Mobile and Wireless Communication Networks

T0189309

IFIP – The International Federation for Information Processing

IFIP was founded in 1960 under the auspices of UNESCO, following the First World Computer Congress held in Paris the previous year. An umbrella organization for societies working in information processing, IFIP's aim is two-fold: to support information processing within its member countries and to encourage technology transfer to developing nations. As its mission statement clearly states,

> *IFIP's mission is to be the leading, truly international, apolitical organization which encourages and assists in the development, exploitation and application of information technology for the benefit of all people.*

IFIP is a non-profitmaking organization, run almost solely by 2500 volunteers. It operates through a number of technical committees, which organize events and publications. IFIP's events range from an international congress to local seminars, but the most important are:

- The IFIP World Computer Congress, held every second year;
- Open conferences;
- Working conferences.

The flagship event is the IFIP World Computer Congress, at which both invited and contributed papers are presented. Contributed papers are rigorously refereed and the rejection rate is high.

As with the Congress, participation in the open conferences is open to all and papers may be invited or submitted. Again, submitted papers are stringently refereed.

The working conferences are structured differently. They are usually run by a working group and attendance is small and by invitation only. Their purpose is to create an atmosphere conducive to innovation and development. Refereeing is less rigorous and papers are subjected to extensive group discussion.

Publications arising from IFIP events vary. The papers presented at the IFIP World Computer Congress and at open conferences are published as conference proceedings, while the results of the working conferences are often published as collections of selected and edited papers.

Any national society whose primary activity is in information may apply to become a full member of IFIP, although full membership is restricted to one society per country. Full members are entitled to vote at the annual General Assembly, National societies preferring a less committed involvement may apply for associate or corresponding membership. Associate members enjoy the same benefits as full members, but without voting rights. Corresponding members are not represented in IFIP bodies. Affiliated membership is open to non-national societies, and individual and honorary membership schemes are also offered.

Mobile and Wireless Communication Networks

IFIP TC6 / WG6.8 Conference on Mobile and Wireless Communication Networks (MWCN 2004)
October 25-27, 2004, Paris, France

Edited by

Elizabeth M. Belding-Royer
University of California, Santa Barbara, USA

Khaldoun Al Agha
University of Paris XI, France

Guy Pujolle
University of Paris XI, France

 Springer

Elizabeth M. Belding-Royer
University of California,
Santa Barbara, USA

Khaldoun Al Agha
University of Paris XI,
France

Guy Pujolle
University of Paris XI,
France

Library of Congress Cataloging-in-Publication Data

A C.I.P. Catalogue record for this book is available from the Library of Congress.

Mobile and Wireless Communication Networks / Edited by Elizabeth M. Belding-Royer, Khaldoun Al Agha, Guy Pujolle

p.cm. (The International Federation for Information Processing)

ISBN 978-1-4419-3580-9   e-ISBN 978-0-387-23150-1   Printed on acid-free paper.

Printed in the United States of America.

9 8 7 6 5 4 3 2 1
springeronline.com

Contents

UNDERSTANDING THE INTERACTIONS BETWEEN UNICAST AND GROUP COMMUNICATIONS SESSIONS IN AD HOC NETWORKS

Lap Kong Law, Srikanth V. Krishnamurthy and Michalis Faloutsos *
Department of Computer Science & Engineering
University of California, Riverside
Riverside, California 92521
{lklaw,krish,michalis}@cs.ucr.edu

Abstract   In this paper, our objective is to study and understand the mutual effects between the group communication protocols and unicast sessions in mobile ad hoc
networks. The motivation of this work is based on the fact that a realistic wireless networks would typically have to support different simultaneous network
applications, many of which may be unicast but some of which may need broadcast or multicast. However, almost all of the prior work on evaluating protocols
in ad hoc networks examine protocols in isolation. In this paper, we compare
the interactions of broadcast/multicast and unicast protocols and understand the
microscopic nature of the interactions. We find that unicast sessions are significantly affected by the group communication sessions. In contrast, unicast
sessions have less influence on the performance of group communications due
to redundant packet transmissions provided by the latter. We believe that our
study is a first step towards understanding such protocol interactions in ad hoc
networks.

1.   Introduction

Most routing protocol evaluations assume implicitly that only the protocol
under consideration is deployed in the network. However, ad hoc networks are
likely to support many types of communication such as unicast, broadcast, and
multicast at the same time. Although the performance evaluation of a protocol
in isolation can lend valuable insights on its behavior and performance, the
protocol may have complex interactions with other coexisting protocols. These
interactions may significantly alter the behavior of the protocols.

---

*This work was supported from grants from Telcordia Technologies and ARL no: 100833196 and from
DARPA FTN Grant no: F30602-01-2-0535.

In this paper, we consider two families of protocols that are typically invoked by applications: unicast routing protocols and group communication protocols such as broadcast and multicast. While an assessment of the behavioral interactions between a large representative set of unicast routing and group communication protocols is beyond the scope of this paper, our attempt is to provide a fundamental understanding of such interactions by considering suitable candidate protocols from each set. Towards this, we consider a representative protocol from the unicast, broadcast and multicast routing. We choose the Ad-hoc On-Demand Distance Vector Routing (AODV) [10] as the representative unicast routing protocol. We consider the Simple Broadcast Algorithm (SBA) [9] as the candidate broadcast scheme and the On-Demand Multicast Routing Protocol (ODMRP) [5] as the candidate multicast protocol. The chosen protocols have been shown to be the elite members of their respective families [2] [11] [6].

The goal of this paper is to draw the attention of the community to the importance of cross protocol interactions. We discuss the possible effects that may arise when both types of protocols coexist in the network. Finally, we conduct extensive simulations to quantify the effects of the interactions. We find that the effects are indeed significant and should be considered in realistic simulations of complete systems.

We wish to point out that in a complementary effort [4], we evaluate the behavioral differences between broadcast and multicast considered in isolation in order to determine the suitability of each for performing group communications. In this paper, our objective is to examine the effects of broadcast and multicast on unicast sessions and vice versa.

The rest of the paper is organized as follows: In the next section, we provide a brief description of the candidate protocols chosen. In Section 3, we deliberate on possible interactions that may arise when unicast and group communication protocols coexist in a network; these deliberations provide a basis for the metrics that we choose when performing our simulation experiments. In Section 4, we first describe the simulation scenarios and parameters chosen; then we present and discuss the results from our simulations. Finally we conclude the paper in Section 5.

2.   Background

In the following paragraphs, we will briefly describe the candidate protocols that we choose for our evaluation. A description of the protocols in more detail can be found from the appropriate references cited. In addition, we present background that facilitates the discussions in the later sections.

Unicast: Ad hoc On-Demand Distance Vector Protocol (AODV) [10] is an on-demand routing protocol that builds routes only when needed. When a source has packet to send but does not have a route, it buffers the packet in a temporary buffer and broadcasts a Route Request (RREQ) message. In order to perform the broadcast, the expanding ring search method is used. With this technique, a node iteratively searches for the destination in zones of increasing size (in terms of hop count) until the destination is found. Nodes that do not have the route to the destination, upon the reception of a unique RREQ message, forward it to their neighbors and update their route tables to set up a reverse route back to the source. Interim nodes that have a fresh route to the destination respond by means of a Route Reply (RREP) message directed to the originating source. As the RREP propagates back to the source, intermediate nodes set up forward pointers to the destination. Once the source receives the RREP, it may begin the transmission of packets from the temporary buffer. In AODV, a sequence number is used to prevent routing loops and nodes must use the information with the most up-to-date sequence number while making routing decisions. When a link along the active route breaks, the node upstream of the broken link propagates a Route Error (RERR) message towards the source to inform the source of the link failure.

Multicast: On-Demand Multicast Routing Protocol (ODMRP) [5] is a mesh based multicast protocol. When a multicast source has a packet to send and the multicast group members are yet to be identified, it floods a Join Query message in the network. The Join Query message is also periodically flooded to refresh group membership information and update routes as long as the source still has packets to send. When a node receives a Join Query message, it stores the source id and sequence number indicated in the message in its message cache; duplicate receptions of the same Join Query are discarded. If the message received is not a duplicate instance of a previous message and if the Time-to-live (TTL) value indicated in the message is greater than zero, the recipient node rebroadcasts the Join Query. When the Join Query reaches a multicast receiver, it creates a Join Reply message and broadcasts it to its neighbors. When a node receives a Join Reply, it checks if it is identified to be the next hop entry. If it is, the node is a forwarding node and the forwarding group flag is set. It then rebroadcasts its own Join Reply. Finally, the Join Reply reaches the multicast source and the routes are established. From then on, until information is further updated, a node will forward the packet only if it is in the forwarding group.

Broadcast: Simple Broadcast Algorithm (SBA) [9] is an intelligent broadcast protocol in the sense that it considerably reduces the number of rebroadcasts as compared with flooding. Furthermore, it has been shown in previous work [11] that SBA outperforms most of the other broadcast schemes such as the counter-based scheme and the location-based scheme. It reduces the effects of a broadcast storm [8] by using a simple technique that we discuss below in brief. SBA incorporates the exchange of periodic hello message between neighbors to enable the acquisition of local neighborhood information by each node. Each hello message contains a list of the one-hop neighbors of the broadcasting node and thus, finally, every node in the network will have its two-hop neighborhood information. The collected neighborhood information is used to decide whether or not a received data packet should be rebroadcasted. The decision is made by determining, by means of the neighborhood information table, if there exists any node that is not covered by previous broadcasts. If all the neighbors of the node are already covered, the node will not rebroadcast the packet; otherwise the node will schedule a time to rebroadcast the packet based on the number of neighbors that it has. The higher the number of neighbors, the sooner the node will rebroadcast the packet. This would therefore make nodes with higher degrees broadcast earlier than lower degree nodes. Thus, this can potentially enable the coverage of a large fraction of nodes with relatively few broadcasts.

We present a brief discussion on the IEEE 802.11 MAC and specifically the *Distributed Coordination Function* (DCF) since all protocols under our evaluation are affected by this protocol.

IEEE 802.11 Medium Access Control (MAC) [1] is the de-facto MAC layer standard for wireless networks and has been popularly considered for ad hoc networks. The fundamental access method with the DCF is known as *carrier sense multiple access with collision avoidance* (CSMA/CA). This protocol basically defines how each station may access the shared wireless channel. *Physical* and *virtual* carrier sensing are used to provide the status of the medium. In order for a host to transmit, it must sense that the medium is idle for a minimum specified duration before attempting to transmit. If the medium is busy, the host must defer its transmission and select a random backoff interval. The backoff interval counter decrements when a host senses that the medium is idle. As soon as the backoff counter reaches zero and the medium is determined idle, a host may proceed with its transmission. The aforementioned access mechanism defines the **basic access** method which is used for all broadcast, multicast and some of the unicast control packets transmissions. IEEE 802.11 MAC DCF also defines an additional mechanism which is used in conjunction with the basic access method to further avoid collisions. This additional mechanism is the **Request to send (RTS)/Clear to send (CTS)** exchange mechanism. Before any data packets are transmitted, a sender host should transmit a short RTS frame to announce its intention to transmit data. All the neighbors of the sender are expected to receive the RTS frame. The destination host upon the reception of the RTS replies with a CTS frame. Since both the RTS and CTS frame contain information that defines the period of time that the medium is reserved, the other neighbors of either the sender or the receiver preclude the initiation of any transmission in this period of time and thus avoid causing collisions. It is assumed as in prior work [2], that all

Table 1.   Common Simulation Parameters (Fixed)

Number of nodes	100
Node speed	5m/s (constant)
Pause time	0 second
Simulation area	500m x 500m
Node transmission range	100m
Simulation time	60 seconds
Broadcast/Multicast group size	40% (40 group members)
AODV temporary buffer size	64 packets
Interface queue size	50 packets

the unicast data packets and some of the control packets are transmitted using this RTS/CTS exchange mechanism.

3.   Issues that may arise when unicast and group communications protocols coexist

In this section, we qualitatively examine the various interactions between coexisting unicast and multicast sessions in an ad hoc network. The sessions compete with each of the network resources and in particular the wireless bandwidth. The exact nuances of the interactions do depend on the actual protocol design; we attempt to characterize certain generic aspects of the interactions via a few chosen metrics of interest.

3.1   Degradations in Packet Delivery Performance

Since, the coexisting protocols attempt to access the shared wireless channel at the same time, a higher number of collisions may be expected. This is especially the case with the control messages that are generated from both protocols since they do not employ the RTS/CTS reservation mechanism. Collisions happen when multiple hosts sense the medium to be idle and try to transmit at the same time or if there are hidden terminals that cause collisions at a receiver. Since all of the packets forwarded using the broadcasting and multicasting protocols are transmitted via the aforementioned **basic access** method, from the perspective of the unicast protocol, there is an increase in the number of broadcast packets that collide primarily with its control messages. Packet drops may occur since routes might not be established. For broadcast and multicast sessions, data packets may be lost directly due to collisions.

To elaborate on the effects of broadcast on unicast protocols, typically when a route to a destination is needed a protocol such as AODV invokes the flood (perhaps a modified version) of an RREQ message as mentioned earlier. In the presence of a group communication session these broadcast RREQ messages typically collide and are hence lost. In some cases, a route is not discovered although it actually exists. This would trigger additional route query attempts and thus would further increase the generated congestion. With AODV if four

consecutive route query attempts were to fail, the sending source would abort
the session and drop the packets.

3.2   Increased Latency Effects

Note that a secondary effect of the congestion is that the MAC layer interface queues fill up quickly. Since the control messages have priority over actual
data, the increase in the volume of control messages (repeated RREQ floods)
may be expected to cause the actual data packets to experience increased (and
potentially large) latencies in each queue.

The unicast packets are likely to be affected to a higher extent than the group
communication packets. Because of the higher levels of redundancy built into
the group communication protocols, they may be less affected by congested
areas; packets that may actually be able to go through less congested areas to
reach their intended destinations more quickly. Furthermore, since there are
no explicit RTS/CTS exchanges, it is enough that only the sender perceive the
channel to be free (as opposed to requiring both the sender and receiver to
be free as in the case of the unicast connections). Thus, the group communication packets have a tendency to depart from their interface queues more
quickly (once they get to the head of the queue) as compared with their unicast counterparts. Finally, note that there are no explicit attempts to retransmit
a group communication session's packets. However, the MAC layer makes
seven retransmission attempts for every unicast packet. Each attempt is made
after an exponential back-off period from the previous retransmission attempt.
Thus this retransmission procedure is expected to contribute immensely to the
latency evolved by the unicast connections.

3.3   Increased Control Overhead

An consequence of congestion is an increase in control overhead. Control overhead includes all packet transmissions excepts the actual data packet
transmissions. In other words, all the control messages transmitted for assist-
ing with the functionality of a protocol account for control overhead. In an
environment where contention levels are high, a great portion of the control
messages are lost due to collisions. These control messages may or may not
be retransmitted (depending on the type of the message). With a broadcasting
protocol such as SBA, the control messages (Hello messages) are not retransmitted upon loss. For protocols such as ODMRP and AODV, control messages
are retransmitted. If a route is not found, several additional attempts are made
before the node aborts the query attempt. We call this type of protocols as
routing-dependent protocols and the broadcasting protocol such as SBA as
routing-independent protocols. Routing-dependent protocols are more sensitive to congestion since a complete route must be established before the route

Table 2.   Parameters for the first part of the simulation

Number of unicast sessions	1,2,4,8
Unicast rate per session	5, 10, 20 packets/second
Unicast packet size	512 bytes
Broadcast/Multicsat packet size	512 bytes
Number of broadcast/multicast source	1
Broadcast/Multicast rate	5 packets/second

Table 3.   Parameters for the second part of the simulation

Number of broadcast/multicast sessions	1,2,3,4
Unicast rate per session	2, 4, 8 packets/second
Unicast packet size	256 bytes
Broadcast/Multicsat packet size	256 bytes
Number of unicast sessions	2
Unicast rate	5 packets/second

discovery cycle ends. Therefore, the control overhead will increase when the contention level increases. On the other hand, routing-independent protocols do not require the construction of any forwarding structure and thus the control overhead is independent of the contention level.

4.   Simulation Study

We conduct extensive simulations to quantify the effects of coexistence of two different types of protocols on the same network.

We use ns-2 simulator [7] in our evaluation. We vary parameters of interest to study the effects of interest under various scenarios. Other system parameters have been chosen to be fixed but the behavioral results from sample simulations for other values of these parameters were similar. The set of generic parameters used are listed in TABLE. 1. We divide the simulations into two parts. In the first part, we simulate scenarios to study the effects of unicast traffic on group communication performance. In the second part, we try to capture the effects of group communication on unicast session performance. The parameters that we vary in the first part of the simulation are listed in TABLE 2, and the parameters that we vary for the second part of the simulation are in TABLE 3. We repeat each simulation 30 times and for each run, we use different scenarios. In order to allow both types of protocols to run concurrently, we have made modifications to the ns-2 source code.

4.1   Simulation results

In the following, we present and discuss the simulation results that we obtain. In order to understand the mutual effects between protocols, we also conducted baseline simulations such that only the protocol of interest is deployed in isolation. We can thus compare the performance of each protocol in the presence of other coexisting protocols with that of the protocol in iso-

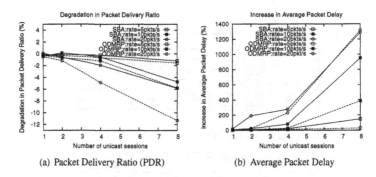

(a) Packet Delivery Ratio (PDR)   (b) Average Packet Delay

Figure 1.   The differences in the performance of SBA/ODMRP in the presence of unicast traffic

lation. Most of the graphs presented reflect the comparisons and are usually represented as a percentage difference.

4.2   The effects of unicast protocol on the performance of group communication protocols

We first present the performance results that capture the effects of the group communication protocols in the presence of unicast protocol.

Packet Delivery Ratio.   As seen from Fig. 1(a), the performance of both SBA and ODMRP degrade when the number of unicast sessions and the unicast rate increases. This is expected since the collisions on the physical channel increases when the background unicast traffic increases. Note that the packet delivery ratio of ODMRP decreases to a greater extent than that of SBA. The reason for this is that SBA has a higher rebroadcast redundancy than ODMRP and therefore the chance of successful packet delivery is higher in spite of congestion as discussed earlier in Section 3.1. The maximum offered load by all the unicast sessions together is 0.625Mb/s. This is a significantly high load given that the channel data rate is assumed to be 2Mbps. However, most of the results that we obtain demonstrate less than a 10% decrease and thus we do not regard these as a significant degradation.

Average Packet Delay.   Although the packet delivery ratio with both SBA and ODMRP only decreases slightly, the average packet delay increases drastically as seen from Fig. 1(b). When the amount of unicast traffic is low, the average packet delay remains almost the same as that of the case wherein the group communication protocol is considered in isolation. However, when the unicast traffic increases, the delay increases significantly. When the number of unicast sessions is 8 and the unicast rate is 20 pkts/second, the average packet delay of both SBA and ODMRP data packets jumps by almost 1300% as com-

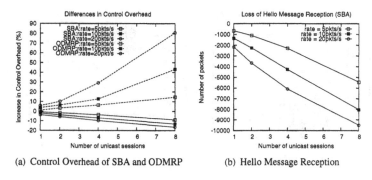

(a) Control Overhead of SBA and ODMRP   (b) Hello Message Reception

Figure 2.   The effect of hello message loss on the control overhead of SBA

pared to the case where no unicast traffic is present. The reason for such a drastic increase in average packet delay is that the data packets are caught up at the lower layer interface queue as discussed earlier in Section 3.2. The situation becomes worse when the amount of unicast traffic further increases as one might expect.

Control Overhead. The control overhead of SBA and ODMRP are affected differently under the presence of unicast traffic. From Fig. 2(a), we see that the control overhead of SBA decreases as the amount of unicast traffic increases. In contrast, the control overhead of ODMRP increases drastically. Upon further investigation, we find that the decrease in control overhead of SBA is caused by the loss of hello messages. From Fig. 2(b), we see that the number of successful hello packet receptions decreases as the amount of unicast traffic increases. These losses of hello packets are mainly due to the collisions. As a result, the two-hop neighborhood information maintained at each node is incomplete and inaccurate. Therefore, on average, the size of the hello message (containing the one-hop neighborhood information) is smaller. This decrease in control overhead is significant, since state information is important for the effective functioning of the SBA protocol. For ODMRP, the control messages are retransmitted for a limited number of times. Because of this, the multicast structure is constructed successfully. Due to an increase in such retransmissions, the control overhead increases in ODMRP as the amount of unicast traffic increases, as we see in Fig. 2(a).

4.3   The effects of group communications protocols on the performance of the unicast protocol

In this subsection, we present the performance of AODV in the presence of broadcast or multicast traffic.

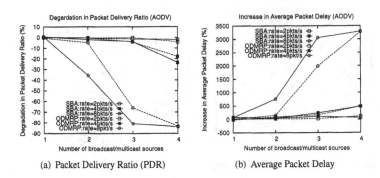

(a) Packet Delivery Ratio (PDR)   (b) Average Packet Delay

Figure 3.   The differences in the performance of AODV in the presence of group communication traffic

Packet Delivery Ratio.   From Fig. 3(a), we see that when the broadcast or
multicast traffic rate is low (2 packets/s), the packet delivery ratio of AODV
remains almost same as that in the case when it is considered in isolation.
However, when the broadcast or multicast traffic increases, the packet delivery
ratio of AODV decreases significantly. When there are 4 broadcast/multicast
sources with a data rate of 8 packets/s, the packet delivery ratio of AODV
differs by almost 90% as compared to the case where no group communication
sessions are present. This tells us that the group communication traffic even at
moderate amounts, degrade AODV performance significantly.

We investigate the reasons for the loss of packets with AODV. In Fig. 4(a)
and Fig. 4(b), we represent the distribution of the AODV data packet drops
when a varying number of broadcast/multicast sources with different data transmission rates are present in the network. At a transmission rate of 2 packets/s,
only a small fraction of the unicast packets are being dropped (compared to
the total of 600 packets) and more than half of them are due to link failures.
We regard these as a normal packet drops since the link failures in this case
are mainly caused by the node mobility. However, as the amount of broad-
cast/multicast traffic increases, the number of packet drops increases drastically. We notice that the number of packet drops at the interface queue upon
link failure increases and accounts for a major portion of the total packet losses.
This happens when there is a link failure and a large number of packets waiting in the queue rely on that link. Note that when the congestion increases, the
number of link failures are also seen to increase. These are primarily due to
"false link failures" and artifact of the IEEE 802.11 MAC protocol [3]. Here if
the intended recipient of an RTS packet is within the sending range (interference range) of some other node, it does not respond to the sender of the RTS
message with a CTS message. After seven consecutive attempts the sender
deems the link to have failed although in reality it still exists. Notice that
the aforementioned packet losses only happen with high amounts of broad-

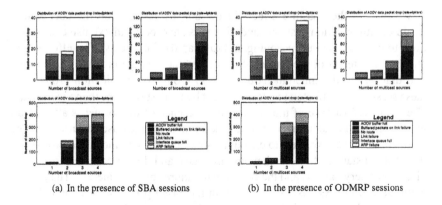

(a) In the presence of SBA sessions   (b) In the presence of ODMRP sessions

Figure 4.   The distribution of AODV data packet drop

cast/multicast traffic. This is in line with our discussion in the previous section that a high amount of traffic increases the contention level at the lower layer interface queue. At very high amounts of broadcast/mutlicast traffic (at a rate of 8 packets/s and $\geq$ 3 sources), packet losses increase drastically. This implies that at this level of background broadcast/multicast traffic, AODV route discovery attempts begin to encounter failures. A lot of packets are buffered in the AODV buffer temporarily without a route to the destination.

Average Packet Delay.   In addition to the poor packet delivery ratio, the average packet delay also increases significantly (see Fig. 3(b)). When the background broadcast/multicast traffic rate is low (2 packets/s), the average packet delay remains steady with only a small delay increase as compared with the case of AODV in isolation. However, when the background broadcast/multicast rate increases, the delay increases drastically. When the broadcast/multicast rate is 4 packets/s, the delay rises by up to 500% and at a rate of 8 packets/s the delay increase is even up to 3250%! This effect is a combination of the delay incurred by the data packet while waiting in the interface queue as well as the delay due to waiting for the route to be found.

5.   Conclusions

In this paper we attempt to understand the impact of coexisting unicast and group communication protocols. Specifically, we attempt to quantify the extent of performance degradations on one due to the other and discuss the underlying effects that cause such degradations. This is motivated by realistic networks wherein multiple applications, some of which may require unicast sessions while others require multicast/broadcast sessions, are likely to exist. To the best of our knowledge this is the first attempt to undertake such a study.

We discuss the possible effects of the unicast sessions on the coexisting group communication sessions and vice versa and perform extensive simulations to corroborate our reasoning and to quantify the effects. The effects are quantified in terms of the packet loss rate and the average incurred delay. We find that due to the inherent redundancy in group communication protocols, these sessions are affected to a lesser extent by unicast sessions than vice versa. The poor performance of the latter in the presence of the former is attributed to the increase in the number of false link failures, the increase in buffer delays which in turn cause packets to encounter link failures with increased probability (at the instance of which they are dropped), and due to collisions of route discovery query packets and the consequent failures of such queries. The group communication sessions also suffer from increased packet losses and delays. Broadcast schemes are likely to be more robust to route failures than multicast schemes since they are independent of routing.

References

[1] ANSI/IEEE std 802.11, 1999 edition.

[2] Broch, Josh, Maltz, David A., Johnson, David B., Hu, Yih-Chun, and Jetcheva, Jorjeta (1998). A performance comparison of multi-hop wireless ad hoc network routing protocols. In *Proceedings of the ACM/IEEE MobiCom'98*, pages 85–97.

[3] Klemm, Fabius, Ye, Zhenqiang, Krishnamurthy, Srikanth V., and Tripathi, Satish K. (2004). Improving TCP performance in ad hoc networks using signal strength based link management. *Ad Hoc Networks Journal*. (to appear).

[4] Law, Lap Kong, Krishnamurthy, Srikanth V., and Faloutsos, Michalis. On evaluating the trade-offs between broadcasting and multicasting in ad hoc networks. (submitted to IEEE MILCOM 2004).

[5] Lee, Sung-Ju, Su, William, and Gerla, Mario (2002). On-demand multicast routing protocol in multihop wireless mobile networks. In *ACM/Baltzer Mobile Networks and Applications, special issue on Multipoint Communication in Wireless Mobile Networks*, volume 7, pages 441–453. Kluwer Acadmic Publishers.

[6] Lee, Sung-Ju, Su, William, Hsu, Julian, Gerla, Mario, and Bagrodia, Rajive (2000). A performance comparison study of ad hoc wireless multicast protocols. In *Proceedings of the INFOCOM'04*, pages 565–574.

[7] McCanne, S. and Floyd, S. Ns-2 simulator.

[8] Ni, Sze-Yao, Tseng, Yu-Chee, Chen, Yuh-Shyan, and Sheu, Jang-Ping (1999). The broadcast storm problem in a mobile ad hoc networks. In *Proceedings of the ACM/IEEE MobiCom'99*, pages 151–162, Seattle, Washington.

[9] Peng, Wei and Lu, Xi-Cheng (2000). On the reduction of broadcast redundancy in mobile ad hoc networks. In *Proceedings of the ACM MobiHoc'00*, pages 129–130, Boston, Massachusetts.

[10] Perkins, Charles E. and Royer, Elizabeth M. (1999). Ad hoc on-demand distance vector routing. In *Proceedings of the IEEE WMCSA'99*, pages 90–100, New Orleans, LA.

[11] Williams, B. and Camp, T. (2002). Comparison of broadcasting techniques for mobile ad hoc networks. In *Proceedings of the ACM MobiHoc'02*, pages 194–205.

CROSS-LAYER SIMULATION AND OPTIMIZATION FOR MOBILE AD-HOC NETWORKS

André-Luc Beylot[1], Riadh Dhaou[1], Vincent Gauthier[2], and Monique Becker[2]

[1] *ENSEEIHT, IRIT/TeSA Lab*
Toulouse, France
{Andre-Luc.Beylot, Riadh.Dhaou}@enseeiht.fr

[2] *GET/INT, Samovar Lab.*
Evry, France
{Vincent.Gauthier, Monique.Becker}@int-evry.fr

Abstract   This paper introduces cross-layer implementation with a multi-rate aware routing scheme and shows that SNR is an important information to use in a routing protocol. The existing routing protocol attempts to minimize the number of hops between source-destination pairs. We use a new metric definition to route the packets and to select the best available link along the path in a multi-rate protocol senario. The new metric is created with information coming from inter-layer interaction between the routing layer and the MAC layer. We use SNR as an information about link quality. We show through simulation that for communications using muti-rate protocol in ad hoc networks, throughput is highly affected as soon as the route goes through low-rate link.

Keywords: Ad hoc networks, AODV, Cross-Layers, IEEE 802.11, Inter-layer interactions, Performance, QoS

Introduction

Ad hoc wireless network are self organizing multi-hop wireless networks where all the nodes take part in the process of forwarding packets. Ad Hoc networks are very different from conventional computer networks. First, the radio resource is rare and time varying. Second, the network topology is mobile and the connectivity is unpredictable. Third architecture-based 802.11 WLAN, is further complicated due to the presence of hidden stations, exposed station, "capturing" phenomena, and so on. Fourth, many current and proposed wireless networking standards have this multi-rate capacity (802.11b, 802.11a, 802.11g, and HyperLan2). The interaction between these phenomena make the behavior of ad hoc network very complex to predict and are really different from wired network architecture.

The aim of Cross-Layer concept is to improve the performance of all layers and share key information between these layers. The goal of this technique is to take benefit of informations about the channel quality to develop a more powerful routing technique. The inter-layer interaction will be managed by the network status. The network status will act as information repository and it will give on demand to each layer, the information about other layers. The inter-layer interaction enables us to use the information on the channel to define a new cost metric in ad hoc network as a function of link quality.

Our proposed schemes use a cross-layer interaction between MAC and network layer. The objective is to create a new QoS cost metric (cf Fig. 1). The proposed metric is a function of SNR (Signal Noise Ratio) and of the number of hops.

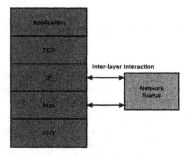

Figure 1.   Inter-Layer Interaction

1.   Related Work

In [BAR04] the authors proposed a new network metric the *medium Time Metric* (MTM), which is derived from a general theoretical model of the reachable throughput in multi-rate ad hoc wireless networks. The MTM avoids using the long range link favored by shortest path routing in favor of shorter, higher throughput, more reliable links.

In [GCNB03] the authors propose a new power-aware routing technique for wireless ad hoc networks (PARO) where all nodes are located within the maximum transmission range of each other. PARO uses a packet forwarding technique where immediate nodes can elect to be *redirector* on behalf of source-destination pairs with the goal of reducing the overall transmission power needed to deliver packet in the network, thus, increasing the operational lifetime of network devices.

In [DACM02] the authors show that the minimum hop path generally contains links which exhibit low reliability. In [DRWT97] and [HLT02] the authors present routing protocols which are based on signal stability rather then on only a shortest path in order to provide increased path reliability.

Based on the IEEE 802.11 protocol, the Receiver Based Auto Rate (RBAR) protocol was presented in [GHB01]. RBAR allows the receiving node to select the rate. This is accomplished by using the SNR or the RTS packet to choose the most appropriate rate and to communicate that rate to the sender using the CTS packets. This allows much faster adaptation to the changing channel conditions than ARF, but requires some modifications to the 802.11 standard.

The Opportunistic Auto Rate (OAR) protocol which is presented in [BSK02], operates using the same receiver based approach, but allows high-rate multi-packet burst to take advantage of the coherence time of good channel conditions. The bursts also dramatically reduce the overhead at high rates by smoothing the cost of contention period and RTS CTS frames over several packets.

2.   IEEE 802.11 MAC Layer Approach

The IEEE 802.11 technology is a good platform to implement single-hop ad hoc network because of its extreme simplicity. But in a multi-hop ad hoc networks environment, the IEEE 802.11 protocol works inefficiently. There are two main effects that reduce the *effiency* of the protocol. First the 802.11b standard extends the 802.11 standard by introducing a higher-speed Physical Layer in the 2.4 Ghz band still guaranteeing the interoperability with 802.11 cards. The 802.11b standard enables multi-rate transmission at 11 Mbps and 5.5 Mbps in addition to 1 Mbps and 2 Mbps. To ensure the interoperability, each WLAN defines a basic rate set that contains the data transfer rate that must be used by all the stations in a WLAN. The overhead due to the use of the basic rate between all the stations in a WLAN is very important and it affects the throughput.

- T_{ctr} is the time required to transmit all the control frame $T_{ctr} = T_{rts} + T_{cts} + T_{ack}$.

- T_{ack} is the time required to transmit MAC ACK frame which includes Physical header and MAC header.

- T_{data} is the time required to transmit a MAC data frame which includes Physical header, MAC header, MAC payload

- T_{st} is the Slot Time.

- $T_{payload}$ is the time required to transmit only the m bytes generated by the application; $T_{payload}$ is therefore equal to $m/data\ rate$ where

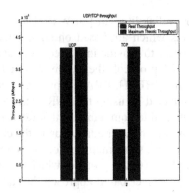

Figure 2.   real throughput vs theoretic throughput with constant size packet of 1024

$data\ rate$ is the data rate used by the NIC to transmit data, i.e., 1, 2, 5.5 or 11 Mbps.

- $\frac{CWmin}{2} * T_{st}$ is the average backoff time

$$Th = \frac{T_{payload}}{T_{difs} + T_{ctr} + T_{data} + 3 * T_{sifs} + \frac{CWmin}{2} * T_{st}}$$

However, even with large packet size (eg., m=1024 bytes) the bandwidth utilization is lower than 39% (cf Fig. 2). This theoretical analysis corresponds to the measurement of the actual throughput at the application level. Two typical "applications" have been considered: FTP and CBR. The experimental results related to the UDP traffic are very close to the maximum throughput computed analytically. As expected, in the presence of TCP traffic the measured throughput is lower than the theoretical maximum throughput. Indeed, when using the TCP protocol overhead related to the TCP-ACK transmission has to be taken into account.

In the second graph (cf Fig. 3), qualnet simulations have been run for which one "CBR" application have been considered (packets size = 1024 bytes). The throughput has been studied as a function of the number of hops and for different available rate using the IEEE 802.11b protocol. We could see how the transmission throughput decreases as a function of the number of hops.

Effect of the SNR

Many current and proposed wireless network standards such as IEEE 802.11a, IEEE 802.11b or HyperLan 2 present a multi-rate capacity. The IEEE802.11b has different adaptive modulations which were investigated with the dynamic

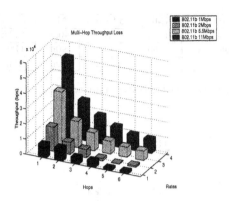

Figure 3. Throughput vs number of hop

channel allocation technology. All of them are trying to improve the effective data rate given the specified bit error rate (BER). Due to the physical properties of communication channels, there is a direct relationship between the rate of communication and the quality of the channel required to support that communication reliably. Since the distance is one of the primary factor that determines wireless channel quality, there is an inherent trade-off between high transmission rate and effective transmission range. The SNR is a very interesting information to monitor because it reflects the link quality. In a multi-rate protocol, each available link may operate at a different rate. The most important challenge is to choose a good trade-off between the link quality and the number of hops. As a short link can operate at high rate, more hops are required to reach the destination.

The Figure 4 and 5 show the Bit Error rate is represented as a function of SNR and the Throughput as a function of the SNR (cf Fig. 5). We can see the effect of the SNR on the transmission performance. But the distance of each link is the primary factor that determines channel quality. Long links have low quality, and thus operate at low rate (cf Fig. 4). Nevertheless, it is difficult to measure the link quality, so we propose to use the Smoothed value of Signal-to-Noise-Ratio since SNR could change dynamically with a high frequency due to electro-magnetic effect. This Smoothed SNR (SSNR) value can be computed as follows:

$$ssnr = (1 - \alpha) * old\_snr + \alpha * cur\_snr$$

where cur_snr and old_snr denote the value of the SNR on receipt of a packet and the previously computed $ssnr$, respectively. The constant value α is a filtering factor and it is set between 0.7 and 0.9 as a function of changing speed of the signal.

Range(meters)	11Mbps	5.5Mbps	2Mbps	1Mbps
Open	160m	270m	400m	550m
Semi-open	50m	70m	90m	115m
Closed	25m	35m	40m	50m

Figure 4.   Relation between the distance of two nodes and the available data rate

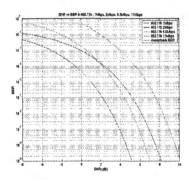

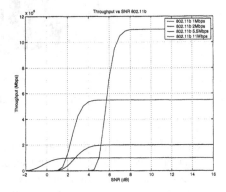

Figure 5.   Bits error rate as a function of SNR

Figure 6.   Throughput as a function of SNR

3.   Network Layer Approach

Ad-Hoc networks require a highly adaptive routing scheme to deal with the frequent topology changes and low performance. In this paper, we propose a routing protocol that utilizes the ad hoc network characteristics to select the route which has a better compromize between the number of hops, the theoretical available bandwidth, and the stability of the route. This protocol is new because it uses the signal strength and SNR available at the MAC layer of an individual host as a route selection criteria. The trade-off between the number of hops and the SNR of each individual route defines a new network metric. The new metric available at the network layer allows to have a global overview of the best available path. In this protocol, a host initiates route discovery on-demand (only when a route is needed to send data). The source broadcasts a route-search packet which will be propagated to the destination, allowing the destination to choose a route and return a route reply.

This paper describes an implementation of the AODV protocol based on a cross layer mechanism in which we use SNR information to obtain better routing techniques. To do so, we use SNR of each node to determine the route which will have globally the best SNR along the path. For this purpose we have added SNR information in each RREQ packet which is used as QoS in-

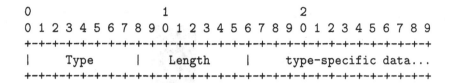

```
0                          1                          2
0 1 2 3 4 5 6 7 8 9 0 1 2 3 4 5 6 7 8 9 0 1 2 3 4 5 6 7 8 9
+-+-+-+-+-+-+-+-+-+-+-+-+-+-+-+-+-+-+-+-+-+-+-+-+-+-+-+-+-+-+
|     Type      |     Length    |     type-specific data...
+-+-+-+-+-+-+-+-+-+-+-+-+-+-+-+-+-+-+-+-+-+-+-+-+-+-+-+-+-+-+
```

Figure 7.   Cross-Layer Extension in AODV Frame

formation. Each node that forwards these packets adds its own SNR information, thus updating the SNR of each link along the route. When the destination
node receives the RREQ packet it directly has the information about the quality
of the route. The destination node then determines the best route and replies
by sending a RREP packet so that each node on the route can save the QoS
information in its routing table. With a route decision mechanism, we take
advantage of this QoS information, the route quality and the global throughput
are improved.

The RREQ and the RREP AODV frame carry the new extension field (cf Fig.
7). Each host along the path picks back the new metric in the extension field of
the RREQ frame. In the case of RREP frame, each host along the path reads
the new metric value and stores it in its routing table.

As we always find the link with the best SNR, we obtain a path with small
transmission range but it may increase the number of hops. Consequently, the
channel access overhead (e.g, backoff time) could be increased in proportion
with the hop count. However it can reduce the link-level transmission time
($\simeq$ Packet Size / Bandwidth), which is highly affected by the packet size. By
reducing the transmission time, we can achieve a better throughput and always
reduce the total energy consumption in the network wide.

This protocol have been tested with ns-2 and is available. It is developed with
the source code of AODV-UU[LN] with cross-layer extension[Gau04].

4.   Expected Result

In the first part of this paper we have presented different effects of SNR on
link quality. First, the throughput and the SNR are directly correlated to the
distance between the sender and the receiver (cf Fig. 9). Second, the throughput is directly correlated to the SNR (cf Fig. 8).

A good link quality is defined by a good SNR. The SNR is a good indicator
of the quality of service of the link. But it is more convenient to aggregate
all SNR informations available on each link into one metric indicator for each
path.

We chose this new metric to find the best available path and to globally improve the network performance. Each link of the best path will have a low Bit

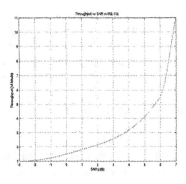

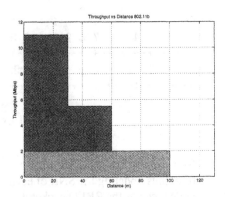

Figure 8.   Throughput vs SNR   *Figure 9.*   Distance vs Throughput in
                                          802.11b

Error Rate and the selected path has the best available throughput between the
sender and the receiver. The measuring method of SNR helps us to have a good
overview over the time of the SNR information and not just the SNR at a given
selected time. It helps us to determine which link has a good stability over the
time and which link has the lower probability to shutdown.

5.   Future Works

The new challenge is to develop a mechanism to monitor the link quality in
real-time during the communication. In this case when the quality of service
of a link falls down, a mechanism should be implemented to find a new path.
It will also be necessary to propose a trade-off between the number of hops
and the quality of each link. It will not lead to select the best available route
but to the selection of a route which presents the best compromise between the
number of hops and the available data rate.
This mechanism could help us to develop a new method of load-balancing
because each node could monitor the number of path search demands and determine a trade-off not to be crossed.

6.   Simulation Issues

We run simulations in order to test the performance and to validate cross-
layer protocols We have written modules simulating a layered stack, which includes MAC layer and network layer. Each layer module communicates with
the upper and the lower layer modules and each layer module communicates
also with the other modules of the same layer in other stacks (see Fig. 1). Modules inside the stack are supposed to communicate with the cross-layer stack.

To simulate these communications, a design solution has been implemented. Small experience maybe hand maneged but for longer experiments, simulation should have to be automatically and dynamically run. So it is necessary to design dynamic and autonomic simulations, which are not easy. This part of the work is not completed yet.

7. Conclusion

In this paper, we started to investigate several cross-layer protocols and we have presented implementation of the AODV protocol which includes cross-layer extension. The concept of cross-layer provides a wide field of information exchange between layers. We focused on SNR which is a useful information to exchange because a low SNR level impacts throughput on the path. A low SNR level leads to a high bit error rate and consequently to a low link throughput.

This protocol uses SNR information in the calculating of the network metric to choose the link with the best available quality (low bit error rate and high throughput). In wireless networks major criteria are the radio channel quality and the energy consumption. These elements cannot be only managed at a local level but have to be managed in a distributed way in the network. Consequently a new network metric has to be created, instead of looking only for the number of hops between the transmitter and the receiver. The quality of the radio channel along the route will also considered.

In order to validate cross-layer methods it is necessary to run a lot of very long simulations in an autonomic and a dynamic way. This work is not completed yet. Nevertheless, cross-layer will probably lead to get very useful results about ad-hoc networks optimization.

The design of dynamic and autonomic simulations may be of general use in order to solve a large set of problems.

References

[BAR04]   D. Holmer B. Awerbuch and H. Rubens. High throughput route selection in multi-rate ad hoc wireless networks. In proceedings of Wireless On-demand Network Systems WONS 2004, Madonna di Campiglio, Italy, 2004.

[BBD03]   M. Becker, AL. Beylot, and R. Dhaou. Aggregation methods for performance evaluation of communication networks, performance evaluation. In *proceedings of International Symposium of Performance evaluation - Stories and Perspectives, Vienna, Austria, Eds: G. Kotsis, pp 215-230*, 2003.

[BSK02]   A. Sabharwal B. Sadeghi, V. Kanodia and E. Knightly. Opportunistic media access for multirate ad hoc networks. In proceedings of ACM MOBICOM 2002, Atlanta, GA, 2002.

[DACM02]   D. De Couto, D. Aguayo, B. Chambers, and R. Morris. Performance of multi-hop wireless networks: Shortest path is not enough. In *The First Workshop on*

Hot Topics in Networks (HotNets-I), Princeton, New Jersey, October 2002. In proceedings of ACM SIGCOMM.

[DRWT97] R. Dube, C. Rais, K. Wang, and S. Tripathi. Signal stability based adaptive routing (ssa) for ad hoc mobile networks. *IEEE Personal Communication*, 1997.

[GAG03] M. Conti G. Anastasi, E. Borgia and E. Gregori. IEEE 802.11 ad hoc networks: Performance measurements. In proceedings of Distributed Computing Systems Workshops ICDCSW'03, Rhode Island, USA, May 2003.

[Gau04] Vincent Gauthier. Adov-uu with cross-layer, 2004. French National Institute of Telecommuncation, http://www-rst.int-evry.fr/~gauthier/.

[GCNB03] J. Gomez, A .Campbell, M. Naghshineh, and C. Bisdikian. Paro: Supporting dynamic power controlled routing in wireless ad hoc networks. *ACM/Kluwer Journal on Wireless Networks (WINET)*, 9(5), 2003.

[GHB01] N. H. Vaidya G. Holland and P. Bahl. A rate-adaptative mac protocol for multi-hop wireless networks. *In proceedings of Mobile Computing and networking*, 2001.

[GK99] P. Gupta and P. Kumar. Capacity of wireless networks. Technical report, University of Illinois, Urbana-Champaign, 1999.

[GW02] A. J. Goldsmith and S. Wicker. Design challenges for energy-constraind ad hoc wireless networks. *IEEE Wireless Communication*, 2002.

[HLT02] E. Nordstrom H. Lundgren and C. Tschudin. Coping with communication gray zone in IEEE based ad hoc networks. In proceedings of WoWMoM, 2002.

[KH95] R. Knopp and P. A. Humblet. Information capacity and power control in single-cell multiuser communication. In proceedings of IEEE International Communications Conference. Seattle, 1995.

[KK03] V. Kawadia and P. R. Kumar. A cautionary perspective on cross layer design. *Submitted to IEEE Wireless Communication Magazine*, July 2003.

[LN] Henrik Lundgren and Erik Nordstrem. Adov-uu. Uppsala University, http://user.it.uu.se/~henrikl/aodv/.

[MBTB98] M. Moulki, AL. Beylot, L. Truffet, and M. Becker. An aggregation technique to evaluate the performance of a two-stage buffered atm switch. In *Annals of Operations Research 79, pp 373-392, Baltzer*, 1998.

[PR99] C. E. Perkins and E. Royer. Ad-hoc on demand distance vector routing. *In proceedings of 2nd IEEE Workshop Mobile Comp. Sys. and Apps. (WMCA), New Orleans, LA*, Feb 1999.

[PRDM00] C. E. Perkins, E. Royer, S. Das, and M. Marina. Performance comparaison of two on-demand routing protocol for ad hoc networks. *In proceedings of IEEE INFOCOM*, 2000.

[SATT04] M. Saito, H. Aida, Y. Tobe, and H. Tokuda. proximity-based dynamic path shortening scheme for ubiquitous ad hoc. In proceedings of Distributed Computing Systems, 2004.

[SS03] J. Chen S. Sheu, Y. Tsai. Mr2rp: The multi-rate and multi-range routing protocol for ieee 802.11 ad hoc wireless networks. *Wireless Networks The Journal of Mobile Communication, Computation and Information*, Volume 9(Issue 2), 2003.

[YSC03] J. Park Y. Seok and Y. Choi. Multi-rate aware routing protocol for mobile ad hoc networks. *IEEE VTC 2003-Spring, Jeju, Korea*, 2003.

IMPROVING TCP PERFORMANCE OVER WIRELESS NETWORKS USING LOSS DIFFERENTIATION ALGORITHMS

Fabio Martignon
Politecnico di Milano, Dept. of Electronics and Information, Piazza L. Da Vinci 32
20133 Milano, ITALY, Tel.: +39-02-2399-3691 - Fax: +39-02-2399-3413
martignon@elet.polimi.it

Abstract   The use of loss differentiation algorithms within the congestion control scheme of TCP was proposed recently as a way of improving TCP performance over heterogeneous networks including wireless links affected by random loss. Such algorithms provide TCP with an estimate of the cause of packet losses. In this paper, we propose to use the Vegas loss differentiation algorithm to enhance the TCP NewReno error-recovery scheme, thus avoiding unnecessary rate reduction caused by packet losses induced by bit corruption on the wireless channel. We evaluate the performance of this enhanced TCP, showing that it achieves higher goodput over wireless networks, while guaranteeing fair share of network resources with classical TCP versions over wired links. Finally, by studying the TCP behavior with an ideal scheme having perfect knowledge of the cause of packet losses, we provide an upper bound to the performance of all possible schemes based on loss differentiation algorithms. The proposed TCP enhanced with Vegas loss differentiation algorithm well approaches this ideal bound.

Keywords: TCP, Wireless Networks, Internet, Protocols.

1.   Introduction

The Transmission Control Protocol (TCP) has proved efficient in classical wired networks, proving an ability to adapt to modern, high-speed networks and new scenarios for which it was not originally designed. However, modern wireless access networks, such as cellular networks and wireless local area networks, pose new challenges to the TCP congestion control scheme.

The existing versions of TCP, like Reno or NewReno, experience heavy throughput degradation over channels with high error rate, such as wireless channels. The main reason for this poor performance is that the TCP congestion control mechanism cannot distinguish between random packet losses due to bit corruption in wireless channels and those due to network congestion.

Therefore, the TCP congestion control mechanism reduces, even when not necessary, the transmission rate. To avoid such limitation and degradation, several schemes have been proposed and are classified in [H.Balakrishnan et al., 1997]. A possible way to reduce the throughput degradation due to transmission errors is to use loss differentiation algorithms that try to estimate the cause of packet losses [S.Cen et al., 2003].

In this paper we propose to use the Vegas loss differentiation algorithm, also known in literature as Vegas Loss Predictor (LP) [Brakmo and Peterson, 1995] to enhance the TCP NewReno error-recovery scheme, as proposed in [S.Cen et al., 2003, Fu and S.C.Liew, 2003], thus avoiding unnecessary rate reduction caused by packet losses induced by bit corruption on the wireless channel.

We evaluate the performance of this enhanced TCP (TCP NewReno-LP), showing that it achieves higher goodput over wireless networks, with both long-lived and short-lived TCP connections, while guaranteeing fair share of network resources with current TCP versions over wired links. TCP NewReno-LP can be implemented by modifying the sender-side only of a TCP connection, thus allowing immediate deployment in the Internet.

We also evaluate the behavior of TCP enhanced with ideal loss prediction, assuming perfect knowledge of the cause of packet losses, thus providing an upper bound to the performance of all possible schemes based on different loss differentiation algorithms. The TCP enhanced with Vegas loss predictor well approaches this ideal bound.

The paper is structured as follows: Section 2 presents TCP NewReno-LP. Section 3 presents the simulation network model. Section 4 analyzes the accuracy of TCP NewReno-LP in estimating the cause of packet losses under several realistic network scenarios. Section 5 measures the performance of TCP NewReno-LP in terms of achieved goodput, fairness and friendliness, and its performance is compared to existing TCP versions, like TCP NewReno and TCP Westwood [Wang et al., 2002], over heterogeneous networks with both wired and wireless links affected by independent and correlated packet losses. Finally, Section 6 concludes the paper.

2.   TCP NewReno Enhanced with Vegas Loss Predictor

The Vegas loss predictor [Brakmo and Peterson, 1995] decides whether the network is congested or uncongested based on rate estimations. This predictor estimates the cause of packet losses based on the parameter V_P, calculated as

$$V_P = (\frac{cwnd}{RTT_{min}} - \frac{cwnd}{RTT}) \cdot RTT_{min} \tag{1}$$

where $cwnd/RTT_{min}$ represents the *expected* flow rate and $cwnd/RTT$ the *actual* flow rate; $cwnd$ is the congestion window and RTT_{min} is the minimum Round Trip Time measured by the TCP source.

Given the two parameters α and β [segments], when $V_P \geq \beta$, the Vegas loss predictor assumes that the network is congested; when $V_P \leq \alpha$, possible losses will be ascribed to transmission random errors. Finally, when $\alpha < V_P < \beta$, the predictor assumes that the network state is the same as in the previous estimation.

We propose to use this predictor within the congestion control of a TCP source as follows: when the source detects a packet loss, i.e. when 3 duplicate acknowledgements are received or a retransmission timeout expires, the Vegas predictor is asked to estimate the cause of the packet loss.

If the loss is classified as due to congestion, the TCP source reacts exactly as a classical TCP NewReno source, setting the slow start threshold ($ssthresh$) to half the current flight size. This allows TCP NewReno-LP to behave as fairly as the standard TCP protocol in congested network environments.

On the contrary, if the loss is classified as due to random bit corruption on the wireless channel, the $ssthresh$ is first updated to the current flight size value.

Then, if the packet loss has been detected by the TCP source after the receipt of 3 duplicate ACKs, the TCP sender updates the $cwnd$ to $ssthresh + 3$ Maximum Segment Sizes (MSS) and enters the fast retransmit phase as the standard TCP NewReno. This allows the source to achieve higher transmission rates upon the occurrence of wireless losses, if compared to the blind halving of the transmission rate performed by current TCP implementations.

If the packet loss has been detected by the TCP source after a retransmission timeout expiration, the congestion window is reset to 1 segment, thus enforcing a friendly behavior of the TCP source toward current TCP implementations.

3.   Simulation Network Model

The TCP NewReno-LP scheme described in the previous Section was simulated using the Network Simulator package (ns v.2 [ns-2 network simulator (ver.2).LBL,]), evaluating its performance in several scenarios as proposed in [S.Floyd and V.Paxson, 2001].

We assume, as in the rest of the paper, that the Maximum Segment Size (MSS) of the TCP source is equal to 1500 bytes, and that all the queues can store a number of packets equal to the bandwidth-delay product. The TCP receiver always implements the Delayed ACKs algorithm, as recommended in [M.Allman et al., 1999].

The network topology considered in this work is shown in Fig. 1. A single TCP NewReno-LP source performs a file transfer. The wired link $S \longleftrightarrow N$ has capacity C_{SN} and propagation delay τ_{SN}. The wireless link $N \longleftrightarrow D$ has capacity C_{ND} and propagation delay τ_{ND}.

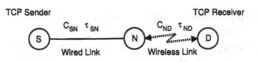

Figure 1.   Network topology in simulations for TCP performance evaluation.

We considered two different statistical models of packet losses on the wireless link: independent and correlated losses. To model independent packet losses, the link drops packets according to a Poisson process, causing a packet error rate (PER) in the 10^{-5} to 10^{-1} range.

To account for the effects of multi-path fading typical of wireless environments, we also considered links affected by correlated errors. From the existing literature [A.A.Abouzeid et al., 2000], we modeled the wireless link state (*Good* or *Bad*) with a two-state Markov chain. The average durations of the *Good* and *Bad* states are equal to 1 and 0.05 seconds, respectively. In the *Good* state no packet loss occurs, while we varied the packet error rate in the *Bad* state from 0% to 100%, to take into account different levels of fading.

4.   Accuracy Evaluation

The key feature of Loss Predictor schemes (LP) is to be accurate in estimating the cause of packet losses, as the TCP error-recovery algorithm we introduced in Section 2, based on the Vegas Predictor, reacts more gently or more aggressively than existing TCP sources depending on the LP estimate. Evidently, when the packet error rate is low and most of packet losses are due to congestion, LP accuracy in ascribing losses is necessary to achieve fairness and friendliness with concurrent TCP flows. On the other hand, when the packet error rate is high such as on wireless links, LP accuracy is necessary to achieve higher goodput, defined as the bandwidth actually used for successful transmission of data segments (payload).

TCP sources detect *loss events* based on the reception of triple duplicate acknowledgements or retransmission timeout expirations. We define *wireless loss* a packet loss caused by the wireless noisy channel; a *congestion loss* is defined as a packet loss caused by network congestion.

The overall *accuracy* of packet loss classification achieved by a loss predictor is thus defined as the ratio between the number of correct packet loss classifications and the total number of loss events.

We measured the accuracy of the Vegas predictor in the network topology of Fig. 1, with $C_{SN} = 10$ Mbit/s, $\tau_{SN} = 50$ ms and $C_{ND} = 10$ Mbit/s, $\tau_{ND} = 0.01$ ms. We considered both the scenarios with and without cross traffic on the wired link and both uncorrelated and correlated errors on the wireless link.

As explained in Section 2, the Vegas predictor detects congestion and wireless losses based on two thresholds, α and β. We tested several values for the parameters α and β and we found the best performance for the accuracy of the Vegas predictor for $\alpha = 1$ and $\beta = 3$. We presented a detailed analysis of the accuracy of the Vegas predictor and other loss differentiation algorithms in [S.Bregni et al., 2003]. In this paper, we summarize only some of the most significant results.

Fig. 2(a) shows the accuracy of packet loss classifications of the Vegas predictor with these parameters as a function of the packet error rate in the scenario with no cross traffic and independent packet losses. Each accuracy value has been calculated over multiple file transfers, with very narrow 97.5% confidence intervals [K.Pawlikowski et al., 2002]. The vertical lines reported in all Figures represent such confidence intervals for each accuracy value.

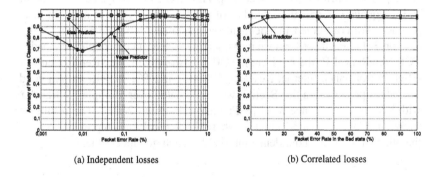

(a) Independent losses   (b) Correlated losses

Figure 2.   Accuracy of classification of packet losses for the Vegas loss predictor as a function of PER in two scenarios: (a) independent packet losses (b) correlated packet losses

Fig. 2(b) shows the accuracy of the Vegas predictor when transmission errors are correlated and modeled as described in Section 3. The Vegas predictor provides high accuracy and approaches an ideal estimator for the whole range of packet error rates.

We have also extended our analysis to more complex network scenarios, with a varying number of TCP connections, multiple hops and various patterns of cross traffic on the wired link. For the sake of brevity we do not report these results. In all the scenarios we examined, the accuracy of the Vegas predictor has always been higher than 70%.

5.   TCP Performance over Wireless Links

So far, this paper has shown that TCP NewReno-LP performs an accurate estimation of the cause of packet losses in various network scenarios. However, as this algorithm is mainly designed to achieve high goodput in the presence of links affected by random errors, a study was made of the performance of this algorithm over wireless links.

To measure TCP NewReno-LP performance, and compare it with other TCP versions, we considered several network scenarios with two different types of connections: the long-lived TCP connections, typical of FTP file transfers, and short-lived connections, typical of HTTP connections. In the following we discuss the results obtained by simulation.

Uncorrelated Losses

Following the guidelines proposed in [S.Floyd and V.Paxson, 2001], we considered the topology shown in Fig. 1. We analyzed three network scenarios with different capacity of the wired and wireless link: $C_{SN} = 2, 5$ or 10 Mbit/s and $C_{ND} = 10$Mbit/s. The Round Trip Time (RTT) is always equal to 100 ms and the queue can contain a number of packets equal to the bandwidth-delay product. We considered independent packet losses, modeled as described in Section 3. For each scenario we measured the steady state goodput obtained by TCP NewReno-LP (the bold line), TCP Westwood with NewReno extensions [High Performance Internet Research Group,] and TCP NewReno. All goodput values presented in this Section were calculated over multiple file transfers with a 97.5% confidence level [K.Pawlikowski et al., 2002]. The results are shown in Figures 3(a), 3(b) and 4, where the vertical lines represent, as in all the other Figures, the confidence interval for each goodput value.

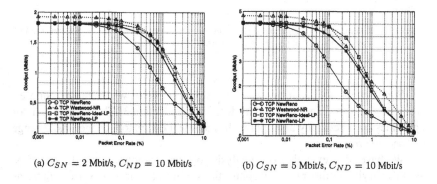

(a) $C_{SN} = 2$ Mbit/s, $C_{ND} = 10$ Mbit/s   (b) $C_{SN} = 5$ Mbit/s, $C_{ND} = 10$ Mbit/s

Figure 3.   Goodput achieved by various TCP versions in the topology of Fig. 1 as a function of PER

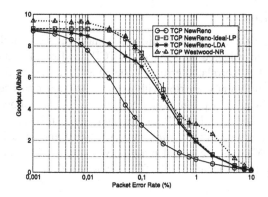

Figure 4. Goodput achieved by various TCP versions in the topology of Fig. 1 with $C_{SN} = 2$ Mbit/s and $C_{ND} = 10$ Mbit/s as a function of PER

It can be seen that for all packet error rates and at all link speeds TCP NewReno-LP achieves higher goodput than TCP NewReno. This is due to the Vegas loss predictor that prevents, most of the time, confusion between real network congestion signals, due to queue overflow, and signals due to link errors.

Note that for packet error rates close to zero, when congestion is the main cause of packet losses, TCP NewReno-LP achieves practically the same goodput as TCP NewReno. This allows TCP NewReno-LP sources to share friendly network resources in mixed scenarios with standard TCP implementations, as it will be shown in Section 5.

In all the considered scenarios, we also measured the goodput achieved by a TCP Westwood source with NewReno extensions (TCP Westwood-NR). In all simulations this source achieved higher goodput than the other TCP versions, especially when the packet error rate was high. However, we believe that there is a trade-off between achieving goodput gain in wireless scenarios and being friendly toward existing TCP versions in mixed scenarios where the sources use different TCPs. In fact, if a TCP source is too aggressive and achieves a goodput higher than its fair share over a wired, congested link, its behavior is not friendly toward the other competing connections. This behavior will be analyzed in Section 5.

To provide a comparison, Figures 3(a), 3(b) and 4 also report the performance achieved by a TCP NewReno based on an ideal estimator that always knows the exact cause of packet losses (TCP NewReno-Ideal-LP). This scheme provides an upper bound on the performance achievable by every scheme based on loss predictors. Note that our scheme approaches this bound for all the considered scenarios.

Correlated Losses

To account for the effects of multi-path fading typical of wireless environments, we also investigated the behavior of TCP NewReno-LP in the presence of links affected by correlated errors, modeled as described in Section 3. We considered two different scenarios with wireless link capacities equal to 2 and 5 Mbit/s, and a Round Trip Time equal to 100 ms. Fig. 5(a) shows the steady-state goodput achieved by the TCP versions analyzed in this paper as a function of the packet error rate in the *Bad* state. TCP NewReno-LP achieves higher goodput than TCP NewReno and practically overlaps to the goodput upper bound achieved by the ideal scheme TCP NewReno-Ideal-LP.

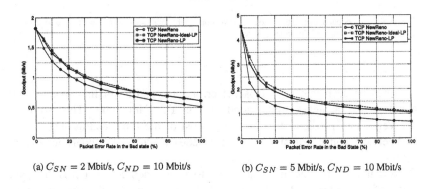

(a) $C_{SN} = 2$ Mbit/s, $C_{ND} = 10$ Mbit/s   (b) $C_{SN} = 5$ Mbit/s, $C_{ND} = 10$ Mbit/s

Figure 5.   Goodput achieved by various TCP versions in the topology of Fig. 1 as a function of PER in the *Bad* state

A similar behavior was observed in Fig. 5(b) where we reported the goodput achieved by the analyzed TCP versions in the topology shown in Fig. 1 with a 5 Mbit/s link capacity as a function of the packet error rate in the *Bad* state. Note that in this scenario the performance improvement of TCP NewReno-LP over TCP NewReno is higher than in the 2 Mbit/s scenario, as wireless losses affect more heavily TCP NewReno goodput when the bandwidth-delay product of the connection is higher [T.V.Lakshman and U.Madhow, 1997].

Impact of Round Trip Time

Packet losses are not the only cause of TCP throughput degradation. Many studies [J.Padhye et al., 1998] have pointed out that TCP performance also degrades when the Round Trip Time (RTT) of the connection increases. TCP NewReno-LP allows to alleviate this degradation to improve performance. Fig. 6(a) and 6(b) report the goodput achieved by TCP NewReno and TCP NewReno-LP sources transmitting over a single link with capacity equal to 2 Mbit/s and a 5 Mbit/s, respectively, as a function of the Round Trip Time of the connection. The link drops packets independently with a loss probability constantly equal to 0.5%.

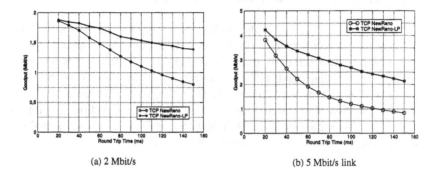

(a) 2 Mbit/s   (b) 5 Mbit/s link

Figure 6.   Goodput achieved by TCP NewReno-LP and TCP NewReno over a single link as a function of the RTT of the connection

We point out the high goodput gain of TCP NewReno-LP over TCP NewReno. This behavior is more evident when the Round Trip Time of the connection increases.

Short-Lived TCP Connections

We also studied the performance of TCP NewReno-LP with short-lived TCP connections. We considered, in line with the literature [N.Cardwell et al., 2000], a typical HTTP connection involving the transfer of a 10 kbyte file over a 5 Mbit/s link affected by a 5% random packet loss, with 100 ms Round Trip Time. We simulated 500 transfers and measured the duration of each file transfer.

The average time to complete the transfer was 0.79 seconds for TCP NewReno-LP and 0.81 seconds for TCP NewReno. Hence, also for short file transfers, TCP NewReno-LP achieves the same results as the current TCP version.

Friendliness and Fairness

So far we have shown that the TCP NewReno-LP scheme estimates accurately the cause of packet losses and that achieves higher goodput than existing TCP versions over wireless links with both uncorrelated and correlated losses.

Following the methodology proposed in [Wang et al., 2002], we evaluated friendliness and fairness of TCP NewReno-LP in a variety of network scenarios and we compared them by those achieved by TCP Westwood-NR. The term *friendliness* relates to the performance of a set of connections using different

TCP flavors, while the term *fairness* relates to the performance of a set of TCP connections implementing the same algorithms.

This section shows how the proposed scheme is able to share friendly and fairly network resources in mixed scenarios where the sources use different TCPs.

To this purpose, we first evaluated TCP NewReno-LP *friendliness* by considering two mixed scenarios: in the first one 5 TCP connections using either TCP NewReno-LP or TCP NewReno share an error-free link with capacity equal to 10 Mbit/s and RTT equal to 100 ms; in the second one the TCP NewReno-LP sources were replaced by TCP Westwood-NR sources.

By simulation we measured the goodput, for each connection, and for all cases. The average goodput of n TCP NewReno-LP and of m TCP NewReno connections, with $n + m = 5$, is shown in Fig. 7(a).

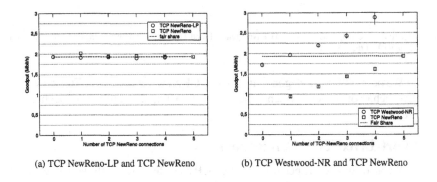

(a) TCP NewReno-LP and TCP NewReno   (b) TCP Westwood-NR and TCP NewReno

Figure 7.   Average goodput of (a) n TCP NewReno-LP and m TCP NewReno connections and (b) n TCP Westwood-NR and m TCP NewReno connections, with $n + m = 5$, over a 10 Mbit/s link with RTT equal to 100 ms

The goodput achieved by both algorithms is very close to the fair share for the full range of sources.

The same experiment was performed with TCP connections using either TCP Westwood-NR or TCP NewReno, and the results are shown in Fig. 7(b). In this scenario TCP Westwood-NR sources proved more aggressive toward TCP NewReno sources than TCP NewReno-LP, and achieved a goodput higher than the fair share practically in every case. This behavior evidences the trade off that exists between achieving high goodput gain in wireless scenarios and being friendly in mixed network scenarios.

To measure the level of *fairness* achieved by TCP NewReno-LP we considered the same scenario described above first with 5 TCP NewReno-LP connections and then with 5 TCP NewReno sources sharing a 10 Mbit/s link with RTT equal to 100 ms. In this scenarios congestion is the only cause of packet losses. The Jain's fairness index of 5 TCP NewReno-LP connections was equal to 0.9987, and that achieved by 5 TCP NewReno sources was equal to 0.9995. These results confirm that TCP NewReno-LP achieves the same level of fairness of TCP NewReno.

We also extended our simulation campaign to more complex scenarios with a varying number of competing connections. The results obtained confirm that TCP NewReno-LP achieves an high level of friendliness toward TCP NewReno, thus allowing its smooth introduction into the Internet.

6.   Conclusions

In this paper, we have discussed and analyzed issues related to the use of Loss Differentiation Algorithms for TCP congestion control. We proposed to use the Vegas loss predictor to enhance the TCP NewReno error-recovery scheme, thus avoiding unnecessary rate reductions caused by packet losses induced by bit corruption on the wireless channel. The performance of this enhanced TCP (TCP NewReno-LP) was evaluated by extensive simulations, examining various network scenarios. Two types of TCP connections were considered, namely long-lived connections, typical of file transfers, and short-lived connections, typical of HTTP traffic. Moreover, we considered two different statistical models of packet losses on the wireless link: independent and correlated losses. We found that TCP NewReno-LP achieves higher goodput over wireless networks, while guaranteeing good friendliness with classical TCP versions over wired links. Moreover, we found that the Vegas loss predictor, embedded in TCP NewReno-LP, proved very accurate in classifying packet losses. Finally, we also defined an ideal scheme that assumes the exact knowledge of packet losses and provides an upper bound to the performance of all possible schemes based on loss differentiation algorithms. The TCP enhanced with Vegas loss predictor well approaches this ideal bound.

References

[A.A.Abouzeid et al., 2000] A.A.Abouzeid, S.Roy, and M.Azizoglu (Tel Aviv, Israel, March 2000). Stochastic Modeling of TCP over Lossy Links. In *Proceedings of INFOCOM 2000*.

[Brakmo and Peterson, 1995] Brakmo, L. and Peterson, L. (October 1995). TCP Vegas: End-to-End Congestion Avoidance on a Global Internet. *IEEE Journal on Selected Areas in Communications*, 13(8):1465–1480.

[Fu and S.C.Liew, 2003] Fu, C. P. and S.C.Liew (Feb. 2003). TCP Veno: TCP enhancement for transmission over wireless access networks. *IEEE Journal on Selected Areas in Communications*, 21(2):216–228.

[H.Balakrishnan et al., 1997] H.Balakrishnan, V.N.Padmanabhan, S.Seshan, and R.H.Katz (December 1997). A Comparison of Mechanisms for Improving TCP Performance over Wireless Links. *IEEE/ACM Transactions on Networking*, 5(6):759–769.

[High Performance Internet Research Group,] High Performance Internet Research Group, U. TCP Westwood Home Page, URL: http://www.cs.ucla.edu/NRL/hpi/tcpw.

[J.Padhye et al., 1998] J.Padhye, V.Firoiu, D.Towsley, and J.Kurose (1998). Modeling TCP Throughput: A Simple Model and its Empirical Validation. In *Proceedings of ACM SIG-COMM '98*.

[K.Pawlikowski et al., 2002] K.Pawlikowski, Jeong, H., and Lee, J. (Jan. 2002). On Credibility of Simulation Studies of Telecommunication Networks. *IEEE Communications Magazine*, pages 132–139.

[M.Allman et al., 1999] M.Allman, V.Paxson, and W.Stevens (April 1999). TCP Congestion Control. *RFC 2581*.

[N.Cardwell et al., 2000] N.Cardwell, S.Savage, and T.Anderson (2000). Modeling TCP Latency. In *Proceedings of INFOCOM 2000*, pages 1742–1751.

[ns-2 network simulator (ver.2).LBL,] ns-2 network simulator (ver.2).LBL.   URL: http://www.isi.edu/nsnam.

[S.Bregni et al., 2003] S.Bregni, D.Caratti, and F.Martignon (San Francisco, 1-5 Dec. 2003). Enhanced Loss Differentiation Algorithms for Use in TCP Sources over Heterogeneous Wireless Networks. In *Proceedings of IEEE GLOBECOM'03*.

[S.Cen et al., 2003] S.Cen, P.C.Cosman, and G.M.Voelker (Oct. 2003). End-to-end Differentiation of Congestion and Wireless Losses. *IEEE/ACM Transactions on Networking*, 11(5):703–717.

[S.Floyd and V.Paxson, 2001] S.Floyd and V.Paxson (August 2001). Difficulties in Simulating the Internet. *IEEE/ACM Transactions on Networking*, 9:392–403.

[T.V.Lakshman and U.Madhow, 1997] T.V.Lakshman and U.Madhow (1997). The performance of TCP/IP for networks with high bandwidth-delay products and random loss. *IEEE/ACM Transactions on Networking*, 5(3):336–350.

[Wang et al., 2002] Wang, R., Valla, M., Sanadidi, M., and Gerla, M. (2002). Adaptive Bandwidth Share Estimation in TCP Westwood. In *Proceedings of Globecom'02*.

TCP PERFORMANCES IN A HYBRID BROADCAST/TELECOMMUNICATION SYSTEM

Davy Darche[1], Francis Lepage[2], Eric Gnaedinger[3]

TDF 1, rue Marconi 57000 Metz[1], CRAN, CNRS UMR 7039, Université Henri Poincaré NANCY I BP 239 54506 Vandoeuvre-les-Nancy Cedex, FRANCE[2,3]

Email: davy.darche@tdf.fr[1] ,francis.lepage@cran.uhp-nancy.fr[2] ,eric.gnaedinger@cran.uhp-nancy.fr[3]

Abstract   High speed cable/ADSL connections are rapidly becoming the standard for Internet at home. Conversely, mobile network operators only offer low bit rate data services to their 2.5G users, and 3G will not be in full swing for another couple of years. In this context, the cooperation between a mobile telecommunication network and a broadcast network can be a suitable alternative to enhance this offer with high speed e-mail, web browsing, file download or even peer-to-peer services. This paper presents a network architecture based on the coupling of a GPRS uplink with a DVB-T downlink, to provide Internet connectivity for unicast and multicast services. We first study the GPRS network and show the issues raised by specific use of this return channel. Then, we analyse, using simulations, how to tune the TCP parameters to increase the GPRS/DVB-T hybrid network performances. Finally, we describe the network architecture of the deployed system, and evaluate its performances.

Keywords: Hybrid network, cooperation, asymmetry, TCP, GPRS, DVB-T

1.   Introduction

In a hybrid network, the uplink and downlink are different. In our case we use DVB-T (Digital Video Broadcasting Terrestrial) for the downlink and GPRS (Generalised Packet Radio Services) for the uplink. The request (*Req.*) is sent by GPRS to the HNIS (Hybrid Network Interconnection System). Then the request is forwarded to the appropriate server on the Internet, whose reply is routed on the DVB-T network [1].

DVB-T is a robust, unidirectional, broadcast network with high bandwidth, whereas GPRS is a bidirectional telecommunication network, with a low bandwidth, and a considerable bit error rate [2]. The cooperation between these two networks creates important asymmetries between the two links of communication as shown by experimental measures (Tab. 1). Thereafter we will study the impact of this asymmetry on the TCP mechanisms. Indeed, the TCP pro-

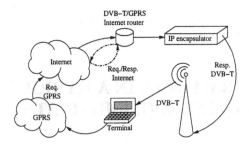

Figure 1.   Hybrid network architecture

tocol, without specific tuning, has poor performances in this hybrid network architecture (Fig. 1).

QoS parameters	GPRS (uplink)	DVB-T
Latency	700ms-4s	50-60ms
Jitter	4s	10ms
Packet error rate	1-2%	$<10^{-6}$
Available bandwidth	20kbits.s^{-1}	20Mbits.s^{-1}

Table 1.   GPRS/DVB-T QoS parameters

Measures in table (Tab. 1) are performed for several bit rates (0.6kbits.s^{-1} to full bandwidth).

2.   Issues raised by the GPRS return channel

In this section we study the GPRS return channel issues. The following experimentations are made on real GPRS networks. We use several GPRS subscriptions of Finnish and French operators. All the results being very similar with the different operators, we only analyse the test results of one operator, RADIOLINJA. All the GPRS subscriptions have two timeslots for upload. The experimental tests consist of sending IP packets (ICMP packet, 84 bytes) from a terminal to a server, and receiving these packets. The sent packets generate a constant bit rate. Then we study the RTT (Round Trip Time) related to each packet. In one case the packets are sent by GPRS and are received by GPRS. We call this transmission mode *bidirectional*, since we use the uplink and the downlink of GPRS. In another case, the packets are sent by GPRS and are received by LAN. We call this transmission mode *unidirectional*. As the latency of the transmission on the LAN is very low compared to the latency on GPRS (LAN<1ms, GPRS$\simeq$1s), we do not take into account the latency of the LAN.

So, we can evaluate the behavior of the GPRS latency as a function of the uplink load, and thus the consequences of an exclusive use of the GPRS uplink.

2.1   GPRS Bidirectional mode

We can notice in the figure 2 that the latency of the GPRS network is around 1s (despite many oscillations), for throughputs ranging from 5kbits.s^{-1} to 20kbits.s^{-1}. At 25kbits.s^{-1}, we reach the bandwidth limit of the GPRS connection. The latency increases strongly due to buffer mechanisms in the GPRS network.

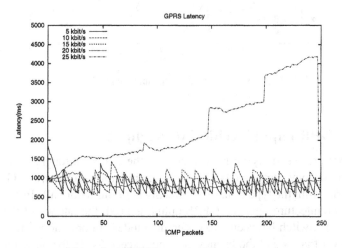

Figure 2.   GPRS latency in bidirectional mode

For lower bit rates, the throughput oscillates presumably because of the time division of the bandwidth in the transmission mode of GPRS.

2.2   GPRS Unidirectional mode

We note that the behavior of the GPRS network (latency according to the load of the uplink) in the unidirectional mode (Fig. 3) is totally different from the bidirectional mode. The latency increases faster for small throughputs. A detailed analysis of the GPRS latencies shows a brutal jump from 2.4s to 4s for the traffic of 15kbits.s^{-1}, 20kbits.s^{-1}, 25kbits.s^{-1}. The latency increases quasi-linearly until reaching a critical threshold of 2.4s (Fig. 3). There are therefore two different behaviors of the GPRS latency:

- A stationary, oscillating mode for throughput lower than 10kbits.s^{-1}.

- A quasi-lineary, increasing mode involving a latency jump from 2.4s to 4s, for throughput greater than 10 kbits.s^{-1}.

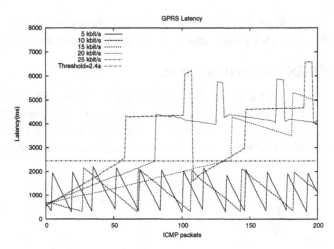

Figure 3.   GPRS latency in unidirectional mode

2.3   GPRS uplink critical throughput

A more precise study permits to specify the critical throughput γ beyond which the latency jumps to 4s. As shown in previous sections, the GPRS latency varies with the type of use (unidirectional or bidirectional). In our hybrid network architecture only the GPRS uplink is used, the latency of the GPRS network can switch between two differents modes. In one mode the throughput is lower than γ and the latency does not exceed 2s. In the other mode, the throughput is greater than γ and the latency increases to 4s-5s.

If we consider the GPRS network as a black box model, we can conclude that the GPRS uplink quality of service is probably related to the traffic on the GPRS downlink. It is therefore necessary to evaluate the throughput needed on the uplink for a specific service when we use GPRS as a return channel. In the GPRS/DVB-T hybrid network, we must limit the uplink traffic to a value smaller than γ to prevent the jump of the latency. This brutal increase of the latency can otherwise be considered by TCP as a timeout. We will solve the problem at the transport layer, as we have limited control on the GPRS network.

3.   Simulation studies of the hybrid network performances

3.1   Simulation model of the hybrid network

In this section, we use the NS2 (Network Simulator v2) simulation tool to study the hybrid network architecture, to solve the GPRS return channel issues, and assess the impact of the various asymmetries of this system. We only simulate OSI layers three and above. In particular the encapsulation mechanisms inherent to the DVB-T and GPRS transmission are not simulated. DVB-T is represented by a unidirectional link with 20Mbits.s^{-1} bandwidth and 50ms of delay. GPRS is modeled by a unidirectional link with 20kbits.s^{-1} bandwidth and 700ms of delay (Fig. 4). We simulate a data transfer from a server **S** to a client **C** trough a TCP connection (FTP). We add a node **n** and a new link to introduce the packet loss rate. This link has the same bandwidth as GPRS and very low delay (1ms). The node **n** has a very large buffer to simulate the GPRS network buffer whereas the server and the client have minimal buffers.

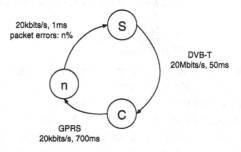

Figure 4. The simulation model

3.2 Asymmetries

Asymmetry of delays. The difference of delay between GPRS and DVB-T is very important, with respectively 50ms on DVB-T and 700ms on GPRS. However, this difference does not affect the TCP functions. It is the large RTT (Round Trip Time), around 750ms, that hinders the throughput of the TCP communication.

$$Throughput_{max} = \frac{TCPWndSize}{RTT} \qquad (1)$$

$$\rho = \frac{RTT_{GPRS}}{RTT_{DVB-T}} \qquad (2)$$

For the simulations, the global RTT is fixed to 750ms ($RTT_{GPRS} + RTT_{DVB-T}$), and we vary the ratio ρ. The ratio ρ has absolutely no effect, the simulation results show that the throughput on the downlink is always 626kbits.s^{-1}. However to increase the TCP performances with this large RTT, we have to increase the TCP window sizes.

Bandwidth asymmetry. The maximal size of a packet, that can be sent from the terminal to the server and from the server to the terminal, without fragmentation is 1500 bytes ($MTU_{max} = 1500$). The TCP protocol generates a maximum of one Ack (40 bytes) per packet of data. A data transfer from the server to the terminal produces, on the uplink, a traffic with a throughput that cannot exceed 2.67% of the downlink throughput.

$$\frac{Ack_{size}}{MTU_{max}} = 0.0267 \tag{3}$$

TCP uses a cumulative acknowledgement mechanism, and thus one Ack can acknowledge more than one packet of data. We have always:

number of Ack $n\leq$ number of data packets m

$$\frac{n.Ack_{size}}{m.MTU_{max}} \leq 0.0267 \tag{4}$$

By using the full bandwidth capacity of the uplink (20kbits.s^{-1}), and without the other constraints (RTT, packet error rate...), we can estimate a *minimum* throughput of 750kbits.s^{-1} for the download traffic. The TCP option *Delayed Ack* permits to defer the sending of the Acks for a duration δ. We can thus increase the number of data packets acknowledged by one Ack. So, it is possible to reduce the bandwidth used on the uplink, or to increase the throughput on the downlink.

TCP option	GPRS throughput	DVB-T throughput
—	19.9kbits/s	750kbits/s
Delayed Ack 5ms	19.9kbits/s	1200kbits/s
Delayed Ack 20ms	19.9kbits/s	1500kbits/s

Table 2.   Bandwidth asymmetry, GPRS latency 1ms, DVB-T latency 1ms

We have run two series of simulations. In the first series, we do not take into account the temporal dimension of the system by fixing the DVB-T and GPRS latencies to 1ms (Tab. 2). We notice that the traffic on the GPRS link uses all the bandwidth capacity (20kbits.s^{-1}) and the throughput on the downlink increases.

In the second simulation, we set the GPRS latency to 700ms and the DVB-T latency to 50ms and we tune the TCP DelayedAck options (Tab. 3). The throughput of the downlink increases slightly more while the traffic on the uplink decreases strongly. In this case the throughput of the system is limited by the temporal dimension (Expr. 1). The value of the TCP window size is 50MSS (Maximum Segment Size) [4] [6] [5]. One MSS is equal to the MTU without the TCP and IP headers, 1460 bytes. Consequently the maximal theorical throughput is 778.6kbits.s^{-1}. This theorical value is confirmed by the simulations. By increasing the value of the TCP window size to 100MSS, we can

TCP option	GPRS throughput	DVB-T throughput
—	19.9kbits/s	747.2kbits/s
Delayed Ack 5ms	12.5kbits/s	757.0kbits/s
Delayed Ack 10ms	12.4kbits/s	752.3kbits/s
Delayed Ack 20ms	10.1kbits/s	761.1kbits/s

Table 3. Bandwidth asymmetry, GPRS latency 700ms, DVB-T latency 50ms

theorically double the throughput of the system. Once again the full bandwidth of the uplink is used, and the downlink throughput reaches $1490kbits.s^{-1}$.

The asymmetry of the packet error rate. In most networks, uplink and downlink have the same packet error rate. In this case, the DVB-T communication link has a very low packet error rate, lower than 10^{-6}, whereas the GPRS link has a large packet error rate, around 1%. By varying the error rate on the two links, we study how these differences affect the TCP performances. We use the standard TCP implementation and options for these simulations.

GPRS error rate	DVB-T error rate	Throughput on DVB-T
0%, 1%, 2%	0%	749kbits/s
0%	0.01%	707.4kbits/s
0%	0.1%	488.4kbits/s
0%	1%	142.8kbits/s
2%	0.01%	704.6kbits/s

Table 4. Error rate asymmetry

The simulations show that the hybrid architecture is more sensitive to the errors on the data download link than to the errors on the return channel. This behaviour is inherent to the cumulative aspect of the TCP acknowledgement mechanism. Conversely, when the loss of a data packet occurs, TCP has to wait for a timeout to detect the packet loss and resend it. This timeout is similar to the RTT, 750ms. This decreases strongly the throughput of the system. The reliability of the DVB-T transmission to transfer the data packet is appropriate for the service we want to provide to the end user (data transfer to the client). Moreover, the error rate of GPRS has very limited influences on the quality of the service. We consider the system in degraded mode, with 2% of error rate on GPRS and 0.01% of error rate on DVB-T. We compare the different TCP implementations with differents options (Tab. 5).

TCP implementation	GPRS throughput	DVB-T throughput
Tahoe TCP	18.7kbits/s	704.6kbits/s
Tahoe TCP, DelAck	9.7kbits/s	693.6kbits/s
Reno TCP	19.1kbits/s	718.7kbits/s
Reno TCP, DelAck	9.9kbits/s	710.2kbits/s
NewReno TCP	19.1kbits/s	718.7kbits/s
NewReno TCP, DelAck	9.9kbits/s	710.2kbits/s
Sack TCP	19.1kbits/s	717.2kbits/s
Sack TCP, DelAck	9.7kbits/s	700.6kbits/s

Table 5. TCP implementation in degraded mode

We notice that the use of the TCP *Delayed Ack* option slightly decreases the performances. As there are fewer Ack to acknowledge packets with the *Delayed Ack* option, the loss of one Ack increases the probability to not acknowledge a data packet. However we can see that we only use half of the uplink bandwidth. This way, we can limit the throughput on GPRS below the critical GPRS throughput, $10kbits.s^{-1}$, and we can solve the GPRS return channel issue.

3.3 Hybrid routing

With the delayed ack option, the upload traffic remains under $10kbits.s^{-1}$. However, if the terminal initiates a data transfer to the server, a new traffic will be added to the $10kbits.s^{-1}$ of acknowledgment traffic. A way to prevent the latency jump on the GPRS link, is to adapt the traffic policy routing on the HNIS router. The Ack related to the upload traffic is routed to the GPRS downlink. Thus the GPRS behaviour switches to a bidirectional mode and the latency is fixed to a value near 700ms. We use a traffic control module to regulate the data traffic on the downlink and to prevent bursts of traffic. This contributes to enforce the architecture stability. Another functionality is to ensure the bandwidth for a single user.

4. Experimentations

For the experimentations, we use a VPN to connect the terminal to the DVB-T/GPRS Internet router. The VPN offers the compression and encryption capabilities. We use a VPN connection to pass through firewall and/or NAT router.

The tests consist of a FTP transfer from the router/server to the terminal. In one case we use the TCP DelayedAck option (set to 600ms) and in the other case the terminal acknowledges all the packets.

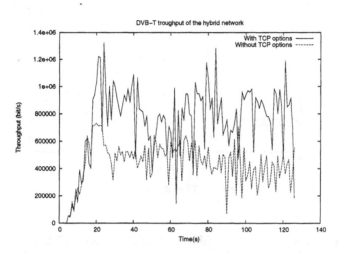

Figure 5. DVB-T throughput in the hybrid architecture

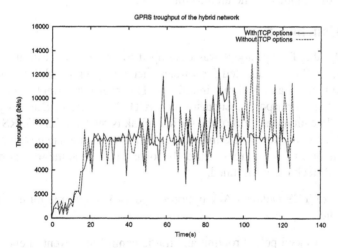

Figure 6. GPRS throughput in the hybrid architecture

We notice that the GPRS throughput is around 6kbit.s^{-1} for the two FTP transfers. With the TCP DelayedAck option, the throughput of the DVB-T traffic can reach 1.3Mbits.s^{-1} whereas without the option, the throughput is limited to 700kbit.s^{-1}. The average throughput is 747.8kbit.s^{-1} with the TCP

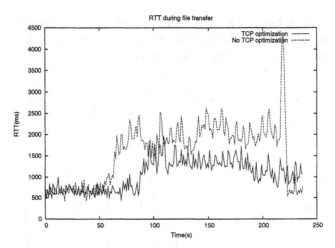

Figure 7.   RTT in the hybrid network

option, against 435.3kbit.s^{-1} without the TCP DelAck. With the TCP optimizations, the RTT has a value of 1185ms, whereas without the RTT has an
average of 1881ms (Fig 7). A low RTT contributes to increase the bit rate
(Expr. 1). Not only the adaptation of TCP increases the performances, but it
increases the stability of the architecture as well.

5.   Conclusion

Initially, the TCP protocol was developed to prevent congestion in wired
networks, TCP is not adapted for wireless networks, with high bit error rate,
high asymmetries and multiple interfaces. However for current Internet services, a reliable transport mechanism such as TCP is required. Due to dynamic
resource allocation, the QoS of the GPRS uplink is related to the GPRS downlink traffic. This behavior is a major issue as it impacts the latency in a DVB-
T/GPRS hybrid network architecture. To solve this issue at the transport level,
the TCP/IP stack has been tuned.

- Use of TCP Delayed Ack option to reduce strongly the return channel
  traffic.

- An advanced policy routing and traffic control to prevent latency peaks
  during upload data transfer.

With these tunings, on the hybrid network architecture, it becomes possible
to provide stable high bit rate services (HTTP, FTP, peers to peers, streaming...)
in a wireless environment.

References

[1] Hybrid Mobile Interactive Services combining DVB-T and GPRS, 2001, Christian Rauch Vodafone Pilotentwicklung GmbH (formerly Mannesmann Pilotentwicklung GmbH) Wolfgang Kellerer, Peter Sties Munich University of Technology (TUM), Institute of Communication Networks Arcisstr.

[2] Multi-Layer Protocol Tracing in a GPRS Network, 2002, Andrei Gurtov, Matti Passoja, Olli Aalto, Mika Raitola Cellular Systems Development Sonera Helsinki, Finland

[3] W. Richard Stevens TCP/IP Illustrated, Volume 1 Ed. Addison Wesley, 1994 ISBN 0-201-63346-9

[4] RFC 793 - Transmission Control Protocol

[5] V. Jacobson, R. Braden and D. Borman, RFC 1323 - TCP Extensions for High Performance.

[6] M. Mathis, J. Mahdavi, S. Floyd and A. Romanow, RFC 2018 - TCP Selective Acknowledgement Options

HANDOFF NOTIFICATION IN WIRELESS HYBRID NETWORKS

Guillaume Chelius
Inria Ares Team, Laboratoire Citi, Insa de Lyon,
21, avenue Jean Capelle, 69621 Villeurbanne Cedex, France
Guillaume.Chelius@insa-lyon.fr

Claude Chaudet
Inria Ares Team, Laboratoire Citi, Insa de Lyon,
21, avenue Jean Capelle, 69621 Villeurbanne Cedex, France
Claude.Chaudet@insa-lyon.fr

Abstract   Hybrid networks composed of a wireless infrastructure network providing Internet access to an underlying ad hoc network are more and more attractive due to their low installation cost. In these all-wireless environments, performance is a key issue as radio bandwidth is scarce. Handoffs management is particularly important as these networks are likely to be highly mobile. Mobility notification should therefore be optimized in order to limit signaling overhead while keeping a good reactivity against terminals mobility. This article presents and studies by simulation different level optimizations applied to a modified Cellular IP protocol.

Keywords: ad hoc networks, hybrid networks, micro-mobility

1.   Introduction

Wireless communications have to play a crucial role in computer networks. They offer open solutions to provide mobility and services where the installation of a complex wired infrastructure is not possible. With the exponential growth of wireless communications, a wide range of wireless devices has been released. In the same time, the number of cellular phones has significantly increased. The Internet becomes pervasive and is now bound to cellular networks. Research on wireless networks has roughly been concentrating on two distinct themes. The first one aims at extending the edge of infrastructure networks by the integration of a last wireless hop. The radio connectivity is provided by Base Stations at the edge of the network. The second theme concerns infrastructure-free and auto-organized wireless networks: Mobile Ad hoc Networks (MANet).

Research efforts aiming at merging cellular wireless and ad hoc networking have recently increased [4, 10, 5, 7–1, 6]. Hybrid networks, the extension of cellular networks using ad hoc connectivity, offer obvious benefits. On one hand, they the extend cellular network coverage using ad hoc connectivity and on the other hand they provide a global Internet connectivity to ad hoc nodes. However, deployment of a wired cellular infrastructure still induces a high cost as well as a lot of constraints. Both costs and constraints can be reduced if we replace the wired infrastructure network by a fully wireless one. The infrastructure network becomes a collection of static wireless nodes acting both as Base Stations and infrastructure routers. Infrastructure communications become wireless and multi-hop. As the wireless medium slightly differs from the wired one, the design of classical micro-mobility protocols must be rethought and if necessary modified.

In this article, we study how ad hoc node mobility/handoffs must be notified in the wireless infrastructure of a wireless hybrid network in order to achieve the best performances. Several strategies are proposed. Section 2 presents the routing protocol as well as the testbed that was used for simulations. Several strategies for mobility notification frame transmission are compared in section 3. Finally in section 4, we propose and compare several mechanisms to reduce the signaling overhead in the wireless hybrid network.

2.   Wireless Hybrid Network

The global architecture is composed of a wireless infrastructure network extended by a general Mobile Ad-Hoc Network (MANet). This differs from the work done in [4] since the infrastructure network is wireless and the ad hoc connectivity may extend further than two hops. In this article, we focus on the ad hoc nodes mobility notification process within the wireless infrastructure network and do not compare routing strategies in hybrid networks. Such a study may be found in [10].

Several micro-mobility protocols such as Cellular IP, Hawaii, Hierarchical Mobile IP or Edge Mobility have been proposed for wire infrastructure networks. Their main tasks are to efficiently manage intra-domain routing, enabling mobiles to perform fast handoffs between Base Stations, as well as to provide a paging service. Several surveys and comparative studies have been published in [2, 3]. In the rest of this article, we will only consider Cellular IP [9] as it is largely deployed and has been the subject of an ad hoc extension in many proposals [8, 1]. The protocol is presented here in a version that has been slightly modified for use in a wireless infrastructure network and in interaction with a separate ad hoc routing protocol which is used to extend the infrastructure coverage.

Every 0.2 s, mobile nodes broadcast ad hoc packets which contain the list of the mobile neighbors, other mobiles and Base Stations, and the identity of the Base Station the mobile has chosen to attach to. These packets play both the role of control packets for the ad hoc routing protocol and route update for the CIP-like protocol.

Infrastructure nodes also act like Base Stations as they communicate through a wireless medium and participate to ad hoc routing through the broadcast of ad hoc packets every 0.2 s. These ad hoc packets contain a list of mobile neighbors and advertise them as a Base Station. Here again, these ad hoc packets play both the role of control packets to the ad hoc routing protocol and BS advertisement for the CIP-like protocol.

Mobility notification is implicitly initiated by a mobile. As a Base Station receives an ad hoc packet from an attached mobile, it transforms the ad hoc packet into a CIP-like route update packet and forwards it to the infrastructure network Gateway through its up-link neighbor. In an infrastructure node, reception of a mobile ad hoc packet updates the route entry for this mobile and reception of a route update packet updates the down-going route to the mobile with the last forwarder as next hop. An infrastructure route has a lifetime of 0.5 s. By default, if no specific route is known, a data packet is forwarded toward the infrastructure Gateway. Mobile nodes always transmit their data packet to their Base Station.

The up-link neighbor of an infrastructure node is the father's node in the CIP-like routing tree. This tree, routed at the gateway and spanning the infrastructure network, is created by diffusion of a gateway advertisement packet, periodically emitted by the gateway and forwarded by all infrastructure nodes.

We performed our simulations using the network simulator NS-2[1]. The topology network used consists of 9 wireless infrastructure nodes with the addition of 2 to 64 mobile nodes which move according to the Random Waypoint Mobility model. The maximum speed of the mobile nodes has been set to 50m/s. Constant Bit Rate data flows of 5 packets of 500 bytes per second are simulated between the mobile nodes.

3. Comparing the Route Update strategies

The radio medium is far from being similar to a classical wire medium such as the Ethernet one and differs in several aspects. Wireless links are pervasive and not isolated due to the broadcast nature of the medium. In consequence, the topology of the infrastructure network is not efficiently mapped into a routing tree as it is in Cellular IP. Another difference between wired and wireless links is the medium transmission quality and efficiency as bandwidth is smaller and latency is greater in air. These differences lead to the fact that some design aspects chosen during the development of Cellular IP may no longer be appropriate. For example, the choice to transmit `route update` packets in unicast increases reliability but also medium occupancy and prevents from routing along efficient paths.

3.1   Acknowledged broadcast

The IEEE 802.11 distributed coordination function (DCF) basically provides two MAC level modes for frame transmissions: unicast and broadcast modes. Unicast frames require acknowledgment from the receiver. The lack of acknowledgement reception causes the retransmission of the frame until the transmission succeeds or the maximum retransmission count is exceeded. Unicast frames can also be protected, especially against hidden node situations, by using a *Request to send - Clear to send* (RTS-CTS) exchange prior to frame transmission. Broadcasted frames are neither protected by RTS-CTS, nor acknowledged. Therefore, correct reception cannot be guaranteed. But, if the same data rate is used, broadcasted frames are far more efficient when transmitting information to a set of neighbor nodes. Both strategies present advantages for `route update` messages transmission. Unicast mode favors reliability and can help maintaining an accurate view of the network topology. Broadcasted frames favor speed and allow the spread of the topology updates faster. Mixing the two approaches to obtain a more reliable broadcast could be profitable. Nevertheless, acknowledging broadcasted frames is not straightforward as multiple acknowledgements from multiple receivers would collide. Therefore, acknowledging broadcasted frames requires the selection of one particular neighbor to acknowledge frames, as if the message was transmitted in unicast mode and every other mode were in a promiscuous reception mode. Wireless hybrid networks provide a hierarchical organization removing the need for dynamic election of the only neighbor acknwoledging frames that arises in a pure ad hoc context. When a node emits a `route update` message, it is destined to its father in the routing tree. If other infrastructure nodes can overhear this message, they will also benefit from this information, adding a route to the mobile and enabling a shortcut in the tree routing scheme of Cellular IP. In the following paragraphs, we will compare the results we obtained for the three possible strategies for `route update` transmission: using unicast transmission, broadcast or acknowledged broadcast. In order to correctly study the differences between this three transmission modes, we will keep the tree routing scheme of Cellular IP and avoid taking the advantages offered by the broadcast and acknowledged broadcast modes.

3.2   Simulation Results

First of all, we will compare three strategies at the MAC-level for `route update` messages transmission. Unicast mode includes RTS-CTS excahnge and acknowledgements. Acknowledges broadcast suppresses RTS-CTS exchange and broadcast suppresses both mechanisms.

With Cellular IP, each node will regularly emit `ad hoc` packets, each Base Station will regularly forward `route update` packets and relay `gateway advertisement` packets. Signaling can represent a high load when the network gets dense. Signaling packets nevertheless carry useful information and should neither be lost, nor be delayed too much. Losing route updates will result in many routing table inconsistencies and delaying these packets too much will result in outdated information in the routing tables. We need to find the correct balance between network load and informations accuracy.

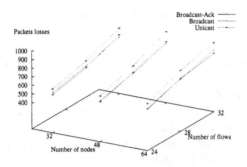

Figure 1.   Losses of CBR packets: route disappearance

Figure 1 presents the losses of data packets due to the absence of route towards the destination in the whole network for configurations where the medium is overloaded (more than 16 CBR data flows). This situation arises when a route has been deleted due to timeout and the new route has not yet been discovered or propagated. Data packets are forwarded to the Gateway, which is the root of the infrastructure network, that drops the packet. Unicast transmission of `route update` messages leads to the highest data packet loss rate, due to the delay introduced by the protection of signaling frames. Routes are not refreshed in time and routing tables entries disappear.

On the opposite, Figure 2 presents the losses of data packets due to retransmissions by the infrastructure nodes. This situation happens whenever a mobile has moved but the routing entry in the infrastructure network still references the old base station. In this situation, broadcast transmission of `route` `update` frames leads to the highest loss number due to the low reliability of the signaling messages transmission.

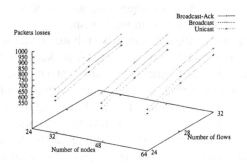

Figure 2.   Losses of CBR packets: wrong routing entry

Both strategies present advantages as well as drawbacks and it is difficult to determine which method will yield optimal results. Acknowledged broadcast is an "in the middle" approach and could lead to better overall results in the end. Figure 3 represents the total number of packets that have been correctly transmitted in each of the simulations performed. As soon as the data flows saturate the medium, broadcast mode outperforms unicast mode by 25% in the best case. Unicast transmission of `Route Update` messages always results in the poorest performance, followed by acknowledged broadcast and broadcast.

The number of packets successfully delivered is highly dependent on the network load. When the network is not overloaded, the performances of the three strategies are equivalent. Then, when the network capacity is exceeded, the overhead introduced by the `route update` transmission mode results in a difference in the number of packets successfully transmitted. As transmitting a packet in broadcast mode requires less time than transmitting the same packet in unicast mode, the medium capacity is exceeded later with broadcast `route update` packets. Finally, performances become equivalent again when the network is overloaded regardless of the transmission mode.

4.   Optimization of the mobility notification

From the results of sections 3, we can deduce that the main challenge to improve data traffic delivery is to reduce the radio medium utilization. We have to reconsider the experimental protocol described in section 2 in order to lower the number of control packets its use requires.

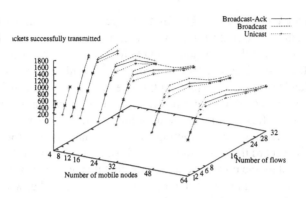

Figure 3. Number of CBR packets correctly received

4.1 Differential Route updates

Frequent route updates are necessary while the mobile performs a handoff. A new route has to be set up in the infrastructure network as fast as possible in order to avoid misrouting and losing of data packets. A first `route update` packet must be sent after the mobile's handoff to create the route. This sending must be repeated in short successive intervals of time in order to prevent the loss of the previous `route update` packets as their delivery is not reliable. After the route setup and while the mobile remains connected to the Base Station, frequent updates are no longer needed. The time interval between two consecutive updates may be increased in order to lower the number of control packets. However, the frequency of `ad hoc` packet emission may not be reduced as cellular stillness is far from meaning ad hoc stillness. In consequence, only a subset of the `ad hoc` packets may be forwarded by Base Stations as `route update` packets.

We introduce a flag in `ad hoc` packets to notify the Base Station whether or not the packet must be forwarded in the infrastructure as a `route update`. The flag is set by the mobile as it is the one which initiates the handoff. After a handoff, the first period between two consecutive *route-update* is 0.2s and this period is increased by a factor 1.5 for each consecutive *route-update*. We call this mechanism *differential route update*.

4.2 Nack route

While using differential route update mechanism, route update emission intervals may be larger than the `ad hoc` packets period. This means that the link between the mobile and its Base Station is refreshed at a higher rate than the infrastructure down-going route to the mobile. In consequence, the Base Station may notice a mobile's handoff long before the mobile's route times out in the infrastructure. This route reminiscence may lead to incorrect routing and data packet loss. In order to prevent this phenomenon, a Base Station may, as soon as it has detected a mobile has left its cell, discard the route towards this mobile in the infrastructure. As routing incoherence between different infrastructure routers may lead to routing loops, it is not enough for the Base Station to only discard the route on its own. It should notify all infrastructure routers concerned by the now out-dated route, that it is no longer valid. This is realized by the Base Station emitting a `route delete` packet which is forwarded to the infrastructure Gateway along the same path as `route update` packets. This packet discards the mobile's route in the infrastructure routers on its path. We call this mechanism *nack route*.

4.3   Nack only

If we carry on trying to reduce control traffic to its extreme, we can completely avoid multiple `route update` emissions after a mobile handoff and send only one. This strategy is optimistic in the sense that it makes the supposition that `route update` packets may not be lost in the infrastructure. If it is lost, there will be no infrastructure route to the mobile and data packets will be dropped. Since only one `route update` is sent for each handoff and no refreshment is further performed, infrastructure routes have an infinite lifetime. To invalid an old route after a mobile handoff, Base Stations send a `route delete` packet, as explained in the previous section. This strategy sounds far from reliable as only one `route update` loss has catastrophic consequences, but it has the advantage of drastically reducing the control traffic.

4.4   Simulation Results

Simulations have been carried out for broadcast, unicast as well as acknowledged broadcast route update transmission modes. Usually, using the optimizations described above in these three transmission modes provides similar results when evaluating the optimizations performances. Even if the numbers of packets successfully transmitted are not the same, the phenomenons described below are the same for the three modes. Therefore we will only present figures for one single transmission mode. Simulations show that the number of nodes in the network has a much lower influence on overall performance when compared to the number of data flows in the network. Therefore, for readability, we will only present results for networks of 64 mobile nodes, as results are also similar when considering fewer nodes.

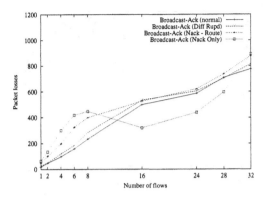

Figure 4. Amount of data packets lost due to route disappearance (64 mobiles; acknowledged broadcast)

To evaluate the performance of the different routing strategies, we will look at the influence of the different optimizations on the routing tables validity. Figure 4 represents the number of data packet losses due to route disappearance, *i.e.*, no route exists to reach the destination mobile, neither in the Base Station to which the sender is attached to, nor in the Gateway. These drops occur when a route expires in the whole network before the new route to the mobile has been propagated. When the medium is lightly loaded, optimizations seem to increase the number of drops at the Gateway. But as soon as the medium gets overloaded, the *Nack Only* optimization which reduces the load due to signaling packets, is highly efficient. As `route update` messages represent a high load, the medium saturation point is postponed. However, losing a `route update` message has a much greater impact with *Nack Only* optimization, that's why performance collapse again under a high data load.

However, Figure 5 shows the number of data packets lost due to outdated entries in the routing tables. The Base Station in charge of the receiver tries to forward the data frame to the mobile, gets no acknowledgment in return, concludes there has been a collision and retries to forward the frame until the retransmission counter is exceeded. These losses are also due to repeated collisions resulting in retransmissions but analysis of the trace files show that this cause is marginal compared to mobility-related losses. These results show that the more optimizations we add, the more the routes are outdated. This is due to the increasing delay between two `route update` packets sending, resulting in a increasing route timeout.

Optimizations described here lead to the same kind of discussion as the one on the different ways to transmit `route update` messages. One one hand, we will try to send as few `route update` frames as possible, but

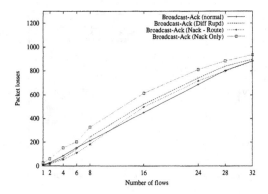

Figure 5.   Amount of data packets lost due to errors in routing tables (64 mobiles; acknowledged broadcast)

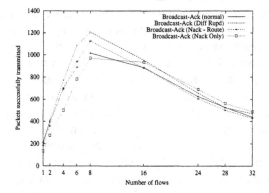

Figure 6.   Amount of data packets successfully transmitted (64 mobiles; acknowledged broadcast)

we will be less reactive to mobility and on the other hand, over-occupying the medium will delay data packets and raise the number of packets lost due to collisions. Figure 6 represents the total amount of data packets successfully delivered. If simple Differential Route Update mode always shows good performances, other mode performances are load-dependent. Nack Only optimization is rather good when the medium is overloaded but represents a loss of performance when there is no real need for saving bandwidth.

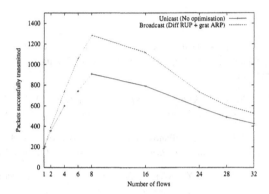

Figure 7.   Number of packets correctly transmitted in normal mode and with optimisations

To conclude this study, Figure 7 shows the total number of data packets correctly transmitted by a regular Cellular IP compared to a modified version in which `route update` packets are broadcasted and the *differential route update* optimization is activated. This optimized Cellular IP leads to the best performances, enhancing the overall data throughput by up to 40%.

5.   Conclusion

In this article, we presented several possible modifications of the Cellular IP protocol for enhancing its performance in a wireless hybrid context. These modifications, concerning routing as well as MAC layer, show that network performances can be increased by up to 40 %. These results can still be enhanced, for example by implementing optimizations of the routing scheme related to the broadcast transmission of signaling frames. These results especially show how micro-mobility protocols derived from wired protocols are inadequate in a wireless context. Mechanisms that have proved themselves worthy in regular networks such as reliable unicast transmission should be left aside in most situations.

Performance should not be expected to get similar to those obtained in wired networks. Nevertheless, the efficiency of wireless hybrid networks can be increased. This work shows that the key issue regarding network performance is the network load. This load can be decreased on one hand by reducing the global signaling volume as studied here and on the other hand by designing suited radio interfaces and medium access protocols. The separated signaling channel mechanism, allocating a particular frequency, time slot or CDMA code to control traffic, widely used in cellular telephony networks might lead to further performance enhancements. Nevertheless, actual wireless hardwares do not allow this due to the long channel switching delay.

References

[1] R. Bhan, A. Croswell, K. Dedhia, W. Lin, M. Manteo, S. Merchant, A. Pajjuri, and J. Thomas. Adding ad hoc Network Capabilities to Cellular IP. Technical report, www.columbia.edu.

[2] A. Campbell, J. Gomez, S. Kim, Z. Turanyi, C-Y. Wan, and A. Valk. Comparison of ip micromobility protocols. *IEEE Wireless Communications*, 9(1):72–82, February 2002.

[3] A. Campbell and J. Gomez-Castellanos. Ip micromobility protocols. *ACM SIGMOBILE Mobile Computing and Communications Review*, 4(4):45–53, October 2000.

[4] R-S. Chang, W-Y. Chen, and Y-F. Wen. Hybrid wireless network protocols. *IEEE Transactions on Vehicular Technology*, 52(4):1099–1109, July 2003.

[5] H-C. Chao and C-Y. Huang. Micro-mobility mechanism for smooth handoffs in an integrated ad-hoc and cellular ipv6 network under high-speed movement. *IEEE Transactions on Vehicular Technology*, 52(6):1576–1593, November 2003.

[6] G. Chelius and É. Fleury. Design of a hybrid routing architecture. In *IEEE MWCN*, Singapore, November 2003. IEEE Communications Society.

[7] Y-Z. Huang. Dynamic adaptive routing for heterogeneous wireless network. Master's thesis, National Central University, West Lafayette, USA, 2001.

[8] V. Typpö. Mobility within Wireless Ad Hoc Networks: Towards Hybrid Wireless Multihop Networks. Master's thesis, VTT Electronics and University of Oulu, Finland, 2001.

[9] A. Valkó. Cellular ip - a new approach to internet host mobility. *ACM Computer Communication Review*, 29(1):50–65, January 1999.

[10] C. Wijting and R. Prasad. Evaluation of mobile ad-hoc network techniques in a cellular network. In *IEEE VTC*, pages 1025–1029, 2000.

SELECTIVE ACTIVE SCANNING FOR FAST HANDOFF IN WLAN USING SENSOR NETWORKS

Sonia Waharte, Kevin Ritzenthaler and Raouf Boutaba
University of Waterloo, School of Computer Science
200, University Avenue West, Waterloo, ON, CA
{swaharte,kmritzen,rboutaba}@bbcr.uwaterloo.ca

Abstract

Seamless connectivity in wireless access networks is critical for time-sensitive applications requiring Quality-of-Service guarantees with bounded data transmission delay. Currently, the latency inherent in the hand-off process can preclude the successful delivery of such applications by introducing delays up to several seconds. In this paper, we address this issue by proposing an improvement of the access points discovery process at the data link layer. We present a novel WLAN architecture using an overlay sensor network as a control plane. By distributing the information on the current network status to the sensor nodes, we show that handoff latency can be significantly reduced.

1.   Introduction

Wireless communications have gained over time great importance with the rapid growth in data transmission rate. Whereas cellular networks still remain constrained by low offered throughput (GPRS enables a theoretical 171.2kbps and UMTS only permits a maximum of 2Mbps for indoor or low range outdoor communications), a major breakthrough has been achieved in Internet access technologies with data rates up to 54Mbps. It is now feasible to envision supporting a wide range of QoS applications encompassing fields as diverse as video streaming or gaming. However, this is contingent on the ability to provide appropriate QoS guarantees through bounded end-to-end transmission delay. In addition, the diversity of terminal devices (laptop, personal digital assistant, etc.) renders the need for mobility support a critical feature.

Several characteristics inherent in wireless networks contribute to the
increase of data transmission delays. In wireless LAN, the de facto
Standard IEEE 802.11 [1] defines a fair random deferred access to the
transmission medium, which introduces unbounded transmission delay
due to idle time periods and retransmissions due to collision. Significant delays also occur during the handoff process. Moving from one
access point to another involves time-consuming mechanisms which may
be detrimental to applications requiring seamless connectivity and QoS
guarantees.

In this paper, we address the problem of handoff latency at the MAC
layer. As this delay mainly results from the exhaustive scan of every
channel in the frequency band, we believe that improvements can be
achieved by distributing the information concerning the surrounding
access points (channel used, supported rate, etc.) to external agents.
By consulting these agents, the mobile nodes can limit the number of
scanned channels and take informed decision about the most appropriate
access point to be associated with.

Based on this idea, we propose a novel architecture using an overlay
sensor network on top of a WLAN. The sensor network is in charge of
maintaining a knowledge base on the network status. By contacting
only the surrounding sensor nodes, a mobile node can obtain precise
information on the network status.

The rest of the paper is organized as follows. Section 2 describes the
mechanisms involved in a handoff process in WLAN and related works.
Section 3 provides a description of our architecture. Validations through
simulations are presented in Section 4. Conclusion and future research
directions are given in Section 5.

2.   Layer 2 Handoff Process and Related Works

In WLAN, a handoff can be defined as the process of leaving the basic
service set of an access point to enter a new one. A handoff is triggered
by a degradation of the signal quality which falls below a predefined
threshold. The handoff can be the result of either excessive noise and
interference or user mobility (decrease of the signal intensity due to the
increasing distance to the associated access point).

At the MAC layer, the handoff process as defined in the IEEE 802.11
Standard [1] can be decomposed into three phases: scanning, authentication and reassociation (Figure 1).

1 Scanning: In order to discover on which channel the surrounding
access points are transmitting, a mobile node needs to scan all the
channels. Two scanning methods are described in IEEE 802.11.

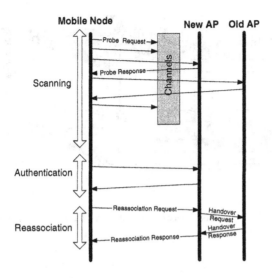

Figure 1. Handoff Process at the MAC Layer

Passive Scanning entails determining the presence of access points by successively listening to all the channels and waiting for the reception of beacon messages identifying the access point. This method, while offering the advantage of low overhead, presents the drawback of introducing a significant delay. In order to alleviate this problem, an active scanning method has been defined. The mobile node broadcasts a Probe Request on each channel and waits a minimum period MinChannelTime for any Probe Response. After the scan of all the channels and the processing of all the beacon messages or Probe Responses received (according to the implemented scanning process), the mobile node can take an informed decision on the most appropriate access point (with the best channel quality).

2 Authentication: The Authentication process involves establishing the identity of the mobile node and authorizing its access to the basic service set of the access point.

3 Reassociation: The Reassociation process consists in transferring an association between an access point and a mobile node to another access point. The operations between the old AP and the new AP are defined by the Inter-Access Point Protocol [2].

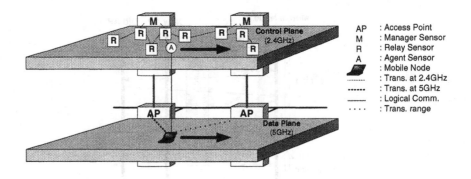

Figure 2.   Overlay sensor network architecture for handoff management

The scanning process has been identified as the principal source of delay in the handoff mechanism [3] [4]. Few works have been conducted aiming at reducing the latency at the MAC Layer [3] [4] . They essentially adopt the same approach by optimizing the waiting time of the mobile node during the active scanning process. Indeed, IEEE 802.11 defines two parameters: MinChannelTime, the minimum waiting period before considering that the channel is idle; and MaxChannelTime, the maximum waiting period after a Probe Response has been successfully received. However, no exact value of these parameters has been explicitly set. The suggested optimal values deduced from previous experiments are approximately 6.5ms for MinChannelTime and 11ms for MaxChannelTime.

Thanks to their sensing, computation and transmission capabilities [5], sensor nodes can serve as an effective monitoring and data gathering technology for WLANs. Few works used sensor networks for managing wireless networks. MeshDynamics [6] uses sensor networks to manage connectivity and routing in ad hoc wireless mesh networks. AirMagnet [7] implements a distributed sensor network in WLAN for security monitoring and intrusion detection. These works demonstrate the practicality and effectiveness of sensor networks as a monitoring infrastructure for wireless networks.

3.   Architecture Design

3.1   Architecture overview

Our access architecture augments the existing IEEE 802.11 access protocol (data plane at 54Mb/s) with an overlay sensor network (control plane at 2Mb/s). As the two planes use different access frequencies (5GHz and 2.4GHz), communications can occur in parallel within each respective plane (Figure 2). This characteristic allows us to lift the control overhead from the data plane.

The overlay network is composed of three types of sensor nodes: the manager, the relays, and the agents. The manager node is a sensor attached to the access point. First, it initializes the sensor relays by conveying information about the associated access point. Second, it serves as the data aggregation server where reassociation requests from mobile nodes are gathered, and reassociation responses are sent back. The relays are fixed sensors uniformly placed throughout the coverage area of an access point. Because of a sensor's short transmission range, the relays are used to route messages between the manager and the sensor agents. They are also responsible for sensing the frequency bands according to a procedure described in the following section. The agents are sensors attached to the mobile nodes. They communicate with the relay sensors upon entering the transmission area of an access point or if a handoff process is initiated due to a degradation of the signal quality.

3.2   Selective Active Scanning for Fast Handoff

In order to determine the presence of surrounding access points, the common method is to successively scan all the channels and therefore detect if a channel is busy by receiving Probe Responses or Beacons. We believe that this time-consuming method can be drastically improved by limiting the number of channels scanned to the ones in which the mobile node is interested, i.e. the access points with which it can potentially be associated. This can be achieved by maintaining a distributed database stored by the relay sensors.

The proposed handoff process is illustrated in Figure 4. The steps involved in the process are the following:

1 The mobile node broadcasts an AP_List Request on the control plane.

2 The neighboring relay nodes reply with an AP_List Response if they satisfy some criteria detailed in a subsequent section.

3 The mobile node processes all the received messages, builds a list
   of the neighbor access points and initiates a scanning process solely
   based on this list.

The remaining steps of the handoff procedure remain unmodified compared to the specifications of the standard, the only difference being that
the process occurs at the control plane instead of the data plane.

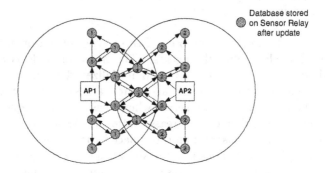

Figure 3.   Example of initialization with two access points[1]

3.2.1   Sensor Network Initialization Process.   Before being
able to handle any handoff, the overlay sensor network needs to be initialized in order to learn about the presence of the surrounding access
points. This process, initiated by the access point, occurs only once
during the network lifetime. Its impact is thus minimum and does not
affect the subsequent handoff mechanisms of the mobile nodes.

In the control plane, each access point sends an initialization packet
containing information about itself using a flooding protocol (Figure 3).
In order to avoid flooding the entire network upon addition of a new
access point, the forwarding process stops two hops after the initialization packet has been forwarded by a relay node that has at least another
entry (meaning that this node is in the transmission zone of another access point). The explanation for this restriction comes from the following
observation: if we consider that for an access point and a relay sensor,
the transmission range can not exceed 150m and 50m respectively, we
are guaranteed to cover the whole transmission area of the access point
within three hops. Thus, if we assume that two access points do not

---

[1]For clarity, not all the relay sensors and not all the messages exchanged during the initialization process are represented.

Table 1. Local Database maintained by the Relay Sensor

AP BSSID	Channel Used	Supported Rate	Signal Strength
xxx-xxx	34 (5170MHz)	54Mbps	65%
yyy-yyy	38 (5190MHz)	24Mbps	No signal

cover exactly the same geographical area, the aforementioned restriction of two hops is sufficient. Nonetheless, this parameter can be increased if necessary without impacting our model.

An example of the database maintained by a relay node is depicted in Table 1. This database is updated on a regular basis with information such as signal strength by only scanning the listed channels. As during the initialization process, a relay node may have received information about an out of range access point, the relay node is allowed to remove this entry if no signal is detected on the corresponding channel.

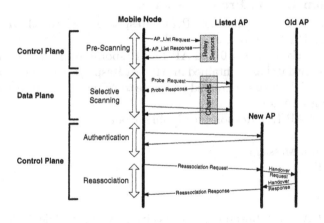

Figure 4. Proposed Handoff Process

3.2.2   Selective Scanning Method.   The proposed method is illustrated in Figure 4. Compared to a traditional handoff process, we add a pre-scanning phase which proactively determines the presence of access points before a mobile node initiates a handoff. As previously mentioned, when the mobile node experiences a degradation of the received signal strength in the current associated basic service set, the mobile node initiates a handoff by first trying to discover new access points which can offer a better service quality. The mobile node then first broadcasts an AP_List Request and waits for the surrounding relay nodes to reply with an AP_List Response. Priority is given to the closest relay sensor by defining a backoff time based on the received signal strength such that:

$$\text{backoff\_time} = (1 - \frac{P_r}{P_t})CW$$

where P_r is the received signal strength, P_t is the maximum signal strength and CW is a random congestion window size.

A relay sensor sends back an AP_List Response by following the specifications of Algorithm 1. This algorithm guarantees that only useful information is sent back to the mobile node. We consider as useful information the announcement of the presence of an access point not previously advertised by another relay sensor.

Algorithm 1: send_Probe_Response
1: **if** (backoff time = 0 and no AP_List Response received) **then**
2:   Send AP_List Response
3: **else if** (backoff time = 0 and AP_List Response received) **then**
4:   Compare database contained in AP_List Response with local database
5:   **if** (Local database contained reference to AP not in received AP_List Response) **then**
6:     Send AP_List Response to Mobile Node
7:   **else**
8:     Ignore Message
9:   **end if**
10: **end if**

If several AP_List Responses are received by the mobile node, the mobile node should process them all. It is worth noticing that the inaccuracies introduced by the difference of positions between the mobile node and the relay sensors (for instance variations in link quality) are overcome by the reception of multiple AP_List Responses. The mobile node should wait for incoming AP_List Responses for a given period of time. According to the AP_List Responses received, the mobile node initiates a traditional scanning process over the listed channels (containing only active channels) to select the most suitable access point in terms of link quality. To provide even faster handoff and to meet user expectations, it is also envisioned to allow the user to limit this scan to channels with a specific Data Rate or a minimum signal strength threshold.

To provide compatibility with existing hardware (which may not be equipped with sensor agents), a traditional handoff process can still be initiated.

3.3   Benefit of the overlay sensor network

The advantages of using sensor nodes as a control plane are two-fold:

- By monitoring the environment, the sensor nodes can obtain information on the presence of access points and the associated quality of their transmission channels. Based on this information, mobile nodes can subsequently make informed decisions about the access point they are willing to be associated with. By directly requesting this information from the nearest sensor node, significant improvement in terms of delay can be achieved. The mobile nodes thus scan only the channels of interest (i.e. the ones used by the access points the mobile node can actually be associated with).

- By transmitting control messages at the sensor plane, in parallel with data transmission at the data plane, bandwidth wastage is reduced.

The decision of using relay nodes instead of contacting directly the access points is motivated by the following factors: less interference occurs as the transmission distance is shortened, thus several mobile nodes can contact different relay nodes without interfering with each other; energy consumption at the mobile node is minimized by contacting a closer relay node; and relay nodes provide more accurate information on the surrounding access points.

It is worth mentioning that the flexibility and ease of deployment of our architecture allow further enhancements such as providing guarantees for QoS applications [8].

4.   Evaluation

In order to assess the benefits of our architecture, we perform simulations in which a mobile node initiates a handoff process with either an active scanning or a selective scanning (using an Overlay Sensor Network architecture) with both standard and optimized MinChannelTime / MaxChannelTime parameters. The simulation parameters are directly taken from [3] and summarized in Table 2. The transmission models rely on IEEE 802.11a and IEEE 802.11b standards for the data plane and the control plane respectively. The relay nodes are uniformly scattered over the coverage area of the access points according to a cellular topology.

Table 2. Simulations Parameters

	Reference	Optimized
MinChannelTime	17ms	6.5ms
MaxChannelTime	38ms	11ms

The simulations are performed using QualNet 3.6.1 [9] and the results are averaged over 50 runs.

We consider that 8 channels can be used by the access points, with data sent at the lowest possible rate (6Mbps) on the data plane in order to maximize the transmission distance.

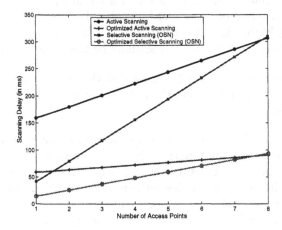

Figure 5. Scanning delay for one user with an increasing number of access points

Figure 5 shows the delay pertaining to the different scanning approaches. In the overlay sensor network architecture, the scanning also includes the pre-scanning delay. We observe that reducing the number of scanned channels brings significant improvements, especially when the number of surrounding access points is limited compared to the number of channels available.

In Figure 6, we aim at evaluating the impact of interfering transmissions on the scanning process of a mobile node. An increasing number of users, randomly placed in the BSS, generate CBR traffic to the access point at 10Mbps. We can observe that the delay introduced to access the medium is still negligible compared to the time wasted to wait for Probe Responses.

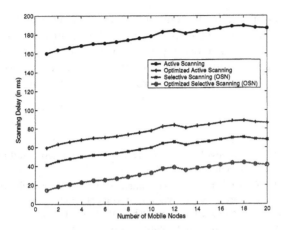

Figure 6. Scanning delay for several users with one access point

5. Conclusion

Supporting user mobility in WLAN remains a challenging task, especially with the QoS requirements of applications necessitating bounded transmission delay. The handoff process is a complex mechanism, involving significant delay, which can be detrimental to QoS guarantees.

Recent work on applying sensor networks (monitoring network connectivity [6], security, intrusion detection [7], etc.), demonstrates the practicality and effectiveness of sensor networks as a monitoring infrastructure for wireless networks.

Based on these observations, this paper aims at reducing the handoff delay in WLANs using a two-tier access architecture consisting of a sensor overlay control plane over an IEEE 802.11 data plane. Using the distributed monitoring and processing capabilities of the sensor network, we shift the burden of transmission control and coordination into the control plane, preserving the data plane solely for data transmission in parallel with the control plane. By maintaining information on the surrounding access points and by delivering this information to the mobile nodes upon request, significant improvements can be achieved.

In this paper, emphasis lies on the handoff process. However, one can envision a number of extensions for our proposed architecture. First, our architecture can be extended to support service class differentiation and QoS interworking between cellular networks and WLANs. Second, an application-adaptive scheduling algorithm can be devised to support per-traffic QoS guarantees (minimum bandwidth and maximum delay) where each mobile node transmitting to the manager node its QoS requirements. Third, better communication techniques could also be incorporated in the sensor overlay network to improve the efficiency of control messages exchange. We believe that the application of sensor networks as a monitoring and control infrastructure for WLANs holds great promise.

References

[1]   IEEE. Wireless LAN medium access control (MAC) and physical layer (PHY) specifications, 1999.

[2]   IEEE. IEEE trial-use recommended practice for multi-vendor access point interoperability via an inter-access point protocol across distribution systems supporting IEEE 802.11 operation, 2003.

[3]   Arunesh Mishra, Minho Shin, and William Arbaugh. An empirical analysis of the IEEE 802.11 MAC layer handoff process. SIGCOMM Comput. Commun. Rev., 33(2):93102, 2003.

[4]   H. Velayos and Karlsson G. Techniques to reduce the IEEE 802.11b handoff time. In Swedish National Computer Networking Workshop, 2003.

[5]   I.F. Akyildiz, Weilian Su, Y. Sankarasubramaniam, and E. Cayirci. A survey on sensor networks. In IEEE Communications Magazine, volume 40 of 8, pages 102114, Aug. 2002.

[6]   Meshdynamics. http://www.meshdynamics.com/index.html.

[7]   Airmagnet.   http://www.fe-solutions.com/support/airmagnet   distributed.html.

[8]   S. Waharte, J. Xiao, and R. Boutaba. Overlay Wireless Sensor Networks for Application-Adaptive Scheduling in WLAN . In 7th IEEE International Conference on High Speed Networks and Multimedia Communications (To appear), 2004.

[9]   Qualnet. http://www.scalable-networks.com/.

AN ANALYSIS OF MOBILE IPv6 SIGNALING LOAD IN NEXT GENERATION MOBILE NETWORKS

Sandro Grech[1], Javier Poncela[2], Pedro Serna[1]
[1] *Nokia Networks, P.O. Box 301, FIN-00045 Nokia Group, Finland (sandro.grech@nokia.com; pedro.serna@nokia.com).*
[2] *Department of Ingeniería de Comunicaciones, Universidad de Málaga, Málaga, 29071 Spain (javier@ic.uma.es).*

Abstract: To date, mobile communication has been dominated by voice services. This is likely to be valid also for the foreseeable future, but at the same time a multitude of data services are also emerging. This trend in mobile communication has fueled the introduction of packet-switched mobile networks, thus introducing the IP suite into the field of mobile communications. This technological shift can be already observed in today's 2.5 and 3G networks. In the first phases, however, mobile devices have an IP point of attachment which seldom changes throughout the lifetime of a communication session. Mobility management is handled below this point of attachment by means of access-specific mechanisms. A unified mobility management mechanism at the IP layer may enable streamlined network architectures, for example as complementary access technologies emerge in next generation mobile networks. Mobile IPv6 represents a key candidate mechanism to fulfill this vision of unified IP-based mobile communication networks. This paper analyses and quantifies the signaling overheads in a mobile communication network that uses Mobile IPv6 for mobility management.

Key words: IP signaling; Mobile IPv6; signaling load analysis; localized mobility management.

1. INTRODUCTION

IP is one of the key enablers of the anticipated widespread adoption of mobile multimedia and data services. The challenges that arise in this new environment are rather different from the ones that IP has traditionally faced. The main issue is related to mobility, which is inherent to wireless and cellular systems. Although mobility may be transparently handled using access-specific mechanisms, mobility may require the mobile device to change its IP address in order to maintain IP connectivity. Without Mobile IP, this change in IP address is exposed to the layers above IP, resulting in disruptions in the mobile communication. The analysis in this paper is specific to Mobile IPv6 [1], since IPv6 is assumed as an underlying enabler of widespread deployment of IP-based mobile communication.

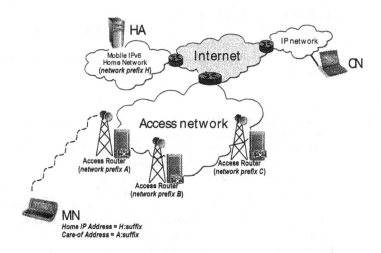

Figure 1. High level reference architecture.

In order to introduce mobility support in IPv6, Mobile IPv6 introduces several new concepts, which will be outlined next with reference to Figure 1. The Mobile Node (MN) acquires a static (or semi-static) IP address, known as the Home Address, from its Home Network. When the Mobile Node resides away from its Home Network it also acquires a Care-of Address (CoA) that matches the prefix of the visited link. The Home Agent (HA) is a router on the home network that maintains a mapping (called binding cache entry) of the Mobile Node's Home Address with the Mobile Node's current Care-of Address. Correspondent Nodes (CN) are any nodes with which a Mobile Node is communicating. The Access Router (AR) is the router that the Mobile Node uses to obtain IP connectivity to the network. In this paper, it is assumed that IP base stations implement the AR functionality.

When a Mobile Node changes its network point of attachment it will require an IP address that matches the network prefix of the visited link. In order to maintain its reachability, the Mobile Node needs to announce this IP address to its Home Agent so that packets which reach the Mobile Node's home IP address can be re-directed by the Home Agent. The Mobile Node should also announce its Care-of Address to any Correspondent Nodes so that they can deliver the packets directly to the Mobile Node (this process is known as Route Optimization). These announcements are performed using messages known as Binding Updates (BU), which create a binding between a Mobile Node's Home Address and its Care-of Address at the receiving node. The bindings created by these messages expire after a pre-defined lifetime.

A Mobile Node announces its Care-of Address to the Home Agent and preferably also to its Correspondent Nodes in the following occasions:
a) a Mobile Node changes its Care-of Address,
b) the lifetime of an existing binding is about to expire (this can be triggered by a Correspondent Node by sending a Binding Refresh Request message),
c) additionally, a Mobile Node should also announce its Care-of Address to its Correspondent Nodes when a Mobile Node starts to communicate with a new Correspondent Node that is not aware of the Mobile Node's Care-of Address.

This study focuses on the Mobile IPv6 signaling load which is generated by a) in the list above. It assumes that the binding lifetimes (b) are long enough so that the signaling load caused in refreshing these bindings is negligibly low when compared to the load caused by IP address changes. The Mobile IPv6 signaling due to c) may or may not be substantial depending mainly on what type of applications the terminal is engaged in. During some web-browsing sessions, for example, the user may be following a set of hyperlinks that point to different web servers (and thus different Correspondent Nodes). In this example the Mobile IPv6 signaling due to change of Correspondent Node may be higher than the Mobile IPv6 signaling due to mobility. In some other applications, such as a speech call, the Correspondent Node is unique for the duration of the call. The overall result depends on the traffic mix, and on other application/user characteristics such as how often does a user using a web-browsing application follow a hyperlink that points to a new web server, etc. The signaling resulting from c) is not considered in the calculations in this study.

2.   BINDING UPDATE PROCEDURE

The Binding Update procedure is a core component of Mobile IPv6. This procedure is used by the Mobile Node to inform the Home Agent and any

Correspondent Nodes that a new Care-of Address has been assigned to the Mobile Node.

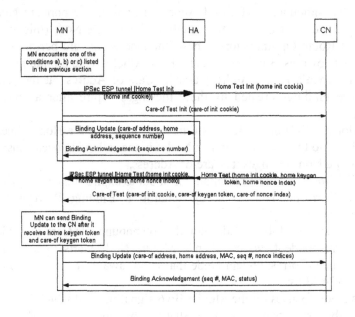

Figure 2. Mobile IPv6 Binding Update signaling.

The sequence of messages exchanged between Mobile Node, Home Agent and Correspondent Node when a Binding Update is required is illustrated in Figure 2. First, the Mobile Node sends two challenges to the Correspondent Node, one of them (Home Init cookie) secured via the Home Agent and the other (Care-of Init cookie), unprotected, directly to the Correspondent Node; these messages are known as Home Test Init (HoTI) and Care-of Test Init (CoTI), respectively. The Mobile Node can perform the binding procedure with the Home Agent in parallel to this operation.

The Mobile Node can send a valid Binding Update message to the Correspondent Node only after receiving two tokens (home and care-of keygen tokens) from the Correspondent Node in response to the Home Test Init and Care-of Test Init messages. The Mobile Node combines these tokens in order to generate the Message Authentication Code (MAC), which the Correspondent Node uses to validate the Binding Update. This procedure is called Return Routability. It enables the Correspondent Node to obtain some reasonable assurance that the Mobile Node is in fact addressable at its claimed Care-of Address and at its Home Address, before accepting any Binding Update from a Mobile Node.

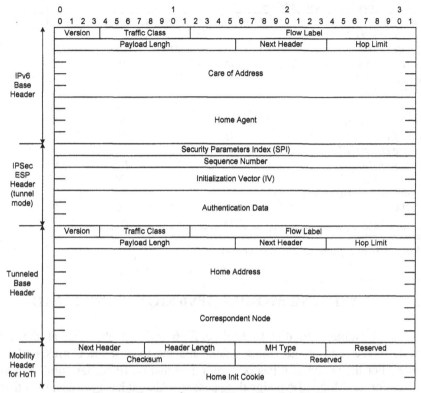

Figure 3. Format of message Home Test Init (HoTI).

The format of message Home Test Init (HoTI) sent from the Mobile Node to the Correspondent Node through the Home Agent (using tunneling) is shown in Figure 3. It is composed of an IPv6 header and a security header that encapsulates the message directed to the Correspondent Node. The source and destination addresses in the base header may be the Home Agent address, the Correspondent Node address or the Care-of Address, depending on the type of message [2]. The encapsulated message consists of an IPv6 header plus a mobility header that carries the Home Test Init message. Note that the 'Next Header' field in the Mobility header is named 'Payload Protocol' in [1].

The IPsec ESP header is used only for messages between the Mobile Node and the Home Agent. For all messages the format is similar but each message type will include its own fields after the Checksum of the Mobility Header and thus its size may vary [1]. Taking into account these variations, the size, in bytes, of the messages used in Figure 2 is provided in Table 1. The Initialization Vector and the Authentication Data fields depend on the Security Association [2].

Table 1. Mobile IPv6 Message Sizes

Message	Fixed Headers	IPsec	Message Data	Total Size
Home Test Init (HoTI)				
(MN → HA)	46	72	10	128
(HA → CN)	46	---	10	56
Home Test (HoT)				
(CN → HA)	46	---	18	64
(HA → MN)	46	72	18	136
Care-of Test Init (CoTI)	46	---	10	56
Care-of Test	46	---	18	64
Binding Update (BU)				
(MN → HA)	46	72	18^1	136
(MN → CN)	46	---	26^2	72
Binding Acknowledgement (BAck)				
(HA → MN)	46	72	10	128
(CN → MN)	46	---	26^2	72

[1] With no options.
[2] Assume Binding Authorization Data Option.

3.   BASELINE MOBILE IPv6 SIGNALING LOAD

The signaling load due to Mobile IPv6 includes the load at the air interface, the load at the Home Agent and the load at the Correspondent Nodes. For the analysis, it has been assumed that no IP header compression is used for the Mobile IPv6 signaling packets. Also, it has been assumed that the home and care-of keygen tokens expire before the Mobile Node sends the next Binding Update to the same Correspondent Node. Due to this, the Mobile Node must perform the complete Binding Update procedure every time, going through the 'Home Test Init – Home Test' and 'Care of Test Init – Care of Test' exchanges every time that a Binding Update needs to be sent. The above assumptions clearly represent a worst-case scenario.

Table 2. Subscriber Density and Subscriber Mobility Figures

Environment	AR relocations/1000 subs/second (R_S)	Subscribers/AR (S_C)	Relocations/second/AR (R_{AR})
Urban	0.556	339	0.189
Dense Urban	0.500	241	0.120
Suburban	0.476	1442	0.685

The subscriber density and subscriber mobility parameters, which model the network, are shown in Table 2. Environments are classified depending on the cell size, which is smaller in dense urban (approx. 0.25 km^2) and larger in a suburban environment (approx. 7 km^2). These reference scenarios are based on typical values for density of data users utilized in equivalent cellular reference models.

The air interface signaling overhead per AR can be obtained from the messages shown in Figure 2. The overhead in the uplink direction is the size of the messages (HoTI, CoTI, and both Binding Updates) sent by a Mobile Node due to an AR relocation, multiplied by the number of relocations per second per AR (R_{AR}); for the downlink direction the calculation is similar.

The total size of the messages exchanged in the air interface (see Figure 2) is: 128 (HoTI) + 56 (CoTI) + 136 (BU to HA) + 72 (BU to CN) + 136 (HoT) + 64 (CoT) + 128 (BAck from HA) + 72 (BAck from CN) = 792 bytes. Thus, the total overhead per Access Router for an urban environment would be 149.69 bytes/s ($\approx$1.2 kbit/s/AR), which represents around 0.6 kbit/s in each direction. This overhead is negligible, even considering bandwidth-limited interfaces.

The air interface signaling load for various environments is shown in Figure 4.a. For suburban ARs the amount of Mobile IPv6 signaling overhead tends to be higher. This is mainly due to the high number of Mobile Nodes per AR in the suburban area (1442 MNs/AR). Results show that, even with multiple simultaneous CNs, the air interface signaling overhead per AR is not substantial.

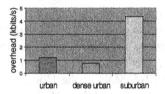

(a) Air interface signaling overhead per AR.

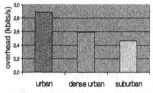

(b) Signaling overhead at Home Agent per 1000 subscribers.

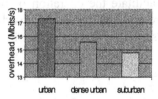

(c) Signaling overhead at Home Agent for 6 million subscribers.

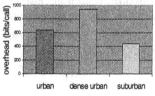

(d) Signaling overhead at Correspondent Node per Mobile Node per call.

Figure 4. Baseline Mobile IPv6 signaling overheads.

The load at the Home Agent is given by the size of the messages involved in the Home Agent signaling: 128 (HoTI) + 64 (HoT) + 136 (BU to HA) + 56 (HoTI) + 136 (HoT) + 128 (BAck from HA) = 648 bytes. To obtain the total number of bytes which are processed (either received or sent) by the Home Agent, this number must be multiplied by the number of subscribers served by the Home Agent (N) and the number of relocations per

1000 subscribers per second: 648 x N x R_s. In an urban environment, the signaling overhead at the Home Agent per 1000 subscribers would be 360 bytes/s ($\approx$2.89 kbit/s) (see Figure 4.b for estimations on various environments).

Moreover, the processing overhead of a Mobile IPv6 BU-BAck pair has been measured [4] to be approximately as reported in Table 3. The Home Agent must also initiate and terminate the IPsec ESP tunnel to the Mobile Node. The additional processing can be roughly estimated to be double. From these results, one can infer that a Home Agent based on a 1 GHz processor can process approximately 5,000 'Binding Update – Binding Acknowledgement' pairs per second.

Table 3. Overhead of a BU-BAck Message Pair in the Home Agent

	No Piggyback	Piggybacked
Sending cost	*0.20 ms*	*0.23 ms*
Receiving cost	*0.29 ms*	*0.23 ms*

A Home Agent serving 6×10^6 subscribers would need to process approximately 3,300 'Binding Update – Binding Acknowledgement' pairs per second. This means that, in theory, if a Home Agent processor does not handle any other traffic, current state of the art processors easily manage this amount of processing [4]. The signaling overhead at a Home Agent serving 6 million mobile subscribers is shown in Figure 4.c. This shows that the bandwidth introduced by 6 million Mobile IPv6 MNs at the Home Agent could be easily supported on a 100 Mbit/s interface.Mobile Nodes should also send Binding Updates to any Correspondent Nodes with which they have ongoing communication, in order to facilitate the operation of route optimization for Mobile IPv6. For a Correspondent Node the overhead caused by the binding messages for this server would be given by the equation:

$$OH_{BU} = \frac{(S_{BU} + S_{CoTI} + S_{HoTI} + S_{CoT} + S_{HoT} + S_{BAck}) \times R_{MN}}{R_u \times (S_d + S_u)}$$

where S_d is the average downlink payload size is, is S_u is the average uplink payload size, R_u is the average uplink packet rate and R_{MN} is the average AR relocations/second/MN. Taking a voice service as an example and using the values given in Table 4, the signaling overhead at the Correspondent Node would be $OH_{BU} = 0.196 \times R_{MN}$.

For voice calls, assuming an average duration of 90 seconds we obtain a mean value of 2.46 relocations/call. Thus, R_{MN} = 2.46/90, which produces 0.027 relocations per second per Mobile Node in a dense urban environment. This amounts to an overhead due to Binding Updates of 945 bits per call, which turns to be less than 0.6% overhead in the Correspondent Node per

Mobile Node, relative to user bandwidth. Figure 4.d shows the signaling overhead at the Correspondent Node per call for the different environments.

Table 4. Typical Traffic Characteristics

	Voice	Video	Streaming	Web-browsing	File transfer
Packtes/second/direction (R_u)	*32.5*	*30*	*6*	*25*	*50*
Mean payload length uplink (S_u) (bytes)	*30.1*	*250*	*25*	*125*	*200*
Mean payload length downlink (S_d) (bytes)	*30.1*	*250*	*500*	*500*	*800*
Average User Bitrate (downlink) (kbps)	*7.8*	*60*	*24*	*50*	*80*
Average User Bitrate (uplink) (kbps)	*7.8*	*60*	*12*	*12.5*	*20*

As can be seen from the results presented, the signaling load and processing overhead introduced by Mobile IPv6 are not critical for any of the affected parties, i.e. the air interface, the Home Agent, and the Correspondent Node. Results show that systems could cope with this overhead with slight increases of their actual capacity, which should be considered in the planning stage. Thus, the introduction of Mobile IPv6 is not hindered by the overheads studied in this section.

4.   ANALYSIS OF INBAND SIGNALING

Apart from signaling due to specific messages required by Mobile IPv6, overheads must also consider the additional signaling included in user packets due to Mobile IP. There are four components that must be considered: Home Address Destination option, Routing Header, inband signaling when the Correspondent Node is also a Mobile Node and IP header compression; their analysis is provided below. For this analysis, traffic estimations per user are required, and the results are dependent on the specific mix of traffic assigned to one user.

In order to avoid ingress filtering of packets by the Access Router, the Mobile Node uses its Care-of Address as IP Source Address when sending packets (uplink). To make the use of the Care-of Address transparent to the higher layers (above IP), each packet must also include a 20 byte Home Address Destination Option. This destination option contains the Home Address of the Mobile Node, which will be used by the Correspondent Node to replace the Care-of Address in the IP Source Address field before being passed to the higher layers.

The overhead associated with the Home Address Destination Option is represented in Table 5(a). For example, for voice service, as there are on average 30.1 bytes per packet, adding 20 bytes for the Home Address

Destination Option would represent an overhead of 66% relative to the payload. If we include the size of the headers (around 60 bytes), without compression mechanisms, the overhead goes down to 21.7%. As there are 32.5 packets per second in each direction this amounts to around 5 kbps per voice flow in the uplink direction. The average subscriber uplink bitrate in the Busy Hour, due to the combined use of all services, experiences an increase from 0.818 kbps to 0.962 kbps. This represents an increase of 17% in uplink network traffic.

Table 5. Mobile IPv6 overhead in (a) uplink traffic due to Home Address Destination Option and (b) downlink traffic due to Routing Header

	Overhead relative to payload (%)		Overhead (with headers) (%)		Overhead per flow (kbps)	
	(a)	(b)	(a)	(b)	(a)	(b)
Voice	66	80.0	21.7	26.0	5.2	6.2
Video	8	9.6	6.1	7.4	4.8	5.8
Streaming	80	4.8	20.2	4.1	1	1.1
Web-browsing	16	4.8	10.5	4.2	4	4.8
File-transfer	10	3.0	4.3	2.7	8	9.6

In the downlink direction, packets sent from the CN to the MN carry the Mobile Node's Care-of Address in the Destination Address field of the IPv6 header. Mobile IPv6 uses a 24 byte Routing Header (type 2) to carry the Mobile Node's Home Address in every packet. This enables the Mobile Node to replace its CoA in the IPv6 header with the Home Address in the Routing Header before handing the packet over to the layers above IP. These higher layers are therefore only aware of the Mobile Node's static Home Address.

The overhead associated with the Routing Header is represented in Table 5(b). In this case, downlink traffic parameters are concerned. For example, for voice service, adding 24 bytes for the Routing Header Option would represent an overhead of 80% relative to the payload. If we include the size of the headers, the relative overhead goes down to 26%. This amounts to slightly more than 6 kbps per voice flow in the downlink direction. The average subscriber downlink bitrate, due to the combined use of all services, in the Busy Hour experiences an increase, due to the use of the Routing Header Option, from 1.14 kbps to 1.31 kbps. This represents an increase of 15% in downlink network traffic.

If the Correspondent Node is also a Mobile Node using Mobile IPv6, then each packet between the two nodes needs both a Routing Header and a Home Address Destination Option. This translates to 44 bytes of overhead in every packet between the two nodes. If we assume that MN to MN communication will be either voice or video (see Table 4 for typical packet sizes) Mobile IPv6 causes an overhead of around 25% in the uplink network traffic and around 35% in the downlink network traffic.

Since the contents of the Home Address Destination Option and Type-2 Routing Header within a Route Optimized Mobile IPv6 user packet are fairly constant, it is expected that IP header compression algorithms, such as the IETF Robust Header Compression (ROHC) [5], will be capable of compressing these headers introduced to the user packets by Mobile IPv6. If IP header compression is applied over the air interface, the impact of the Mobile IPv6 headers will be minimized over this resource limited link. In the wired part of the network, however, the full headers will be exposed and thus the overheads calculated in the previous sections are valid for this part of the network.

5.   CONCLUSION

The results of the baseline Mobile IPv6 signaling load show that the overhead caused by the introduction of Mobile IPv6 in next generation mobile systems is rather small. For example, the air interface signaling load per AR, with more than 300 subscribers, in an urban environment is around 1.2 kbit/s. In the Correspondent Nodes the increase of load amounts to less than 1% per Mobile Node. Also, the additional signaling and processing loads in the Home Agent are also in reasonably low limits.

The inband signaling caused by Mobile IPv6 due to the use of the Home Address Destination Option in uplink messages and the Routing Header Option in downlink messages is, however, bigger, having its highest impact on voice services. The total overhead depends on the exact user traffic mix, increasing as the voice services represent a bigger share of the total traffic. Our results show that this overhead would result in an increase of more than 15% on the traffic, both in the uplink and downlink directions. The impact of this overhead could be reduced using header compression mechanisms in the air interface. If the Correspondent Node is also a Mobile Node using Mobile IPv6, then the overhead caused by Mobile IPv6 is around 25% in the uplink direction and 35% in the downlink.

From the standpoint of mobility signaling, Mobile IPv6 scales well by maintaining sustainable levels of overhead at the AR, intra-domain links, Home Agent and Correspondent Nodes. This is mainly accredited to the relatively low relocation frequencies characteristic of cellular environments.

Several proposals for introducing hierarchical mobility to Mobile IP have appeared in the past years ([6], [7]), in order to address the issue of signaling load, signaling overhead, and handover speed in Mobile IPv6. These mechanisms are however aimed towards the reduction of signaling load and signaling overhead associated with the Mobile IPv6 outband signaling (i.e. BU, BAck, CoTI, HoTI, CoT and HoT). The analysis in this paper shows these are not so critical when compared to the inband signaling and user plane processing. If coupled with a mechanism such as [8], which reduces the Mobile IPv6 handover latency, we thus believe that a hierarchical extension to Mobile IPv6 is not necessary for the architecture considered in this paper.

ACKNOWLEDGEMENTS

This work was performed as part of a co-operation agreement between Nokia and the University of Málaga.

REFERENCES

1. D. B. Johnson, C. E. Perkins, "Mobility support in IPv6", RFC 3775.
2. J. Arkko, V. Devarapalli, F. Dupont, "Using IPsec to Protect Mobile IPv6 Signaling between Mobile Nodes and Home Agents", RFC 3776.
3. S. Kent, R. Atkinson, "IP Encapsulating Security Payload (ESP)" RFC 2406, November 1998.
4. R. Wakikawa, "The Design and Implementation of Mobile IPv6 with multiple network interface support", Master's Thesis, Keijo University - Japan, June 2000.
5. C. Bormann, et al., "Robust Header Compression (ROHC)", RFC 3095.
6. H. Soliman, C. Castelluccia, K. El-Malki, L. Bellier, "Hierarchical Mobile IPv6 Mobility Management (HMIPv6)", draft-ietf-mipshop-hmipv6-02.txt, June, 2004.
7. A. Campbell, J. Gomez, S. Kim, A. Valko and C.Y. Wan, "Design, Implementation, and Evaluation of Cellular IP", IEEE Personal Communications, pp. 42-49, August 2000.
8. R. Koodli, et al., "Fast Handovers for Mobile IPv6", draft-ietf-mipshop-fast-mipv6-01.txt, January, 2004.

PEER-TO-PEER BASED ARCHITECTURE FOR MOBILITY MANAGEMENT IN WIRELESS NETWORKS

Shou-Chih Lo[1] and Wen-Tsuen Chen[2]

[1]Dept. of Computer Science & Information Engineering, National Dong Hwa University, Hualien, Taiwan, R.O.C.; [2]Dept. of Computer Science, National Tsing Hua University, Hsinchu, Taiwan, R.O.C.

Abstract: Mobility management is an important task in wireless networks. The Mobile IP protocol provides a basic solution to the mobility management. However, Mobile IP suffers from several problems. In this paper, we propose an enhancing version of Mobile IP by using the Peer-to-Peer (P2P) network technology. We organize home agents into P2P networks and use the Domain Name System (DNS) to provide the universal telephone number that can uniquely identify one person regardless of the type of equipped device. We claim that our proposed version can provide the advantages of update locality, scalability, load balancing, fault tolerance, and self-administration.

Key words: Mobile IP; Domain Name System; Peer-to-Peer; Wireless Networks.

1.   INTRODUCTION

Mobility management in wireless networks is an important task in order to keep connectivity with roaming users at anytime. Mobile IP[1], which is a standard proposed by the Internet Engineering Task Force (IETF), can serve as the global mobility management in the future heterogeneous wireless networks[2].

Mobile IP uses the *home agent* (HA) and *foreign agent* (FA) to maintain the mobility of a mobile node (MN). The HA maintains the address binding of an MN, and the address binding is a mapping between the permanent home address to the care-of address (CoA) temporally borrowed from an FA.

Mobile IP suffers from several problems[3-5] such as the triangular routing, frequent and long distant registration updates, and single point of failures.

In this paper, we propose some mechanisms to solve these problems experienced in Mobile IP, and most of importance, we introduce the emerging technique of *Peer-to-Peer* (P2P) networks[6-8] into Mobile IP. P2P networks are overlay networks whose topologies are fully independent of physical networks. P2P networks are mostly designed for the data sharing applications. One user can publish its shared data items such as songs or pictures into the P2P network. The developed P2P lookup mechanisms enable one user to efficiently locate the desired data item in logarithmic time. Also, P2P networks with self-organizing and self-configuring features can provide load balancing and fault tolerance.

We take the address binding as a shared data item, and organize a set of HAs into a P2P network. We develop a mechanism to distribute the address binding of an MN to a selected HA with low update cost from the P2P network. We allow each system operator to organize its own P2P network and use the *Domain Name System* (DNS) to provide access to the various P2P networks. Moreover, we provide a universal identifier converted from a typical telephone number to reach a user regardless of the type of equipped device.

The rest of this paper is organized as follows. In Section 2, we give a brief survey on mobility management using Mobile IP. In Section 3, we present the design of our proposed architecture. Section 4 compares the difference and performance of a variety of approaches. Finally, we give a conclusion in Section 5.

2.   RELATED WORK

Mobile IP specified a mechanism to enable an MN to change its point of attachment without changing its IP address. Both Mobile IPv4 and Mobile IPv6 are discussed in the IETF. In this paper, we explain our main idea based on Mobile IPv4. The same idea can be deployed in the IPv6 framework.

Mobile IP has the problem of frequent registration updates particularly for an MN with high mobility. The regional registration[3-5] is commonly used to reduce the registration cost. When an MN moves within the same domain (or region), the registration update is locally handled by a domain-level agent (called Gateway FA, GFA). We called this approach region-based Mobile IP.

As an MN moves far away from its permanent HA, the long distant registration update to the HA would cost high. The dynamic HA assignment becomes the potential solution to this problem. In the approachs[9,10], the GFA is used to be a temporary HA. The association of a temporary HA to an MN is recorded in DNS. In the approach[11], an FA can select a near and light loaded HA to register for an MN, and a redirection link is created between the permanent and temporary HAs. However, how to select a proper HA for an MN is not discussed.

Another problem raised in Mobile IP is the triangular routing. The straightforward solution is to bypass the HA and directly establish the connection to the currently visited FA of the MN. In the approachs[9,10,12,13], the DNS is used to support the query of address binding of any MN. The address binding is stored in DNS as a resource record, and can be refreshed by the dynamic update[14]. We call this approach DNS-based Mobile IP.

In Mobile IP, the HA or FA is sensitive to the single point of failure. The fault-tolerant issue becomes important. In the approachs[15,16], an HA or FA has some other redundant ones as its backup set. Once the HA or FA is failed, another one would be dynamically selected from its backup set.

3. PEER-TO-PEER BASED ARCHITECTURE

In this paper, we propose a P2P-based Mobile IP architecture to efficiently manage the MN's mobility. We combine the advantages of region-based and DNS-based Mobile IPs in our proposed architecture. Moreover, our design takes advantage of the load-balancing and scalability characteristics of P2P networks.

3.1 System Overview

We use subnet-level granularity to explain the basic operations of our mobility management. The detailed descriptions will be given in Section 3.5. Suppose that each subnet is associated with an FA. Several subnets would constitute a domain which is associated with a GFA. The functional overview of our proposed architecture is depicted in Fig. 1.

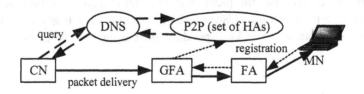

Figure 1. P2P-based Mobile IP.

To avoid the problems of single point of failure and overloaded traffic on the HA, we organize all the existing HAs into a P2P network. With the essential feature of P2P networks, one HA can freely join (when newly installed) and leave (when failed) the network. We would dynamically select an HA for an MN from the P2P network, which is close to the MN. If the selected HA is heavy loaded, we would seek another light loaded one in the neighborhood. With a little modification on the existing P2P lookup mechanism, we can efficiently locate the HA that is selected for a particular MN in the P2P network.

When an MN moves within the same domain, the registration update is locally performed to the GFA. Only whenever the MN moves to another domain, the registration update to the HA is performed. Meanwhile, we may select a new HA that is close to the MN for reducing the long distant registration update.

When a CN (Corresponding Node) would like to connect to an MN, it issues a query to DNS where the P2P lookup mechanism is triggered to locate the MN's HA. The found HA would return the location of GFA the MN is currently located in to the CN. As a result, the CN can directly establish a connection to the GFA and this connection would be further redirected to the MN.

We claim that this architecture can have the following advantages:

Update locality. The frequent registration updates due to the MN's movement of small scope will be partially localized by the regional registration technique. Moreover, the periodical registration updates to the HA during the binding renewal period would be cost saving, because we have selected a near HA to the MN.

Load balancing. The DNS does not perform the complex name resolution for an MN. Instead, the DNS only provides the entry point to the P2P network and triggers the P2P lookup mechanism to find the MN's HA. We put the burden of the complex name resolution on the P2P network where the actual execution would be distributed to the nodes involved in the P2P network. The set of HAs in the P2P network will work together and can migrate the workload with each other.

Self-administration. Each system operator can administer its own P2P network, which facilities the prevention of binding data from the revelation to other system operators. Also, a system operator can freely increase or decrease the number of HAs depending on the amount of users that are served.

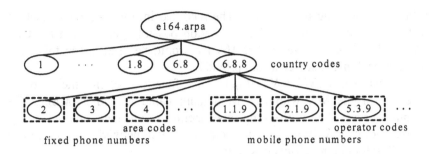

Figure 2. ENUM domain name space.

3.2 DNS Structure

Assume each MN is associated with and identified by a unique telephone number (called ENUM[17] throughout the paper). As illustrated in the IETF RFC 2916[18], a typical E.164 telephone number like +886-3-5741234 can be transformed to the domain name with format:

"4.3.2.1.4.7.5.3.6.8.8.e164.arpa".

"e164.arpa" is the suggested root of the ENUM domain names. A possible portion of the ENUM domain name space is shown in Fig. 2. The second-level domains include one entry for every country code, and the third-level domains include one entry for every area code or every operator code.

Assume the ENUM domain name space is divided into non-overlapping zones according to the ENUMs administered by different system operators. We indicate each zone in Fig. 2 by using a dotted rectangle. A zone will have one primary name server and several secondary name servers for the fault-tolerant reason.

A system operator can install a set of HAs for each of its service zones. These HAs are not necessary to be associated with network routers as done in Mobile IP, and can be artificially distributed into the service coverage of the corresponding zone. These HAs are responsible for storing the resource records of MNs having their ENUMs in that zone. The resource record might contain the IP address binding for Internet applications that need mobility support and the service binding which specifies the preferred means (e-mail, telephone, etc.) to be reached at a particular period of time.

The name server in the zone performs the name resolution, given an MN's ENUM domain name, by locating the HA which stores the MN's resource record in the P2P network. The system operator has the responsibility to keep the list of HAs for each of its service zones up to date in the corresponding name server. The recorded information might include the HA's IP address and the HA's location possibly gotten from the GPS (*Global Positioning System*). The HA list would be used in the following functions each name server would provide.

- get_host_name($ENUM\_NAME_{MN}$). This function returns the resource record of the MN with ENUM domain name $ENUM\_NAME_{MN}$. The name server when performing this function would randomly select one HA from its HA list as the entry point to the P2P network, and from there the P2P lookup mechanism is activated.
- get_neighbor_HA(IP_{target}, k). This function returns the k nearest HAs from the HA list that are close to the node with IP address IP_{target}. The actual measurement of locality is beyond the scope of the paper and GPS is one of the possibilities.

3.3   P2P Structure

The emerging P2P networks have potential to support large data sharing applications. Some system protocols, such as Pastry[6], CAN[7], and Chord[8], have been proposed for building large P2P networks. These protocols are based on a *distributed hash table* (DHT), which allows shared data items to be uniformly distributed into the nodes in the P2P network.

Each node participated in the P2P network has part of index information to shared data items. In Chord, for example, one user can issue a data lookup query to any node in the P2P network, and from there the query would be subsequently forwarded with at most $\log N$ hops till to the target node containing the desired data item. Also, Chord can efficiently support the node's join and leave with $\log^2 N$ messages.

Our goal is to construct an individual P2P network with the HAs in each zone. Within this network, the MN's address binding would be considered as a shared data item. An MN can publish its address binding (or service binding) identified by an ENUM to the P2P network. We call the node the binding data is hashed to a *destined* HA for an MN. A user can locate the destined HA of a particular MN by sending a lookup query, carrying the MN's ENUM, to the P2P network.

Note that the destined HA is not artificially selected but is determined by the DHT. The registration cost would be high if the destined HA is far away from the MN's current location. To support update locality, we artificially select a near HA called *assigned* HA to the MN, and the actual binding data is stored in the assigned HA. Since we can only locate the destined HA through the P2P lookup query, we establish a redirection link from the destined HA to the assigned HA.

Sometimes the near HA to the MN would be heavy loaded. In this case, we select another HA in the neighborhood, which is light loaded as an assigned HA. To support this function, we construct a neighbor list for each HA, which records the state (alive and heavy loaded or not) of its neighbors. We can get the k nearest neighbors to an HA X by asking the name server of the local zone via function get_neighbor_HA(IP_X, k).

The HA in our P2P network provides the following functions:

- locate_HA($ENUM\_NAME_{MN}$): This function locates the destined HA of the MN with ENUM domain name $ENUM\_NAME_{MN}$ by using the P2P lookup mechanism.
- redirect_HA(IP_{target}, $ENUM\_NAME_{MN}$): This function creates a redirection link from the destined HA of the MN with ENUM domain name $ENUM\_NAME_{MN}$ to the node with IP address IP_{target}.

3.4 Region Structure

We construct a GFA in each domain, and this GFA can provide a global CoA (GCoA) to a registered MN under the domain. By contrast, the FA can provide a local CoA (LCoA) to a registered MN under the subnet. Packets, which are sent from a CN and are destined to an MN, are tunneled to the GFA by the GCoA and then tunneled to the MN by the LCoA. During the Mobile IP session, the CN after querying the DNS would directly deliver packets to the GFA the MN is currently located in. If the MN makes a movement and changes to another GCoA and/or LCoA, the GFA known by the CN has the responsibility to redirect the packets to the MN.

If the movement is within the same domain, we can either establish a redirection path between the old FA and the new FA (path 1 in Fig. 3) or between the GFA and the new FA (path 2 in Fig. 3). The choice is depending on the update and packet delivery costs.

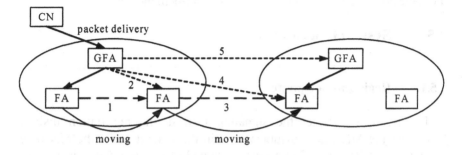

Figure 3. Possible redirection paths.

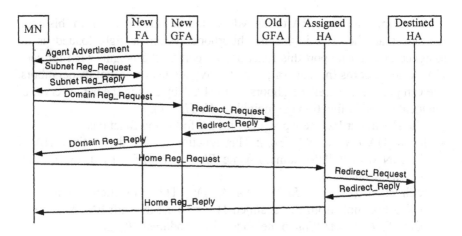

Figure 4. Registration flow during inter-domain movement.

If the movement is across different domains, we have three choices on the redirection paths. One is between the old FA and the new FA (path 3 in Fig. 3). Another is between the old GFA and the new FA (path 4 in Fig. 3). The other is between the old GFA and the new GFA (path 5 in Fig. 3) after the MN has registered to the new GFA. The choice is depending on the update and packet delivery costs and the IP address space used. The first two have to use the LCoA of global IP address space; while the last one can use the LCoA of private IP address space.

The GFA has another job of maintaining a list of available HAs, which are located in the same zone and are close to the GFA, for each of service zones. The GFA can select an HA from the corresponding list as an assigned HA for an MN coming from a certain service zone. The GFA can send the get_neighbor_HA(IP_{GFA}, k) request to the name server of a particular zone to construct this list. Those MNs with their ENUM domain names belonging to the same zone would share a common list of HAs in the GFA.

3.5   System Operations

3.5.1   Registration Update

In Fig. 4, we depict the signaling flow during registration update. Whenever the MN changes subnets within the same domain, the MN only communicates its new LCoA to the serving GFA (we use path 2 in Fig. 3). Whenever the MN changes domains, it first obtains an LCoA by performing a subnet-specific registration update to the serving FA. The serving FA assigns the MN a designated GFA. Then the MN performs a domain-specific

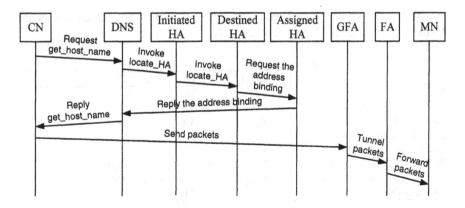

Figure 5. Packet delivery flow.

registration update by communicating its current LCoA to the designated GFA. The designated GFA replies to the registration with a GCoA and an assigned HA. If the MN has an active session, the redirection path (we use path 5 in Fig. 3) between new and old GFAs is created. Then the MN performs a home registration update by communicating its current GCoA to the assigned HA.

The assigned HA can either accept or reject the MN's registration according to its current capacity. If rejected, the assigned HA would reply the MN with another HA selected from its neighbor list. The MN would subsequently attempt to register to a different HA till accepted. If the accepted HA is different from the destined HA of the MN, the accepted HA would call the redirect_HA($IP_{assigned\_HA}$, $ENUM\_NAME_{MN}$) function. The subsequent binding renewal would be only performed to the assigned HA.

In our registration scheme, any new connection from a CN is directed to the new GFA since the address binding has been updated. Those old connections would be guaranteed to be deliverable via the redirection path.

3.5.2   Packet Delivery

In Fig. 5, we depict the signaling flow during packet delivery. The CN first sends the get_host_name($ENUM\_NAME_{MN}$) request to DNS. The DNS would contact the name server of the zone to which $ENUM\_NAME_{MN}$ belongs. The name server randomly selects an HA (called initiated HA) from the HA list to initiate the P2P lookup operation. The destined HA of the MN can be found by calling function locate_HA($ENUM\_NAME_{MN}$). Then, we follow the redirection link to reach the assigned HA and from there the address binding (i.e., MN's GCoA) is retuned to the CN. As a result, the CN establishes a connection and sends packets to the GFA, and these packets are further tunneled to the FA, and to the MN.

Table 1. Performance comparison

	Standard	Region Based	DNS Based	P2P Based
Triangular Routing	Yes	Yes	No	No
Frequent Registration	Yes	No	Yes	No
Single Point of Failure	Yes	Yes	Yes	No
Load Balancing	No	No	No	Yes
Update Locality	No	Yes	No	Yes
Registration Update	2*2 hops	2*[3, 2] hops	2*2 hops	2*[3+log N, 2] hops
Connection Setup	3 hops	4 hops	2*d+2 hops	2*(d+log N+2)+3 hops
Packet Delivery	3 hops	4 hops	2 hops	3 hops

4.   PERFORMANCE EVALUATION

We have mentioned three categories of enhancements to Mobile IP: region_based, DNS_based, and P2P_based ones. Here we make a comparison of these different enhancing mechanisms to the standard Mobile IP. In Table 1, we summarize the advantages/disadvantages and costs of these mechanisms. Our proposed P2P_based mechanism essentially inherits the advantages of region_based and DNS_based ones, so we have no triangular routing (due to DNS) and frequent registration update (due to regions) problems. The self-configuring characteristic of P2P networks makes our mechanism having no single point of failure on the HA.

Moreover, our proposed P2P_based mechanism has good load balancing due to the following reasons:

1. In the DNS: The domain name hierarchy of DNS can naturally distribute the workload to different name servers. Moreover, the random selection of the entry point to a P2P network from a name server can distribute the P2P lookup overhead to different nodes.

2. In the P2P network: The operations of a P2P lookup query are naturally distributed to the nodes involved. The neighbor list associated with each HA can be a reference to migrate the registration related jobs from a heavy loaded HA to a light loaded one.

3. In the GFA: The assigned HA is randomly selected from a list maintained by the GFA, which can avoid a certain HA to become heavy loaded.

Next, we analyze the registration update, connection setup, and packet delivery costs in terms of hop distances for these mechanisms. The standard Mobile IP follows the path MN-FA-HA during the registration update, hence the cost is twice (for round trip) the hops from the MN to the HA. During the connection setup and packet delivery, the path CN-HA-FA-MN is followed.

In the region_based one, the inter-domain registration update follows the path MN-FA-GFA-HA; while the intra-domain registration update follows the path MN-FA-GFA. We use the bracket in the table to denote the maximal and minimal values of the cost. The connection setup and packet delivery follow the path CN-HA-GFA-FA-MN.

In the DNS-based one, the path MN-FA-DNS is followed during the registration update, and the path CN-DNS-CN-FA-MN is followed during the connection setup. The path segment CN-DNS-CN is to iteratively locate the proper name server in the domain name hierarchy. Assume the average number of iterations is denoted by d. The packet delivery follows the path CN-FA-MN.

In the P2P_based one, the path MN-FA-GFA-Assigned HA-Destined HA is followed during the inter-domain registration update; while the path MN-FA-GFA is followed during the intra-domain registration update. The path CN-DNS-Initiated HA-Destined HA-Assigned HA-CN-GFA-FA-MN is followed during the connection setup. The P2P lookup from an initiated HA to a destined HA would take $\log N$ hops in a typical P2P network like Chord. The packet delivery follows the path CN-GFA-FA-MN. As can be seen, the connection setup is longer than that in other mechanisms. It is one of our future work to reduce this setup delay by using data replication in the P2P network.

5. CONCLUSION

In this paper, we introduce the emerging P2P network technology into Mobile IP to efficiently support mobility management. In our proposed architecture, we provide the dynamic HA assignment and the capability of load balancing and fault tolerance on the HA. We overcome the problems in Mobile IP such as triangular routing and frequent registration update at the expense of the delay on connection setup. In the future, we will incorporate the AAA (Authentication, Authorization, and Accounting) server into our architecture to enhance the capability of security.

ACKNOWLEDGMENTS

The authors would like to thank the anonymous referees for their helpful suggestions. This research was partially supported by the National Science Council of the Republic of China under Contract No. NSC 93-2752-E-007-003-PAE.

REFERENCES

1.   C. Perkins, "IP Mobility Support for IPv4, Revised," *RFC3220*, IETF, Jan. 2002.
2.   L. Morand and S. Tessier, "Global Mobility Approach with Mobile IP in All IP Networks", *IEEE Int'l Conf. on Communications* (ICC), pp. 2075-2079, May 2002.
3.   Campbell, J. Gomez, S. Kim, Z. Turanyi, C-Y. Wan, and A. Valko, "Design, Implementation and Evaluation of Cellular IP," *IEEE Personal Communications Magazine*, vol. 7, no. 4, pp. 42-49, Aug. 2000.
4.   S. Das, A. Mcauley, A. Dutta, A. Misra, K. Chakraborty, and S. K. Das, "IDMP: An Intradomain Mobility Management Protocol for Next-Generation Wireless Networks," *IEEE Wireless Communications*, vol. 9, no. 3, pp.38-45, Jun. 2002.
5.   R. Ramjee, K. Varadhan, L. Salgarelli, S. R. Thuel, S. Y. Wang, and T. L. Porta, "HAWAII: a Domain-Based Approach for Supporting Mobility in Wide-Area Wireless Networks," *IEEE/ACM Trans. on Networking*, vol. 10, No. 3, pp. 396-410, Jun. 2002.
6.   Rowstron and P. Druschel, "Pastry: Scalable Distributed Object Location and Routing for Large-Scale peer-to-Peer Systems," *Proc. IFIP/ACM Int'l Conf. on Distributed Systems Platforms* (*Middleware*), Nov. 2001.
7.   S. Ratnasamy, P. Francis, M. Handley, R. Karp, and S. Shenker, "A Scalable Content-Addressable Network," *ACM SIGCOMM*, pp. 161-172, Aug. 2001.
8.   Stoica, R. Morris, D. Karger, M. F. Kaashoek, H. Balakrishnan, "Chord: A Scalable Peer-to-Peer Lookup Service for Internet Applications," *IEEE/ACM Trans. on Networking*, vol.11, no. 1, pp. 17-32, 2003.
9.   Y. Chen and T. Boult, "Dynamic Home Agent Reassignment in Mobile IP," *IEEE Wireless Communications and Networking Conference*, pp.44-48, 2002.
10.   R. Zheng, Y. Ge, J. C. Hou, and S. R. Thuel, "A Case for Mobility Support With Temporary Home Agents," *ACM SIGMOBILE Mobile Computing and Communications Review*, vol. 6, no. 1, pp. 32-46, Jan. 2002.
11.   M. Kulkami, A. Patel, and K. Leung, "Mobile IPv4 Dynamic Home Agent Assignment," *IETF*, draft-ietf-mip4-dynamic-assignment-00.txt, Jan. 2004, Work in Progress.
12.   M. Conti, E. Gregori, and S. Martelli, "DNS-based Architecture for an Efficient Management of Mobile Users in Internet," *15th Int'l Symposium on Parallel and Distributed Processing*, pp. 1957-1964, Apr. 2001.
13.   C. Snoeren and H. Balakrishnan, "An End-to-End Approach to Host Mobility," *6th Int'l Conf. on Mobile Computing and Networking* (*MOBICOM*), pp. 155-166, Aug. 2000.
14.   P. Vixie, S. Thomson, Y. Rekhter, J. Bound, "Dynamic Updates in the Domain Name System (DNS UPDATE)", *RFC2136*, IETF, Apr. 1997.
15.   J. W. Lin and J. Arul, "An Efficient Fault-Tolerant Approach for Mobile IP in Wireless Systems," *IEEE Trans. on Mobile Computing*, vol. 2, no. 3, Jul.-Sept. 2003.
16.   J. H. Ahn and C. S. Hwang, "Efficient Fault-Tolerant Protocol for Mobility Agents in Mobile IP," *15th Int'l Symposium on Parallel and Distributed Processing*, pp. 1273-1280, 2001.
17.   C. Mctaggart, "Telephone Numbers, Domain Names, and ENUMbers," *IEEE Communications Magazine*, pp. 26, Sept. 2002.
18.   P. Faltstrom, "E.164 Number and DNS," *RFC2916*, IETF, Sept. 2000.

SUPPORTING GROUPWARE IN MOBILE NETWORKS

Nadav Lavi
Israel Cidon
Idit Keidar
Department of Electrical Engineering, Technion - Israel Institute of Technology

Abstract   We present MaGMA (Mobility and Group Management Architecture), an architecture for groupware support in mobile networks. MaGMA's main objective is enabling mobile users to use real-time group applications over the IP infrastructure. Our solutions address group management as well as support for QoS and seamless handoff. We illustrate the advantages of MaGMA using mathematical analysis and simulations.

1.   Introduction

The widespread availability of the Internet has enabled the use of many groupware and collaborative computing applications (e.g., chat, ICQ, Net-Meeting, Exchange, Lotus Notes, Webex, desktop video conferencing, etc.). With the advance of wireless personal communication, such groupware applications are becoming popular in cellular and mobile networks [16]. For example, major cellular providers (Verizon, Nextel, Orange) offer, or plan to offer soon, group services such as push-to-talk (PTT) [7, 18]. The PTT cellular revenue, which was $84 million in 2003, is expected to reach $10.1 billion by 2008; and the 2.3 million PTT cellular subscribers community of 2003 is expected to grow to 340 million by 2008 [19]. While traditional PTT is limited to voice, the emerging convergence is expected to merge real time and non-real time aspects of group communication.

The converged Internet infrastructure is starting to provide the required support for real-time applications, such as voice over-IP (VoIP) and video-conference over-IP (VCoIP), which require quality of service (QoS) among stationary endpoints. This has led to the emergence of many QoS standards and technologies, e.g., Diffserv, RSVP, and MPLS, as well as real-time protocols such as RTP, H.323, MGCP, and SIP.

At the same time, wireless access to the global Internet is becoming widely supported, and WLAN access points are ubiquitously available. Given the trends predicted in wireless standard forums such as the Open Mobile Alliance

(OMA) [9] and 4G, it is expected that the next phase in the evolution of converged group services will be their integration with wireless mobile devices. These networks are most likely to adopt the TCP/IP architecture including its related convergence standards.

In this paper, we propose MaGMA an architecture for group management in mobile networks interconnected via the global Internet. MaGMA provides an initial comprehensive solution for the mobile world, addressing aspects such as scalable group management, mobility, handoff, and QoS provision. MaGMA's architecture consists of a collection of mobile group managers (MGMs), which manage group membership and also implement a multicast overlay for data delivery. Each mobile node (MN) interacts with an MGM proximate to it. MaGMA supports a subscription model in which nodes can request to be notified of other nodes' mobility.

We propose a number of group management protocols. We have implemented MaGMA in the ns2 network simulator [8]. We present simulation results and validate them through mathematical analysis.

This paper proceeds as follows: Section 2 gives background on current mobility solutions. Section 3 describes the network model and our proposed architecture. Section 4 presents solutions for mobile group management. Section 5 evaluates the proposed solutions through simulations and analysis. Section 6 addresses transport issues, and Section 7 concludes.

2.   Related Work

We are not aware of any previous comprehensive solution for mobility support in groupware applications. We now overview leading mobility solutions for IP in general, and group communication systems in particular.

Mobility Solutions   Mobile IP [10] is the current standard for seamless mobility in the IPv4 based Internet. Mobile IP uses a simple method of encapsulation and tunneling. Every MN is associated with a home domain, in which resides a proprietary server named *home agent*. While at its home domain, the MN receives packets as a regular stationary node. When the MN moves to a foreign domain and changes its IP address, it notifies its home agent of its new IP address. Thus, the home agent can forward to the MN packets destined to the MN's home IP address through a tunnel it creates to the new location. This forwarding scheme, called *triangle routing*, generally leads to routes that are longer than the direct path, and therefore suffers from poor performance. Moreover, Mobile IP applies only to unicast sessions between the MN and a corresponding node and does not include QoS support.

In [11], a route optimization to Mobile IP that avoids triangle routing is proposed. In this approach, the MN sends *binding* information to the corresponding node, thus enabling direct communication between the two. It is unclear whether Mobile IP with route-optimizations can support simultaneous movements of both endpoints of a communication. Moreover, this solution requires a modification of the host IP stack, and a home agent in each MN's home domain, and is therefore difficult to deploy. In addition, creating new connections to a MN must always involve its home domain, even if the MN is distant from it for an extensive period. In contrast, MaGMA provides a flexible architecture in which the communication infrastructure is deployed in the network, and communication with an MN is independent of its home domain.

Balakrishnan and Snoeren [14] propose a DNS-based solution to IP mobility. Similar to Mobile IP, every node has a home domain. When an MN moves and changes its IP address, it registers a secure DNS update at its home domain DNS server. In order to avoid the use of stale binding information, DNS caching is minimized (by setting TTL=0) and direct binding is used. The major drawbacks of this approach are that both endpoints cannot move simultaneously, that DNS standards do not support user self-configuration, and that operating systems and DNS servers often do not comply with DNS TTL directions. Finally, DNS caching elimination will overload the DNS system.

Mysore and Bhaghavan [6] propose to use the IP multicast infrastructure for mobility support. While this solution can potentially provide good performance, unfortunately, IP multicast is not widely deployed. Therefore, this scheme cannot provide seamless mobility in today's Internet.

Group Communication A closely-related group-management protocol is CONGRESS [1], which was designed for ATM environments. Like MaGMA, CONGRESS uses an overlay among servers. However, in contrast to MaGMA, the overlay is hierarchical and restricted to membership management, and does not support QoS multicast. Moreover, CONGRESS was not designed with mobility in mind and does not incorporate a handoff solution.

Prakash and Baldoni [12] propose protocols for group communication support in virtual cellular networks where base-stations can move, and for ad-hoc networks. In contrast, we consider a network with stationary base-stations. Bartoli [2] proposes a totally-ordered multicast protocol for a dynamic membership in wireless networks. In contrast to MaGMA, it assumes a failure-free environment and focuses on reliability and ordering rather than QoS support.

3. Model and Architecture

3.1 Network model

Similarly to mobile IP, we model the network as a collection of autonomous domains. Every MN has a unique ID (UID), which identifies the MN in all of its locations. Upon moving to a new domain, the MN obtains a new local IP address, e.g., using DHCP. We do not address intra-domain handoff, i.e., micro mobility. Rather, we assume that a micro-mobility mechanism is in place (e.g., [5, 13, 17]), and that an adequate routing protocol exists in each domain. We assume that hosts can crash, and that such crashes are detectable by other hosts.

3.2   Design goals

Our main goal is to provide support for managing and keeping a coherent up-to-date view of each group in a highly dynamic mobile environment. The solution we seek should address the following issues:

- Mapping group names to their current subscribers.

- Mobility support with seamless handoff.

- QoS support for real time applications.

- Transport efficiency, including the avoidance of triangle routing, and minimizing the number of duplicates of multicast messages sent.

- Low control overhead and a scalable control plane.

- Support for incremental deployment.

3.3   Architecture

Our architecture consists of a collection of MGMs positioned in different domains. For simplicity's sake, through most of this paper we assume that MGMs are static and well known. In Section 4.5, we discuss possible extensions of the basic architecture in which MGMs can be added on-the-fly. The role of MGMs is twofold: managing group membership and forwarding packets in order to facilitate QoS multicast. Our architecture calls for the use of multiple servers for the following reasons:

- to offer scalability in the number of groups and the number of group members;

- to efficiently support groups with geographically dispersed members, as well as localized ones;

- to facilitate QoS reservation among domains;

- to reduce traffic overhead; and

- to provide fault-tolerance in the presence of network partitions (where a node may not be able to communicate with a remote server) as well as server failures.

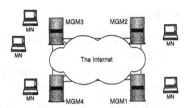

Figure 1. The MaGMA network architecture.

Our architecture is depicted in Figure 1. Ideally, one MGM is located in each domain, although this is not mandatory. Each MN is served by the closest MGM to its domain. The MGMs provide the following services to MNs:

- *Joining* or *leaving* a group.

- *Moving* to a new location - the moving node sends a *move* message to its new MGM.

- *Retrieving* the current membership view (list of current subscribers) of a given group and *multicasting* to a given group.

The MGMs form an overlay network among them. The overlay is used both for membership management and QoS multicast. The overlay construction can be employ known techniques for building efficient QoS-aware overlays, e.g., [15, 4], and its discussion is beyond the scope of this paper. We assume that MNs are likely to remain in the groups they join for extensive periods. Therefore, move messages dominate the control traffic.

4. MGM Protocols

We now present protocols for managing group membership of mobile users. The first two protocols are based on existing mobility solutions. We then propose two additional protocols, which handle mobility explicitly.

4.1 Exploiting Mobile IP

Since IP Mobility is the de facto standard for mobility on the Internet, we first consider a naïve solution based on this approach. One can delegate the responsibility for mobility management to Mobile IP, and have the MGMs only map group names to nodes' home addresses. This eliminates the need to handle *move* requests. The main drawback of this solution is the uncontrolled delay and QoS degradation resulting from Mobile IP's triangle routing.

4.2   DNS based solutions

As an alternative to Mobile IP, DNS-based mobility [14] can be used when DNS support is available and user mobility is limited. We now examine possibilities for extending this solution to support group management.

A simple approach, similar to the Mobile IP based solution suggested above, can delegate mobility handling to the nodes' home domain DNS servers, and have the MGMs map each group to its subscribed MNs in a domain name format. With this approach, a retrieve operation first gets from the MGM a list of MN names and then employs DNS queries to translate the MN names to actual IP addresses. This solution is simple and can be gradually implemented in today's Internet. However, the retrieve-translate procedure can take a substantial amount of time. Moreover, this solution suffers from the previously described problems associated with dynamic DNS resolution.

An alternative DNS-based solution replaces the MGM with a proprietary group DNS server used only for group translations. This DNS server maps group names directly to their subscribed MNs' IP addresses, thus reducing the translation delay. An additional drawback of this suggestion is that current standards do not support DNS server updates by hosts from foreign domains.

4.3   MGMFlood

We now turn to solutions in which mobility is handled by the MGMs and not delegated to other services. In our first such scheme, MGMFlood, each MGM forwards (floods) to all other MGMs all control messages (*join/leave/move*) received from MNs in its domain. When an MN crashes, its local MGM detects the crash and sends an appropriate *leave* message to all other MGMs.

Note that the protocol must guarantee view consistency in the presence of rapid mobility. When an MN frequently changes its location, it updates a different MGM each time it moves. Thus, different MGMs can receive the flooded *move* messages in different orders. In order to ensure consistency, each MN keeps an increasing *Domain Hop* (DH) counter, counting the number of times it moves between domains. This counter is sent to the MGM in every *join/leave/move* message, and is stored along with the MN's UID at the MGMs. An MGM that receives a move message with a lower DH than the one associated with the UID in its current view ignores this message.

The MN stores the DH as long as it is up. In case the MN re-joins a group after a crash, it registers with DH=0, which indicates to the MGM to send back, when available, the highest DH value associated with this MN.

MGMFlood is simple and allows for seamless handoff due to its prompt reaction to mobility updates. However, it entails high control message overhead, as all MGMs keep views of all groups, including groups not residing in their domains. This solution may not scale well, especially if there are many small groups and localized memberships.

4.4   MGMLeader

Our second solution reduces the overhead by propagating updates only to those MGMs that have group members in their domains. When an MGM receives an MN's message regarding a group that is represented in its domain, it extracts the MGMs that have members in the group from its local view, and forwards the message only to those MGMs.

If an MGM receives a control message (*join* or *move*) for a group that does not yet exist in its domain, then it needs to discover the group's up-to-date view, and to forward the event to the appropriate MGMs. The challenge is preserving a coherent view at all MGMs in the presence of concurrent operations without inducing excessive overhead.

In order to minimize the control overhead and ensure view consistency, only one of the participating MGMs sends the view to the new MGM. To this end, one MGM is designated as the *coordinator* of the group. Every active group has a *coordinator*, and a single MGM can be the coordinator of multiple groups. If the *coordinator* fails or leaves the group (because all the MNs in its domain leave) then a new *coordinator* is elected, as explained below.

When a new MGM joins a group due to a *move* event, it extracts the moving MN's former MGM from the *move* message, and sends the event message to that MGM. The former MGM, in turn, forwards the message to the *coordinator*. When the *coordinator* receives a *move* message originating from an MGM that is not already in the group, it sends the group's view to the new MGM and forwards the message to all the group's MGMs. This message flow is illustrated in Figure 2.

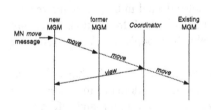 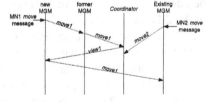

Figure 2.   Move message flow.

Figure 3.   Potential view inconsistency in over-simplified leader-based solution.

This communication between the two MGMs also facilitates establishing a tunnel from the former MGM to the new one, so that the former MGM can forward data packets destined to the moving MN via its new MGM, to guarantee smooth handoff; such tunneling is suggested in [11].

When a new MGM joins a group due to a *join* message, it broadcasts the message to all the MGMs. As before, when the *coordinator* receives this message from the new MGM, it sends the group's view to the MGM.

As with the previous solution, MGMs may receive certain event messages out of order. The ordering of events related to the same MN is solved, as in the MGMFlood protocol, using the DH counter. However, this does not suffice to ensure view consistency when MGMs dynamically join and leave the group. Figure 3 illustrates a problematic scenario that can occur if concurrent joins are handled carelessly. In this example, while a new MGM retrieves the group's view from the *coordinator*, an existing MGM sends another event to the group's MGMs. The existing MGM is unaware of the new MGM and thus does not forward the message to it. This causes the new MGM to an inconsistent view of the group.

In order to address this difficulty, each MGM maintains an increasing *Local Event Counter* (LEC) for every group. Whenever an MGM receives a *join*, *leave*, or *move* message from a local MN, it increments the appropriate LEC. The group's LEC is included in every message pertaining to this group sent by the MGM. When an MGM joins a group, it initiates the group's LEC to 1. In addition, the MGM keeps, for every group, a *LECvector*, holding the highest known LEC for each MGM in this group.

In every message sent from one MGM to another, both the sender's LEC and the receiver's latest known LEC (from the *LECvector*) are included. When an MGM receives a packet, it checks the LECs. If its local LEC is higher than the one known to the sender it sends back its local view and LEC. If it discovers that the sending MGM's LEC is higher than the one it knows, it retrieves the local view of the sending MGM. When the *coordinator* forwards *move* messages of new MGMs, it includes the LECs corresponding to the view it is sending to the new MGM. In case some events are not reflected in this view, the receiving MGMs forward their local views to the new MGM. In addition, we ensure consistency using periodic updates, whereby the group's *coordinator* sends the current view all to the group's MGMs.

We now discuss coordinator election. When the last MN in the *coordinator*'s domain leaves the group or moves to another domain, the *coordinator* appoints a new MGM as the new *coordinator* of the group and informs the group's MGMs of the new *coordinator* in the forwarded *move* or *leave* message. Subsequently, the leaving *coordinator* forwards control messages that it still receives to the new *coordinator*. In order to avoid appointing an MGM that has already left, an MGM can not leave the group until it receives the *coordinator*'s permission. If the *coordinator* notices, after receiving a *leave* or *move* message, that an MGM has no members in the group, it sends a permission-to-leave message to that MGM. The only scenario where the *coordinator* does not permit the MGM to leave is if the *coordinator* appoints the MGM to be the new *coordinator*. In this case the MGM needs to find a new MGM to replace it as the group's *coordinator*.

In case the *coordinator* crashes, the surviving MGMs start an election procedure by flooding their local views to all MGMs in the system. The MGM with the lowest ID is elected as the new *coordinator*, and it sends to the group's MGMs an up-to-date view computed using the local views sent during the election procedure. Throughout the election procedure, all new events are buffered by the MGMs, and are disseminated only after the new view is received.

4.5   Dynamic MGMs

Thus far, we have assumed that MGMs are static and well-known. However, our solution can be extended to support a dynamic architecture, where MGMs can join and leave. The MGMLeader is a natural choice for such an architecture, since it accommodates for a dynamic set of MGMs maintaining the membership of a single view.

We plan to extend the architecture to support a delegation mechanism, where MGMs can grant MNs permission to operate as membership servers. This will allow the use of our group services in wireless hybrid networks, i.e. islands of ad-hoc network interconnected via access-points [3].

5.   MGM Protocol Evaluation

5.1   Packet delay evaluation

Mobile IP may exhibit poor performance due to its use of triangle routing. With the MGM architecture, on the other hand, packets are sent directly to the MN's current location. To illustrate the advantage of this approach, we simulate a single constant bit rate (CBR) UDP session, and measure the end to end delay with both approaches (MGM and Mobile IP).

We simulate a network of four domains. The transmission source is in Domain 0 throughout the simulation. The receiver is initially in its home domain (Domain 1) and then moves towards Domain3 through Domain2, as depicted in Figure 4. The domains are connected via 5Mb links with 20ms delay. The simulated wireless interface is 914MHz Lucent WaveLAN DSSS.

Figure 5 shows the average measured packet delays for three architectures: Mobile IP, a centralized architecture with a single MGM (in Domain 0) servicing all MNs, and a distributed architecture, with an MGM in each domain. The MGM solutions transmit messages via the optimal route, whereas Mobile IP uses the triangle route, degrading performance by a factor of 3.

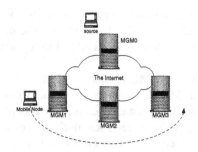

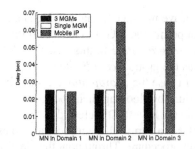

Figure 4.   Scenario simulated in Figure 5.   *Figure 5.*   Average end-to-end packet delay for system depicted in Figure 4: simulation.

5.2   Control plane evaluation

We now evaluate the overhead associated with control messages. We measure the average control overhead associated with a single *move* event. We simulate the following uniform network model:

- 11 domains (Domains 0-11), 1 MGM in each domain;

- 10-100 receiving MNs, initially uniformly distributed in Domains 1-10, then moving among these domains;

- 8 groups, where every receiving MN participates in a single group chosen uniformly at random;

- a fixed number of sources in Domain 0.

We simulate the MGMFlood and MGMLeader protocols in this setting, and measure the average control overhead associated with a single *move* message. The average is calculated over 300 events for each number of MNs. In each event, a random MN moves to a new random domain. The results are depicted in Figure 6, with 95% confidence intervals for MGMLeader. We also mathematically analyze the expected control overhead. For MGMFlood, this is straightforward. Since each control message is sent to all MGMs, and there are ten MGMs, the overhead is exactly ten messages per *move* event. Not surprisingly, the analysis and simulation results for this protocol accurately match each other (see Figure 6).

For MGMLeader, our analysis provides an upper bound and a lower bound (both depicted in Figure 6). Recall that a new MGM joining a group communicates with the MN's former MGM, which forwards the message to the *coordinator*. The lower bound (*coordinator* case) occurs if the former MGM is the *coordinator* of the retrieved groups. The upper bound (non-*coordinator* case) occurs when the former MGM is not the group's *coordinator*. In this

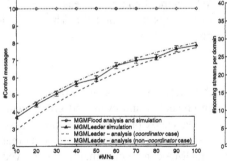 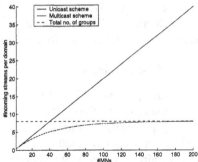

Figure 6. Average number of control messages per movement, uniform system, varying number of MNs, fixed number of domains: analysis vs. simulations.

Figure 7. Average number of incoming streams per domain in uniform system with 8 groups, varying number of MNs, fixed number of domains: analysis.

case, one more message is sent compared to the *coordinator* case: from the former MGM to the *coordinator*. Due to space consideration, the detailed analysis derivation is given in the full paper.

We observe that with MGMLeader, the overhead increases with the number of MGMs that have members in the group. In sparse groups, few MGMs are involved, and hence few control messages are sent. MGMLeader outperforms MGMFlood in all the simulated situations, but its advantage is less significant in dense groups. We conclude that MGMLeader is preferable for sparse groups, whereas the much simpler MGMFlood may be adequate for dense groups in which all or most MGMs participate.

6. Transport Issues

We suggest two solutions for *multicast*. The first solution uses unicast: the transmitting MN *retrieves* the group's view and sends data directly to all the group members using multiple unicast streams. This solution incurs minimal load on MGMs, but suffers from poor scalability as domains with many users will have many incoming streams.

The second solution uses multicast over the MGMs' overlay. The transmitting MN sends the data to its MGM, which forwards the data to all the group's MGMs, which in turn forward the data to their locally subscribed MNs. This solution can exploit IP Multicast where available. It is more scalable, and enables each MN to use a single stream, thus extending its battery life.

Figure 7 depicts the number of incoming streams per-domain in both solutions, analyzed for the uniform network model of Section 5. We assume that all groups are active. In the unicast scheme, the average number of incoming data streams is the average number of MNs in the domain, whereas in the multicast scheme, it is the average number of groups in the domain. The detailed analysis derivation is given in the full paper.

7.   Conclusions

We have presented MaGMA, an architecture for supporting group services in the emerging wireless networks. We presented and evaluated group management protocols for this architecture. MaGMA provides a comprehensive solution for seamless mobility with QoS support, important requirements that are not addressed in current solutions. MaGMA can be incrementally deployed since an MN may communicate with any MGM located in its vicinity.

References

[1]   T. Anker, D. Breitgand, D. Dolev, and Z. Levy. CONGRESS: connection-oriented group address resolution services. *SPIE on Broadband Networking Technologies*, 1997.

[2]   A. Bartoli. Group-based multicast and dynamic membership in wireless networks with incomplete spatial coverage. *Mobile Networks and Applications*, 3(2):175–188, 1998.

[3]   E. M. Belding-Royer, Y. Sun, and C. E. Perkins. Global Connectivity for IPv4 Mobile Ad hoc Networks. Internet draft, IETF, draft-royer-manet-globalv4-00.txt, Nov. 2001.

[4]   Y. Chawathe. Scattercast: an adaptable broadcast distribution framework. *Multimedia Syst.*, 9(1):104–118, 2003.

[5]   E. Gustafsson, A. Jonsson, and C. E. Perkins. Mobile IPv4 Regional Registration. Internet draft, IETF, draft-ietf-mobileip-reg-tunnel-06.txt, Mar. 2002.

[6]   J. Mysore and V. Bharghavan. A new multicasting-based architecture for internet host mobility. In *ACM/IEEE Conf. on Mobile Computing and Networking*, pp. 161–172, 1997.

[7]   Nextel. http://www.nextel.com/services/directconnect.shtml.

[8]   ns2 homepage. http://www.isi.edu/nsnam/ns.

[9]   Open Mobile Alliance. http://www.openmobilealliance.org.

[10] C. E. Perkins. IP Mobility Support for IPv4. RFC 3344, IETF, Jan. 2002.

[11] C. E. Perkins and D. B. Johnson. Route Optimization in Mobile IP. Internet draft, IETF, draft-ietf-mobileip-optim-11.txt, Sep. 2001.

[12] R. Prakash and R. Baldoni. Architecture for Group Communication in Mobile Systems. In *17th IEEE Symposium on Reliable Distributed Systems*, pp. 235–242, 1998.

[13] R. Ramjee, T. La Porta, S. Thuel, K. Varadhan, and L. Salgarelli. IP micro-mobility support using HAWAII. Internet draft, IETF, draft-ietf-mobileip-hawaii-01.txt, Jan. 2000.

[14] A. C. Snoeren and H. Balakrishnan. An end-to-end approach to host mobility. In *6th Int'l Conf. on Mobile computing and networking*, pp. 155–166, 2000.

[15] L. Subramanian, I. Stoica, H. Balakrishnan, and R. H. Katz. OverQoS: offering internet QoS using overlays. *SIGCOMM Comput. Commun. Rev.*, 33(1):11–16, 2003.

[16] The Yankee Group. NG Push-to-Connect Provides Simple UI and Enriches User Experience, Sep. 2003.

[17] A. G. Valkó. Cellular IP: a new approach to internet host mobility. *SIGCOMM Comput. Commun. Rev.*, 29(1):50–65, 1999.

[18] Verizon Wireless. http://news.vzw.com/news/2003/08/pr2003-08-14.htm.

[19] WinterGreen. Push To Talk (PTT) Market Opportunities, Market Forecasts, and Market Strategies 2003-2008, Dec. 2003.

RSM-WISP: ROAMING AND SERVICE MANAGEMENT IN HOTSPOT NETWORKS THROUGH A POLICY BASED MANAGEMENT ARCHITECTURE

Idir FODIL[1,2], Vincent JARDIN[1], and Guy Pujolle[2]

[1] 6WIND, Research and Development, {Idir.fodil, Vincent.Jardin}@6wind.com

[2] LIP6, University of Paris6, { Idir.fodil, Guy.Pujolle}@lip6.fr

Abstract: This paper describes RSM-WISP, a new management architecture designed for WISPs to facilitate the implementation and management of the services they offer at the access side of the WLAN, and to manage roaming contracts between WISPs. Our architecture is based upon the policy based management principles as introduced by the IETF, combined with more intelligence at the network edge. RSM-WISP adopts an architecture that is composed of two elements: a WISP management center (MC) that deploy policies and monitors all the WLANs, and programmable access router (CPE) located in each WLAN. The CPE ensure service enforcement, service differentiation (access to different service levels) and guarantee, user access management, and dynamic WLAN adaptation according to user's SLA (service level agreement). Concerning roaming management, it is achieved on the CPE through multiple service provider support capabilities.

Key words: WLAN, Hotspot, IEEE802.11, WISPs, Policies, PBM, Management, Services, SLA, Roaming

1.   INTRODUCTION

The recent years have seen expanding advances in new access network technologies which aimed to provide users with high speed access to the internet, and ability to use their network services everywhere and every time. Among these, the IEEE802.11 [1] standard has confirmed that it is the most simple and effective technology for providing network access in public places for users equipped with wireless cards. In order to provide their users with their subscribed service levels, and to benefit from public WLANs deployment, WISPs must be able to efficiently manage their public wireless networks at the wireless side and Internet access side. The wireless management which consists in guaranteeing micro mobility, security and quality of service in the wireless side is actually supported by significant projects in research, industry and standardization community. For the access network management, its main functionalities is to provide means for services specification and deployment, service differentiation, user access management, security guarantee and roaming management [5,6].

In this paper, we propose RSM-WISP, efficient, simple and scalable management architecture for public WLAN, which enables service differentiation between users, network adaptation according to users SLA, heterogeneous access networks support, and roaming management. This architecture is based on the use of IETF policy based management (PBM) [8, 9], enhanced with our improvements that allow more intelligence in access equipments. In RSM-WISP, access management is processed at IP Level in the access routers instead of access points. For policy configuration, some XML schemes have been defined, offering open, easy and customizable management architecture. We have implemented and validated this architecture on 6WINDGate routers, and we will use it in the context of INFRADIO project [19], which aims to deploy large IPv6 WLAN in university campuses with advanced functionalities such as user access control. The rest of the paper is organized as follows. Hotspot management requirements are provided in section 2. The RSM-WISP architecture, policy specification, and implementation are detailed in section 3. And finally, conclusion, actual and future works are overviewed.

2.   HOTSPOT ACCESS NETWORK MANAGEMENT

2.1   Management Objectives

In order to identify the hot spot access network management requirements, some roles need to be identified. The actors can be categorized by being one of the following: (i) Single point WISP: business venue/site owner that offers public WLAN services as value add to other core services. For example, hotels, airports, coffee shop, etc. (ii) Multiple point WISP:

traditional service providers such as ISPs, GSM/GPRS/3G operators that offer hotspot services as part of their offerings.

According to these roles, hotspot access management can be grouped into two families: WISP management, and roaming management.

WISP management: The WISP management is a set of tools enabling efficient operations of the hotspot network within its resources in accordance with WISP goals. It consists on the following points:

- Network Provisioning: setting up suitable quality of service configurations in order to meet user needs, and maintaining effective hotspot operations.
- Reactivity: network monitoring and automatic adaptation when degradations affect services.
- Access Management: authentication and authorization of the users.
- Adaptation: dynamic hotspot configuration according to user service level agreement.
- Accounting: varied billing strategies must been supported like free access, prepaid access for a certain amount of time or volume, pay per use period and differential fees for higher bandwidth.
- Heterogeneous access network support: providing multiple point WISP with ability to include hotspot offer in their services. This means that users can buy an internet access and use it at home (DSL or cable) and also in hotspots of their provider.

Roaming Management: There are two different roaming scenarios:

- Per bandwidth: this roaming contract is between single point WISP and multiple point WISP. Single point WISP rent a specific bandwidth for multiple point WISP, who applies its own user and service management strategies. A single point WISP may have contracts with one or several multiple point providers. For example, in an airport we can find one or several multiple point WISP.
- Per user: this roaming relationship is established between multiple point WISPs, like roaming in GSM networks. When users arrive in foreign network, authentication is done between the user, the foreign WISP and the user WISP. After successful authentication, the user is authorized to access available services. After disconnection, the foreign provider invoices the user WISP.

2.2   Management Challenges

Numerous solutions have been proposed [12, 13, 14, 15, 16], but most of them don't address the whole access management paradigm. Some, provide AAA functionalities (authentication, authorization, accounting), others provide security, and others mobility management. Moreover, dynamic WLAN adaptation according to users SLA, service differentiation, heterogeneous network support, and roaming management are not achieved.

The first reason is that service differentiation and heterogeneous network support can not be achieved using layer 2 based solutions, because they are link layer specific and cannot provide means for identifying services. Secondly, management is distributed among access points of the WLAN, which is not optimum network management solution because more than one AP has to be configured and adapted. Thirdly, dynamic network adaptation according to users and services is very difficult and challenging task with currently available network management tools. And finally, roaming management is very complex in such environment, because multiple service provider support on hotspot network still hard task.

Unfortunately, current network management cannot provide suitable tools for achieving the above needs. This is essentially due to the fact that network management is not much automated, and need skilled staffs with accurate knowledge of the various management tools. Moreover, existing tools are closed, service specific and cannot allow new service deployment. These generates extremely complex and very difficult network management, which weighs down and slows down introduction of new services, as well as significantly increase service providers operating costs.

We investigate the use of IETF policy based management approach in wireless LAN networks combined with central management held by access router instead of access points. We have enhanced the IETF architecture, because it is incomplete even though it is worthy foundation, since service providers and users needs have not been translated into suitable policies [11], and intelligence is not distributed among network equipments. Furthermore, we focused on designing an IP level solution, because it's the only way to differentiate services and to provide independent access network support. As result, we designed a policy architecture which provides WISPs with ability to offer innovative and differentiated services to their customers, to manage them in simple easier and more cost effective way, and to have roaming contracts with other WISPs.

3.   RSM-WISP

The main objective of the RSM-WISP architecture is to provide WISPs with suitable tools enabling them to efficiently manage their networks and users, and to establish and manage roaming contracts with other WISPs. Based on the use of policies installed on the access router by WISP and according to users SLA containing allowed services and QoS parameters, the access router configure itself dynamically to ensure the contracted service. For Roaming Management, according to the roaming contract (per user, or per bandwidth), WISPs can install their own policies on the router and manage their users. Policies of different WISP are separated and we assume that no conflict can happen between them since the access router appears as a dedicated router for each WISP.

3.1 Architecture

The RSM-WISP architecture has two main components, the management Center who takes on the WISP sold SLA guarantee and the access router (CPE) linking the public WLAN to the Internet.

The Management Center: The management center is the component of the architecture related to the WISP. The Management Center is responsible for the SLA negotiation, the generation of relevant policies and the application of these policies on the access router (CPE). The management center is a set of five modules: Service Portal (SPo), customer Agreement Database (CAD), Policy Server (PS), Policy Database (PDB) and Management tool (MNT).

The Access Router (CPE): Rather than configuring and managing each access point by itself, we choose to configure access router. Like that, user's re- authentication in the same WLAN is avoided, and handoff delays are reduced. Moreover, access points provisioning and management can be done by the router allowing global view of the network and more efficient resource management. In the RSM-WISP architecture, the CPE is the equivalent of the PEP+PDP (Policy enforcement and policy decision points) [8, 9] in the IETF architecture. The CPE is more "intelligent" than a simple PEP since it has the capability of monitoring events, keeping network states, and providing users the ability to modify their services on the fly. The CPE ensure plays the following roles:
- Enforcement of the policies sent by the PS,
- Translation of these policies in proprietary configurations,
- Auto-adaptation according to the network state,
- Reconfiguration or new PS policies solicitations,
- Response to monitoring requests sent by the PS,
- Periodic delivery of monitoring information up to the PS,
- Storage of policies sent by the PS.

Management Center and Access Routers communication: The policy server is the link between the management center and the access routers. The communication between the Policy Server and the access routers is achieved via 5 exchanges: provisioning (from PS to CPE) through a secured protocol, policy enforcement reports, on demand monitoring (PS send monitoring request to the CPE) periodical monitoring information reports (periodically sent from CPE to PS), and policy solicitation (when an unknown behavior occurs, the CPE sends a request to the PS. The PS deals with the problem, takes the appropriate decisions and sends the relevant policies to the CPE).

3.2   Policy Specification

In order to provide policies those allow appropriate translation of WISPs and users requirements onto access router configurations, we have specified the entire service provisioning and adaptation process. Thanks to this model, we have identified two policy families: WISP Policies and Roaming Policies.

Roaming Policies: point to the subscribed roaming contracts between the WISPs. These policies contain parameters related to foreign WISP, associated roaming model, and AAA parameters. If a foreign WISP has per bandwidth roaming contract, it will insert its own policies for users and services management as described after. But, if the contract is per user, service deployment will be done only when new user connect to the hotspot and according to parameters pushed by the foreign WISP. In other words, when a roaming contract is established on per user model, users coming from foreign WISPs are treated as users of the local WISP.

WISP-Service Policies: These policies define the set of policies chosen by the WISP administrator in order to manage their own services and their users. For foreign WISPs who have per bandwidth contract, they also insert their own WISP-Services policies in order to manage their users and services. We divide these policies into service specification, service update, user access management and on-demand service policies.

- **Services Specification policies (SSP):** These policies represent the full description of service deployment methods adopted by the WISP to manage its services. Since deploying differentiated services consists in specifying IP service parameters (port, protocol, etc) and their quality of service, we divide the SSP policies in two categories: QOSP and FAP.
  - **Quality of service policies (QOSP):** These policies allow WISPs, to specify their own services according to the quality of service strategy adopted in the hotspot network. Obviously, specified strategies are tightly depending on the home WISP quality of service strategy. In case where DiffServ is applied, each service will be assigned to specific class of service (example: VoIP → EF, Web → BE) with associated parameters. In case where Not DiffServ strategy, each service will be assigned a specific queue on the output.

- **Filtering Actions Policies (FAP):** These policies give a description of the services through filtering rules. Parameter of the filtering policies can be static (example: destination port =80) to handle known services or dynamic to handle applications such as VoIP, VoD, etc (pushed when a session is launched). The filtering rules can be either IPv6 or IPv4. In order to provide users with their guaranteed service levels, the filtering policies are applied in coordination with the quality of service policies. This is done thanks to an enhanced filtering engine which combine filtering and quality of services functionalities.
- **Service updates Policies (SMP):** In network management process, the WISP must be able to dynamically change its current services specification. For example, it may change bandwidth or services parameters. For those reasons we have defined the services updates policies which provide WISP with ability to dynamically change its current configuration. Currently, we provide means for changing Bandwidth parameters of existing service or class of service in DiffServ case. This policy is defined as follows:

  On Service update IF request= "change" then service_bandwidth ="new_rate"
- **User Access Management Policies (UAMP):** UAMP policies allow access control management of the users by specifying which types of users have access to certain services, under which conditions, and dynamic network adaptation according to the users SLA. When applying these policies, the access router adapts itself to meet the user's quality of service requirements contained in the service level agreement (SLA). There are two possible types of SLA that a WISP can sell, which led to two possible types of UAMP policies:
- **Per service SLA:** in this SLA, users can choose one or more service among services list, and for each service specify their own quality of service parameters. For example, WISP sells VoIP, FTP, Mail, Web, VoD, and Video Conferencing. User John will buy VoIP and Mail, while Barbara buys VoD, Mail and FTP. Each service of each user has its own quality parameters. In order to give WISP with ability to manage their users and services, the UAMP policies have been defined as follows:

  On New User If (service name) and (conditions)

  Then Authorize service

  Else re-adaptation

Conditions are related to quality of service parameters (available bandwidth, etc), date, time, number of currently running service sessions, etc. Re-adaptation consist in authorizing service, even when conditions are not accepted through quality of service dynamically reconfiguration. For example, the voice over IP service is programmed using the following policy:

    If (service = VoIP and VoIP available bandwidth)
               Then authorize VoIP else Readapt.

If there is no available bandwidth for VoIP service, then the access router evaluate if it can recover bandwidth from other classes or change its configuration thanks to Readapt actions.

- **Packaged SLA:** in this SLA, services are grouped in different packages, and users can buy one among them. Each package has its specific QoS parameter. For example gold package contain VoIP, Mail and Web with 20, 20 and 20 Kbps respectively. Time connection is related to the entire service package. In this SLA, when user buys a package, he/she is given a profile. In order to manage this packages, the WISP will program its access router using the following UAMP policies: On New user If (user_profile) and (conditions) then

            Allow list of services
            Else degrade to other profile

Conditions are related to available bandwidth on the access router, or to number of current connected users. For example, for the precedent gold package, the WISP will install

       if (user=gold) and (available bandwidth) then
          Allow Mail, VoIP, Web
          Else degrade to silver package.

The available bandwidth provides means for checking if there are enough resources for the specified service package. For the both SLA, the UAMP policies provide means for dynamic service deployment thanks to automatic router adaptation.

- **On-Demand Service Policies (ODSP):** Materialize the value added services that a WISP may offer for its customers. For example, user may change its profile from silver to gold, in order to have better quality on voice over IP. The application of service update policies generate a modification of the associated filtering policies that have been applied for the user. These policies provide users with means for service upgrade and are pushed directly from user terminal to the access router (Web interface or some protocols). These policies have two main objectives, provide users with means for dynamically changing their requirements and allow them to configure access equipments according to their SLA which is stored in the user side (smart card).   At present, we have defined the following policy

      On Update if (request="change") Then
      (user_profile = "new_profile")

This policy allows users to dynamically change their profile, thus allowing them to get more services without interruption.

3.3 Architecture Implementation

In this section, we describe the RSM-WISP implementation on the access router. We have used the following access router functionalities: Dual stack (Ipv4 and Ipv6 support), DHCPv4/ DHCPv6 server, Radius Client, Filtering, and Quality of services. Figure 1 shows the elements of the RSM-WISP implementation architecture.

Policy Manager: All policies defined in our architecture are described and validated using XML schemas and installed using: CLI (command line interface), an XML/HTTP connection, or a web interface directly of from remote machine. The Policy manager which is handled by the WISP administrator can receive policies from foreign WISPs when they have roaming contracts. It is responsible of validating the policies XML schemas, storing them in database, sending Add/Delete/Update messages to the appropriate WISP block. The entire policy manager has been developed using C++ language, because it provides more flexibility and scalability in implementing new services.

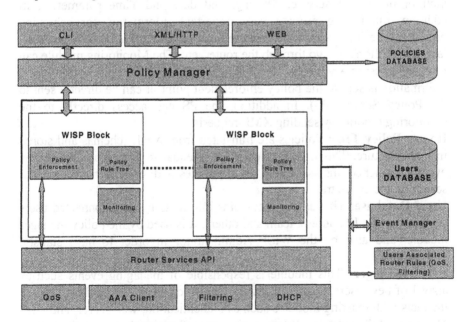

Fig1. RSM-WISP Access Router Implementation

WISP block: When a foreign WISP establish roaming relationship according to per bandwidth model, a new module called WISP block is instantiated and created on the access router. The WISP block contains policy enforcement, policy rule tree and monitoring modules.

Policy Enforcement: The policy enforcement module is the heart of the Policy Architecture on the access router. It ensures the following tasks:

– After reception of the policies from the Policy Manager, it translates these policies into C++ objects and stores them in tree structure, and processes them. The policies which can be directly applied (QOS Policies) are translated to routers rules thanks to the Router Service API Module. For the others, it notifies the "event module" of the events types it is waiting for (UAM policies are launched by arrival of new users).

– Communicate with monitoring module to get local router information. For example, bandwidth use, number of users, ... etc

– Ensure keeping states about users deployed services in order to remove them when the user leaves the network.

– Periodically, or on request, it sends monitoring reports to the Policy manager.

Monitoring: The Monitoring module provides the policy enforcement a global view about all local router parameters and states. Currently, we can monitor quality of service, filtering, and date and Time parameters. In addition, monitoring provide very important information for achieving billing. These information concern amounts of data volume per IP address, last time an IP packet go through the router, etc. The Monitoring module can be acceded using XML requests, or simple function calls. All the monitoring information is sent to the policy enforcement point or can be directly sent to the Policy Server (PS). In addition, the PS can access directly to the monitoring module by sending XML requests.

Router Policy Tree: Policies are translated from XML schemas and stored in tree structure. This tree is of complexity equal to 1, because when new event is launched, the associated set of policies is directly retrieved without searching the entire tree.

Users Database: This database contains information about connected users such as profile, IP address, team and others. It is used by the policy manager module, and also by the WISP Administrator in order to have statistic information.

Event Manager: This module is responsible of managing events such as arrival of new users, new application request, or other events. This module interacts with existing modules such as authentication, web server, and CLI. Moreover, this module allows adding new functionalities on the policy manager such as other authentication mechanisms or new events. For the event manager we have used the C language.

Users Associated Router Rules: This file contains indexes of actual router rules deployed for each user. The index size is low because it contains only single information per user. This file allows removing or updating services for users.

Router Services API: We have designed these API for the following three reasons:

– Provide single and simple way to use router services
– Offer means for dynamically updating router rules.

The API services are of two types: Functions calls and XML requests. The XML request support has been added in order to provide PDP or other advanced equipment with ability to directly monitor the access router, and changes its configuration without requiring other router modules.

Filtering Module: The Filtering Module called PFM is an engine that allows filtering and quality of service deployment at the same time. It works as follows:

– Output interface: implementation of quality of service queuing disciplines. We specify queues parameters (bandwidth, priority, borrow...) and scheduling algorithms (CBQ, WFQ...).
– Input interface: specification of filtering rules, based on IP packet fields such as version, protocol, port...

Quality of Service Module: This module provides traffic conditioning elements such as droppers, markers, shapers... It allows for example traffic limiting for services or users.

4. CONCLUSION

In this paper, new network management architecture for roaming and service management in hotspot networks has been detailed. The lack of solutions that allow multiple service provider support, service guarantee and service differentiation led us to propose this architecture. Our solution allows WISPs to get benefits from the large deployment of public WLANs, by differentiating services offered to their customers, efficient and simple architecture. Moreover, since access network is managed by the access routers, we can extend its functionalities to manage access points and to interact with wireless management solutions. For example, access router may control radio resources, and allow or deny new users that try to associate in busy or congested access points. This approach is currently subject of lot of works in IEEE and IETF [17]. Compared to the classical IETF PBM architecture, our solution offer two major improvements: (1) A Further abstraction level has been added providing the administrator with the possibility to deploy services without having to know which device parameters to configure. (2) A distribute management model where more intelligence is pushed toward the access equipments (access networks). Furthermore, because of the IP based, our solution can work over different air interfaces, across wireless LAN cards from different vendors, and does not require any modification to layer 2 protocols.

5.   REFERENCES

[1] IEEE. 802.11b/d3.0 Wireless LAN Medium Access Control (MAC) and Physical Layer (PHY) Specification, August 1999.

[2] Upkar Varshney and Ron Vetter, "Emerging Mobile and Wireless Networks", Communications of the ACM, Vol. 43, N°. 6, June 2000.

[3] Rajeswari Malladi and Dharma P. Agrawal, "Current and Future Applications of Mobile and Wireless Networks", Communications of the ACM, Vol. 45, N°. 10, October 2002.

[4] A.Mahler and C.Steinfield The Evolving Hot Spot Market for Broadband Access "ITU Telecom World 2003 Forum panel on Technologies for Broadband, Geneva, October 2003"

[5] Donald M. Fye, "Evolution of WLAN Roaming Services", CDG WLAN Technical Forum, Dallas, Texas, October 2, 2003

[6] Michael Kende, "WLAN challenges and opportunities", National Summit on Broadband Deployment , April 28, 2003

[7] Idir Fodil and Vladimir Ksinant "User Service Management in Hotspot network using Policies", European Wireless 2004, the fifth European wireless Conference, February 24-27 2004, Barcelona, Spain

[8] A.Westrinen and al, "RFC 3198: Terminology for Policy Based Management ", IETF, November 2001.

[9] David Kosiur,"Understanding Policy-Based Networking". Wiley Computer Publishing, 2001.

[10] Raouf Boutaba and Jin Xiao, " Network Management State of the Art", WCC, IFIP World Computer Congress, August 2002.

[11] O.Corre, I.Fodil, V.Ksinant and G.Pujolle, " An Architecture for Access Network Management with Policies", MMNS 2003, 6th IFIP/IEEE Conference on Network Management, September 2003.

[12] Junbiao Zhang and al, "Virtual Operator based AAA in Wireless LAN Hot Spots with Ad-hoc Networking Support", Mobile Computing and Communications Review, Volume 6, Number3.

[13] Joseph W. Graham II, *"Authenticating Public Access networking"*, SIGUCCS'02, November 20-23, 2002, Providence, Rhode Island, USA.

[14] IEEE Daft P802.1X/D11: Standard for Port based Network Access Control, LAN MAN Standards Committee of the IEEE Computer Society, March 27, 2001.

[15] Pekka Nikander, "Authorization and charging in public WLANs using FreeBSD and 802.1x", USENIX annual technical conference, June 10-15 2002.

[16] P. Kalhoun and al., "Light Weight Access Point Protocol", Internet Draft, June 2003.

[17] IETF CapWap Working Group, http://www.ietf.org/html.charters/capwap-charter.html

[18] Alper E. Yegin, Yoshihiro Ohba, Reinaldo Penno, George Tsirtsis ,and Cliff Wang, " Protocol for Carrying Authentication for Network Access (PANA) Requirements", Internet Draft , June 2003.

[19] INFRADIO Project : http://rp.lip6.fr/infradio/

INTEGRATED RECONFIGURATION MANAGEMENT FOR THE SUPPORT OF END TO END RECONFIGURATION

Aristotelis Glentis, Nancy Alonistioti
Communication Networks Laboratory, Department of Informatics & Telecommunications, National and Kapodistrian University of Athens, 157 84, Athens, Greece

Abstract: The development, delivery and management of mobile services are the subject of many research activities in both the academia and industry. The ultimate goal of these efforts is a dynamic environment that enables the delivery of situation-aware, personalized multimedia services over heterogeneous, ubiquitous infrastructures, commonly termed as systems beyond 3rd generation (3G). Reconfigurability and adaptability are key aspects of the mobile systems beyond 3G. Reconfigurable mobile systems and networks introduce additional requirements and complexity. Using the existing network control plane is inadequate for the realization of the reconfigurability process. Introducing a Reconfiguration Management Plane is very important for the deployment of network/system wide reconfigurability. In this paper we intend to discuss the basic functionality of RMP and respective interrelations.

Key words: reconfiguration management, mobile systems beyond 3G.

1. INTRODUCTION

1.1 Towards reconfigurability

The evolution of technology has led to the introduction of the Open Business Model in the world of mobile telecommunications. The Open Business model is the model that the Internet is based on: the ISP provides the connectivity and the user access the application/service provider using open APIs and protocols that reside on top of the IP protocol. Mobile telecommunications are based on different business models, since in the mobile world the user is confined to the provider's network and cannot access services outside this domain. The mobile provider is the one responsible for the deployment and maintenance of value added services. With the arrival of 3G networks and UMTS, which offer an all IP network, this fact has changed and the opportunity of the adoption of the Open Business Model in mobile communications is possible, and can be beneficial both to the telecom operator and the application/service providers[1,2]. The telecom operators can benefit since their users can have access to a larger range of applications and services, which they don't have the burden to deploy, manage, maintain themselves. The application/service providers benefit from the larger user base that can access their products. The users can benefit from the plethora of new services and from the competition between telecom providers and achieve optimum ratio of quality of service per price.

In order to achieve this goal the need for end to end reconfigurability rises. The users want to access the applications/services that have registered to, discover new service or applications that are offered, update their software and don't be tied to a certain underlying network infrastructure but can choose the one preferred from the networks that are available in the environment according to their preferences. The users could be able to change environments (i.e. from UMTS to GSM and 802.11b) without loosing the service, if possible. The service, on the other hand, must be able to adapt to the change of the network characteristics, or to the request of the user for having better or worse quality of the service, etc. However, in the mobile world there are two issues that have to be solved, the different capabilities of the mobile devices to execute applications or services, and the mobility of the user who comes across different networks with different characteristics.

1.2   Related work

The issue of reconfigurability on these two axes has been tackled in the past mainly in the two edges of the OSI layer model: the physical and the application. On the physical layer research has been carried out so that devices can detect the networks that are available and use them to communicate. However, the research was limited to the use of different physical layers to carry the information and no provision was made for the interoperability with application's requirements. Furthermore, several attempts have been made for the introduction of adaptive protocols and respective design[3]. Building on the knowledge from early software radio projects in the military domain[4,5], SDR Forum has pioneered in exploring reconfigurability concepts in the United States. However, being the vanguard of reconfigurability developments and the first to define a software radio architecture[6,7], seems to have come at the expense of a rather restricted view on reconfigurability that focuses primarily on the radio domain (RF processing, down-conversion, RF processing, A/D conversion, etc)[8]. On the application layer, research has been carried on the adaptation of the application or service according to the predefined profiles of the user and the service in the MOBIVAS platform[9,10]. The user can discover different instances of the service according to the profile and the terminal capabilities of his device. However no input from the underlying network is used in the service provision decision. The tackling of the problem in the two edge layers, physical and application, is not efficient and sufficient since it creates a lot of difficulties to the network devices, and to the application developers and providers.

Based on the above discussion, it is apparent that in the design of fully reconfigurable networks and systems, the introduction of advanced reconfiguration management functionality is necessary. In this paper we introduce a holistic solution for addressing reconfiguration management across all layers, namely, the Reconfiguration Management Plane (RMP). RMP enhances reconfigurability control in order to address end-to-end reconfiguration management aspects.

2.   RECONFIGURATION MANAGEMENT ASPECTS

In order to address reconfiguration management it is important to tackle reconfiguration in two levels:
- The local level (addressing network node and mobile device reconfiguration)
- The system level (addressing network wide reconfiguration and service adaptation).

One simplistic approach (addressing only the local level of reconfiguration) would be to assume that each of the reconfigurable devices has a local reconfiguration manager (LRM), who is responsible for the reconfiguration plane of the local device. It keeps track of the state of the device and performs the necessary actions needed for reconfigurability. The actions vary, it can be downloading components and installing software to offer new protocol stacks, changing the QoS values of the protocol stack in use, ensuring that the requirements of the application running are met, choosing the optimum combination of protocol stacks and routing according to the policy that is defined by the user or operator, triggering reconfiguration on user or application request, etc. In order to achieve these goals the LRM should have a clear picture of the network topology and be able to contact different servers and services, as well as finding the optimal network path. Furthermore, the application can reside on a server that is not in the region controller by the telecom operator, so the LRM should be able to construct the path to the remote application server. This requires that the LRM uses a lot of CPU power and that each device has the routing info for all the networks that it participates. Provided that network topologies change often in reconfiguration environments, the LRM will be flooded with control messages. The processing and space complexity becomes a major issue considering the limitations on the mobile devices.

On the other hand, another simplistic approach would be to delegate the reconfiguration responsibility to the application developer and provider. In this case, the value added service developers should be able to easily access the network in order to provide the parameters needed for the service to operate smoothly. The parameters needed (for example QoS settings) should be propagated in all the devices that are in the network path from the server that provides the VAS and the user terminal. As a result the service should have knowledge of the network topology and the application developer should cope with different network infrastructures and provide the methods to communicate with them. The application also should be able to access all the internal nodes of the network, something that is not acceptable from the telecom providers view, since this might reveal the internal structure. One possible solution for this problem is to have the necessary functions packaged in a library. This solution is quite cumbersome since all applications and services should be linked with the library, and changes to the interfaces or addition of new network infrastructure would mean the need to upgrade both the library and the application or service. This introduces a lot of overhead to the application developer. The application developers want a clean interface between the application and the network, and shouldn't be forced to cope with specific network functions.

From the above the need for an entity that controls the reconfiguration process on the network level and provides a layer of abstraction both to the terminal and the service application comes to the surface.

3.   RECONFIGURATION MANAGEMENT PLANE ARCHITECTURE

3.1   General architecture

The entity identified in the previous context, is the Reconfiguration Management Plane (RMP), which is introduced and described in the current section. The RMP can be viewed as another control plane that is operating on all OSI layers and runs along with the network control plane. Its main task is to provide layer abstractions to the applications and services on the one hand and to the terminals and network devices on the other. Furthermore the RMP is responsible to coordinate the reconfiguration process and provide the required resources in order to be completed. The RMP is comprises different components that are illustrated in figure 1.

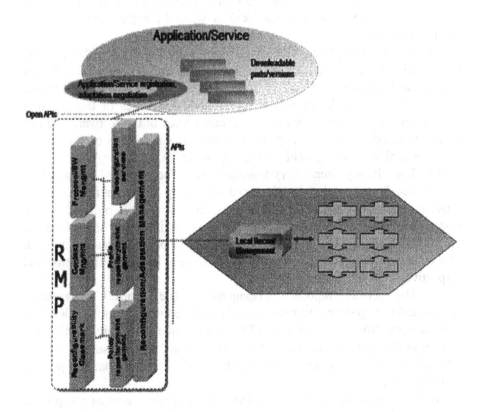

Figure 1. The RMP entity

3.2   Architectural components

The components are:

1) The protocols/SW management. The reconfiguration procedure is based on the ability of the mobile devices, network nodes to download and install software that makes possible the support of new protocol stacks. The reconfiguration procedure encompasses the triggering of certain protocol and software to be downloaded on the mobile terminal or other network nodes/ devices in order to support efficient user connectivity and optimal service provision (e.g., downloading of certain protocols that are not installed in the mobile device). Therefore this functional entity is responsible to identify, locate and trigger the suitable protocol or SW for download.

2) the Context management. The context management is responsible to manage the context information of the network. In the RMP domain, context doesn't concern the applications that run over the network, but the network data themselves, i.e. the nodes, the state of the nodes, the congestion information etc. The context manager is responsible to have a picture of the network. This is necessary in the decision of feasibility of a reconfiguration event, or the choice of an alternate path between the application server and the user terminal.

3) the Reconfigurability Classmark. This component has a duplex role. On one side it keeps track of the different nodes of the network and their state regarding reconfigurability, i.e. the protocol versions that are installed. On the other side it keeps a database of the capabilities of the different network nodes that exist. For example it keeps track of the software that can be downloaded and run on each device, and the capabilities of the mobile device regarding reconfigurability and upgrading.

4) The Policy repository/management. The policy management component is the main decision making entity for reconfiguration. It provides the entry point for the reconfiguration related policies of the system. Furthermore, it undertakes the merging of different profiles from the profile management and combines them with the policies that are defined. The output is the final decision about the feasibility of a reconfiguration and respective actions to be triggered.

5) The profile repository/management. The profile management component is responsible to manage and combine the different profiles. The profiles come from different parts of the system since they refer to different entities of the system. The profiles can be classified in: the user profile, the network profile, the application profile, the terminal profile, charging profile, security profile etc.

6) The reconfiguration services. The reconfiguration services component is responsible for the communication between the RMP and the

application/service. It accepts and processes reconfiguration requests for the network in order to provide the necessary environment for an application and service to execute. It also provides feedback to the application for the feasibility of the request, or can also initiate a reconfiguration on the application in case for example of network configuration changes or selection of different settings by the users, mobility etc..

7) The reconfiguration Adaptation/Management. The reconfiguration adaptation management component is responsible for initiating the triggering of service adaptation based on network capability restrictions or reconfiguration policies. It triggers and coordinates the reconfiguration actions, exchanges messages with the LRM of the devices and coordinates the reconfiguration process. Internally, it accesses the components of the RMP and provides the necessary tools and information for the reconfiguration to take place.

Although there is no central security management component, security should be considered and tackled in all the RMP components. The authorization, authentication and integrity of communications among the RMP components should be assured by using proper security schemas. The selection of the security schema depends on the network infrastructure and the communication channels that are used between the components. However special consideration should be taken for the communication of the RMP and the external entities (end user terminals, network nodes, applications), most notably with the protocols/SW management component, since the need of a security schema that supports public key cryptography is needed for authentication purposes. Furthermore, solutions that use digital signatures could be considered to protect the network nodes and user terminal from downloading unauthorized software.

3.3   Communication between RMP and external entities

The communication between the RMP and the external entities is based on open APIs. The RMP communicates primarily with the local reconfiguration managers and with the Application/Service. The open APIs provide the infrastructure for applications from different vendors to communicate with the networks of different providers. Since the providers are not willing to reveal the inner structure of the network to the external applications, they are able to provide context and adaptation related information to the application/services through the use of open APIs in a controlled way. The open APIs can guarantee the construction of the communication path between the application and the reconfigurable network, so that the applications can provide the requirements and trigger reconfiguration when needed, and the network can give proper feedback without revealing internal structural information.

The local reconfiguration managers are internal components of the devices, and as a result they can have proprietary characteristics. However the need to communicate with different nodes and with the RMP makes the use of open APIs essential. The RMP has to communicate with the LRMs in order to trigger a reconfiguration procedure, to collect essential data (like protocols installed, node capabilities, etc), query about the status of the node, etc. The LRMs need to communicate with the RMP in order to trigger a reconfiguration procedure, to download software and answer to requests from the RMP. The use of open APIs is essential to provide the communication path between the devices of different manufacturers and enhancing interoperability.

3.4   Case studies

In order to clarify further the use and functionality of the RMP the following sequence diagrams that depict an overview of the main functionalities and operations performed by the RMP in the event of a reconfiguration, are also presented. The two cases that are illustrated are: the first case (illustrated in figure 2) is when the user terminal initiates the reconfiguration procedure, and the second (illustrated in figure 3) is when the reconfiguration procedure is initiated due to the provision requirements of a Value added service after a user selection for application download. In either case the RMP is the one that steers the reconfiguration process. It triggers the reconfiguration in the network nodes that are among the path from the Value Added Service Provider (VASP) to the user terminal, and provides the software for protocol stack reconfiguration in the network nodes and the user terminal. The RMP communicates with the LRM of the network nodes and the user terminal on the network side and with the application/service on the application side. However, the user terminal might have to communicate directly with the VASP in order to download extra software components (for example it might be essential to download codecs that are needed for the service the user is currently downloading).

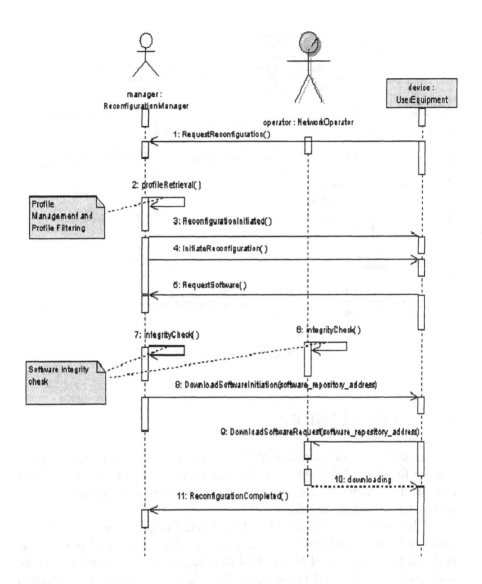

Figure 2. Terminal Initiated

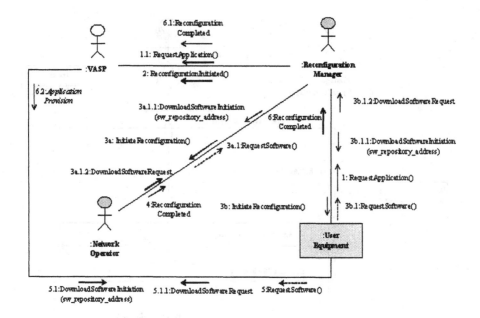

Figure 3. Network Initiated

4.   CONCLUSIONS

The evolution of reconfigurability notion has been heralded as main concept for 4G mobile communications. In order to reach its full potential, a consistent framework that deals with reconfigurability challenges and control has to be introduced.

In this paper we have introduced a generic framework to cope with the complexity of reconfigurability management. This work will provide the basis for the evolution of End-to-End Reconfigurability notions that will be addressed and enhanced inside the E2R architecture design. The proposed architecture for reconfiguration management addresses the effective policy based reconfiguration triggering towards the network nodes and the combination of adaptation triggering towards the end-user services in order to achieve the optimal service provision and perception to the end user in a transparent way.

ACKNOWLEDGEMENTS

This work has been performed in the framework of the EU funded project E2R. The authors would like to acknowledge the contributions of their colleagues from E2R consortium.

REFERENCES

1. UMTS Forum Report No. 9, "The UMTS third generation market - structuring the service revenues opportunities"; http://www.umts-forum.org/ .
2. J. Pereira, "Beyond third generation", Wireless Personal Mobile Communications (WPMC), 22 September 1999, Amsterdam, The Netherlands.
3. M. Dillinger, K. Madani, N. Alonistioti, 2003, "Software Defined Radio, Architectures, Systems and Functions", John Wiley, England.
4. M. C. Cox, "Joint tactical radio system (JTRS)", presentation available from http://www.jtrs.sarda.army.mil/.
5. The GloMo project, "Global mobile information systems (GloMo)", DARPA ITO; http://www.janet.ucla.edu/glomo/.
6. P. G. Cook, "Software architecture in software defined radios", contribution to SDR Forum meeting, February 24, 1999; http://www.sdrforum.org/.
7. J. Bickle et al., "Software radio architecture (SRA) 2.0 technical overview", presentation to OMG TC meeting, December 11, 2000, Orlando, Florida.
8. S. M. Blust et al., "SDR definitions", contribution to SDR Forum Plenary & Technical Committee meeting, September 1, 2000; http://www.sdrforum.org/ .
9. The MOBIVAS project; http://mobivas.cnl.di.uoa.gr .
10. N. Houssos, V. Gazis, A. Alonistioti, " Application-transparent adaptation in wireless systems beyond 3G", 2nd International Conference on Mobile Business (M-Business 2003), Vienna, Austria, 23-24 June 2003.

REPLICA ALLOCATION CONSIDERING DATA UPDATE INTERVALS IN AD HOC NETWORKS

Hideki Hayashi, Takahiro Hara, and Shojiro Nishio

Dept. of Multimedia Eng., Grad. Sch. of Information Science and Tech., Osaka Univ.
{hideki, hara, nishio}@ist.osaka-u.ac.jp

Abstract   In ad hoc networks, since network division occurs frequently, it is effective to replicate data items. This paper proposes effective replica allocation methods in ad hoc networks where each data item is updated at irregular intervals. The proposed methods allocate replicas based on probability density functions of the update intervals of data items. Also, they invalidate replicas that have been updated with high probability because accesses to old replicas impose extra computational overhead and roll backs. As a result, the proposed methods not only improve data accessibility but also reduce the number of accessing old replicas.

Keywords: ad hoc networks, replica allocation, data accessibility, data update, mobile computing environment

1.   Introduction

As one of the research fields in mobile computing environments, there has been increasing interest in *ad hoc networks* that are constructed of only mobile hosts that play the role of a router. Disconnections occur frequently in ad hoc networks, since mobile hosts move freely, and this causes frequent network division. If network division occurs due to the migration of mobile hosts, mobile hosts in one of the two divided networks cannot access data items held by mobile hosts in the other network. In Figure 1, if the radio link between two mobile hosts is disconnected at the central part, the mobile hosts on the left-hand side and those on the right-hand side cannot access data items D_1 and D_2, respectively. A key solution to this problem is to replicate data items on mobile hosts that are not the owners of the original data item.

In ad hoc networks, there also be many applications in which mobile hosts access data held by other mobile hosts. A good example is when a research project team constructs an ad hoc network and the team members refer to data obtained by other members for efficiency. Recently, ad hoc networks have attracted much attention as an infrastructure of next-generation computer environments, e.g., wearable computing en-

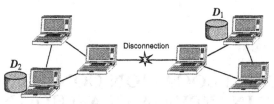

Figure 1.   Network division.

vironments and sensor networks. Therefore, it will be more and more important to improve data accessibility in ad hoc networks.

In [2, 4], we proposed three replica allocation methods for improving data accessibility in ad hoc networks where a data item is not updated. These methods periodically determine replica allocation based on the access frequency from each mobile host to each data item and the network topology at that moment. In [3], we extended the three methods proposed in [2] to adapt to an environment where each data item is periodically updated. These extended methods replicate data items on mobile hosts based on the access frequency, the time remaining until each item is next updated, and the network topology.

In a real environment, it is more likely that data items are updated at irregular intervals. In this paper, we propose three replica allocation methods to improve data accessibility based on probability density functions of the update intervals of data items. In an assumed environment, mobile hosts may access invalid replicas that have been updated. Such invalid accesses cause roll backs when the hosts later connect to the mobile hosts holding the originals. Invalid accesses and roll backs consume the power of mobile hosts, causing a serious problem since mobile hosts typically have poor power resources. Thus, the proposed methods also invalidate replicas that have been updated with high probability. As a result, the proposed methods not only improve data accessibility but also reduce the number of accessing invalid replicas.

The remainder of the paper is organized as follows. In Section 2, we show some conventional works related to our work. In Section 3, we present our assumed environment. In Section 4, we propose three replica allocation methods. In Section 5, we show the results of simulation experiments. Finally, in Section 6, we summarize this paper.

2.   Related Works

In the research field of ad hoc networks, a few studies have been made to improve data accessibility [7–12].

In [7] and [9], the authors have proposed methods in which replicas are allocated to a fixed number of mobile hosts which act as servers and the consistency among the replicas is keep using a strategy based on the quorum system which has been proposed for distributed database. These are considered similar to our methods because replicas are allocated to mobile hosts. However, these methods assume that replicas are allocated to only mobile hosts selected as servers and their storages have unlimited memory space. Our methods effectively allocate replicas to all mobile hosts with limited memory space.

In [8, 11], the authors have proposed caching methods in the Internet based mobile ad hoc network. This method is considered similar to our methods because each mobile host allocates replicas. However, this method differs from our methods with the point that the authors assume that some of the mobile hosts are connected to the Internet. We assume that all mobile hosts are not connected to the Internet.

In [10], the authors have defined two new consistency levels among replicas and proposed methods that disseminate updated data to keep the consistency. These methods are considered similar to our methods because replicas are allocated to all mobile hosts. However, these methods are different from our methods in which each mobile host invalidate replicas considering update intervals of the data items.

In [12], the authors have proposed a method that predicts time when a network division occurs and allocates replicas to mobile hosts before the network division occurs. This is similar to our proposed methods because the authors assume frequent network divisions. However, this method differs from our methods with the point that the authors assume a specific mobility model. Our methods work for any mobility models.

3. Assumptions and Approach

The system environment is assumed to be an ad hoc network where mobile hosts access data items held by others. In this paper, mobile hosts connected to each other by one-hop/multihop links are simply called *connected mobile hosts*. We make the following assumptions:

- We assign a unique *host identifier* to each mobile host. The set of all mobile hosts in the system is denoted by $M = \{M_1, M_2, \cdots, M_m\}$, where m is the total number of mobile hosts and M_j $(1 \leq j \leq m)$ is a host identifier. Each mobile host moves freely.

- Data are handled as a data item which is a collection of data. We assign a unique *data identifier* to each data item. The set of all data items is denoted by $D = \{D_1, D_2, \cdots, D_n\}$, where n is the total number of data items and D_j $(1 \leq j \leq n)$ is a data identifier. The original of each data item is held by a particular mobile host.

- Each mobile host has memory space of C data items for creating replicas, excluding the space for the original data item. Replicas are relocated in a specific period called *relocation period*.

- The access frequencies to data items from each mobile host are known. In a real environment, the access frequencies can usually be known by recording the log of access requests at each host.

- Each data item is updated by the mobile host holding the original at irregular intervals. The update interval of D_j is represented by a probability density function, $f_j(t)$. After a data item is updated, the replicas become invalid. In a real environment, $f_j(t)$ can usually be known by recording the log of the update at mobile hosts holding the originals.

- Each mobile host holds a table in which the information on the update times (time stamps) of all data items in the entire network is recorded. This information table is called the *time stamp table*.

In this environment, a request for a data item is successful only when the request issuing host accesses the original or its replica with the same time stamp as the original. That is, replicas with a time stamp different from the original are invalid. The request succeeds immediately if the request issuing host holds the original or connects to the mobile host holding the original. Otherwise, if the request issuing host or at least one connected mobile host holds the replica, the request issuing host tentatively accesses the replica. After the tentative access, when the request issuing host connects to the host holding the original, the tentative access is determined as having either succeeded or failed. This can be achieved by comparing the update logs at the host holding the original with the information on the time stamp of the accessed replica and the access time at the request issuing host. If the tentative access fails, the roll back occurs so that the request issuing mobile host recovers its state before accessing the replica. If the request issuing host and connected mobile hosts do not hold the original/replicas, the request fails.

4.   Replica Allocation Methods

In this section, we first propose three replica allocation methods. Then, we explain the cache invalidation in our proposed methods.

4.1   Replica allocation

In this paper, we propose three new replica allocation methods, which are extensions of the methods proposed in [3], to adapt to an environment where data items are updated at irregular intervals.

First, we define a *PTT value* for each replica of D_j at M_i as follows:

$$p_{ij} \cdot \int_0^{to_j - t_j} f_j(t + t_j) \cdot t \, dt \tag{1}$$

Here, p_{ij} denotes the access frequency of M_i to D_j; to_j denotes the lifetime of replicas of D_j ($to_j \geq 0$); t_j denotes the time that has passed since D_j's most recent update, which is the time interval between the current time and the time stamp of D_j. If D_j has already timed out ($to_j < 0$), the PTT value is defined as 0.

The PTT value represents the average number of successful access requests until the mobile host discards the replica of D_j. Thus, by allocating replicas with high PTT values at a relocation period, not only data accessibility is expected to be higher but also the number of accessing invalid replicas is expected to be lower. The detailed decision process of to_j is explained in the next subsection.

Based on this idea, we extend the three methods proposed in [3] by mainly changing their algorithms to use PTT values instead of PT values in order to adapt to an environment where each data item is updated at irregular intervals. We call the three extended methods the *E-SAF (Extended-Static Access Frequency)*α method, the *E-DAFN (Ex.-Dynamic Access Frequency and Neighborhood)*α method, and the *E-DCG (Ex.-Dynamic Connectivity based Grouping)*α method. In the following, we describe the details of the three extended methods.

4.1.1   E-SAFα method.   Each mobile host allocates replicas of data items in descending order of PTT values within the limit of its own memory space. If a mobile host issues an access request for a data item whose replica at the host has become invalid and the request is satisfied, the request issuing host again allocates the valid replica, i.e., refreshes the replica. This operation is also done in the other two methods.

4.1.2   E-DAFNα method.   In the E-SAFα method, since mobile hosts with the same access characteristic allocate the same replicas and there are many replica duplications, the data accessibility is low. To solve this problem, after allocating replicas with the E-SAFα method, the E-DAFNα method eliminates replica duplications between two neighboring mobile hosts. The algorithm is as follows:

1 At a relocation period, each mobile host broadcasts its host identifier and information on access frequencies to data items. After all mobile hosts complete their broadcasts, from the received host identifiers, every host knows its connected mobile hosts.

2 Each mobile host preliminary determines the replica allocation with the E-SAFα method.

3 In each set of connected mobile hosts, starting from the mobile host with the lowest suffix (i) of host identifier (M_i), the following procedure is repeated in the order of the breadth first search. When there is duplication of a data item (original/replica) between two neighboring mobile hosts, and if one of them is the original, the host which holds the replica replaces it with another replica. If both of them are replicas, the host with the lower PTT value replaces the replica with another replica. When replacing the replica, from among data items whose replicas are not allocated at either of the two hosts, a different replicated data item is selected whose PTT value is the highest.

4.1.3   E-DCGα method.   The E-DCGα method shares replicas in larger groups of mobile hosts than the E-DAFNα method. This method creates groups of mobile hosts as *biconnected components* [1] and then allocates replicas of data items on mobile hosts in each group in descending order of PTT values in the group. By grouping mobile hosts as a biconnected component, the group is not divided even if one mobile host disappears from the network or one link is disconnected in the groups. Thus, the group has high stability. The algorithm is as follows:

1 At a relocation period, each mobile host broadcasts its host identifier and information on access frequencies to data items. After that, every host knows its connected mobile hosts.

2 In each set of connected mobile hosts, starting from the mobile host with the lowest suffix (i) of host identifier (M_i), an algorithm to find biconnected components is executed. Then, each biconnected component is put in a group. If a mobile host belongs to more than one biconnected component, it can belong only to the group in which the corresponding biconnected component was found earlier.

3 In each group, the PTT value of each mobile host in the group to each item is calculated. Then, the PTT value of the group to each item is calculated as a summation of PTT values of mobile hosts in the group to the item. These calculations are done by the mobile host with the lowest suffix of host identifier in the group.

4 In descending order of PTT values in each group, replicas are allocated until the memory space of all mobile hosts in the group becomes full. Here, replicas of data items which are held as originals by mobile hosts in the group are not allocated. Each replica

is allocated at a mobile host whose PTT value to the data item is the highest among hosts that have free memory space to create it.

5 After allocating replicas of all data items that have no original in the group, if there is still free memory space at mobile hosts in the group, replicas are allocated in descending order of PTT values until the memory space is full. Each replica is allocated at a mobile host with the highest PTT value to the data item among hosts that have free memory space and do not hold the original/replica. If there is no such mobile host, the replica is not allocated.

4.2 Cache invalidation

In our proposed methods, a lifetime is assigned to each data item based on its probability density function of update intervals. Since each mobile host discards replicas whose most recent update is a long time ago, the number of accesses to invalid replicas can be reduced. In the following, we explain the procedure of cache invalidation in our methods.

The probability that D_j is updated within time τ_j since D_j's most recent update is expressed by the following expression:

$$\int_0^{\tau_j} f_j(t) \, dt. \tag{2}$$

In our proposed methods, a constant threshold α is assigned to all data items in the entire network. If the value of expression (2) becomes equal or more than α, each mobile host holding a replica of D_j discards the replica from its own cache. More specifically, when each mobile host allocates a replica of D_j or records the time stamp of D_j, the value of τ_j that satisfies the condition $\int_0^{\tau_j} f_j(t) \, dt = \alpha$ is calculated using expression (2). The value of τ_j calculated in this manner is defined as the lifetime, to_j. Then, using expression (2) and the found $\tau_j(= to_j)$, the PTT value is calculated at every relocation period.

In order to reduce the number of accessing invalid replicas, each mobile host monitors its own cache space and discards replicas whose lifetimes have passed. The cache space for the discarded replicas is kept free. When the mobile host accesses the original or a valid replica, the new replica of the data item is allocated again on the free cache space

At the time of discarding replicas, each mobile host discards the information on time stamps of replicas whose lifetimes have passed from its own time stamp table regardless of whether it holds the replica. When it accesses the original or a valid replica, it again records the information on the time stamp of the replica in its own time stamp table.

If α is set to a small value, each mobile host discards replicas whose originals have been updated with low probabilities, and thus the number of accesses to invalid replicas is reduced. However, since a small value of α also discards many valid replicas, the number of successful accesses is also reduced. Thus, the value of α should be chosen carefully according to the system characteristics and the performance requirements.

5.   Simulation Experiments

In this section, we present simulation results from our performance evaluation of the proposed methods.

5.1   Simulation model

The number of mobile hosts in the entire network is 40. Mobile hosts exist in a size 500 [m] × 500 [m] flatland and randomly move in all directions at a speed randomly determined from 0 to 1 [m/sec]. The radio communication range of each mobile host is 80 [m]. The number of kinds of data items in the entire network is 40, and M_i holds D_i $(i = 1, \cdots, 40)$ as the original. The size of each data item is 1 [MB]. Each mobile host creates up to 10 replicas with our proposed methods in Section 4. Replicas are periodically relocated every 400 [sec]. The access frequency of M_i to D_j is $p_{ij} = 0.5 \times (1 + 0.001j)[1/10\text{sec}]$. That is, each mobile host requests data items based on their access frequencies at every 10 [sec]. Each data item is updated with intervals based on the exponential distribution with mean U [sec].

In the simulation experiments, we randomly determine the initial position of each mobile host in the flatland and evaluate the following two criteria for our proposed methods during 1,000,000 [sec].

- *Data accessibility*:
  The ratio of the number of successful access requests to the number of all access requests issued during the simulation time.

- *Rate of accessing invalid replicas*:
  The ratio of the number of tentative data accesses that resulted in failure to the number of all access requests.

5.2   Effects of α value

First, we examine the effects of α value when the average update period U is fixed to 500. Figures 2 and 3 show the results. In both graphs, the horizontal axis indicate the value of α. The vertical axes indicate the data accessibility and the rate of accessing invalid replicas, respectively. For comparison, the performances when data replication is

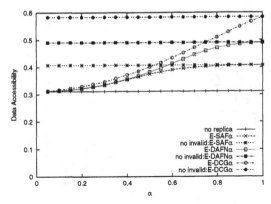

Figure 2. α and data accessibility.

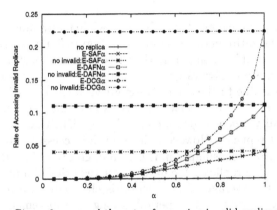

Figure 3. α and the rate of accessing invalid replicas.

not made and when replica invalidation is not made are shown as "no replica" and "no invalid:(replica allocation method)," respectively.

In Figure 2, in the cases where each mobile host invalidates replicas, as α gets larger, the data accessibility of each method gets higher. This is because the lifetime of each data item gets longer, and thus a larger number of replicas are available in the entire network. In the cases where each mobile host does not invalidate replicas, the data accessibility of each method is not affected by α. Comparing the three methods, the E-DCGα method gives the highest data accessibility. This is because mobile hosts can share many kinds of data items in stable groups.

Figure 3 shows that in the cases where each mobile host invalidates replicas, mobile hosts rarely access invalid replicas when α is very small. This is because the lifetime is very short, and thus replicas are discarded in a short time. On the contrary, as α gets larger, the rate of accessing

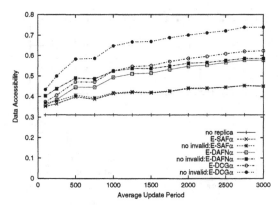

Figure 4.   Average update period and data accessibility.

invalid replicas gets higher. This is because while many valid replicas are accessible in the entire network as shown in Figure 3, many invalid replicas are also accessible. Comparing the three replica allocation methods, the E-DCGα method always gives the highest rate of accessing invalid replicas. This is due to the same reason given in the case of Figure 2.

The above results show that, in this experiment, the increment of data accessibility when α gets larger is larger than that of rate of accessing invalid replicas. However, since invalid accesses and roll backs impose extra power consumption, it is meaningful to reduce the number of accessing invalid replicas. We have conducted other simulation experiments where the average update period is changed. The results show that when the average update period is long, the increment of rate of accessing invalid replicas is larger than the case in this subsection.

5.3   Effects of average update period

Next, we examine the effects of the average update period, U, on our proposed methods when α is fixed to 0.7. Figures 4 and 5 show simulation results. In both graphs, the horizontal axis indicates the average update period. The vertical axes indicate the data accessibility and the rate of accessing invalid replicas, respectively.

In Figure 4, as the average update period gets longer, the data accessibility gets higher in all cases. This is because replicas held by each mobile host are valid for a longer time. In each replica allocation method, the data accessibility in the case where each mobile host does not invalidate replicas is larger than that in the case where it invalidates replicas. This is because when each mobile host discards replicas whose lifetimes have passed, it may discard valid replicas that have not yet been updated.

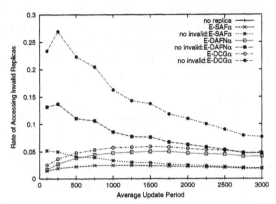

Figure 5.   Average update period and the rate of accessing invalid replicas.

Figure 5 shows that in the cases where mobile hosts invalidate replicas, when the average update period is very short, the rate of accessing invalid replicas is very low. In these cases, since replicas are discarded in a very short time, each mobile host can create replicas only when connecting with mobile hosts holding the originals in a relocation period. As a result, it creates only replicas with the same version as the originals and thus rarely accesses invalid replicas. In the cases where each mobile host does not invalidate replicas, as the average update period gets longer, the rate of accessing invalid replicas of all methods gets lower. This is because replicas held by each mobile host are valid for a longer time. The above results show that invalidation of replicas is very effective in each method when the average update period is short. However, when the average update period is long, the effectiveness becomes low since replicas are valid for a long time.

6.   Conclusions

In this paper, we proposed effective replica allocation methods in ad hoc networks where each data item is updated at irregular intervals. The proposed methods allocate replicas based on probability density functions of the update intervals of data items. Furthermore, in the proposed methods, each mobile host discards replicas that are updated with high probability. As a result, the proposed methods not only improve data accessibility but also reduce the number of accessing old replicas.

The results of simulation experiments according to our proposed methods show that data accessibility and the rate of accessing invalid replicas significantly depend on the setting of threshold α.

In [5], we proposed cache invalidation methods to effectively invalidate old replicas by broadcasting invalidation reports. In [6], we also proposed updated data dissemination methods to update old replicas effectively. As part of our future work, we plan to evaluate our methods in an environment where methods proposed in [5] or [6] are used together.

Acknowledgments

This research was partially supported by The 21st Century Center of Excellence Program "New Information Technologies for Building a Networked Symbiotic Environment" and Grant-in-Aid for Young Scientists (A)(1668005) of the Ministry of Education, Culture, Sports, Science and Technology, Japan, and by Tateishi Science and Technology Foundation.

References

[1] A.V. Aho, J.E. Hopcroft, and J.D. Ullman, "The Design and Analysis of Computer Algorithms," *Addison-Wesley*, 1974.

[2] T. Hara, "Effective replica allocation in ad hoc networks for improving data accessibility," *Proc. IEEE Infocom'01*, pp. 1568–1576, 2001.

[3] T. Hara, "Replica allocation methods in ad hoc networks with data update," *ACM-Kluwer Journal on Mobile Networks and Applications (MONET)*, vol. 8, no. 4, pp. 343–354, 2003.

[4] T. Hara, N. Murakami, and S. Nishio, "Replica allocation for correlated data items in ad-hoc sensor networks," *SIGMOD Record*, vol. 33, no. 1, pp. 38–43, 2003.

[5] H. Hayashi, T. Hara, and S. Nishio, "Cache invalidation for updated data in ad hoc networks," *Proc. CoopIS'03*, pp. 516–535, 2003.

[6] H. Hayashi, T. Hara, and S. Nishio, "Updated data dissemination in ad hoc networks," *Proc. Int'l Workshop on Ubiquitous Mobile Information and Collaboration Systems (UMICS'04)*, pp. 29–43, 2004.

[7] G. Karumanchi, S. Muralidharan, and R. Prakash, "Information dissemination in partitionable mobile ad hoc networks," *Proc. Symposium on Reliable Distributed Systems (SRDS'99)*, pp. 4–13, 1999.

[8] S. Lim, W-C. Lee, G. Cao, and C.R. Das, "A novel caching scheme for internet based mobile ad hoc networks," *Proc. ICCCN'03*, pp. 38–43, 2003.

[9] J. Luo, J.P. Hubaux, and P. Eugster, "PAN: Providing reliable storage in mobile ad hoc networks with probabilistic quorum systems," *Proc. ACM MobiHoc'03*, pp. 1–12, 2003.

[10] K. Rothermel, C. Becker, and J. Hahner, "Consistent update diffusion in mobile ad hoc networks," *Technical Report 2002/04, Computer Science Department, University of Stuttgart*, 2002.

[11] F. Sailhan, and V. Issarny, "Cooperative caching in ad hoc networks," *Proc. Int'l Conf. on Mobile Data Management MDM'03*, pp. 13–28, 2003.

[12] K. Wang, and B. Li, "Efficient and guaranteed service coverage in partitionable mobile ad-hoc networks," *Proc. IEEE Infocom'02*, vol. 2, pp. 1089–1098, 2002.

ANOVA-INFORMED DECISION TREES FOR VOICE APPLICATIONS OVER MANETS[*]

Mouna Benaissa,[1] Vincent Lecuire,[1] D.W. McClary,[2] and Violet R. Syrotiuk[2]

[1] *CRAN, CNRS UMR 7039*
Henri Poincare-Nancy I University
Campus Scientifique, BP 239
54506 Vandoeuvre-les-Nancy Cedex, France
{mouna.benaissa,vincent.lecuire}@cran.uhp-nancy.fr

[2] *Department of Computer Science & Engineering*
Arizona State University
P.O. Box 878809
Tempe, AZ U.S.A. 85287-8809
{daniel.mcclary,syrotiuk}@asu.edu

Abstract   Both real-time multimedia and mobile networks present challenges ripe for new analysis techniques. We examine the applicability of statistical design of experiments and inductive learning theory in the prediction of delay for real-time audio transmissions over mobile ad hoc networks. Utilizing analysis of variance methods and simple decision tree agents, we find both significant factor interaction between traffic load and node mobility as well as a dramatic reduction in error percentage in prediction of end-to-end delay.

1. Introduction

Real-time multimedia transmissions and mobile ad hoc networks each provide distinct challenges. Jitter and end-to-end delay are significant factors of interest for multimedia transmissions. Likewise, end-to-end delay is often a primary concern within *mobile ad hoc networks* (MANETs). When these fields meet, the importance of better understanding the domains increases significantly.

The reliability and consistency required by real-time audio transmission becomes much harder to ensure when dealing with the constantly varying condi-

[*]This research is supported in part by the National Science Foundation under grant ANI-0240524.

tions of a mobile ad hoc network. Thus, there is a need to effectively understand and predict end-to-end delay.

In light of this need, we show the validity of techniques from outside the
domain of common networking. Specifically, we examine *statistical design of
experiments* (DoE) and *analysis of variance* (ANOVA) to isolate and quantify
factor interactions among variables in a MANET. Additionally, we adapt artificial intelligence techniques from basic decision-tree learning to provide a
mechanism for predicting delay within a mobile environment. By fusing these
two techniques, we find significant measures of predictability for end-to-end
delay for audio transmissions over MANETs.

We first present a performance analysis of end-to-end delay for audio transmissions over MANETs followed by an ANOVA analysis for accurate prediction of end-to-end delay. We begin with a discussion of related works and an
introduction to the simulation methods used. From this, we present the simulation results followed by an introduction to DoE and ANOVA analysis, learning
theory and decision trees, and the results produced by these techniques.

2.   Simulation Analysis of Audio Packet Delays

The results in this section are based on simulations using the network simulator `ns-2`, version 2.26 [6, 9]. This package includes extensions for mobile ad
hoc network simulation, including a set of routing protocols, an IEEE 802.11
MAC layer, a radio propagation model, and a node mobility model. However,
we extended this package to generate voice traffic over the wireless environment.

Technical Considerations

All nodes communicate with a wireless radio based on the IEEE 802.11
standard [7]. The radio propagation range for all nodes is 250 meters and the
channel bandwidth is $11\,Mbps$. The specific *medium access control* (MAC)
scheme is CSMA/CA with acknowledgments. At the link layer, we leave most
of the 802.11 parameters set to default values. Thus, the RTS threshold is set
to 250, the short retry limit (SRL) is set to 7, and long retry limit (LRL) is set
to 4.

At the network layer, we use the *Ad hoc On Demand Distance Vector* (AODV)
[13] protocol for routing. The choice of a reactive protocol such as AODV
rather than a proactive protocol such as OLSR [11] is based on results from
our earlier work [3]. Indeed, the proactive protocol consists of every node
emitting `hello` messages periodically in order to learn the network topology.
On the other hand, reactive protocols invoke a route discovery procedure on an
on-demand basis.

As mobile ad hoc networks are characterized by intermittent connectivity, the audio stream is interrupted when a route error causes the routing protocol to establish a new route to the destination. These interruptions are typical events that strongly disrupt the speech played on the receiver side. Benaissa et al. [3] have shown that such interruptions can be long, typically a number of seconds, regardless the routing protocol employed. However, OLSR causes more interruptions than AODV, thus AODV is better adapted to the deployment of audio applications over ad hoc networks.

The AODV parameter values are set as recommended in [14] such that they minimize network congestion and allow the protocol to operate as quickly and accurately as possible. As such, the HELLO interval is set to 1.0, the route reply wait time is set to 1.0, the reverse route life is set to 3.0, and the active route timeout is set to 3.0.

The network layer maintains a send buffer of 64 packets. This buffer contains (only) data packets waiting for a route. All packets (both data and control) sent by the routing layer are queued at the interface queue until MAC layer transmission. We set the maximum size of the interface queue to 50 packets and maintain it as a priority queue with two priorities, each served FIFO. Control packets receive higher priority than data packets.

Network Environment and Methodology

Our network model consists of 50 nodes in a 1000×1000 meter square flat area. Our results are based on the average packet delay of 20 scenarios, or patterns. Patterns are randomly generated by different seeds. Each simulation executes for $250s$. Input parameters for the simulation are speed s in m/s and the network load ℓ in kbps.

In order to avoid large variation in successive patterns, some seed number effects are cancelled. We make the following assumptions:

- We define the same initial position and heading of nodes for all patterns. Thus, when the seed changes, the initial network topology and traffic peers remain identical. All movements are different (see Figure 1).

- For different patterns at a given network load ℓ, enabled traffic peers are identical. When ℓ is increased, additional traffic peers are set.

- During the simulation, the effective speed of nodes and the network load are constant and equal to the input parameters s and ℓ.

Mobility

Nodes move according to the random waypoint mobility model. In this model, each node x chooses a random destination within the simulation area

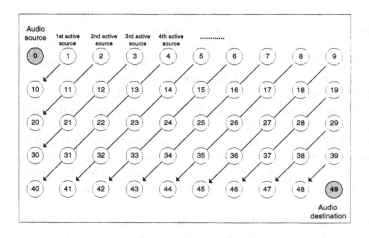

Figure 1. The initial MANET topology.

and travels toward its destination in a straight line at a given speed. Upon node x's arrival at its destination, it pauses, chooses a new random destination, and continues its motion. Node speed s varies from $1m/s$ to $10m/s$ and we use a pause time of 10 seconds.

Traffic Pattern

We develop a traffic generator to simulate unicast voice traffic as well as background traffic.

Voice traffic is generated between a source node (0) and a destination node (49) (see Figure 1). A voice flow is typically divided into *talkspurts* (periods of audio activity) and *silent periods* (periods of audio inactivity, during which no audio packets are generated). We consider an average talkspurt of 30.83% and an average silent period of 61.47% as recommended by the ITU-T specification for conversational speech [10]. Alternating periods of activity and silence are exponentially distributed with an average duration of $1.004s$ and $1.587s$, respectively. We consider the PCM codec (Pulse Codec Modulation — see Recommendation G.711 in [8]) at $64kbps$ as audio traffic. Then, 320 byte voice packets are generated periodically during activity periods, each voice packet representing a sample of $40ms$.

Background traffic is generated by *constant bit rate* (CBR) data sessions with selected sources and destinations. One CBR session transmits packets of 2048 bytes payload every $0.3s$ (i.e., $57.28kbps$ load per session). To increase the load in the network, the number of active sessions is increased. In our simulation, network load ℓ varies from $1 \times 57.28kbps$ to $10 \times 57.28kbps$.

Simulation Results

Figure 2 illustrates average end-to-end delay as a function of the number of active sessions for three different node speeds: $2m/s$, $5m/s$ and $7m/s$. We find that from 1 to 5 active sessions, delays are nearly equal (less than $40ms$), regardless of the speed. For higher numbers of active sources, the increases in delay are stronger when node speed is higher. Delay increases from $50ms$ to $200ms$ for node speed $2m/s$, while it reaches $450ms$ at node speed $7m/s$ under the same load level.

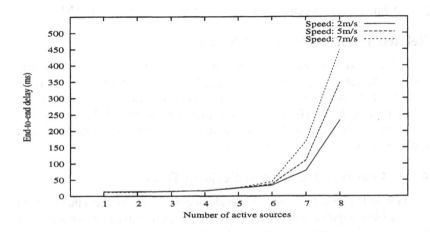

Figure 2.   End-to-end delay as a function of load for different node speeds.

Our results illustrate that node speed does not have a significant impact on end-to-end delay under light load traffic. However, when load traffic increases, end-to-end delay increases rapidly when node speed is high.

3.   Designed Experiments and ANOVA Analysis

Though often treated as isolated, the factors affecting a transmission in any network interact in ways that are often imperceptible when considered by system and protocol designers. Factor interaction is not a new consideration in other engineering domains and techniques have long been in place to effectively unearth and analyze them. Designed experiments, a technique originating from agricultural engineering in the early part of the 20th century, allows the otherwise vast number of experimental runs necessary to detect and quantify factor interactions to be done in a limited number of trials. As defined by Montgomery [12], a *factor* (e.g., load) is an experiment variable taken to have some effect on the *response variable* of interest (e.g., end-to-end delay). A *factor interaction* is the failure of a factor to produce the same resultant value for the response variable when another factor is at a different level.

Key to DoE's ability to effectively analyze factor interactions is analysis of variance (ANOVA). In the simplest case, ANOVA examines the effect of a single factor on a response variable. ANOVA is defined by a sum of squares identity [12]. Using these sums of squares, ANOVA allows calculation of main effects for all factors as well as interactions. Our aim is to utilize DoE and the underlying foundation provided by ANOVA to significantly improve the predictability of end-to-end delay for audio-transmissions over MANETs.

Related Work — DoE and ANOVA

Designed experiments and ANOVA have rarely been applied to mobile networking. Vadde and Syrotiuk [17], inspired by the work of Barrett et al. [1, 2], used ANOVA analysis to identify main effects and factor interactions on service delivery in MANETs. Along similar lines, Perkins et al. [15] examined 2^k factor interaction particularly among node speed, network size, and number of traffic sources.

4.   Learning Theory and Decision Trees

Machine learning theory aims to develop agents and algorithms that are able to effectively improve future action by learning from both its environment and its own decision-making processes. Among the simplest learning algorithms to implement is that of *decision tree learning*. Decision tree learning examines a set of properties and returns a decision. Typically decisions are boolean, but are easily extended to numerical results. Thus, decision trees are naturally suited to learning parameter optimization.

In functionality, the inductive learning provided by a decision tree begins with a heuristic from which an initial decision tree structure is extrapolated. The branching conditions of this tree are then tuned via the inductive process to produce better optimized results. Thus, for the purpose of optimizing the predictability of end-to-end delay, decision trees allow us to quickly produce a proof-of-concept argument for inductive learning theory for MANETs as well to evaluate the potential benefit of integrating a decision mechanism into our simulation model.

Related Work – Learning Theory and Decision Trees

Within artificial intelligence and other fields, learning theory has evolved into a sophisticated and rich field in its own right. However, much in the same manner as ANOVA analysis, learning theory methodologies have not been significantly explored within mobile ad hoc networks. Farago et al. [5] developed a meta-MAC protocol utilizing computational learning theory. We take our direction in learning agent construction from Russell and Norvig [16].

5.   DoE and Learning Methodologies

DoE and ANOVA Methodologies

In order to validate the effectiveness of DoE and ANOVA techniques within our chosen application domain, we choose those variables from Benaissa et al. [3] that provide both the most sound basis for factor interactions as well as reasonable extension into a decision tree learning mechanism. We choose CBR traffic load and node mobility as our factors of interest and end-to-end delay as our response variable. We then use `Design Expert` [4], a software suite designed to aid in the construction and processing of DoE, to develop an appropriate model to fit our data set.

We design our statistical experiment as a 2^2 response surface populated by our data points, where 2^k is indicative of a two-level (high and low) model with k variables of interest or *main effects*. From this, we generate two models for ANOVA analysis. The first, an untransformed model of the data, is used to develop certain of our decision tree agents. The second, a model transformed such that any factor interactions present are effectively quantified, is used to verify the ANOVA validity.

Decision Tree Methodologies

To leverage learning theory techniques, we choose decision tree inductive learning for a number of reasons: ease of implementation, small spatial requirements and a solid modeling of a top-down traversal of protocol stacks. For our proof-of-concept, we examine multiple means of constructing the inductive learning aspects of the tree, both with and without statistical influence. As is standard for decision tree tests, average-guessing or *strawman* algorithms are constructed to provide a baseline for judgment of improvements via the learning mechanism. Given our implementations of both statistically-influenced and standard inductive learning, we implement both statistically-influenced and standard mean-guessing strawman algorithms.

Our first step in developing decision tree agents is to create a heuristic that serves as the basis for construction of the initial tree. For purposes of conceptual proof, we do not consider factor interactions in the construction of this heuristic, though such interactions are responsible for much of the power of this manner of inductive learning. Instead, we consider the simplest case, in which we assume end-to-end delay is primarily dependent on traffic load. Examination of the data set yields a decision tree with three distinctly different relations to end-to-end delay. As illustrated in Figure 3, we choose a simple, one-level tree branching at what initial observation indicates are significant differences in the load vs. end-to-end delay relationship.

Figure 3.   A simple one-level decision tree estimates delays based on noticeable variances of load vs. mobility. More refined decision trees take into account multiple factors and their interaction.

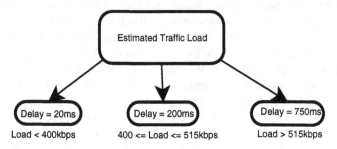

Having constructed an initial tree, we implement the inductive learning process as a number of agents, each utilizing different means of hypothesis correction during the learning and testing phases. These agents can be broken into two sets — those that do not use statistical influence in their correction, and those that do. For each of these categories, we create agents which employ incremental correction techniques and agents that employ averaging. Additionally, we design an agent that uses no correction following its learning phase, as well as one that employs only statistical means of self-correction.

Agent correction in both learning and testing phases is structured such that those agents without statistical influence utilize their particular means of correction, be it incremental or average-based, from the point of the value of the initial heuristic. Statistically-influenced agents utilize averaging based on the ANOVA-generating equation during the learning phase and their particular correction method during the testing phase to further refine the agent's accuracy. The agent with no testing-phase adaptation employs averaging in its learning-phase correction, and the completely statistically-driven agent averages against the ANOVA-generated equation during both phases.

6.   DoE and Learning Theory Results and Discussion

DoE Results and Discussion

As described in section 5, we create a pair of designed experiments. One uses no data transformation so as to yield equations suitable for adaptation into our decision tree agents and the other transformed to most accurately assess any factor interaction present. In each case, we utilize `Design Expert` to process our data sets. We specify our factors as L for traffic load, and M for node mobility, with respective low and high levels ($57.28kbps$, $572.80kbps$) and ($1m/s$, $8m/s$). We then choose as our response variable for end-to-end delay for the real time audio flow with low and high levels ($4.10ms$, $1663.96ms$).

From initial sum of squares analysis, we choose to design the experiments as a cubic model analyzing factors L, M, L^2, M^2, LM, L^3, M^3, L^2M, and LM^2 where, e.g., LM denotes the interaction effect between traffic load and node mobility. In keeping with standard practice, we choose the cubic model because it yields the lowest F-value and $P(X > $ F-value$)$. On this cubic model we then use `Design Expert` to perform ANOVA analysis of the data.

As expected, the initial untransformed model uncovers significant factor interaction. We find significant factors to be L, M, L^2, LM, L^3, and L^2M. Thus, consistent with our hypothesis, the interaction between traffic load and node mobility within MANETs is significant with respect to end-to-end delay on real-time audio transmissions. Likewise, in this case the square and cube of traffic load holds some significant relationship to end-to-end delay. For the purposes of enhancing our decision tree agents, ANOVA yields the relationship for the delay, D:

$$D = 70.40 + 0.17L - 52.54M - 3.3e^{-3}L^2 + 13.82M^2 + \\ 0.024LM + 9.85e^{-6}L^3 - 1.17M^3 - 2.40e^{-4}L^2M + 0.01LM^2.$$

By using a Box Cox test [12], we determine that by applying an inverse square root transform ($y' = \frac{1}{\sqrt{y+\lambda}}$) with $\lambda = -0.5$. we achieve a more precise measurement of the principal factor interaction, that between traffic load and node mobility. As shown in Figure 4, we find that the two factors interact with one another in at least two key points. From the generated interaction plots as well as the lack of fitting error, there exists a definite factor interaction between traffic load and node mobility for audio transmission over MANETs. Specifically, we notice significant interactions when traffic load is in the ranges $114.56 < L < 171.84$ and $458.4 < L < 515.52$. Thus, we say load-mobility interaction displays its most prominent effects when traffic load is in the more extreme regions of its possible values.

Figure 4.   A factor interaction graph for traffic load and node mobility. Path intersections indicate levels at which interaction is most significant.

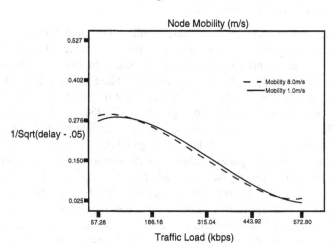

Decision Tree Results and Discussion

For testing of both the validity of decision tree agents in reducing uncertainty in end-to-end delay in MANETs, as well as the integration of the technique with statistically-influenced heuristics, we measure the testing-phase performance of our previously designed agents in three categories over n estimates:

- Correct guess percentage: $\dfrac{\text{correct estimates}}{n}(100)\%$

- Mean error percentage: $\dfrac{|\text{delay}_{\text{estimated}} - \text{delay}_{\text{actual}}|}{n}(100)\%$

In order to establish a baseline for evaluating our agents, we run trials with both a "dead" strawman, which simply guesses the mean delay found in the learning phase for all testing-phase estimates, and a statistical strawman, which guesses the result of our ANOVA equation for each testing-phase estimate. The "dead" strawman produces correct guess percentage of 4.25%, and a mean error of 572.10%. With similar inaccuracy, our statistical strawman reports a correct guess percentage of 2.38% and mean error percentage of 1010.78%. Additionally, we consider our non-adapting decision tree agent, which produces results: correct guess percentage of 4.40%, and a mean error of 140.25%.

These baselines established, we find that both categories of decision-tree agents perform well, with certain techniques outshining others. The decision tree agents that adapt incrementally both perform with correct percentages of

15.14% in the statistical case and 19.65% for the standard model. In our normal and statistically-influenced agents, we find that those that employ testing-phase heuristic refinement via averaging produce correct percentages of 33.04% and 32.69% while maintaining mean error percentages less than 100%. Specifically, the statistically-influenced agent with difference-based testing phase adaptation produces a correct percentage of 32.69% while lowering mean error percentage to 42.00%. In similar fashion, our non-statistical model produces a correct percentage of 33.04% and mean error percentage of 39.84%.

In consideration of these results, we find that decision tree agents that utilize difference-based averaging perform with dramatic improvement over simple averaging means. Additionally, the lack of adverse effects produced when utilizing ANOVA-based heuristics leads us to believe that ANOVA-produced equations provide just as sound if not more sound a basis for decision heuristics as currently exist. That said, given the significance of the factor interaction detected by the DoE phase of our work, we believe that this interaction can be exploited to produce far more accurate decision trees than our current one-level model. In such a case, we expect the learning to be far richer and the results even more promising.

7.   Conclusions and Future Work

Our results in examination of real-time audio transmission over mobile ad hoc networks serve as an initial proof-of-concept for the validity of both DoE as a means of effectively verifying the results of the simulation, and detecting factor interaction among network parameters as well as decision tree learning as a method of increasing the predictability of end-to-end delay. The work is currently in a preliminary state. Further investigation of the interaction between traffic load and mobility is certainly required. Likewise, we expect deeper consideration and refinement of the decision tree model, based on the results of our ANOVA analysis of the data, will improve even present levels of predictability and error percentage. In the same vein, consideration of other factors involved in real-time audio transmission, particularly routing and MAC protocol configurations (e.g., RTS threshold, short and long retry length) may yield yet more information about the complex interactions between these components. We plan to embed such a learning mechanism within an application, such that delay may be minimized on the fly. As such, we have begun consideration of these directions and expanding our research as the domain demands.

References

[1] Barrett, C.L, A. Marathe, M.V. Marathe, and M. Drozda (2002). "Characterizing the Interaction Between Routing and MAC Protocols in Ad-Hoc Networks," Proceedings of the Third ACM International Symposium on Mobile Ad Hoc Networking and Computing (MobiHoc'02), pp. 92–103.

[2] Barrett, C.L., M. Drozda, A. Marathe, and M.V. Marathe (2003). "Analyzing Interaction Between Network Protocols, Topology and Traffic in Wireless Radio Networks," Proceedings of the IEEE Wireless Communications Networking Conference (WCNC'03)", pp. 1760–1766.

[3] Benaissa M., V. Lecuire, F. LePage, and A. Schaff (2003). "Analysing End-to-End Packet Delay and Loss in Mobile Ad hoc Networks for Interactive Audio Applications," Proceedings of the Workshop on Mobile Ad Hoc Networking and Computing (MADNET'03), pp. 27–33.

[4] "Design Expert Software," Stat Ease Inc. lOnlinel. Available: http://www.statease.com

[5] Farago A., A.D. Myers, V.R. Syrotiuk, and G.V. Zaruba (2000). "Meta-MAC Protocols: Automatic Combinations of MAC Protocols to Optimize Performance for Unknown Conditions," IEEE Journal on Selected Areas in Communications, Vol. 18, No. 9, Sept. 2000, pp. 1670–1681.

[6] Greis, M. (2001). "Tutorial for the Network Simulator ns," lOnlinel Available: http://www.isi.edu/nsnam/ns/tutorial

[7] IEEE Standards Department (1999). "Part 11: Wireless LAN Medium Access Control (MAC) and Physical Layer (PHY) specifications: Higher-Speed Physical Layer Extension in the 2.4 GHz Band," Technical Report, European Telecommunications Standards Institute.

[8] International Telecommunication Union ITU-T (2001). "Pulse Code Modulation (PCM) of Voice Frequencies," Recommendation G.711, Telecommunication Standardization Sector of ITU.

[9] ISI (2001). "The Network Simulator ns," lOnlinel Available: www.isi.edu/nsnam/ns/

[10] ITU-T (2004). "List of ITU-T Recommendations," lOnlinel Available: http://www.itu.int/publications/itu-t/itutrec.htm

[11] Jaquet, P., P. Muhletaler, A. Quayyum, A. Laouiti, T. Clausen, L. Viennot, and P. Minet (2002). "Optimized Link State Routing Protocol," Internet Draft draft-ietf-manet-olsr-06.tx, Internet Engineering Task Force.

[12] Montgomery, D. (2001). *Design and Analysis of Experiments*, John Wiley & Sons, Inc.

[13] Perkins, C.E. (2001). *Ad hoc Networking*, Addison Wesley Longman, Inc.

[14] Perkins, C.E. and E. Royer (1999). "Ad hoc On-demand Distance Vector AODV Routing," Proceedings of the IEEE Workshop on Mobile Computing System and Applications, pp. 90–100.

[15] Perkins, D.P., H.D. Hughes, and C.B. Owen (2002). "Factors Affecting the Performance of Ad Hoc Networks," Proceedings of IEEE International Conference on Communications (ICC'02), Vol. 4, pp. 2048–2052.

[16] Russell, S. and P. Norvig (1995). *Artificial Intelligence: A Modern Approach*, Prentice Hall, Inc.

[17] Vadde, K.K and V.R. Syrotiuk (2004). "Factor Interaction on Service Delivery in Mobile Ad Hoc Networks," to appear in IEEE Journal on Selected Areas in Communications, (accepted March 2004).

ROUTE STABILITY TECHNIQUES FOR ENHANCED VIDEO DELIVERY ON MANETS

Carlos T. Calafate, Manuel P. Malumbres and Pietro Manzoni *
Department of Computer Engineering
Polytechnic University of Valencia (UPV)
Camino de Vera S/N, 46022 Valencia, Spain
{calafate, mperez, pmanzoni}@disca.upv.es

Abstract   One of the main problems associated with MANETs is that mobility and the associated route discovery and maintenance procedures of reactive routing protocols cause interruptions on real-time video streams. Some of these interruptions are too large to be concealed using any sort of video technology. In this work we argue that increased bandwidth and QoS strategies do not solve problems associated with mobility. We present a solution for enhanced video transmission that increases route stability by using an improved route discovery process based on the DSR routing protocol, along with traffic splitting algorithms and a preventive route discovery mechanism. We achieve improvements in terms of goodput, and more important, we reduce the video gap durations up to 97% for high mobility scenarios, improving the user viewing experience dramatically.

Keywords: Multipath routing, video streaming gaps

1.   Introduction

Mobile Ad-hoc networks (MANETs) are composed by a set of independent mobile nodes which "cooperate" without any type of infrastructure. The low cost and ease of deployment of this kind of networks makes them extremely attractive for applications ranging from disaster relief situations to small home environments. However, the current performance of MANETs can hardly be accepted for real-time multimedia.

The IEEE 802.11 [IEEE 802.11, 1999] technology is the most deployed and used. The IEEE 802.11a/g standards enable MANETs to use more bandwidth for multimedia applications, allowing various simultaneous video flows per

* This work was supported by the *Ministerio de Ciencia y Tecnología* under grant TIC/2003/00339, and the *Junta de Comunidades de Castilla La-Mancha* under grant PBC/03/001.

node to exist. The medium access issues concerning QoS, included in the IEEE 802.11e working group, are still not standardized but are of extreme importance to provide service differentiation.

The purpose of this work is to present the problems that still persist after applying QoS techniques at the MAC level. We found [Carlos T. Calafate et al., 2004b] that even when a video flow does not have to face competition with other flows, and when the routing protocol operates in optimal conditions, video performance is still not optimal due to mobility.

We show that there is a close relationship between video gaps and route discovery events. We then describe a traffic splitting approach based on multi-path, which uses a route selection mechanism that optimizes the use of disjoint routes. The use of disjoint routes reduces video gaps occurrence generated by node mobility thus improving the quality of the received video. Finally, in order to prevent possible route losses we supply a preventive route discovery mechanism activated when a video flow has not at least two disjoint routes available.

To measure the effect of video gaps over the final video quality delivered to the user, we propose a metric called "video annoyance" (VA). This metric helps in evaluating the behavior of the proposed route discovery and traffic splitting mechanisms.

Concerning the structure of this paper, in section 2 we give a brief description of the related work in this area. In section 3 we propose enhancements to DSR's route discovery technique. Section 4 presents the effects of applying enhanced route discovery procedures over the delivered video quality, performing a detailed study of the video loss pattern. In section 5 we present a multipath routing algorithm that introduces traffic splitting as a mechanism to improve final video quality when node mobility is significant. In section 6 we perform a global evaluation of all mechanisms presented in this paper. Finally, in section 7 we make some concluding remarks, along with some references to future work.

2.   Related work

The use of multiple routes in MANETs has recently become a promising solution for multimedia data transmission.

[L. Wang et al., 2001] use a probing technique to assess the quality of available routes, so that the traffic is forwarded based on the delay of each route. Their objective is also to achieve a fair load distribution as well as improved throughput, end-to-end delay and queue utilization. [Nasipuri et al., 2001] proposed a strategy for quick route recovery through packet re-direction in intermediate nodes to reduce the frequency of query floods.

[Wu, 2002] proposes a more selective route discovery procedure to DSR to increase the degree of disjointness of routes found without introducing extra overhead. It allows the source to find a maximum of only two paths (node disjoint paths) per destination.

[Lee and Gerla, 2001] show that the paths found by DSR's route discovery mechanism are mainly overlapped. They therefore propose a route discovery technique that provides nodes disjoint paths.

In [Marina and Das, 2001] the AODV protocol [Perkins and Royer, 1999] has been extended in order to provide multi-path capabilities, though no new route discovery mechanism was proposed. Both node disjoint and link disjoint approaches are presented.

3. Route discovery extensions to DSR

The Dynamic Source Routing (DSR) protocol is an efficient routing protocol for MANETs. DSR, by default, finds only a small number of routes. By extending its route discovery mechanism we increase the average number of routes found per node. This extra information alone offers to each node extra possibilities when a route is lost, requiring on average less route discovery processes. Lee and Gerla proposed in [Lee and Gerla, 2001] a route discovery technique based on altering the route discovery process. So, during the "RREQ" propagation phase, packets with the same route request ID can be forwarded if they arrive "through a different incoming link than the link from which the first RREQ is received, and whose hop count is not larger than the first RREQ".

With our route discovery proposal we significatively reduce the routing overhead when compared to Lee and Gerla's proposal (see [Carlos T. Calafate et al., 2004a]). From now on we shall refer to our solution as "Super Restrictive" mode (SR). In SR mode, we add a list (*SRlist*), to the already existing route discovery table structure in all nodes. This list is used to store the intermediate hops that forwarded the route request. The cost in terms of memory is very small - only some bytes per source. The main enhancement of the SR mode is that it discontinues the propagation of a route request if some of the previous hops (except the source) is the same. With this method we assure that the discovered paths are node disjoint, increasing therefore the usefulness of the routes found.

The size of the *SRlist* can be controlled and limited. When a route request arrives and the list is already full, it is not propagated. This means that only a pre-defined number of route requests are forwarded. If the size of the *SRlist* is very high we obtain the basic SR mode; if it is equal to one the behavior is similar to the DSR's propagation mode. The size of the *SRlist* is a new parameter and it will be referred to as *PNC* (Previous Node Count).

The SR solution restricts the route request forwarding process to route sizes not superior to the first one arriving. To increase the flexibility of the approach we accept routes with an extra size up to a certain value. We call this parameter *flexibility*. In [Carlos T. Calafate et al., 2004a] we demonstrated that only small values for this parameter, though, make sense in terms of route size and stability.

In the following sections we test three different *Flexibility / PNC* pairs. In mode 1 ($Flex. = 0, PNC = 2$) the propagation using the SR technique is restricted to the maximum, so that only one extra route per node is allowed relative to the default DSR behavior. Modes 2 and 3 maintain one of the parameters of mode 1, but in mode 2 ($Flex. = 2, PNC = 2$) we increase flexibility and on mode 3 ($Flex. = 0, PNC = 4$) we increase the number of RREQs propagated per node. In all modes the use of cache on route propagation is turned off, maintaining the rest of DSR's behaviors unchanged.

4.   Effects of route stability on real-time video streams

In this section we show how the use of a standard routing protocol, such as DSR, can affect a real-time video stream in terms of video communication disruptions (video gaps). We will prove that video gaps are intimately related to route discovery procedures, and how the SR mode presented in Section 3 can considerably alleviate this problem. The evaluations are done using the ns-2 simulator [K. Fall and K. Varadhan, 2000]. Each obtained value is averaged over 5 simulation runs. Concerning node movement, it was generated using the random waypoint mobility model. A filter was applied to accept only scenarios without network partitioning (i.e., with no unreachable destinations), in order to obtain a connected graph.

We first evaluate a 1000x1000 m squared scenario with 80 nodes. The traffic load consists of a single H.264 [H.264, 2003] video stream obtained from the well known *Foreman* sequence at 10 frames-per-second. Each video frame is split into 7 RTP packets, resulting in a target bit-rate of 186 kbit/s. Our purpose is to observe the performance of the different routing protocols independently from other traffic flows.

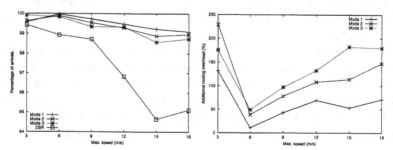

Figure 1.   Packet arrivals and routing overhead when streaming an H.264 video flow

In figure 1 we observe how modes 1 to 3 always perform slightly better that the original DSR implementation in terms of packet arrivals. The best performing mode is mode 1, where the improvements over DSR reach 4.5% in packet arrivals. Mode 1 also generates less routing control packets than the other two modes, with a small relative increase compared to DSR.

Loss pattern analysis

H.264 standard offers a wide range of tools to reduce the effects of degradation in the presence of losses. Different types of intra macroblock updating strategies and error concealment tools are available, which aims at estimating parts of frames which are not received. Thus, observing the results of figure 1, and taking into account these facts we could conclude that the difference in terms of peak signal-to-noise ratio (PSNR) between receiving 99% or 95% of the packets is very slight.

However, packets dropped in bursts long enough to cause video gaps affect H.264 video decoders in a drastically different manner. When no information relative to consecutive frames arrives to the decoder, this will freeze the last decoded frame until communication is resumed. When communication is resumed the decoder's effort is also increased since it must resynchronize and recover from losses as quickly as possible. We therefore argue that the PSNR is not a representative factor and we propose a new metric which we called *VA* (*Video Annoyance*) parameter, defined as:

$$VA = \frac{\sum_{i=0}^{N}(G_i)^2}{NF^2}, \ 0 \leq VA \leq 1, \tag{1}$$

where NF is the total number of frames in the sequence, N is the total number of independent video gaps occurring in a video sequence and G_i is the size of the i-th video gap. We define a video gap $G_i = \frac{B_i}{PPF}$ as the number of video frames lost sequentially; B_i is the number of consecutive packets lost for gap i ($B_i \geq 2$) and PPF is the number of packets per frame.

This number does not need to be an integer, since for example $4\frac{1}{2}$ frames can be lost in a single burst, being communication resumed with information relative to some position inside a frame. What must be defined, though, is the minimum number of consecutive lost packets to create a video gap. In this work we set that threshold to one entire frame, that is 7 packets ($B_i \geq 7$). When $VA = 0$, there are no video frame gaps, though losses can occur. When $VA = 1$ the entire sequence was lost. The quadratic relation takes into account the fact that many distributed 1-frame gaps are almost imperceptible to the viewer, while a single 50 frames gap, a 5 seconds interruption at 10Hz, is quite disturbing for the user.

Analyzing a typical packet drop pattern on a simulation using DSR and a single H.264 video flow, we can observe that some of the packet loss events are bursty. Bursts of packets lost cause the video flow to be stopped, so that several entire frames are lost.

Table 1 presents a comparison between the different routing protocols concerning the VA parameter. As it can be seen, the VA associated with modes 1 to 3 is only a small fraction (1-2%) of the VA achieved with the original DSR implementation. This result proves how the different SR modes improve the video experience in terms of video disruptions.

Table 1. Video annoyance statistics

Protocol	VA (10^{-6})	VA % towards DSR
DSR	38,1	-
Mode 1	0,502	1,3169
Mode 2	0,708	1,8575
Mode 3	0,768	2,0167

To further analyze and validate the improvements shown with the VA parameter, we now consider the video gap histogram for all protocols at maximum node speed (see figure 2). We observe that DSR performs much worse than any of the SR modes for all gap intervals, being mode 1 the one that achieves the best results.

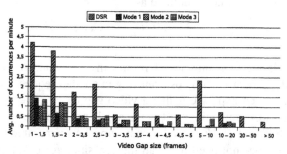

Figure 2.   Video frame gaps histogram at a maximum node speed of 18 m/s

Another important factor shown in figure 2 is that SR modes 1 to 3 present gap sizes of no more than 20 frames, contrarily to DSR. In fact, DSR is prone to lose as much as 217 consecutive frames, equivalent to more than 20 seconds of video interruption at 10Hz, while in mode 1 the maximum gap size experienced is of only 13 frames (1.3 seconds of video interruption at 10Hz).

Gap causes and solutions

In scenarios like the one under evaluation, where congestion is not a problem, packet losses directly depend on link breaks and subsequent route failures. Since DSR uses link layer information to detect broken links, the interval between the detection of a broken link and the reception of the associated notification by the source is, in terms of video streaming, not excessively long. In fact, we find that when a route breaks the number of packets lost can be estimated by:

$$N = 2 \cdot S_{rate} \cdot T_{RERR}, \qquad (2)$$

where S_{rate} is the source's packet generation rate and T_{RERR} is the time that the "Route error" packet takes to arrive at the source. This phenomena can sometimes be alliviated by DSR's packet salvaging mechanism.

To clarify the results of table 1 and figure 2 we compared the behavior of the different SR modes with the behavior of DSR in terms of RREQs. We calculated the number of route requests generated by the video source at different speeds (see figure 3.) The results are presented using the values obtained for the DSR as a reference. We observe that modes 1 to 3 present less video gaps for all speeds as expected due to the higher number of routes found. The relationship between the number of RREQs and video gaps also explains the improvements in terms of VA. The mode with best overall performance (mode 1) shows an average reduction of 68% on the number of RREQs generated in relation to DSR, and so we will use it as a basis for the improvements performed in the following sections.

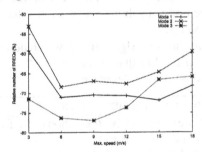

Figure 3. The behavior of the different SR modes towards DSR in terms of RREQs

5. Multipath routing

In this section we will present further improvements by analyzing how the use of simultaneous paths on data transmission effectively reduces the downtime of H.264 video flows, making the communication experience smoother and more pleasant to the user. We should point out that the sequence numbers included in RTP headers allow the receiver to reconstruct the sending order. Also, the video decoder can use sequence numbers to determine the proper location of a packet without necessarily decoding packets in sequence. Therefore, we consider that out-of-order delivery provoked by multipath routing is not a primary issue.

Traffic splitting strategies

Traffic splitting in the context of multipath routing refers to the technique of distributing the packets of a certain stream through different paths. Concerning the paths themselves, we can talk about their degree of disjointness and also make a distinction between link disjoint and node disjoint paths [Marina and Das, 2001]. Node disjoint paths are those where none of the intermediate nodes are in common. Link disjoint paths are those where all links differ, though common nodes may exist.

An optimal strategy in terms of traffic splitting would be one where the shortest disjoint path is used. In general, node disjoint paths are preferable since they achieve the best trade-off in terms of both bandwidth and node resources. There are some cases where no node disjoint paths are available and, therefore, link disjoint paths are used. In fact, the link disjointness condition is enough to reduce the effect of mobility on ad-hoc networks.

We define a metric, see Equation 3, that demonstrates the exact gains in terms of path dispersion using the average degree of path disjointness.

$$Dispersion = 1 - \frac{CL}{NL},\tag{3}$$

where CL is the number of common links relative to the previous path, and NL is the number of links of the current path. The objective is to obtain values close to 1, which is the optimal solution; dispersion values close to 0 reflect a bad traffic dispersion policy.

We propose an algorithm (see Algorithm 1), which describes how to get the maximum disjointness path. We shall use the term "*Disjoint solution*" to the mechanism that makes use of this algorithm.

Algorithm 1 Maximum route disjointness algorithm
---
if (no path has been chosen previously) *then* choose the first shortest path;
else { find the shortest node disjoint path;
  if (not found) *then* find the shortest link disjoint path;
  if (not found) *then* find the shortest path with least common links;
  if (not found) *then* choose first shortest path; }

The action of finding the disjointness of one route is always done relatively to the previously used route. This technique easily adapts to extra routes found through the forwarding or interception of routing packets, as well as to routes which were considered lost. Though this algorithm aims at finding the best choices in each situation, it could be considered computationally expensive for small embedded systems. Therefore we also propose a much simpler solution which consists of randomly choosing routes which are not larger than a certain size (s) relative to the first one. This alternative solution, referred as R_s, is used as a reference for the Disjoint solution.

Using the same simulation setup that in section 4, we evaluated the Disjoint algorithm comparing it with the R_s solution. Table 2 shows the average results by setting $s = 2$ and setting the maximum allowed node speed to 18 m/s. As it can be seen, the maximally disjoint solution always achieves the best results.

Table 2. Comparison of traffic splitting strategies

Mode	Video arrivals(%)	Routing overhead	End-to-end delay (ms)	Dispersion
Disjoint	99,70	6759	39,54	0,71
R_2	97,60	11346	51,19	0,32

In terms of end-to-end delay, the Disjoint solution always performs quite better than the R_2 one, which means that the paths used are shorter. In what refers to the routing overhead, again the Disjoint solution performs much better.

If we observe the results concerning the dispersion achieved with both methods, we verify that the Disjoint solution presents dispersion values that more than double the ones from solution R_2. We also verify that the dispersion value almost does not vary with speed.

This analysis allows us to conclude that the results achieved justify the extra computational complexity required by the Disjoint solution, being the R_2 a possible solution for environments with few resources, though the performance suffers some degradation. From now on we will always use the Disjoint solution when performing traffic splitting.

Preventive route discovery

In order to improve the traffic splitting *Disjoint* strategy presented in previous section, we propose a mechanism to perform preventive route discovery processes. Its objective is to minimize the video gap effects on the video quality delivered to the user.

We also have to consider the possibility that, after completing a preventive route discovery cycle, no disjoint routes are found. In this case, we have to start another route discovery process to avoid video flow stall if the current route is lost. The rate at which we generate preventive route discovery processes must be evaluated in order to be useful and not to overload the network. By varying the preventive route discovery period among values: 0.5, 1, 2, 4, 8, 16 seconds and never (default) we calculated the routing overhead differences among the various inter-request values, see figure 4. Remind that when all routes to the destination are lost, a new route discovery process is started and the probability of producing a video gap is high.

As expected, the routing overhead is higher than the default solution in all cases. The lower the inter-request interval, the higher the routing overhead

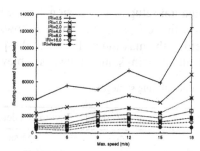

Figure 4.   Effect of inter-request interval on routing overhead

becomes. We consider that for values under 4 seconds the routing overhead becomes prohibitive. In terms of packets arrivals, around 99% of packets arrive for all solutions under test at all speeds. Considering the VA parameter, we achieve the best average results for an inter-request interval of 8 seconds, which also offers very good results in terms of packet arrivals and routing overhead.

6.   Overall evaluation

In this section we perform a global evaluation of the SR and Disjoint solutions. The simulation setup is configured with a 1000x1000 meters scenario and 80 nodes. As before, the mobility pattern is generated using the random waypoint mobility model and we consider only scenarios where node topology forms a connected graph. Only one video flow conforms the injected traffic with the same characteristics described in section 4. The Disjoint solution uses a multipath routing algorithm with a preventive route discovery mechanism set to minimum period of 8 seconds.

We can clearly see on the left side of figure 5 how DSR performs worse than the remaining modes for moderate/high speeds. The Disjoint mode is the best for all speeds, but the SR mode alone can provide very good enhancements.

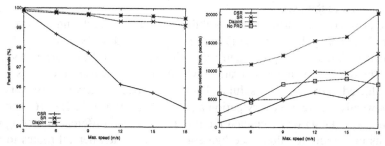

Figure 5.   Comparison in terms of video packet arrivals (left) and routing overhead (right)

In terms of routing overhead, we can see in figure 5 that the SR mode does not generate an excessive number of control packets. The Disjoint mode causes more routing overhead since this protocol is performing preventive route discoveries frequently. Also notice that the rate of growth between the three solutions is quite different. Comparing the routing overhead for the minimum and maximum speeds we can see that while DSR control packets have increased by a factor of 9, SR and Disjoint mode have increased by factors of 5 and 2 respectively. This shows the better adaptation and appropriateness of these two to high mobility scenarios. We also show the routing overhead achieved with the Disjoint mode when turning off the Preventive Route Discovery mechanism (No PRD in figure 5). We can see that the routing overhead is highly reduced when that mechanism is turned off, showing an overhead similar to the SR mode.

If we now focus on our main goal - reducing video streaming gaps - we find that there has been a gradual improvement, and that the SR mode plus the multipath Disjoint solution are able to significantly reduce the video gap occurrence. Table 3 shows the improvements in terms of global gap percentage and the VA parameter. The difference in terms of VA between SR and Disjoint modes is not greater due to the fact that both approaches avoid large video gaps, being large video gaps those that provoke significant differences in terms of VA metric. For video gaps of less than 3 frames, though, we find that the disjoint mode is very effective.

Table 3. VA parameter comparison

Protocol version	DSR	SR	Disjoint
Gap (%)	2,41	0,303	0,0776
Avg. VA (10^{-7})	32,3	2,14	0,619

Finally, relative to the benefit of including or not the preventive route discovery mechanism, we achieve a slight increase in routing overhead and a reduction of 60% in terms of VA and of 50% in gap percentage by turning it on. The main reason for this slight difference are the few situations when preventive routing saves us of video gaps.

7. Summary

We presented several enhancements to the DSR protocol in order to provide a better support to H.264 video stream delivery. The proposals focused on the route discovery process, the packet splitting strategy, and the preventive route discovery process.

We showed that video gaps are intimately related to route discovery procedures and that this problem can not be solved through conventional QoS mechanisms. We also proposed an alternative metric to PSNR, called video annoyance, in order to measure video gaps in a clear and straightforward manner. We extended DSR's route discovery mechanism to increase the average number of routes found per node. This extra information alone offers to each node extra possibilities when a route is lost, requiring on average less route discovery processes. We introduced a dynamic algorithm for maximizing the degree of disjointness of consecutive paths for a same stream, and evidenced the goodness of the algorithm against a more relaxed solution.

By comparing the standard DSR protocol against our proposals we showed that the enhancements lead to a very significative reduction of video gaps. We also verified that the routing overhead is maintained low, even when applying packet splitting and performing preventive route discovery, showing that the effectiveness of the presented strategy does not come at the cost of too many additional control packets.

References

(1999). International Standard for Information Technology - Part 11: Wireless Medium Access Control (MAC) and Physical Layer (PHY) Specifications, IEEE 802.11 WG, ref. no. ISO/IEC 8802-11:1999(E) IEEE Std. 802.11.

(2003). Draft ITU-T Rec. and Final Draft Int. Standard of Joint Video Specification (ITU-T Rec. H.264 | ISO/IEC 14496-10 AVC).

Nasipuri, A., Castaneda, R., and Das, S. R. (2001). Performance of multipath routing for on-demand protocols in mobile ad hoc networks. *ACM/Baltzer Mobile Networks and Applications (MONET) Journal, vol. 6*, pages 339–349.

Carlos T. Calafate, M. P. Malumbres, and P. Manzoni (2004a). A flexible and tunable route discovery mechanism for on-demand protocols. *12-th Euromicro Conf. on Parallel, Dist. and Network based Proc.*, La Coruña, Spain.

Carlos T. Calafate, M. P. Malumbres, and Pietro Manzoni (2004b). Performance of H.264 compressed video streams over 802.11b based MANETs. In *International Conference on Distributed Computing Systems Workshops (ICDCSW '04)*, Hachioji - Tokyo, Japan.

Johnson, David B and Maltz, David A (1996). Dynamic source routing in ad hoc wireless networks. In Imielinski and Korth, editors, *Mobile Computing*, volume 353. Kluwer Academic Publishers.

K. Fall and K. Varadhan (2000). ns notes and documents. The VINT Project. UC Berkeley, LBL, USC/ISI, and Xerox PARC.

Lee, S. and Gerla, M. (2001). Split multipath routing with maximally disjoint paths in ad hoc networks. IEEE ICC, pages 3201–3205.

L. Wang, Y. Shu, M. Dong, L. Zhang, and O.W.W. Yang (2001). Adaptive multipath source routing in ad hoc networks. ICC 2001. Page(s): 867-871

Marina, M. and Das, S. (2001). On demand multipath distance vector routing in ad hoc networks. In Proceedings of IEEE International Conference on Network Protocols (ICNP), pages 14–23.

Perkins, Charles E. and Royer, Elizabeth M. (1999). Ad hoc On-Demand Distance Vector Routing. In *Proceedings of the 2nd IEEE Workshop on Mobile Computing Systems and Applications, New Orleans, LA*, pages 90–100.

Wu, Jie (2002). An Extended Dynamic Source Routing Scheme in Ad Hoc Wireless Networks. 35th Annual Hawaii International Conference on System Sciences (HICSS'02)-Volume 9, Big Island, Hawaii.

A NEW SMOOTHING JITTER ALGORITHM
FOR VOICE OVER AD HOC NETWORKS

Mouna Benaissa and Vincent Lecuire

CRAN, CNRS UMR 7039, Henri Poincaré-Nancy I University
Campus Scientifique, BP 239, 54506 Vandoeuvre-les-Nancy Cedex, France
{mouna.benaissa,vincent.lecuire}@cran.uhp-nancy.fr

Abstract   Voice over IP applications require playout buffer at the receiver side to smooth network delay variations. Existing algorithms for dynamic playout adjustment used in Internet do not operate correctly in wireless ad hoc networks because they estimate end-to-end delay based on set of previous received audio packets. Mobility in ad hoc networks leads to topology changing and estimate based on past history is not appropriate. In this paper, we propose a new algorithm for playout delay adjustment based on *Route Request* AODV control messages to provide more accurate delay estimation. The performance evaluation shows that this algorithm outperforms existing playout delay adjustment algorithms. Performance criteria are the loss late percentage (reliability criterion), averaged playout delay (interactivity criterion) and playout delay variation (stability criterion).

Keywords: packet audio, playout delay, delay estimation, ad hoc network, AODV.

1.   Introduction

One of the challenges of transmitting real-time voice on packet networks is how to overcome the variable inter-packet delay -the jitter- encountered as packets move on the transmission path through the network. In order to compensate these variable delays, packets are buffered at the receiver side and their *playout* is delayed for a period of time. Thus, most of the packets will be received before their scheduled *playout* times [Moon et al., 1998][Clark et al., 1992]. The playout delay must take into account three constraints. The first one is the interactivity constraint which requires playout delay below a certain value considered to be quite acceptable in human conversation (less than 300ms but 100ms is recommended to obtain excellent interactivity [ITU-T, 2001]). Second, the reliability constraint which requires little packet loss rate (generally less than 5%). Third, the stability constraint requires no large playout delay variation (this constraint is effective when playout delay is adjusted dynamically).

Extensive research work has been done to adjust dynamically the playout delay according to delay variation through the duration of an audio session. Existing algorithms estimate the end-to-end delay using collected delays measures of the more recent received audio packets. These measures can involve one packet (autoregressive algorithms)[Ramjee et al., 1994][Kansal and Karandikar, 2001], L packets [Moon et al., 1998][Leon and Sreenan, 1999][Agrawal et al., 1998][Liang et al., 2001], M talkspurts [Pinto and Christensen, 1999][Ramos et al., 2003] or all received packets [Fujimoto et al., 2002]. Generally, the playout delay is computed per-packet but adjusted per-talkspurt. However, these algorithms do not operate correctly in wireless ad hoc networks: when the topology changes, end-to-end delay estimation based on past history (delays of audio packets which arrived by obsolete route) becomes inappropriate.

This paper highlights effect of network reconfigurations on audio traffic transfer in ad hoc networks and then presents a new algorithm for playout delay adjustment appropriate to such environnement. End-to-end delay estimation is based on an ad hoc routing event: The *Route request* control message (RREQ) of AODV routing protocol. Of course, we suppose that AODV is adopted for the deployment of voice applications over ad hoc networks. However, it is not a drawback since we show in our prior work given in [Benaissa et al., 2003] that it is more appropriate for such applications.

The paper is organized as follows: In the section II, we present ad hoc network characteristics which can have a particular effect on audio traffic transfer, compared to wireline networks. In section III, we show that RREQ-AODV message provides an accurate estimation for end-to-end delay. Then, we describe our new playout delay algorithm based on RREQ-AODV delay messages. Section IV provides performance evaluation and comparison results obtained by simulation using ns-2. The performance criteria are loss late percentage (reliability criterion), averaged playout delay (interactivity criterion) and playout delay variation (stability criterion). Section V concludes the paper.

2. Ad hoc reconfiguration phases: A typical disturbing event for VoIP

In ad hoc network, mobile nodes communicate with others using multi-hop wireless links. There is no stationary infrastructure such as base stations. Each node in the network also acts as a router to forward data packet to other nodes [Perkins, 2001]. There is two approaches for existing routing protocols in mobile ad hoc networks: the proactive approach such as OLSR [Jaquet et al., 2002] and the reactive approach such as AODV[Perkins, 2001]. The proactive approach consists in every node emitting hellos messages periodically in order to learn the network topology. Reactive protocols invoke a route determination procedure on demand only.

Packets audio streaming over mobile ad hoc networks distinguishes clearly communication phases and network reconfiguration phases in an audio session. During a reconfiguration phase, the audio stream is interrupted because of the delay caused by a routing protocol to establish a new route towards the destination. The receiver identifies this phase when a sudden interruption occurs on packets arriving, generally followed by series of packets arriving with high end-to-end delays. Indeed, packets waiting for the new route, arrive with long delays at the destination if they are not dropped in the network queues. These interruptions disturb the played out audio speech at the receiver and can be long, generally of some seconds [Benaissa et al., 2003]). After this event, the new route for the new communication phase can present different or similar network conditions (traffic load and number of hops) compared to the previous one. We say *strong*, a reconfiguration which leads to different network conditions. We say *light*, a reconfiguration which leads to similar network conditions. Prior work given in [Benaissa et al., 2003][Jaquet and Viennot, 2000] have shown that:

- OLSR causes more reconfiguration phases than AODV, thus, AODV is more adapted for packet audio applications.

- Delays can vary significantly after a reconfiguration phase: Playout delay adjustement is needed.

- Reconfiguration phase causes disruption on the audio speech: It is possible to benefit from this interruption to adjust playout delay whithout any additionnal disturb.

Based on these works, we propose a new playout delay adjustment algorithm for voice over ad hoc networks. This algorithm presents a new approach for playout delay estimation and considers new event leading to its adjustment, appropriate to ad hoc environment.

3. RREQ-AODV algorithm

Jitter control required by voice applications faces a typical problem of spontaneous changes which occur on ad hoc network topology. When such a change happens, the delay on the new topology must be correctly estimated to be able to adjust the playout delay in accordance with current network conditions. The existing strategies used to estimate future delay are not efficient in ad hoc networks because they are based on past delay measures. The proposed approach is different: it uses AODV routing information to predict network conditions. In the following section, we show that RREQ message generated by AODV during a route discovery process provides pertinent information about future audio packet delay.

3.1 Delay indication using RREQ-AODV messages

The AODV routing protocol generates control messages to establish a new route towards the destination. The source node initializes a route discovery process just before sending data. It is achieved using a RREQ message which is broadcasted across the network. When the destination node receives the first RREQ message, it carries back the route in a RREP message to the source and ignores the next received RREQ messages for this route. The route established is the network path built by the RREQ message. Thus, RREQ message and audio packets use the same path from the source to the destination. This path is known by the receiver because AODV source uses the ones built by the first RREQ message which reaches the destination. So, the end-to-end delay achieved by the RREQ message presents for the receiver a pertinent indicator about delay audio packets to be received through this route.

At the receiver side, this indicator is available and updated dynamically before receiving audio packets of each communication phase. A new RREQ message is received during a reconfiguration phase from a new route discovery cycle, since the source maintains at most one route per destination. This can happen in several cases:

- **The beginning of an audio session**: As any reactive routing protocol, AODV initiates a route discovery process to start an audio session. This involves the sending of the first RREQ message which provides to the receiver a delay indication from the beginning of the session.

■ **Mobility**: When a node moves from an active audio path, a new reconfiguration phase begins. The node which detects the link failure sends a route error (ERR) message to the source. If the audio source node still desires the route, it reinitiates route discovery process. So, the destination receives a new RREQ message which provides a new delay indication appropriated to the new topology.

■ **Long silence period**: AODV maintains a timer-based states in each node, about the usage of individual routing table entries. A routing table entry is expired if not used recently. Thus, the audio route expires during a silence period which is longer than a route expiry time. Due to this long silence period, the source needs to initiate route discovery process at the next talkspurt. This provides a new RREQ message to the audio destination and updates the delay indicator.

■ **High traffic load**: AODV maintains topology information via *HELLO* messages. If a node does not receive any *HELLO* message from its known neighbor, the link is considered broken. This can occur when network traffic load is high, even if there is no mobility. So, a new route discovery process starts and presents a new RREQ message to the receiver. This message provides a new delay indication appropriated to the network load conditions.

In our work, the receiver requires always to be notified about route changing by RREQ messages. It is obtained in the following way: Firstly, the procedure of RREQ message broadcasting is modified. In the initial procedure, the intermediate node receiving the RREQ message may send a RREP message if it has a route to the destination and stop broadcasting. Thus, the audio destination does not receive any *RREQ* message and can not detect this reconfiguration. To notify the receiver about this reconfiguration, this procedure is modified. The intermediate node having a route to the destination unicasts the received RREQ message to the final destination which reply a RREP message to the source. Secondly, a local repair procedure is not used. Using this procedure, AODV attempts to repair localy a failed link instead of informing the source and initiates a new route discovery process. To notify the destination about this reconfiguration, this procedure is not used in our work. Note that in 50 nodes networks, this approach does not have any significant performance advantage while it is recommended in larger networks to increase scalability [Lee et al., 2003].

3.2   RREQ-AODV algorithm description

In this section, we describe a new approach to adjust the playout delay in voice over ad hoc networks. This approach is based on a typical event of AODV routing protocol to estimate correctly the end-to-end delay even if network topology was changed: the RREQ control message arrival on the receiver side.

3.2.1   Playout delay estimation. The playout delay estimation is based on delay indication provided by RREQ messages. Let $Drreq_n$ be the end-to-end delay achieved by the RREQ message received during the nth reconfiguration phase. The playout delay estimation $\hat{d}_{n,k}$ for packets of the kth talkspurt belonging to the nth normal phase is computed as follows:

$$\hat{d}_{n,k} = Drreq_n + \beta_{n,k} \tag{1}$$

where $\beta_{n,k}$ is a safety factor, added to ensure that the estimated end-to-end delay is greater than the actual network delay. To get more accurate playout delay estimation, this factor is adjusted dynamically. This is discussed in the next section.

3.2.2   Playout delay adjustment.   The delay indication $Drreq_n$ is updated at every RREQ message arrival and used at the beginning of a new normal phase (adjusted during reconfiguration phase) or a new talkspurt (adjusted during silence period): If arriving packet i is the first packet of talkspurt k or a normal phase n, the playout delay is computed as given in equation 1. This delay is preserved for each subsequent packet j. The playout times p_i and p_j are computed as below:

$$p_i = t_i + \hat{d}_{n,k} \tag{2}$$

$$p_j = p_i + (t_j - t_i) \tag{3}$$

where t_i and t_j are (respectively) times at which packets i and j are generated at the sender.

3.3   Safety factor adjustment

We distinguishes two events for adaptation of the safety factor $\beta_{n,k}$: the beginning of a new normal communication phase (a RREQ message is received) and the beginning of a new talkspurt (a new talkspurt begins in the same normal phase while no new RREQ message is received). The adaptation of $\beta_{n,k}$ is performed in the following way for each case:

- **RREQ message is received:** The reception of a new RREQ message indicates that a new network topology is established. Then, the new delay indication $Drreq_n$ is updated in equation (1). To identify the type of the occurred reconfiguration (light or strong), the algorithm computes the difference δ between the current delay indication $Drreq_n$ and the previous one $Drreq_{n-1}$ as follows:

$$\delta = |Drreq_n - Drreq_{n-1}| \tag{4}$$

   The algorithm compares this difference to a certain threshold $threshold\_req$ and adjust $\beta_{n,k}$ accordingly:

   1 δ **is large enough** ($\delta > threshold\_req$): The result of this comparison indicates that the network conditions on the new topology have changed significantly; it was a strong reconfiguration. For the new normal phase, the delay estimation is based on the new delay indication $Drreq_n$ and $\beta_{n,k}$ is set to its primary value β_{min}:

$$if \; |Drreq_n - Drreq_{n-1}| > threshold\_req$$

$$then \ \beta_{n,k} \leftarrow \beta_{min}; \qquad\qquad (5)$$

2 δ **is small** ($\delta <= threshold\_req$): In this case, the algorithm considers that the network conditions are similar on the new topology; it was a light reconfiguration. The safety factor $\beta_{n,k}$ preserves its previous value $\beta_{n-1,k}$:

$$if \ |Drreq_n - Drreq_{n-1}| <= threshold\_req$$
$$then \ \beta_{n,k} \leftarrow \beta_{n-1,k}; \qquad\qquad (6)$$

■ **No *RREQ* message is received while a new talkspurt begins**: The absence of RREQ message indicates that no changes happen on the ad hoc network topology. In this case, the delay estimation for this talkspurt can be based on the recent delay past history. We propose to adapt $\beta_{n,k}$ as a function of loss percentage q_{k-1} achieved on the more recent talkpsurt ($k-1$):

$$\beta_{n,k} = f(\beta_{n,k-1}, q_{k-1}) \qquad\qquad (7)$$

To keep a certain stability of the estimated playout delay, the adjustment of $\beta_{n,k}$ in equation (7) is performed in a gradual way as follows:

1 **No loss observed on the previous talkpsurt** ($q_{k-1} = 0$): This means that the safety factor is large and it can be decreased to improve interactivity without degrading reliability. Then, $\beta_{n,k}$ is decreased by a factor r:

$$if \ q_{k-1} = 0 \ then \ \beta_{n,k} = (1-r)\beta_{n,k-1} \qquad\qquad (8)$$

2 **Loss percentage is less or equal than the user tolerable limit** q_{ref} ($q_{k-1} \leq q_{ref}\%$): This means that the algorithm gets a good reliability but there is no margin on the safety factor to improve interactivity. Then, $\beta_{n,k}$ preserves its previous value for the next talkspurt in order to maintain the same level of reliability and interactivity:

$$if \ q_{k-1} \leq q_{ref} \ then \ \beta_{n,k} = \beta_{n,k-1} \qquad\qquad (9)$$

Note that in our work, q_{ref} is set to 3%.

3 **Loss percentage exceeds the user tolerable limit** ($q_{k-1} > q_{ref}\%$): This means that the safety factor is small. It must be enlarged at the next talkspurt to increase reliability; $\beta_{n,k}$ is increased by a multiple of factor r as function of observed loss percentage q_{k-1}, as follows:

$$\begin{aligned}
if \ \ q_{ref.} < q_{k-1} < 10\% \quad & then \quad \beta_{n,k} = (1+2r)\beta_{n,k-1} \\
if \ \ 10 < q_{k-1} < 20\% \quad & then \quad \beta_{n,k} = (1+4r)\beta_{n,k-1} \\
if \ \ 20 < q_{k-1} < 30\% \quad & then \quad \beta_{n,k} = (1+6r)\beta_{n,k-1} \\
if \ \ q_{k-1} > 30\% \quad & then \quad \beta_{n,k} = 2\beta_{n,k-1} \qquad (10)
\end{aligned}$$

The equation 10 is defined as to avoid adjusting $\beta_{n,k}$ with strong value from talkspurt to another. Indeed, we aim to keep a certain stability of the playout delay. For larger value of r, the playout delay adjustment is larger.

To keep acceptable interactivity (end-to-end delay less than $300ms$ including delay required to collect audio samples), the adjustment of $\beta_{n,k}$ is bounded between $\beta_{min} = 40$ and $\beta_{max} = 200$. Parameters *threshold_req* and r are chosen in way they give the better tradeoff between interactivity, reliability and stability. *threshold_req* is set to 80 an r is set to 0.05. Figure 1 shows that the playout

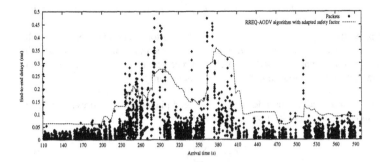

Figure 1. Playout delay using algorithm RREQ-AODV with adapted safety factor β.

delay follows suitably end-to-end delays and presents a good stability.

4. Performance evaluation

In this section, we evaluate and compare the RREQ-AODV algorithm performances to autoregressive based algorithms 1 and 4 reported in [Ramjee et al., 1994] (refered as mean delay algorithm and spike algorithme in this paper) and L packets statistics based algorithm reported in [Moon et al., 1998] (refered as *Moon* algorithm in this paper). The results shown in this section are evaluated on six audio traces obtained by simulation.

4.1 Performance metrics

To measure the obtained audio quality Q at the receiver when applying a playout delay algorithm, we take into account three criteria: Interactivity (averaged playout delay I), reliability (percentage of loss due to late arrivals F) and stability (averaged playout delay jitter S). The E-model predicts the subjective quality Q of a telephone call based on its characterizing transmission parameters. It combines impairment caused by these parameters into a single rating Q. According to the ITU-T recommendations, the rating value range of Q corresponds to a speech transmission category, as follows: Best for range of $[90, 100]$, High for range of

$[80, 90]$, Medium for range of $[70, 80]$, Low for range of $[60, 70]$ and Poor for range of $[0, 60]$. The rating Q is given by:

$$Q = Q_0 - E \qquad (11)$$

where Q_0 takes into account the effects of noise. The default value of Q_0 is 94.2. E combines impairment of different transmission parameters. In our work, E groups the impairment relative to interactivity $E(I)$, the impairment relative to reliability $E(F)$ and the impairment relative to stability $E(S)$. $Q(I, F, S)$: $\mathcal{R}^+ \times \mathcal{R}^+ \times \mathcal{R}^+ \longrightarrow [0, 100]$ is given by:

$$Q(I, F, S) = 94, 2 - E(I) - E(F) - E(S) \qquad (12)$$

Let p_i be the playout delay of packet i, N be the total sent audio packets, A be the total received audio packets, L be the total played out audio packets during the audio session. A packet i, sent at time t_i and received at time a_i, is played out if it arrives before its playout time tp_i, ie: $a_i \leq tp_i$ (where $tp_i = t_i + p_i$).

4.1.1   Interactivity metric.   The averaged playout delay I provides indication about the interactivity level. I is given by:

$$I = \frac{1}{L} \sum_{i=1}^{L} (p_i) \qquad (13)$$

In human conversation, end-to-end delay must not exceeds $110ms$ for a good interactivity but tolerates degraded audio quality for end-to-end delay between $110ms$ and $260ms$ (In our work 40ms are required to collect samples of one audio packets). When end-to-end delay exceeds $260ms$, audio quality is poor. Considering these bounds, $E(I)$ is given by [Boutremans and Le Boudec, 2003]:

$$E(I) = \begin{cases} 0.001\,I & for\ I \leq 110 \\ 18.89\,tanh(0.02\,(I - 185)) + 17.1 & for\ 110 < I \leq 260 \\ 0.01I + 32 & for\ I > 260 \end{cases}$$

$$(14)$$

4.1.2   Reliability metric.   The loss late percentage F indicates the reliability level. F is given by:

$$F = (\frac{A - L}{A})100 \qquad (15)$$

Independently of the codec in use, $E(F)$ is given by [Boutremans and Le Boudec, 2003]:

$$E(F) = 34.3ln(1 + 12.8F) \qquad (16)$$

According to equation 16, when the percentage of loss is less than 3%, the audio quality is good but tolerates degraded audio quality for loss percentage between 3% and 15%. When loss percentage exceeds 15%, audio quality is poor.

4.1.3   Stability metric.   The jitter S on the playout delay during the audio session provides indication about the stability of the playout delay. S is given by:

$$S = \frac{\sum_{i=2}^{L} |p_i - p_{i-1}|}{(L-1)} \tag{17}$$

In our work, $E(S)$ is given as follows:

$$E(S) = 2 \times S \tag{18}$$

As S increases, E(S) increases and playout delay is less stable.

4.2   Reference traces

We selected six reference traces which characterize different ad hoc network conditions (load traffic and mobility speed). These traces were obtained by simulation using ns-2. Our network model consisted of 50 nodes in a 1000 × 1000 meter flat, square area. The nodes moved according to the random way point mobility model. All nodes communicated with 802.11 based wavelan wireless radios, which have a bandwidth of 11Mbps and a propagation radius of 250m (See [Benaissa, 2004] for complete details on the simulation environment and methodology). We use AODV protocol for routing and PCM codec (Pulse Codec Modulation- See Recommendation G.711 in [ITU-T, 2001]) to generate audio traffic. For each trace, we get sending and receiving time of all audio packets and all RREQ messages transfered from the audio source to the audio destination. Principals characteristics of these traces are:

- Reference trace 1, 2 and 3: They present normal node mobility ($1m/s$ to $2m/s$) and normal load traffic conditions. We consider that network conditions on these three traces are favorable to VoIP applications.

- Reference trace 4: It presents high mobility ($8m/s$) and high load traffic. We consider that network conditions on this trace are difficult for VoIP applications. Such a trace is useful to study the behavior of playout delay algorithm in difficult situation.

- Reference trace 5: This trace presents light load traffic and high mobility (6m/s). End-to-end delays are very small during normal phases and very high after reconfiguration phases. When route reconfiguration occurs, packets waiting for new route arrive with very high delays because network queues are not loaded and thus these packets are not discarded. This trace is used to show particularly the effect of mobility on end-to-end delays packets.

- Reference trace 6: This trace presents normal load traffic conditions without any mobility. We consider that this trace is appropriate for mean delay algorithm, spike algorithm and *Moon* algorithm.

4.3   Performance comparison

In this section, we evaluate and compare RREQ-AODV algorithm to mean delay algorithm, spike algorithm and *Moon* algorithm on the six reference audio traces. We give a table 1, summarizing the obtained results: I, F, S, and $Q(I, F, S)$.

4.3.1   Reference trace 1, 2 and 3: Normal mobility with normal load conditions.   Results given in table 1 show that RREQ-AODV algorithm outperforms other algorithms on the three traces. Mean delay algorithm obtains good stability and fiability at the cost of degraded interactivity. Impairment relative to stability is very important with spike algorithm. *Moon* algorithm gives degraded reliability. Algorithm RREQ-AODV obtains the better tradeoff between interactivity, reliability and stability. When observing results in details from the traces, we remark that RREQ-AODV algorithm follows more suitably the delays tendency than the others.

4.3.2   Reference trace 4: High mobility with high load traffic.   The four algorithms lead to poor audio quality. Mean delay algorithm outperforms others when considering reliability ($F = 4.4\%$) but it obtains poor interactivity (I is 10 times higher than with RREQ-AODV algorithm). Indeed, RREQ-AODV algorithm provides good interactivity but at the cost of poor reliability ($F = 20.91\%$). Spike algorithm and *Moon* algorithm give degraded stability ($S > 10ms$) and reliability. When observing results in more details, we remark that most of lost packets are those arriving with great delays, which are not useful to the VoIP application. These packets are played out when using mean delay algorithm which leads to excessive playout delay. In difficult network conditions, playout delay adjustment algorithms cannot give a good tradeoff between interactivity and reliability. In this case, mechanisms for the quality of service must be deployed in the network.

Trace	Algorithm	I	F	S	Q(I,F,S)	Quality
1	Mean delay	241.46	6.01	0.32	41.57	poor
	Spike	194.29	6.72	3.63	45.08	poor
	Moon	161.46	8.75	1.10	58.52	poor
	RREQ-AODV	131.04	5.33	0.32	73.91	medium
2	Mean delay	143.27	4.40	0.21	74.23	medium
	Spike	72.66	6.74	1.45	69.89	low
	Moon	92.45	8.25	0.93	68.46	low
	RREQ-AODV	86.31	3.45	0.24	81.09	good
3	Mean delay	181.35	5.30	0.21	60.29	low
	Spike	117.08	8.73	1.88	64.16	low
	Moon	115.41	10.79	0.86	63.16	low
	RREQ-AODV	124.66	6.43	0.34	71.65	medium
4	Mean delay	1467.28	4.40	2.43	27.33	poor
	Spike	361.42	13.50	11.21	1.74	poor
	Moon	832.55	11.75	10.79	11.60	poor
	RREQ-AODV	148.80	20.91	1.83	40.48	poor
5	Mean delay	958.15	5.35	2.74	29.24	poor
	Spike	208.63	9.53	9.01	23.41	poor
	Moon	568.03	10.06	9.03	19.10	poor
	RREQ-AODV	84.91	6.74	0.28	72.22	medium
6	Mean delay	44.32	3.8	0.04	80.48	good
	Spike	37.55	6.70	0.66	71.60	medium
	Moon	36.17	6.17	0.18	73.99	medium
	RREQ-AODV	61.91	1.3	0.19	88.53	good

Table 1.   Result obtained on the six audio traces

4.3.3 Reference trace 5: high mobility with light load traffic.

RREQ-AODV algorithm provides medium audio quality and then outperforms other algorithms which lead to poor audio quality. Mean delay algorithm, spike algorithm and *Moon* algorithm do not give good stability and interactivity. This is due to the fact that these algorithms consider long delays caused by reconfiguration phases in their estimation, while such delays do not give an appropriate indication about future delays. However, RREQ-AODV algorithm leads to more stable playout delay ($S = 0.28$) and excellent interactivity at the cost of degraded reliability. A more careful analysis of the results reveals that most part of the lost packets are those arriving with great delays. Such packets are considered lost by the audio application. Thus, the algorithm RREQ-AODV does not increase playout delay if there is no additional advantage. These results confirm that algorithms which are based on delay past history are not appropriate to adjust playout delay in the presence of mobility even with light load traffic. In these conditions, algorithm RREQ-AODV reacts correctly.

4.3.4 Reference trace 6: no mobility with normal load conditions.

Results show that all algorithms provide excellent interactivity and stability but lead to different reliability levels. When observing results in more details, we remark that mean delay algorithm and spike algorithm underestimate playout delay and thus lose packets arriving with acceptable delays. As algorithm RREQ-AODV considers a minimum bound for the safety factor ($\beta_{n,k} > 40$), it looses less packets than the others.

5. Conclusion

In this paper, we have proposed a new playout delay algorithm specially designed for voice over ad hoc networks. Its first strength is in the way that it estimates the end-to-end delay in the presence of mobility which leads to route reconfiguration. The algorithm uses RREQ-AODV message delay as a delay indicator. This is appropriate because the receiver is sure that the audio packet will go through the same path. Its second strength is the adaptation strategy which gives the same importance to interactivity, reliability and stability constraints. The performance evaluation results show that our algorithm outperforms existing algorithms in all cases when considering simultaneously: the interactivity, the reliability and the stability criteria, as well as when considering only interactivity and reliability criteria. A drawback of our solution is that the methodology is tied to AODV. However, a general methodology for reactive protocols can perhaps be derived from the proposal. In conclusion, our algorithm will contribute to improve the quality of voice application running on ad hoc networks. Of course, other mecanisms, such as FEC and network differenciated services, should be also used for supply enough QoS for user of voice applications.

6. References

[Agrawal et al., 1998] Agrawal, P., Chen, J., and Sreenan, C. (1998). "use of statistical methods to reduce delays for media playback buffering". In *International Conference on Multimedia Computing and Systems*, pages 259–263.

[Benaissa, 2004] Benaissa, M. (2004). *Dynamic playout delay adjustment for voice over ad hoc network*. PhD thesis, Henri Poincaré - Nancy I University.

[Benaissa et al., 2003] Benaissa, M., Lecuire, V., Lepage, F., and Schaff, A. (2003). "analysing end-to-end delay and loss in mobile ad hoc networks for interactive audio applications". *Workshop on Mobile Ad Hoc Networking and Computing MADNET'2003*, pages 27–33.

[Boutremans and Le Boudec, 2003] Boutremans, C. and Le Boudec, J. Y. (2003). "Adaptive joint playout buffer and fec adjustment for internet telephony". In *Proceedings of IEEE INFO-COM'2003*, San-Francisco, CA.

[Clark et al., 1992] Clark, D., Shenker, S., and Zhang, L. (1992). "Supporting real-time applications in an integrated services packet network architecture and mechanism". *SIGCOMM'92*, pages 14–26.

[Fujimoto et al., 2002] Fujimoto, K., Ata, S., and Murata, M. (2002). "Adaptive playout buffer algorithm for enhancing perceived quality of streaming applications". In *IEEE Globecom*.

[ITU-T, 2001] ITU-T (2001). "List of itu-t recommendations". http://www.itu.int/publications/itu-t/itutrec.htm.

[Jaquet et al., 2002] Jaquet, P., Muhletaler, P., Quayyum, A., Laouiti, A., Clausen, T., Viennot, L., and Minet, P. (2002). "Optimized link state routing protocol". *Internet Draft draft-ietf-manet-olsr-06.tx, Internet Engineering Task Force*.

[Jaquet and Viennot, 2000] Jaquet, P. and Viennot, L. (2000). "Overhead in mobile ad hoc network protocols". *Technical Report, INRIA*.

[Kansal and Karandikar, 2001] Kansal, A. and Karandikar, A. (2001). "Adaptive delay estimation for low jitter audio over internet". *in proccedings of IEEE Globcom*, pages 17–18. San Antonio, USA.

[Lee et al., 2003] Lee, S.-J., Royer, E. M., and Perkins, C. E. (2003). "Scalability study of the ad hoc on-demand distance vector routing protocol". *ACM/Wiley International Journal of Network Management*, 13(2):97–114.

[Leon and Sreenan, 1999] Leon, P. D. and Sreenan, C. (1999). "An adaptive predictor for media playout buffering". In *Proc. of IEEE International Conference on Acoustics, Speech, and Signal Processing (ICASSP)*, pages 3097–3100.

[Liang et al., 2001] Liang, Y., Farber, N., and Girod, B. (2001). "Adaptive playout scheduling and loss concealment for voice communications over IP networks". *IEEE Transactions on Multimedia*.

[Moon et al., 1998] Moon, S., Kurose, J., and Towsley, D. (1998). "Packet audio playout delay adjustment: Performance bounds and algorithms". *ACM/Springer Multimedia Systems*, 6:17–28.

[Perkins, 2001] Perkins, C. (2001). "Ad hoc networking". Addison Wesley Longman.

[Pinto and Christensen, 1999] Pinto, J. and Christensen, K. (1999). "An algorithm for playout of packet voice based on adaptive adjustment of talkspurt silence periods". In *Proceedings of the IEEE 24th Conference on Local Computer Networks*, pages 224–231.

[Ramjee et al., 1994] Ramjee, R., Kurose, J., Towsley, D., and Schulzrinne, H. (1994). "Adaptive playout mechanisms for packetized audio applications in wide-area networks". In *Proceedings of the IEEE Infocom*, pages 680–688.

[Ramos et al., 2003] Ramos, V., Barakat, C., and Altman, E. (2003). "A moving average predictor for playout delay control in voip". Technical report, Nice-Sophia Antipolis University, Mistral and Planète research team, INRIA.

ON THE COMPLEXITY OF RADIO RESOURCES ALLOCATION IN WCDMA SYSTEMS

Emmanuelle Vivier[1], Michel Terré[2], Bernard Fino[2]
[1]*ISEP, 21 rue d'Assas, 75006 Paris*   [2]*CNAM, 292 rue Saint Martin, 75003 Paris*

Abstract-   Recent attention to resource allocation algorithms for multiservice CDMA networks has focused on algorithms optimizing the aggregate throughputs (sum of all individual throughputs) on the uplink and on the downlink. Unfortunately, for a given set of real time (RT) and non real-time (NRT) communications services, such optimal algorithms involve non-integer spreading factors that do not belong to a finite set of spreading length as used in 3G systems. In this paper, we propose four algorithms for power and spreading allocation to RT and NRT services implementable in a real CDMA network like UMTS in the Frequency Division Duplexing (FDD) mode. On the downlink, two algorithms are presented; the first one maximises the aggregate downlink NRT throughput whereas the second one maximises the number of simultaneously transmitted NRT services. On the uplink, an algorithm that maximises the aggregate uplink NRT throughput and a suboptimal one, more easy to implement, are presented. Thanks to power control, both algorithms allow more simultaneous transmitting terminals than the optimal one. In both directions, the resulting aggregate NRT throughputs are very close to the ones obtained by the optimal algorithms. The small difference is the price paid for obtaining truly assignable spreading factors.

1   INTRODUCTION AND SYSTEM MODEL

Wireless multimedia services in 3G networks are characterized by different quality of service requirements. The radio resource management problem in CDMA networks[7] is closely connected to the necessity of power control to maximize the number of terminals supported by such systems, hence for increasing cellular capacity. Unlike TDMA systems, radio resources are not countable but could be seen as different rates allocated to active services. The maximum individual rate for each transmitter is theoretically bounded by the use of one OVSF (Orthogonal Variable Spreading Factor) tree for spreading data. More, the use of such spreading sequences makes data rates belong to a finite set of values. In this context, two main QoS classes, related to (RT) and (NRT) services are considered in a given cell. Resources allocation process aims at detemining a set of OVSF codes that ensures to each communication a correct transmission. Therefore, the process has to be compliant with some constraints:

1) the allocation process occurs very periodically, every 10 ms in the UMTS FDD Terrestrial Radio Accesss Network (UTRAN FDD)[11].
2) priority is given to RT communications[1,2].
3) uplink NRT services use the leftover capacity[3] and downlink NRT services use the remaining transmission power of the base station.
4) each signal has to maintain a minimum target signal to noise-plus-interference ratio in order to be correctly demodulated. This threshold is slightly higher than the minimum operating point in order to take into consideration the random variations of the interference level[3,5].

The following notations are used:

I_{inter}	Intercell interferences
I_{th}	Thermal noise in a 3,84 MHz-large band (I_{th}=-108 dBm)
I	$I = I_{inter} + I_{th}$
P_{max}	Maximum transmission power of the base station
p_{max}	Maximum transmission power of the RT and NRT terminals
M	Number of NRT services uniformly distributed in the cell
Q	Number of RT services uniformly distributed in the cell
p_i, p'_i	Transmission power allocated · for the transmission of the ith NRT, RT service (downlink) · to the terminal transmitting the ith NRT, RT service (uplink)
g_i, g'_i	Channel gain between the terminal transmitting the ith NRT, RT service and the base station, and vice-versa
N_i	Spreading factor for the ith NRT service
N_{RT}	Spreading factor for RT communications
α $0 \leq \alpha \leq 1$	normalized cross-correlation between the spreading codes at the receiver level
Γ_{NRT}	Minimum signal to noise-plus-interferences ratio to be reached for NRT services
$\Gamma_{NRT}(i)$	signal to noise-plus-interferences ratio for the ith NRT service
$\Gamma_{RT}(i)$	signal to noise-plus-interferences ratio for the ith RT service
Γ_{RT}	Minimum signal to noise-plus-interferences ratio to be reached for RT services

It must be noticed that Γ_{NRT} and Γ_{RT} are the same for all the concerned terminals and that, for simplicity reasons, N_{RT} is the same for all RT communications, as well as α for all cross-correlations.

NRT terminals are sorted in decreasing order of g_i: the transmission channel quality is a decreasing function of i, $1 \leq i \leq M$.

The algorithms optimizing the aggregate throughputs on the downlink and on the uplink[3] involve non-integer spreading factors that do not belong to a finite set of spreading length as used in 3G systems. They are described respectively in section 2 and 3. In addition, in the context of an UMTS FDD network, section 2 describes the proposed algorithms for the downlink and is concluded by the comparison of the performances of those algorithms. In the same context, section 3 presents two adaptations of the theoretical optimal algorithm on the uplink: an optimal one and a suboptimal one, and compares both performances. Finally, section 4 presents our conclusions.

2 DOWNLINK

2.1 Algorithms

On the downlink, Γ_{NRT} and Γ_{RT} can be expressed as follows:

$$\Gamma_{NRT}(i) = N_i p_i g_i \left[I + \alpha \left(\sum_{k=1}^{Q} p'_k + \sum_{\substack{k=1 \\ k \neq i}}^{M} p_k \right) g_i \right]^{-1} \geq \Gamma_{NRT} \qquad (1)$$

and

$$\Gamma_{RT}(i) = N_{RT} p'_i g'_i \left[I + \alpha \left(\sum_{k=1}^{M} p_k + \sum_{\substack{k=1 \\ k \neq i}}^{Q} p'_k \right) g'_i \right]^{-1} \geq \Gamma_{RT} \qquad (2)$$

First, the amount of the transmission power dedicated to RT communications must be determined. For this purpose, interferences generated for NRT services must be estimated. In the worst case,

$$P_{max} = \sum_{k=1}^{M} p_k + \sum_{k=1}^{Q} p'_k \qquad (3)$$

Transmission power allocated for the transmission of the ith RT service is therefore straightforward to reach $\Gamma_{RT}(i) = \Gamma_{RT}$:

$$p'_i = \Gamma_{RT} (I + \alpha P_{max} g'_i) [(N_{RT} + \alpha \Gamma_{RT}) g'_i]^{-1} \qquad (4)$$

Then, the remaining transmission power of the base station can be allocated to NRT services. Several allocation policies are conceivable. The aggregate downlink rate is: $\Omega_{NRT}^{\downarrow} = \sum_{i=1}^{M} 7,68 / N_i$, in Mbits/s. Actually, a constant chip rate (including the radio supervision) of 5120 chips per 10/15 ms is performed.

The optimal allocation, maximising $\Omega^{\downarrow}_{NRT}$ consists in allocating all the remaining transmission power for the NRT service benefiting from the highest channel quality: $p_1 = P_{max} - \sum_{i=1}^{Q} p'_i = P_{max} - P_{RT}$

It leads to:

$$N_1 = \Gamma_{NRT} \left(I + \alpha\, g_1 P_{RT} \right) \left[(P_{max} - P_{RT}) g_1 \right]^{-1} \tag{5}$$

and $\Omega^{\downarrow*}_{NRT} = 7,68 / N_1$.

Unfortunately, N_1 has no reason to be in the set of values SF↓={4, 8, 16, 32, 64, 128, 256, 512} that have been normalized for UTRAN downlink[10]. Consequently, $\Omega^{\downarrow*}_{NRT}$ is a theorical upper bound for $\Omega^{\downarrow}_{NRT}$.

In the following, we propose two algorithms. Under the constraint of spreading factors belonging to SF↓, and always considering NRT services in the decreasing order of their channel gain,

1) the first algorithm, named Downlink Discrete Spreading Factor Up ('DownlinkDSF-U'), maximises the aggregate downlink rate by allocating to the considered NRT service the lowest spreading factors that leads to a 'feasible' solution. Once a speading factor is allocated, it is not modified any more even when considering a following NRT service.

2) the second algorithm, named Downlink Discrete Spreading Factor Down ('DownlinkDSF-D'), maximises the number of simultaneous transmitted NRT services by allocating the highest spreading factor of SF↓ (i.e 512) to NRT services as long as it leads to a 'feasible' solution. Then, the number of simultaneous transmitted NRT services being fixed, it rises, while it is feasible, the individual rate of each service one step by one step.

The 'feasibility' of a solution is now defined: when $N_1, N_2, \ldots N_m$ are known, the transmission power allocated for the transmission of the ith NRT service, $1 \le i \le m \le M$ is determined as follows:

From (1) we obtain, to reach Γ_{NRT}:

$$p_i = \Gamma_{NRT} \left(I + \alpha\, g_i\, P_T \right) \left[g_i \left(N_i + \alpha\, \Gamma_{NRT} \right) \right]^{-1} \tag{6}$$

where $P_T = \sum_{k=1}^{Q} p'_k + \sum_{k=1}^{m} p_k = P_{RT} + \sum_{k=1}^{m} p_k \le P_{max}$

Therefore:

$$P_T - P_{RT} = \Gamma_{NRT} \left(I \sum_{i=1}^{m} \left[g_i (N_i + \alpha\, \Gamma_{NRT}) \right]^{-1} + \alpha\, P_T \sum_{i=1}^{m} \left[N_i + \alpha\, \Gamma_{NRT} \right]^{-1} \right)$$

and

$$P_T = \frac{P_{RT} + \Gamma_{NRT} I \sum_{i=1}^{m} [g_i (N_i + \alpha \ \Gamma_{NRT})]^{-1}}{1 - \alpha \ \Gamma_{NRT} \sum_{i=1}^{m} [N_i + \alpha \ \Gamma_{NRT}]^{-1}}$$ (7)

If $\alpha \Gamma_{NRT} \sum_{i=1}^{m} [N_i + \alpha \ \Gamma_{NRT}]^{-1} < 1$ and $P_T \leq P_{max}$, the solution is 'feasible'
and p_i is obtained with (6) and (7).

2.2 Performances

RT and NRT terminals are uniformly distributed in the cell for the distance from the base station from 325 m to 1.2 km. In order to determine the channel gains, we chose Okumura-Hata propagation model in an urban area with f=2 GHz, $h_{Base station}$=40 m and $h_{terminal}$=1.5 m [6]. Let P_{max}=10 W and I_{inter}=-63 dBm (equivalent to 6 base stations situated 2 km far away from the considered base statio and transmitting at P_{max}). Γ_{RT} and Γ_{NRT} are set to 7,4 dB. Q is set to 50 and M varies from 1 to 500. Finally, N_{RT}=256 and α =0,5.

Figure 1 illustrates the variations of $\Omega_{NRT}^{\downarrow *}$, $\Omega_{NRT}^{\downarrow}$ obtained with 'DownlinkDSF-U' and $\Omega_{NRT}^{\downarrow}$ obtained with 'DownlinkDSF-D'. Figure 2 gives the number of simultaneously transmitted NRT services with 'DownlinkDSF-U' and 'DownlinkDSF-D' as a function of the total number of active downlink NRT services in the cell (M). It is recalled that with the theoretical optimal algorithm, only one NRT service is served.

It appears that when 'DownlinkDSF-D' is applied, the number of simultaneous transmitted NRT service is exactly M when M is low (typically lower than 25). All NRT services being transmitted, $\Omega_{NRT}^{\downarrow}$ first increases and then fluctuates, depending on the random distribution of the terminals. Then, as the base station uses all its power to reach more and more terminals benefiting from worse and worse conditions of propagation, it can not transmit information to all NRT services and the individual rates remain minimum. Therefore, the aggregate throughput is nearly proportional to the number of simultaneously served NRT services and never exceeds 550 kbits/s.

On the opposite, the 'DownlinkDSF-U' never simultaneously transmit information to more than 4 NRT services, whose spreading factor is at least 32. More, the probability of having terminals benefiting from higher conditions of propagation increases with M increasing. Hence, $\Omega_{NRT}^{\downarrow}$ and

$\Omega_{NRT}^{\downarrow*}$ are increasing functions. Finally, $\Omega_{NRT}^{\downarrow}$ varies from 340 to 640 kbits/s, i.e. from 73% to 90% of $\Omega_{NRT}^{\downarrow*}$.

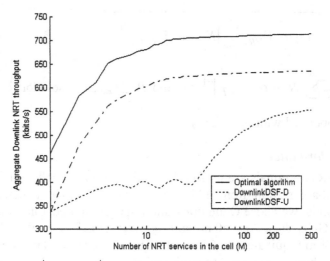

Figure 1: $\Omega_{NRT}^{\downarrow*}$ and $\Omega_{NRT}^{\downarrow}$ obtained with 'DownlinkDSF-U' and 'DownlinkDSF-D'

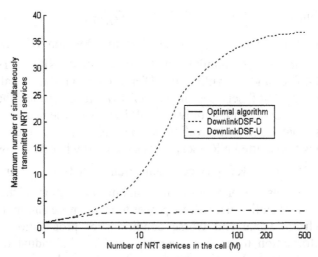

Figure 2: Number of simultaneously transmitted NRT services with 'DownlinkDSF-U' and 'DownlinkDSF-D', function of the total number of active NRT services in the cell (*M*)

The following section focuses on the UTRAN uplink.

3 UPLINK

3.1 Algorithms

On the uplink, Γ_{NRT} and Γ_{RT} can be expressed as follows:

$$\Gamma_{NRT}(i) = N_i p_i g_i \left[I + \alpha \left(\sum_{k=1}^{Q} p'_k g'_k + \sum_{\substack{k=1 \\ k \neq i}}^{M} p_k g_k \right) \right]^{-1} \geq \Gamma_{NRT} \qquad (8)$$

and

$$\Gamma_{RT}(i) = N_{RT} p'_i g'_i \left[I + \alpha \left(\sum_{k=1}^{M} p_k g_k + \sum_{\substack{k=1 \\ k \neq i}}^{Q} p'_k g'_k \right) \right]^{-1} \geq \Gamma_{RT} \qquad (9)$$

As on the downlink, RT communications are still served first. Hence a maximum acceptable total power received by the base station from all NRT services $P_R^{NRT\,max}$ is determined. This threshold represents the maximum value that ensures RT communications not to be blocked by NRT services. Therefore:

$$\sum_{i=1}^{M} p_i g_i \leq P_R^{NRT\,max} \qquad (10)$$

In the same way, P_R^{RT} is the total power received by the base station from all RT services: $P_R^{RT} = \sum_{i=1}^{Q} p'_i g'_i$.

Consequently, in order to reach exactly Γ_{RT}, we obtain from (9) :

$$p'_i g'_i = \Gamma_{RT} \left(I + \alpha \left(P_R^{NRT\,max} + P_R^{RT} \right) \right) [N_{RT} + \alpha \Gamma_{RT}]^{-1}$$

and:

$$P_R^{RT} = Q \, \Gamma_{RT} \left(I + \alpha \left(P_R^{NRT\,max} + P_R^{RT} \right) \right) [N_{RT} + \alpha \, \Gamma_{RT}]^{-1} \qquad (11)$$

Therefore:

$$Q = \left\lfloor P_R^{RT} \left(N_{RT} + \alpha \, \Gamma_{RT} \right) \left[\Gamma_{RT} \left(I + \alpha \left(P_R^{NRT\,max} + P_R^{RT} \right) \right) \right]^{-1} \right\rfloor \qquad (12)$$

and from (11) Q determines the real value of P_R^{RT}. Finally:

$$p'_i = \Gamma_{RT}\left(I + \alpha\left(P_R^{NRT\,max} + P_R^{RT}\right)\right)\left[\left(N_{RT} + \alpha\,\Gamma_{RT}\right)g'_i\right]^{-1}$$

Once transmission powers for RT communications are determined such that $0 \leq p'_i \leq p_{max}$, $P_R^{NRT\,max}$ can be shared between NRT services.

Algorithms maximising the aggregate NRT uplink rate $\Omega_{NRT}^{\uparrow*}$ consider as on the downlink the NRT services in the decreasing order of their channel gain[3,4]. While $\Omega_{NRT}^{\uparrow}(m) = \sum_{i=1}^{m} 3{,}84/N_i$ (in Mbits/s, excluding the radio supervision) increases, they set p_i to p_{max}, except for the last NRT transmitting terminal of the cell because of (10). Once again, this process leads to non-integer numbers for spreading factors: as in the previous section, $\Omega_{NRT}^{\uparrow*}$ is the theoretical upper bound for $\Omega_{NRT}^{\uparrow}$. Actually, for the UTRAN, uplink spreading factors values must belong to $SF_\uparrow = \{4, 8, 16, 32, 64, 128, 256\}$ [10].

Therefore, in this paper, we propose an algorithm named Uplink Discrete Spreading Factor 'UplinkDSF' that ensures compatibility with UTRAN's spreading factors requirements. A perfect power control is considered. For a set of spreading factors N_1, N_2, ...N_m, $1 \leq m \leq M$, the corresponding transmission powers are determined as follows:

From (8) we obtain, to reach Γ_{NRT}:

$$p_i g_i = \left(\Gamma_{NRT}\left(I + \alpha\,P_R^{RT} + \alpha\,P_R^{NRT}\right)\right)\left[N_i + \alpha\,\Gamma_{NRT}\right]^{-1} \qquad (13)$$

where $P_R^{NRT} = \sum_{i=1}^{m} p_i g_i$. Therefore,

$$P_R^{NRT} = \frac{\Gamma_{NRT}\left(I + \alpha\,P_R^{RT}\right)\sum_{i=1}^{m}\left[N_i + \alpha\,\Gamma_{NRT}\right]^{-1}}{1 - \alpha\,\Gamma_{NRT}\sum_{i=1}^{m}\left[N_i + \alpha\,\Gamma_{NRT}\right]^{-1}} \qquad (14)$$

If $\alpha\,\Gamma_{NRT}\sum_{i=1}^{m}\left[N_i + \alpha\,\Gamma_{NRT}\right]^{-1} < 1$ and $P_R^{NRT} \leq P_R^{NRT\,max}$, p_i is obtained with (13) and (14).

Lastly, if $p_i \leq p_{max}$, the solution is feasible. Otherwise, the spreading factor of the lowest path gain of the terminals that do not check $0 \leq p_i \leq p_{max}$ is increased by one step. In this way, a new set of spreading factors is considered at the input of the algorithm. If the solution is feasible, a new NRT service is considered. Otherwise, the cell has reached its capacity.

3.2   Performances

The same uniform distribution of RT and NRT terminals as for the downlink is considered, as well as the determination of the channel gains. Let I_{inter}=-92.3 dBm be the received power when Q=50 RT terminals transmit at p_{max}=23 dBm[9] with an attenuation of 132 dB. Finally, $P_R^{NRT\ max}$ =1,76 pW is equivalent to $3\times I$, Γ_{RT} and Γ_{NRT} are set to 7,4 dB and α=0.5. M varies from 10 to 250. Figure 3 gives the number of simultaneously transmitting uplink NRT services as a function of the number of active uplink services in the cell (M). When M is low, this number is higher with the optimal algorithm because the spreading factors values are not bounded by 256. $P_R^{NRT\ max}$ is reached quite soon with the optimal algorithm because all transmitting terminals transmit at p_{max}. Once $P_R^{NRT\ max}$ is reached, the number of simultaneously transmitting uplink NRT services decreases. Actually, some terminals, well placed in the cell, can transmit at high rate and therefore generate quite a lot of received interference whereas others can not transmit without exceeding p_{max} with a spreading factor set to 256 (UplinkDSF) or without decreasing $\Omega_{NRT}^{\uparrow*}$. Finally, Figure 4 illustrates the variations of $\Omega_{NRT}^{\uparrow*}$ and $\Omega_{NRT}^{\uparrow}$: $\Omega_{NRT}^{\uparrow}$ varies from 69% to 95% of $\Omega_{NRT}^{\uparrow*}$.

This UplinkDSF is easy to implement and gives very satisfactory results. However, it is not optimal. Consequently, the results of an optimal algorithm are also displayed. Of course, this 'Optimized UplinkDSF' algorithm ensures spreading factor values in SF↑. For each expected threshold of interferences lower than $P_R^{NRT\ max}$ and for all NRT service (still in the decreasing order of their channel gains), it identifies individually all the spreading factors generating a level of interferences at the receiver side equal or lower than the expected threshold. Once the truly generated interferences are aggregated and checked lower than the expected threshold, it identifies the generated aggregated level of interferences equal or lower than $P_R^{NRT\ max}$ that maximises the aggregate throughput. This approach is optimal and its complexity is closely connected to the number of analysed expected thresholds. For 25% of our results, a throughput about 3.84/256 Mbits/s higher than with the 'UplinkDSF' is obtained. It leads to a mean increase of 4 kbits/s and corresponds to an 1% gain of throughput.

The number of simultaneously transmitting uplink NRT services is approximately the same for both algorithms. Nevertheless, it can be noticed that 'UplinkDSF' allocates power of transmission to new NRT services as long as the resulting solution is feasible whereas 'Optimized UplinkDSF' stops at the maximum throughput. In conclusion, the 'UplinkDSF' algorithm is quasi-optimal in terms of throughput (it reaches 99% of the optimal one) and, when they are numerous in the cell, ensures more fairness among NRT services.

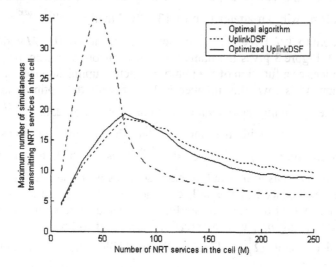

Figure 3: Number of simultaneously transmitting uplink NRT services, function of the number of active uplink services in the cell (*M*)

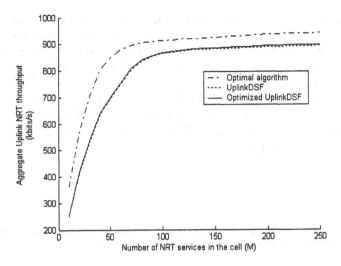

Figure 4: Variations of $\Omega_{NRT}^{\uparrow *}$ and $\Omega_{NRT}^{\uparrow}$

4   CONCLUDING REMARKS

The radio resources allocation that maximises the aggregate NRT throughput of a CDMA network should gives on the downlink all the base station's power for the transmission of only one NRT service: the one benefiting from the best propagation conditions in the cell. On the uplink of such a system, while the aggregate throughput increases and the received interferences do not exceed a maximum threshold, it should allow NRT terminals that benefit from the best conditions of propagation to transmit at their peak power. But both algorithms lead to the determination of non integer values for spreading factors and therefore represent theoretical upper bounds.

In this paper, we proposed four algorithms for power and spreading allocation to RT and NRT services implementable in a real CDMA network like UMTS FDD. As for the determination of the theoretical upper bound of the aggregate rate, NRT users benefiting from the best conditions of propagation are the first served once RT communications are satisfied.

On the downlink, two algorithms were presented, allocating spreading factors in the set {4, 8, 16, 32 64, 128, 256, 512}; 'DownlinkDSF-U' maximises the aggregate downlink NRT throughput whereas 'DownlinkDSF-D' maximises the number of simultaneously transmitted NRT services.

On the uplink, the proposed algorithm 'UplinkDSF' allocates spreading factors in the set {4, 8, 16, 32 64, 128, 256} and gives a very interesting aggregate uplink NRT throughput. The optimal algorithm: 'Optimized UplinkDSF', whose complexity is more difficult to evaluate, leads to a hardly higher gain of the aggregate throughput. Thanks to power control, both algorithms allow more simultaneous uplink transmitting terminals than the theoretical optimal one.

With 'DownlinkDSF-U', 'UplinkDSF' and 'Optimized UplinkDSF', the resulting aggregate NRT throughputs are very close to the ones obtained by the optimal algorithms. The small difference is the price paid for obtaining truly assignable spreading factors.

REFERENCES

1. S. Choi and P. G. Shin, "An uplink CDMA system architecture with diverse QoS guarantees for heterogeneous traffic," IEEE Transactions on Networking, vol. 7, no. 5, oct. 1999.
2. O. Gürbüz, H. Owen, "Dynamic resource scheduling schemes for W-CDMA systems," IEEE Commun. Mag., Oct. 2000.
3. S. J. Oh, D. Zhang, and P. M. Wasserman, "Optimal Resource Allocation in Multiservice CDMA Networks," IEEE Trans. Wireless Commun., vol. 2, July 2003, pp. 811-821.
4. V. Rodriguez and D. J. Goodman, "Power and data rate assignment for maximal weighted throughput in 3G CDMA," in Proc. IEEE WCNC'03, pp. 525-531, vol.1.
5. T. Shu, Z. Niu, "Uplink capacity optimization by power allocation for multimedia CDMA networks with imperfect power control," IEEE J. Select Areas Commun., vol. 21, no. 10, pp. 1585-1594, Dec. 2003.
6. X. Lagrange, P. Godlewski, S. Tabbane, "Réseaux GSM-DCS" Hermès, 4ème édition, pp. 141-142.
7. J. Zander, "Radio resource management in future wireless networks: requirements and limitations," *IEEE Commun. Mag., Aug.* 1997
8. 3GPP TS 05.02, TSG GERAN, "Digital cellular telecommunications system (Phase 2+); Multiplexing and multiple access on the radio path".
9. 3GPP, Technical Specification 25.101 v3.16.0, UTRA (UE) FDD; Radio Transmission and Reception, Dec. 2003.
10. 3GPP, Technical Specification 25.211 v5.3.0, Physical channels and mapping of transport channels onto physical channels (FDD), December 2002.
11. 3GPP, Technical Specification 25.212 v5.3.0, Multiplexing and channel coding (FDD), December 2002.

OPTIMIZATION OF PILOT POWER FOR SERVICE COVERAGE AND SMOOTH HANDOVER IN WCDMA NETWORKS

Iana Siomina and Di Yuan
Department of Science and Technology, Linköping University
SE-601 74 Norrköping, Sweden
iansi, diyua@itn.liu.se

Abstract   In WCDMA networks, the presence of a cell is announced through a common pilot channel (CPICH). The power levels of the pilot channels have a great impact on coverage and service quality. Using mathematical optimization techniques, we address the problem of minimizing the amount of pilot power for providing service coverage and smooth handover. We present numerical results for several realistic planning scenarios of WCDMA networks, and analyze the pilot power solutions found by optimization versus those obtained by ad hoc strategies.

Keywords: WCDMA, pilot power, coverage, handover, mathematical optimization.

1.   Introduction

In a WCDMA network, a cell announces its presence through a common pilot channel (CPICH). Pilot signals provide channel estimation to mobile terminals, and thereby facilitate cell selection and handover. Mobile terminals continuously monitor pilot signals of the network. Typically, a mobile terminal is attached to the cell with the strongest pilot signal.

Pilot power levels strongly affect coverage and service [2]. The pilot power of a cell effectively determines the cell size, and, consequently, the traffic load in the cell. Thus, to optimize the network performance, the pilot power levels should be carefully chosen [12]. Previous work of analyzing the effect of pilot power on network performance can be found in, for example, [4, 6, 8–10, 13].

We study the problem of providing service in a WCDMA network using a minimum amount of pilot power. There are a couple of reasons for minimizing pilot power consumption. First, as the total power available to the network is limited, less amount of pilot power means more power for user traffic. (Typically, the pilot power of a cell lies somewhere between 5% and 10% of the cell power, [5].) A second reason for minimizing pilot power is to reduce pilot pollution and interference (e.g., [1, 6]).

We consider two types of service constraints. The first constraint is full service coverage, that is, a mobile terminal must be able to receive at least one pilot signal anywhere in the network. This constraint is defined by imposing a carrier-to-interference ratio (CIR) target for the pilot signals. Second, the pilot signals are planned to enable smooth handover. Here, by smooth handover, we mean that, when a mobile terminal moves across cell boundaries and changes its home cell, the handover operation can be performed with minimum risk of dropping a call or disrupting a data session. To enable smooth handover, the CIR target is enforced not only in the interior of a cell, but also at its boundaries to other cells. Mathematically, these two types of constraints are of the same characteristic; they are hence formulated using one set of constraints in our mathematical formulations. In our numerical experiments, we study several realistic planning scenarios of WCDMA networks, including two city networks in Europe. We analyze the pilot power solutions found by our optimization technique versus those obtained using two ad hoc approaches. Our numerical study also provides some insights into the impact of the constraint of smooth handover on pilot power consumption.

The remainder of the paper is organized as follows. In Section 2 we describe our system model. The optimization problem is formalized in Section 3, and two ad hoc solutions are presented in Section 4. Mathematical formulations are discussed in Section 5, and a Lagrangean heuristic is described in Section 6. We present our numerical study in Section 7. Finally, in Section 8 we draw some conclusions and discuss forthcoming research.

2.   System Model

2.1   Preliminaries

Consider a WCDMA network consisting of m cells. Let P_i^{Tot} denote the total transmission power available in cell i. This amount of power is shared among the pilot channel, other signaling channels, as well as user traffic. We use x_i to denote the pilot power of cell i.

The service area is represented by a grid of bins, for which predictions (or measurements) of signal propagation are performed. Let n denote the number of bins, and g_{ij} the power gain between the antenna of cell i and bin j. Thus, in bin j, the power of the received pilot signal of cell i is $x_i g_{ij}$. The interference experienced by a mobile terminal in bin j, with respect to cell i, reads $I_{ij} = (1 - \alpha_j)P_i g_{ij} + \sum_{k \neq i} P_k g_{kj} + \nu_j$, where P_i and P_k are the tranmission power of cells i and k, respectively. Parameter $\alpha_j \in (0, 1)$ is the orthogonality factor in bin j, and ν_j represents the effect of the thermal noise in bin j.

We consider planning scenarios with high traffic load, and assume therefore that all base stations operate at full power, i.e., $P_i = P_i^{Tot}, i = 1, \ldots, m$. This corresponds to the worst-case interference scenario. We can then write I_{ij} as

$$I_{ij} = (1 - \alpha_j)P_i^{Tot}g_{ij} + \sum_{k \neq i} P_k^{Tot}g_{kj} + \nu_j. \qquad (1)$$

The strength of a pilot signal is defined by its CIR. For cell i and bin j, the CIR is

$$\gamma_{ij} = \frac{x_i g_{ij}}{I_{ij}} = \frac{x_i g_{ij}}{(1 - \alpha_j)P_i^{Tot}g_{ij} + \sum_{k \neq i} P_k^{Tot}g_{kj} + \nu_j}. \qquad (2)$$

2.2   Service Constraints

Two service constraints are taken into account in our planning problem. The first constraint is full coverage. A necessary condition for a mobile terminal to access any network service is the detection of at least one pilot signal. We assume that, to successfully detect a pilot signal, the CIR must meet a threshold γ_0. Thus, the service of cell i is available in bin j only if the following is true.

$$\gamma_{ij} \geq \gamma_0. \qquad (3)$$

Full service coverage means that for any bin, there are one or more cells, for which (3) holds. Pilot power minimization subject to (3) has been previously studied in [11].

The second service constraint involves smooth handover. Handover occurs when a mobile terminal moves from the service area of one cell to that of another. Full service coverage does not necessarily ensure smooth handover. For example, consider two adjacent bins served by two different cells, for which the CIR of each of the two pilot signals is good in its respective bin, but very poor in the other. A mobile terminal that moves from one bin into the other, crossing the boundary of its home cell, may have difficulties in detecting the pilot signal of the other cell in time. When this occurs, there is a risk of dropping a call or interrupting a data session.

To facilitate smooth handover, a mobile terminal should be able to detect the pilot signal of the cell to which handover will take place, before it leaves its current home cell. For this purpose, the pilot power levels should be set such that, for the above example, the pilot signal of a cell not only covers its own bin, but also provides some coverage in the adjacent bin served by the other cell. One way to provide this kind of coverage is to require that the CIR of a pilot signal is above γ_0 in bins adjacent to the current cell. However, this may lead to an unreasonably (and unnecessarily) large amount of pilot power, and a risk of pilot pollution. Instead, we handle smooth handover by requiring that, if two adjacent bins belong to different cells, both pilot signals have a CIR of at least γ_0 at the boundary of the two bins. (This requirement increases the likelihood of being in soft or softer handover for mobile terminals at cell boundaries.) Modeling this constraint would require prediction of signal propagation at bin boundaries. Such predictions are not available in our system model. (An implicit assumption in Section 2.1 is that, for every bin, the power gain of a cell is identical in the entire bin.) However, it is reasonable to assume that, for two adjacent bins, the power gain at their boundary is somewhere between the gain values of the two bins. In this paper, we use the average value of the two power gain values to represent the power gain at the boundary.

Consider cell i and two adjacent bins j_1 and j_2. If x_i meets the CIR target in both bins, or in none of the two, the aforementioned constraint of smooth handover does not apply. Assume that cell i has a sufficiently high pilot CIR in bin j_1 but not bin j_2, and that $g_{ij_1} > g_{ij_2}$ (because otherwise the CIR in j_2 is at least as good as that in j_1). To enable smooth handover for mobile terminals moving from j_2 into j_1, the strength of the received pilot signal is calculated using the average value of g_{ij_1} and g_{ij_2}. The new CIR formula is as follows.

$$\frac{x_i \frac{(g_{ij_1}+g_{ij_2})}{2}}{I_{ij_1}} = \frac{x_i \frac{(g_{ij_1}+g_{ij_2})}{2}}{(1-\alpha_{j_1})P_i^{Tot}g_{ij_1} + \sum_{k\neq i} P_k^{Tot}g_{kj_1} + \nu_{j_1}}. \qquad (4)$$

Note that, the interference computation in (4) uses the power gain of bin j_1 (i.e., same as in (1)), not the average power gain. The reason for this is simple: The pilot power levels are planned for the scenario of worst-case interference – using the average power gain in the denominator of (4) would lead to less interference and thus jeopardize full coverage.

To formalize the constraint of smooth handover, we use $A(j)$ to denote the set of adjacent bins of bin j. For most bins, this set contains eight elements. If cell i satisfies (3), then the new CIR formula applies for all bins in $A(j)$. For convenience, we introduce the notation $\bar{g}_{ij}$ to represent the new, adjusted power gain for bin j, that is, $\bar{g}_{ij} = \min\{g_{ij}, \min_{l\in A(j)} \frac{g_{il}+g_{ij}}{2}\}$. We can then write the constraint of smooth handover as follows.

$$\bar{\gamma}_{ij} = \frac{x_i \bar{g}_{ij}}{I_{ij}} = \frac{x_i \bar{g}_{ij}}{(1-\alpha_j)P_i^{Tot}g_{ij} + \sum_{k\neq i} P_k^{Tot}g_{kj} + \nu_j} \geq \gamma_0. \qquad (5)$$

Examining the two service constraints, (3) and (5), we observe that the latter is always as least as strong as the former. Therefore, if cell i covers bin j, the pilot power must be at least $\bar{P}_{ij}$, which is derived from (5):

$$x_i \geq \bar{P}_{ij} = \frac{\gamma_0((1 - \alpha_j)P_i^{Tot}g_{ij} + \sum_{k \neq i} P_k^{Tot}g_{kj} + \nu_j)}{\bar{g}_{ij}}. \tag{6}$$

Remark We impose (5) regardless of whether bin j lies on the boundary of cell i or not. Suppose that bin j is in the interior of cell i, and, consequently, (3) is satisfied for bin j as well as for all bins in $A(j)$. In this case, it can be easily realized that constraint (5) is also satisfied and thus redundant. As a result, the impact of (5) on the pilot power of a cell is determined by those bins on the cell boundary.

3.   Problem Definition

Our pilot power optimization problem, which we denote by PPOP, is defined as follows.

- **Objective**: Minimize the total pilot power, i.e., $\min \sum_{i=1}^m x_i$.

- **Constraint one**: Every bin is covered by at least one pilot signal, that is, for any bin j, there exists at least one cell i for which $x_i \geq \bar{P}_{ij}$.

- **Constraint two**: The pilot power of cell i is limited by P_i^{Tot}, i.e., $x_i \leq P_i^{Tot}, i = 1, \ldots, m$.

The following proposition states the computational complexity of PPOP.

Proposition 1 PPOP is $\mathcal{NP}$-hard.
Proof See the first appendix at the end of the paper.

4.   Two Ad Hoc Solutions

One ad hoc solution to PPOP is the one in which all cells use the same level of pilot power, which we refer to as the solution of uniform pilot power. We use P^U to denote the minimum (total) power of uniform-power solutions that satisfy the constraints of PPOP. The value of P^U can be derived quite easily. Let $P_j^U = \min_{i=1,\ldots,m} \bar{P}_{ij}$. To provide service in bin j, the power of at least one pilot signal must be greater than or equal to P_j^U. As this is true for any bin, taking the maximum of P_j^U over the bins gives a lower bound on any feasible uniform power. In addition, setting all pilot power levels to $\max_{j=1,\ldots,n} P_j^U$ yields a feasible solution to PPOP. We have thus shown the following.

$$P^U = m \cdot \max_{j=1,\ldots,n} P_j^U = m \cdot \max_{j=1,\ldots,n} \min_{i=1,\ldots,m} \bar{P}_{ij}. \tag{7}$$

A second ad hoc approach to PPOP is a greedy heuristic. We call the solution generated by this heuristic the power-based pilot power, because for every bin, the heuristic chooses the cell for which the required power is minimal among all the cells. For bin j, we use $c(j)$ to denote the cell that minimizes $\bar{P}_{ij}$, that is, $c(j) = \arg\min_{i=1,\dots,m} \bar{P}_{ij}$. The pilot power of cell i is thus $P_i^G = \max_{j=1,\dots,n:c(j)=i} \bar{P}_{ij}$. The total pilot power of this solution is therefore

$$P^G = \sum_{i=1}^{m} P_i^G = \sum_{i=1}^{m} \max_{j=1,\dots,n:c(j)=i} \bar{P}_{ij}. \tag{8}$$

5.   Mathematical Formulations

5.1   A Cell-bin Formulation

Problem PPOP can be formulated mathematically using the pilot power variables $x_i, i = 1, \dots, m$, and the following set of binary decision variables.

$$y_{ij} = \begin{cases} 1 & \text{if cell } i \text{ covers bin } j, \text{ i.e., } \bar{\gamma}_{ij} \geq \gamma_0, \text{ or, equivalently, } x_i \geq \bar{P}_{ij}, \\ 0 & \text{otherwise.} \end{cases}$$

Because the pilot power of a cell has an upper limit, not all cells are able to cover a bin. Therefore, we only need y-variables for feasible combinations of cells and bins. For this reason, we define a set $C(j)$, which consists of all cells that can cover bin j using a feasible pilot power, i.e., $C(j) = \{i = 1, \dots, m : \bar{P}_{ij} \leq P_i^{Tot}\}$. PPOP can then be stated as follows.

$$[\text{PPOP-CB}] \quad P^* = \min \sum_{i=1}^{m} x_i \tag{9}$$

$$\text{s. t.} \quad \sum_{i \in C(j)} y_{ij} \geq 1, \quad j = 1, \dots, n, \tag{10}$$

$$\bar{P}_{ij} y_{ij} \leq x_i, \quad i \in C(j), j = 1, \dots, n, \tag{11}$$

$$y_{ij} \in \{0, 1\}, \quad i \in C(j), j = 1, \dots, n. \tag{12}$$

Constraints (10) ensure that every bin is covered by at least one cell. By (11), x_i must be at least $\bar{P}_{ij}$, if cell i covers bin j. The non-negativity restrictions on the x-variables are implicitly handled by (11).

5.2   A Refined Formulation

From a computational standpoint, formulation PPOP-CB is not efficient. In particular, its linear programming (LP) relaxation is very weak. Solving PPOP using this formulation is out of reach of a standard problem solver[1]. To avoid this weakness, we derive a second, refined formulation. The refinement is

---

[1]In our numerical study, CPLEX [3] did not manage to find optimal or near-optimal solutions within any reasonable amount of time even for the smallest test network (60 cells and 1375 bins).

based on the rather simple observation that, in an optimal solution to PPOP, the pilot power of cell i attains a value belonging to the discrete set $\{\bar{P}_{ij}, j = 1, \ldots, n : i \in C(j)\}$. In the refined formulation, we use the following set of binary variables.

$$z_{ik} = \begin{cases} 1 & \text{if the pilot power of cell } i \text{ equals } \bar{P}_{ik}, \\ 0 & \text{otherwise.} \end{cases}$$

In Section 5.1, we defined the sets $C(j), j = 1, \ldots, n$, each of which contains the set of cells that can cover a bin. In the refined formulation, it is more convenient to use the notation $B(i)$, which describes the possibility of coverage from the perspective of cells. Specifically, we let $B(i) = \{j : \bar{P}_{ij} \leq P_i^{Tot}\}$. Also, we define a set of indication parameters for the refined formulation:

$$a_{ijk} = \begin{cases} 1 & \text{if bin } j \text{ is covered by cell } i, \text{ provided that } z_{ik} = 1, \\ 0 & \text{otherwise.} \end{cases}$$

Below we present the refined formulation.

$$[\text{PPOP-RF}] \ P^* = \min \sum_{i=1}^{m} \sum_{k \in B(i)} \bar{P}_{ik} z_{ik} \tag{13}$$

$$\text{s. t.} \quad \sum_{k \in B(i)} z_{ik} = 1, \quad i = 1, \ldots, m, \tag{14}$$

$$\sum_{i \in C(j)} \sum_{k \in B(i)} a_{ijk} z_{ik} \geq 1, \quad j = 1, \ldots, n, \tag{15}$$

$$z_{ik} \in \{0, 1\}, i = 1, \ldots, m, k \in B(i). \tag{16}$$

In PPOP-RF, (14) states that exactly one of the possible pilot power levels is selected for every cell. By (15), every bin is covered by at least one cell.

Although it may not be trivial, it can be shown that the LP relaxation of PPOP-RF is always at least as strong as that of PPOP-CB.

Proposition 2 The LP relaxation of PPOP-RF is at least as strong as that of PPOP-CB. In addition, there exist instances for which the former is strictly better than the latter.

Proof See the second appendix at the end of the paper.

6.   A Lagrangean Heuristic

For large-scale networks, it is time-consuming to solve PPOP-RF exactly using a standard solver. We therefore developed a Lagrangean heuristic as an approximate solution method. Due to space limitation, we will not present the algorithm in its full detail. In brief, the Lagrangean heuristic comprises two components. The first component is a Lagrangean relaxation, in which constraints (15) are relaxed using Lagrangean multipliers $\lambda_j, j = 1, \ldots, n$. The relaxation decomposes into one easily-solved subproblem per cell. For cell i, the subproblem is to minimize the function $\sum_{k \in B(i)} (\bar{P}_{ik} - \sum_{j=1}^{n} \lambda_j a_{ijk}) z_{ik}$, subject to the constraint $\sum_{k \in B(i)} z_{ik} = 1$. The Lagrangean dual is then solved using subgradient optimization.

The second component is a primal heuristic, in which the solution of the relaxation, if infeasible, is modified to a feasible solution. The heuristic consists of two phases. The first phase involves covering bins that are not covered by any cell in the solution of the relaxation. Among these bins, the heuristic selects the bin for which the cardinality of the set $|C(j)|$ is minimal. To cover this bin, the cell that needs a minimum amount of incremental power is chosen. This is then repeated until all bins are covered. In the second phase, the heuristic attempts to reduce the total pilot power by examining bins covered by multiple cells. For each of such bins, the heuristic identifies whether any cell can reduce its pilot power if the bin is removed from the coverage area of the cell. The second phase terminates when no improvement of this type is identified.

The Lagrangean heuristic solves the relaxation and applies the primal heuristic for a predefined number of subgradient optimization iterations. At termination, the Lagrangean heuristic yields both an upper bound (the best feasible solution found) and a lower bound (the best value of the Lagrangean relaxation) to the optimum.

7.   Numerical Study

We used three WCDMA networks in our numerical study. The first network was provided by Ericsson Research, Sweden. The other two networks, provided by the MOMENTUM project [7], originate from planning scenarios for Berlin and Lisbon, respectively. Table 1 displays some network characteristics.

Table 1.   Network characteristics.[a]

	Network N1	*Network N2*	*Network N3*
Sites	22	50	52
Cells	60	148	140
Bins	1375	22500	62500
Bin size (m^2)	40×40	50×50	20×20
$P_i^{Tot}, i = 1, \ldots, m$	20 W	19.95 W	19.95 W
γ_0	-18.24 dB	-20 dB	-20 dB
$\nu_j, j = 1, \ldots, n$	-100 dbm	-108.1 dbm	-100 dbm
$\alpha_j, j = 1, \ldots, n$	0.4	$\{0.327, 0.633, 0.938\}$	$\{0.327, 0.633, 0.938\}$

[a]For networks N1 and N3, the orthogonality parameter α_j depends on bin type (urban, rural, or mixed).

For N1, we used a standard solver [3] to find the optimal solution using PPOP-RF. For N2 and N3, we applied the Lagrangean heuristic described in Section 6. We present our main results in Table 2, which displays the optimized pilot power, and the pilot power of the two ad hoc solutions.

Table 2. Pilot power solutions.[a]

Network	Uniform pilot power		Power-based pilot power		Optimized pilot power	
	Total	Average	Total	Average	Total	Average
N1	92.58	1.54	46.87	0.78	41.52	0.69
N2	464.40	3.13	192.51	1.30	160.00	1.08
N3	415.28	2.97	174.20	1.24	147.83	1.06

[a] For the optimized pilot power, its worst-case deviation from optimum, in a relative sense, are 9% and 12% for N2 and N3, respectively.

We observe that the power-based solution offers a substantial improvement over the solution of uniform pilot power. The former is, however, still quite far away from optimum, when compared to optimized pilot power, which corresponds to only a few percent of the total power available. Our results suggest, therefore, that it is possible to use a small amount of pilot power for providing service coverage and smooth handover in WCDMA networks.

The optimized pilot power levels of network N1 are further examined using a histogram in Figure 1. We conclude that the power levels of most pilot signals lie between 0.3 W and 1.0 W, and, in addition, most cells use a pilot power that is less than the average (0.69W in this case).

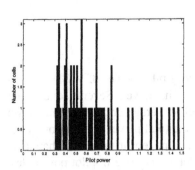

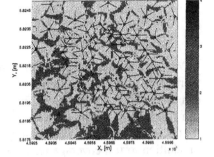

Figure 1. A histogram of the pilot power of network N1.

Figure 2. Pilot signal coverage of network N2 (city of Berlin).

Figure 2 illustrates the pilot signal coverage of network N2 (city of Berlin). For each bin, its color (or darkness) represents the number of cells providing coverage in the bin. The figure also shows the locations of the base stations as well as the antenna directions. We observe that most parts of the service area are covered by one or two pilot signals. (A more detailed examination of the coverage statistics shows that about 32% of the bins are covered by more than one pilot signal.) In many parts of the figure, bins covered by multiple cells form lines that indicate cell boundaries.

In the next part of our numerical study, we examine the impact of the constraint of smooth handover on pilot power. Ignoring smooth handover, the pilot signals need to satisfy constraint (3) only, i.e., if cell i covers bin j, the minimum pilot power is P_{ij}, which is derived from (3): $P_{ij} = \gamma_0 \cdot ((1 - \alpha_j)P_i^{Tot}g_{ij} + \sum_{k \neq i} P_k^{Tot}g_{kj} + \nu_j)/g_{ij}$. We replace $\bar{P}_{ij}$ by P_{ij} in PPOP-RF, which then minimizes the total pilot power without the constraint of smooth handover. The results are presented in Table 3. Comparing the results in this table to those in Table 2, we observe that between 30% and 50% additional pilot power are necessary to support smooth handover.

Table 3. Pilot power solutions without smooth handover.[a]

Network	Uniform pilot power		Power-based pilot power		Optimized pilot power	
	Total	Average	Total	Average	Total	Average
N1	64.24	1.09	31.61	0.53	27.87	0.46
N2	356.44	2.41	146.56	0.99	114.80	0.78
N3	276.79	1.98	133.64	0.95	110.14	0.79

[a]For the optimized pilot power, its worst-case deviation from optimum, in a relative sense, are 6% and 12% for N2 and N3, respectively.

8.   Conclusions

We have studied the problem of minimizing pilot power of WCDMA networks subject to service coverage and smooth handover. Several conclusions can be drawn from our study. First, both full coverage and smooth handover can be achieved using only a few percent of the total power in a network, even for the scenario of worst-case interference. Second, ensuring smooth handover in addition to full coverage results in a moderate increase in pilot power (less than 50%). Moreover, our study shows that optimized pilot power considerably outperforms ad hoc approaches, and, therefore, mathematical models can be very helpful for optimizing power efficiency in WCDMA networks.

An extension of the current research is pilot power optimization for the purpose of load balancing. Because pilot signals influence cell size, pilot power can be adjusted to equalize the load over cells. This topic is to be addressed in forthcoming research.

Acknowledgments

The authors wish to thank the research group at Ericsson Research, Linköping, Sweden, for the technical discussions and the data of the first test network. We also thank the group of the MOMENTUM project [7] for providing us with the test networks of Berlin and Lisbon. This work is partially financed by CENIIT (Center for Industrial Information Technology), Linköping Institute of Technology, Sweden.

Appendix: Proof of Proposition 1

We show that any instance of the minimum-cost set covering problem (which is $\mathcal{NP}$-hard) can be polynomially transformed to an instance of PPOP. Consider an instance of the set covering problem, where $\{S_1, S_2, \ldots, S_m\}$ is a collection of sets, and B is a set of items. A set S_i is associated with a cost c_i, and contains some (possibly all) of the items in B. The objective of the set covering problem is to select a subset of $\{S_1, S_2, \ldots, S_m\}$ at minimum cost, such that all the items in B are included. The corresponding instance of PPOP has m cells and $m + |B|$ bins. We choose the parameters $\bar{P}_{ij}$ and P_i^{Tot} as follows.

- For bin $j = 1, \ldots, m$, $\bar{P}_{jj} = \epsilon$, where ϵ satisfies $0 < \epsilon < \min_{i=1,\ldots,m} c_i$, and $\bar{P}_{ij} > P_i^{Tot}, \forall i \neq j$. (That is, cell j is the only cell that can cover bin j.)
- For cell $i = 1, \ldots, m$ and $j = m + 1, \ldots, m + |B|$, $\bar{P}_{ij} = c_i$ if S_i contains the $(j - m)$th item of set B, otherwise $\bar{P}_{ij} > P_i^{Tot}$.

The above transformation is clearly polynomial. Moreover, a feasible solution to the PPOP instance is also feasible to the set covering instance, and vice versa. Finally, for any such pair of solutions, the two objective functions have the same value. Hence the conclusion.

Appendix: Proof of Proposition 2

We prove the first part of the proposition by showing that, for any feasible solution to the LP relaxation of PPOP-RF, there is a corresponding solution to the LP relaxation of PPOP-CB, and, in addition, the total pilot power of the former is greater than or equal to the latter.

Consider a feasible solution, denoted by $\bar{z} = \{\bar{z}_{ik}, i = 1, \ldots, m, k \in B(i)\}$, to the LP relaxation of PPOP-RF. Consider solution $\bar{y} = \{\bar{y}_{ij}, i \in C(j), j = 1, \ldots, n\}$, where $\bar{y}_{ij} = \sum_{k \in B(i)} a_{ijk}\bar{z}_{ik}$, to the LP relaxation of PPOP-CB. It is easy to verify that $\bar{y}$ satisfies (10). Next, we show that the total power of $\bar{y}$ is at most as that of $\bar{z}$. For $\bar{y}$, the optimal value of x_i in PPOP-CB is obviously $\bar{x}_i = \max_{j=1,\ldots,n:i\in C(j)} \bar{P}_{ij}\bar{y}_{ij}$. According to the definitions of the sets $B(i)$ and $C(j)$, this equation can also be written as $\bar{x}_i = \max_{j\in B(i)} \bar{P}_{ij}\bar{y}_{ij}$. Assume that the maximum occurs for bin j^*, i.e., $\bar{x}_i = \bar{P}_{ij^*}\bar{y}_{ij^*}$. For PPOP-RF, the pilot power of cell i reads $\sum_{k \in B(i)} \bar{P}_{ik}\bar{z}_{ik}$. For cell i, let $\tilde{B}(i) = \{k \in B(i) : a_{ij^*k} = 1\}$. The set $\tilde{B}(i)$ contains all bins in $B(i)$ that, in order to be covered by cell i, require a pilot power of at least $\bar{P}_{ij^*}$. Then, $\sum_{k \in B(i)} \bar{P}_{ik}\bar{z}_{ik} \geq \sum_{k \in \tilde{B}(i)} \bar{P}_{ik}\bar{z}_{ik} \geq \sum_{k \in \tilde{B}(i)} \bar{P}_{ij^*}\bar{z}_{ik} = \bar{P}_{ij^*} \sum_{k \in \tilde{B}(i)} \bar{z}_{ik} = \bar{P}_{ij^*} \sum_{k \in B(i)} a_{ij^*k}\bar{z}_{ik} = \bar{P}_{ij^*}\bar{y}_{ij^*} = \bar{x}_i$. Because this holds for any cell, we have proved the first part of the proposition.

To show the second part of the proposition, it is sufficient to give an example. Consider two cells and four bins, where $\bar{P}_{11} = 1.2$, $\bar{P}_{12} = 0.8$, $\bar{P}_{13} = 0.6$, $\bar{P}_{21} = 0.6$, $\bar{P}_{22} = 0.8$, and $\bar{P}_{24} = 0.3$. Assume also that $\bar{P}_{14}$ and $\bar{P}_{23}$ exceed their limits (and are thus irrelevant to the discussion). In the integer optimum, $y_{11} = y_{12} = y_{13} = y_{24} = 1$, and the total pilot power equals 1.5. The optimal LP solution of PPOP-CB is $y_{11} = 0.5, y_{12} = 0.75, y_{13} = 1, y_{21} = 0.6, y_{22} = 0.25$,

$y_{23} = 1$, with a total power of 0.9. (The relative gap is therefore 40%.) The LP relaxation of PPOP-RF, on the other hand, yields the integer optimum.

References

[1] M. M. El-Said, A. Kumar, and A. S. Elmaghraby, Sensory system for early detection of pilot pollution interference in UMTS networks, in: *Proceedings of the 10th International Conference on Telecommunications (ICT '03)*, Tahiti, Papeete, French Polynesia, February 2003, pp. 1323–1328,

[2] A. Eisenblätter, T. Koch, A. Martin, T. Achterberg, A. Fügenschuh, A. Koster, O. Wegel, and R. Wessäly, Modelling feasible network configurations for UMTS, in: *Telecommunications Network Design and Management*, eds. G. Anandalingam and S. Raghavan (Kluwer Academic Publishers, 2002) pp. 1–24.

[3] *ILOG CPLEX 7.0, User's Manual* (ILOG, August 2000).

[4] D. Kim, Y. Chang, and J. W. Lee, Pilot power control and service coverage support in CDMA mobile systems, *Proceedings of IEEE VTC '99*, Houston, TX, May 1999, pp. 1464–1468.

[5] J. Laiho, A. Wacker, and T. Novasad (eds.), *Radio Network Planning and Optimisation for UMTS*, (John Wiley & Sons, 2002).

[6] R. T. Love, K. A. Beshir, D. Schaeffer, and R. S. Nikides, A pilot optimization technique for CDMA cellular systems, *Proceedings of IEEE VTC '99*, Houston, TX, May 1999, pp. 2238–2242.

[7] MOMENTUM, http://momentum.zib.de (2003).

[8] Y. Sun, F. Gunnarsson, and K. Hiltunen, CPICH power settings in irregular WCDMA macro cellular networks, *Proceedings of IEEE PIMRC '03*, Beijing, China, September 2003, pp. 1176–1180.

[9] K. Valkealahti, A. Höglund, J. Pakkinen, and A. Flanagan, WCDMA common pilot power control with cost function minimization, *Proceedings of IEEE VTC Fall '02*, Vancouver, Canada, September 2002, pp. 2244–2247.

[10] K. Valkealahti, A. Höglund, J. Pakkinen, and A. Flanagan, WCDMA common pilot power control for load and coverage balancing, *Proceedings of IEEE PIMRC '02*, Lisbon, Portugal, September 2002, pp. 1412–1416.

[11] P. Värbrand and D. Yuan, A mathematical programming approach for pilot power optimization in WCDMA networks, *Proceedings of Australian Telecommunications Networks and Applications Conference (ATNAC)*, Melbourne, Australia, December 2003.

[12] J. Yang and J. Lin, Optimization of pilot power management in a CDMA radio network, *Proceedings of IEEE VTC Fall '00*, Boston, MA, September 2000, pp. 2642–2647.

[13] H. Zhu, T. Buot, R. Nagaike, and S. Harman, Load balancing in WCDMA systems by adjusting pilot power, *Proceedings of the 5th International Symposium on Wireless Personal Multimedia Communications*, Honolulu, Hawaii, October 2002, pp. 936–940.

AN ALTERNATIVE METRIC FOR CHANNEL ESTIMATION WITH APPLICATIONS IN BLUETOOTH SCHEDULING

João H. Kleinschmidt, Marcelo E. Pellenz and Luiz A. P. Lima Jr.
Graduate Program in Computer Science, Pontifical Catholic University of Paraná, Curitiba – PR, Brazil. E-mail: {joaohk, marcelo, laplima}@ppgia.pucpr.br

Abstract: Once Wireless Local Networks (WLAN) and Bluetooth devices share the same frequency band (ISM) there is a potential risk of interference if they are supposed to operate close to each other. Additionally, the signal fading effects on mobile Bluetooth networks may deeply affect the overall performance. That is why the use of strategies that minimize transmission on channels with great interference or severe fading is so important. This paper proposes and investigates the use of parameter m of the Nakagami distribution, as the channel estimation metric. We observed that parameter m may provide faster estimates on the channel condition than the bit error rate metric. This metric is applied in a new scheduling algorithm for Bluetooth piconets. Simulation results showing the performance of the algorithm for different traffic conditions are eventually presented.

Key words: Bluetooth; wireless networks; Nakagami-m fading; scheduling.

1. INTRODUCTION

Bluetooth is emerging as an important standard[1] for short range and low-power wireless communications. It operates in the 2.4 GHz ISM (Industrial, Scientific and Medical) band employing a frequency-hopping spread spectrum technique. The transmission rate is up to 1 Mbps, using GFSK (Gaussian Frequency Shift Keying) modulation. The Bluetooth MAC protocol is designed to facilitate the construction of ad hoc networks. The devices can communicate with each other forming a network with up to eight nodes, called *piconet*. Within a piconet, one device is assigned as a master node and the others devices act as slave nodes. Devices in different piconets can communicate using a structure called *scatternet*. The channel is divided in time slots of 625 μs. A time-division duplex (TDD) scheme is used for full-duplex operation. For data transmission Bluetooth employs seven asynchronous packet types. Each packet may occupy 1, 3 or 5 time slots. The throughput of Bluetooth links using asynchronous packets was investigated[2] for the additive white Gaussian noise (AWGN) channel and for the Rayleigh fading channel. In other work[3], we extended the results presented by Valenti[2] looking into the performance of Bluetooth links in Nakagami-*m* fading channels.

The sharing of the same frequency band between WLAN and Bluetooth devices may cause interference, if they are operating close to each other. Additionally, may occur mutual interference between different Bluetooth piconets operating in the same area. In Bluetooth networks with node mobility, like in sensor networks applications, the fading effects in the radio signal may significantly decrease the link performance. The use of strategies that minimize the transmission in channels with great interference or severe fading, may substantially improve the piconet performance. Extensive empirical measurements have confirmed the usefulness of the Nakagami-*m* distribution for modeling radio links[13,14]. The Nakagami-*m* distribution[4] allows a better characterization of real channels because it spans, via the parameter *m*, the widest range of multipath fading distributions. For *m*=1 we get the Rayleigh distribution. Using *m*<1 or *m*>1 we obtain fading intensities more and less severe than Rayleigh, respectively.

This work proposes the use of fading parameter *m* as an alternative channel quality metric. This parameter can be estimated based on the received symbols. In a mobile wireless network, when a node position changes from line-of-sight to non-line-of-sight, for example, the impact in the signal propagation characteristic may be interpreted as a change in the parameter *m*. This model is interesting when Bluetooth devices are applied to ad hoc sensor networks. Power class one Bluetooth devices can cover ranges up to 100 meters, allowing the formation of large area piconets or scatternets. We also propose a new scheduling algorithm for Bluetooth piconets, which uses the channel quality information in the scheduling policy.

This paper is structured as follows: in Section 2 some issues about piconet scheduling and related works are presented. In Section 3 we present and evaluate the performance of the main Nakagami fading parameter estimators found in the literature. Section 4 proposes a new strategy based on channel quality estimation and Section 5 shows the simulation results for different scenarios. Finally, conclusions are drawn in Section 6.

2.   RELATED WORK ON PICONET SCHEDULING

In a Bluetooth piconet, the master controls the channel access. A slave can send a packet only if it receives a polling packet from the master. The master transmits packets to the slave in even slots while the slave transmits packets to the master in odd slots. Thus, Bluetooth is a master driven TDD standard and this poses several challenges in scheduling algorithms since there could be a waste of slots if only the master or the slave has data to send. Recently, many schemes have been proposed in the literature for piconet and scatternet scheduling.

In the study of Capone[5], several polling schemes are compared. In the round robin scheme a fixed cyclic order is defined and a single chance to transmit is given to each master-slave queue pair. The exhaustive round robin (ERR) also uses a fixed order but the master does not switch to the next slave until both the master and the slave queues are empty. The main disadvantage of the ERR is that the channel can be captured by stations generating traffic higher than the system capacity. A limited round robin (LRR) scheme that limits the number t of transmissions can solves this problem. A new scheme called LWRR (limited and weighted round robin) with weights dynamically changed according to the observed queue status is also presented[5]. Other works about piconet scheduling consider QoS issues in Bluetooth[6,7]. The results[5,6,7] do not consider any loss model for the wireless channels.

In other paper[8], a scheduling policy based on slave and master queues is shown. The master-slave pairs are distinguished based on the size of the Head-of-the-Line (HOL) packets at the master and slave queues. Then, the pairs are classified in three classes according to slot waste. This information is used in the HOL K-fairness policy (HOL-KFP) [8]. When the authors introduced channel errors, the HOL-KFP had its performance reduced. An extension for HOL-KFP called wireless adapted-KFP (WAKFP) was proposed and the results indicate that a better performance is achieved in the presence of channel errors[8].

In [9] an algorithm called Bluetooth Interference Aware Scheduling (BIAS) is presented that uses a channel estimation procedure in order to detect the presence of other wireless devices in the same band (such as other Bluetooth or IEEE 802.11b devices). The scheduling algorithm will avoid packet transmission in frequencies that have a high bit error rate (BER), called bad frequencies. This fact reduces the packet loss due to interference of other near devices. Few of the scheduling schemes presented here consider a loss model for the wireless channel. The works [8,9] use a simple error model.

This paper models de wireless channel fading through the Nakagami-m distribution, apply an alternative metric for channel state estimation and propose a new scheduling algorithm using that metric. It is important to point out that the proposed metric has a faster estimation convergence than the bit error rate used in [8,9]. The fading parameter m gives us an indication of the fading severity, which will directly impact on either the bit or the packet error rates.

3.   ESTIMATORS FOR THE NAKAGAMI FADING PARAMETER

The Nakagami probability density function (pdf) is a two-parameter (m, Ω) distribution, where m is the fading parameter and $\Omega = E[r_n^2]$ is the second moment of the received signal samples, r_n. The estimation of parameter m has found recently many applications, as in systems with optimized transmission diversity. In order to use the Nakagami distribution to model a given set of empirical data, one must determine, or estimate, the fading figure m from the data. Knowledge of the fading parameter is also required by the receiver for optimal reception of signals in Nakagami fading. Many estimators have been proposed in the literature. The fading parameter m is defined as

$$m = \frac{\Omega^2}{E\left[\left(r_n^2 - \Omega\right)^2\right]} \quad m \geq 0.5 \cdot$$

Given $\{r_1, r_2, \ldots, r_N\}$ as realizations of N i.i.d. Nakagami-m random variates, the kth moment of the Nakagami distribution is given by

$$\hat{\mu}_k = \frac{1}{N}\sum_{i=1}^{N} r_i^k \cdot$$

The parameter m may be estimated from its definition, using the 2nd and 4th sample moments, $\hat{\mu}_2$ e $\hat{\mu}_4$:

$$\hat{m}_s = \frac{\hat{\mu}_2^2}{\hat{\mu}_4 - \hat{\mu}_2^2} \cdot$$

High order sample moments can deviate significantly from the true moments if the sample size is not large enough because outliers. Cheng and Beaulieu [11] proposed two new estimators, one based in integer moments,

$$\hat{m}_t = \frac{\hat{\mu}_1 \hat{\mu}_2}{2\left(\hat{\mu}_3 - \hat{\mu}_1 \hat{\mu}_2\right)},$$

and other based on real moments,

$$\hat{m}_{1/p} = \frac{\hat{\mu}_{1/p}\,\hat{\mu}_2}{2p\left(\hat{\mu}_{2+1/p} - \hat{\mu}_{1/p}\,\hat{\mu}_2\right)},$$

where $\hat{\mu}_{1/p} = E[r_i^{1/p}]$ and $\hat{\mu}_{2+1/p} = E[r_i^{2p+1}]$. These estimators are efficient for a moderated number of samples and are appropriated for low complexity implementations. We apply these estimation techniques for a Bluetooth piconet transmission using DM1 packets. For a DM1 packet we have a sequence of $N=240$ channel samples available for channel estimation. These samples represent the soft decision information about the received symbols. We simulate the variability of fading intensity (parameter m) every new transmission in the same master-slave link of a piconet, during 10 time slots. The simulated and estimated values of parameter m are presented in Table 1.

Table 1. Simulated and estimated values of parameter m

		1	2	3	4	5	6	7	8	9	10
Simulated value of m		0.5	1	0.5	1.5	1	1.5	0.5	1	0.5	1.5
Estimated value of m	$\hat{m}_s$	0.60	0.93	0.80	1.74	1.27	2.17	0.56	0.99	0.50	2.22
	$\hat{m}_t$	0.63	0.95	0.94	1.87	1.35	2.32	0.63	1.07	0.48	2.33
	$\hat{m}_{1/p}$	0.56	0.95	0.62	1.60	1.25	2.08	0.50	0.96	0.52	2.08

Figure 1 shows the estimators convergence on every time slot of the piconet polling, in a same master-slave link transmission. We may observe that estimator $\hat{m}_{1/p}$ presents a better convergence. At the end of each time slot transmission we obtain the estimated value, $\hat{m}$. This estimative is applied in the scheduling algorithm proposed in the next section as the channel state information. Notice that tracking the channel condition in one time slot based on the bit error rate may be not feasible due to the low number of available bits from DM1 packet.

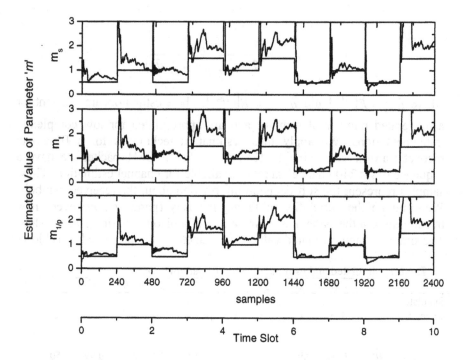

Figure 1. Performance of Nakagami fading parameter estimators in a Bluetooth piconet

4.   PROPOSED SCHEDULING ALGORITHM

The channel condition can greatly affect the performance of the piconet and the polling strategy. In mobile environments, the status of the wireless channel changes very rapidly and this means that a better performance will be achieved if a node is polled at the moment it has a good channel condition and not polled when the conditions are bad. Since Bluetooth is a technology designed for WPANs (Wireless Personal Area Networks), channels errors due to mobility and interference of other devices are very common. A good scheduling algorithm must consider these issues.

We propose an algorithm – called Bluetooth Channel State Scheduling (BCSS) algorithm – that uses the channel state information for piconet scheduling. The values of the fading parameter m can be efficiently estimated as discussed in Section 3. The master will carry out the estimations using the data packets exchanged with the slaves. Every time a master receives a packet, the value of m for that link will be updated. Since this task does not require extra information to be exchanged between the master and the slaves, no extra time is added to the scheduling policy. In the new scheduling policy, the master will poll only the slaves that are above a certain threshold for m, indicating a good channel state. The slaves that are below the threshold, indicating that they are at a bad channel state, will be jumped for at most t_j times. Notice that if the channel state is always good the algorithm is reduced to a round robin policy.

5.   SIMULATION RESULTS

We developed an event driven simulator in C++ to compare the BCSS algorithm with round robin and ERR strategies. The effects of Nakagami fading are simulated using the models described in a previous study[3]. A Poisson traffic source was assumed for the traffic generation in each piconet node. This model can simulate various applications of Bluetooth. In the first simulation scenario we investigate the influence of the fading parameter m in a round robin scheduling. It consists of a piconet with a master and 7 slaves separated by a distance d. The parameter λ is the mean arrival rate in packets per time slot. Fig. 2 shows the average delay for this scenario for three different values of m, using DM1 packets. In this scenario all nodes have the same traffic conditions. We can observe that the state of the channel has great influence in the average delay of the piconet, affecting the performance of the network.

In the second simulation scenario a piconet with the master and 4 slaves is considered. Fig. 3 and 4 compare the average delay for different traffic conditions and DM1 packets using round robin, ERR and BCSS algorithm, for distances of seven and ten meters. In the BCSS algorithm we choose t_j=6 and a threshold m=1. This means that only the slaves with m greater than one will be polled, and the others will be jumped for at most six times. The traffic is the same in the master and the slave queues. In the simulation we assume that the channel conditions are changing every two rounds of the polling scheme. We also consider this scenario with d=10m for different traffic conditions in the master and slave links, as defined in Table 2. The results are shown in Fig. 5. The simulation results show that the BCSS algorithm improves its performance when the traffic is high. For low traffic, ERR has the best performance. Other works [5,12] also concluded that the exhaustive service (ERR) does not have good performance under high traffic.

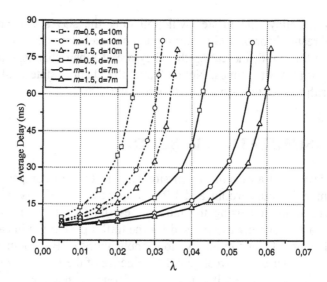

Figure 2. Average delay for different values of *m*

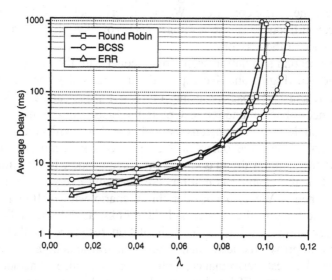

Figure 3. Average delay for d = 7m

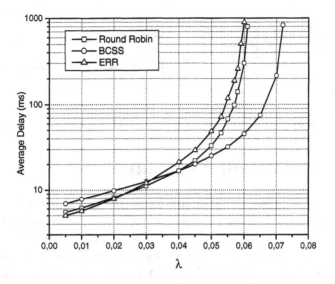

Figure 4. Average delay for d = 10m

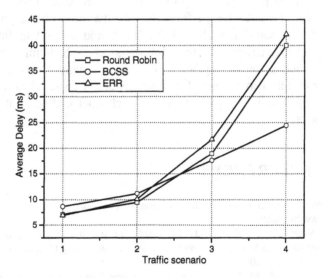

Figure 5. Average delay for different traffic scenarios

Table 2. Traffic conditions for master-slave links

Traffic scenario	λ_1 (master-slave 1)	λ_2 (master-slave 2)	λ_3 (master-slave 3)	λ_4 (master-slave 4)
1	0.01	0.01	0.02	0.02
2	0.01	0.01	0.04	0.04
3	0.03	0.05	0.03	0.05
4	0.03	0.03	0.06	0.06

6.   CONCLUDING REMARKS

This paper proposed an alternative metric for channel state estimation using Nakagami-*m* fading distribution. This metric is applied in the polling strategy of a piconet scheduling algorithm denoted BCSS. The BCSS algorithm is considerably efficient for high traffic loads if the channel conditions change frequently. These variations in channel conditions are present in many applications of the Bluetooth technology in environments with interference and mobility. The BCSS algorithm can be combined with other scheduling policies to improve their performance. This work can be extended to evaluate the performance of the proposed algorithm with different traffic sources, such as FTP, HTTP and voice. Our future works include improvements to the intra-piconet scheduling policy and implementation of an inter-piconet scheduling scheme for scatternets. The parameter *m* can also be used for other important issues in Bluetooth, like scatternet formation, routing and specific channel coding strategies using AUX1 packets.

7.   REFERENCES

1. Bluetooth SIG, Specifications of the Bluetooth system, *Core Version 1.1*, February 2001. http://www.bluetooth.com
2. M. C. Valenti, M. Robert and J.H. Reed, On the throughput of Bluetooth data transmissions, IEEE Wireless Communications and Networking Conference, Orlando, FL, pp. 119-123, March 2002.
3. J. H. Kleinschmidt, M. E. Pellenz, and E. Jamhour, Bluetooth network performance in Nakagami-*m* fading channels, The Fifth IFIP TC6 International Conference on Mobile and Wireless Communications Networks, Singapore, October 2003.
4. J. Proakis, *Digital Communications*, New York, NY: McGraw-Hill, 4th edition, 2001.
5. A.Capone, M. Gerla and R. Kapoor, Efficient polling schemes for Bluetooth picocells, IEEE International Conference on Communications, Helsinki, Finland, June 2001.
6. J. B. Lapeyrie and T. Turletti, FPQ: a Fair and Efficient Polling Algorithm with QoS Support for Bluetooth Piconet, IEEE Infocom, San Francisco, CL, 2003.

7. Y. Liu, Q. Zhang and W. Zhu, A Priority-Based MAC Scheduling Algorithm for Enhancing QoS support in Bluetooth Piconet, IEEE International Conference on Communications, Circuits and Systems, July 2002.

8. M. Kalia, D. Bansal and R. Shorey, MAC Scheduling and SAR Policies for Bluetooth: A Master Driven TDD Pico-Cellular Wireless System, IEEE International Workshop on Mobile Multimedia Communications, San Diego, California, November 1999.

9. N. Golmie, Performance evaluation of a Bluetooth channel estimation algorithm, IEEE International Symposium on Personal Indoor and Mobile Radio Communications, Lisbon, Portugal, September 2002.

10. A. Ramesh, A. Chockalingam and L.B. Milstein, SNR estimation in generalized fading channels and its application to turbo decoding, IEEE International Conference on Communications, Helsinki, Finland, June 2001.

11. J. Cheng and N. C. Beaulieu, Generalized moment estimators for the Nakagami fading parameter, *IEEE Communication Letters*, vol. 6, no. 4, pp. 144-146, April 2002.

12. N. Johansson, U. Körner and P. Johansson, Performance Evaluation of Scheduling Algorithms for Bluetooth, IFIP TC6 International Conference on Broadband Communications, Hong Kong, November 1999.

13. M. Nakagami, The m-distribution, a general formula of intensity distribution of rapid fading, *Statistical Methods in Radio Wave Propagation*, W. G. Hoffman, Ed. Oxford, England: Pergamon, 1960.

14. H. Suzuki, A statistical model for urban radio channel model, *IEEE Trans. on Communications*, vol. 25, pp. 673 –680, July 1977.

DISTRIBUTED PAIRWISE KEY GENERATION USING SHARED POLYNOMIALS FOR WIRELESS AD HOC NETWORKS

Anindo Mukherjee, Hongmei Deng and Dharma P. Agrawal
Center of Distributed and Mobile Computing
University Of Cincinnati, Cincinnati, OH 45229
{mukherao, hdeng, dpa}@ececs.uc.edu

Abstract: The infrastructure-less property of wireless ad hoc network makes the traditional central server based security management schemes unsuitable and requires the use of a distributed key management mechanism. In this paper, we propose a distributed pairwise key establishment scheme based on the concept of bivariate polynomials. In our method, any mobile node in an ad hoc network can securely communicate with other nodes just by knowing their corresponding IDs. The bivariate polynomials are shared in such a manner that the shares depend on the coefficient matrix of the polynomial, the requesting node's ID and the ID of the nodes that respond to the request. We study the behavior of our scheme through simulations and show that our scheme compares well with other schemes and has a much better performance when averaged over the lifetime of the network.

Key words: Security; Ad Hoc Networks; Symmetric Keys; Bivariate Polynomials; Threshold Secret Sharing

1. INTRODUCTION

Security in ad hoc networks is riddled with a constant change in paradigms. Murphy et al. [6] defined ad hoc networks as "A transitory association of mobile nodes which do not depend on any fixed support infrastructure." Thus, an ad hoc network can be either a network of radios in a battlefield or a network of laptops in an office environment etc. This multitude of applications makes the deployment of a common security infrastructure a complex problem. In addition to the diversity in the kinds of applications, security in ad hoc networks is severely constrained due to the dynamic nature of the networks. Participants may join and leave the network at any time. The traditional central server based security management mechanism may not be directly applicable since the incoming participants need not have access to a central trusted server after the network has been deployed. Thus, a distributed key management mechanism is necessary in securing wireless ad hoc networks. Several approaches towards implementing a distributed key management scheme have been proposed in literature. Zhou and Hass [9] use a partially distributed certificate authority system, in which a group of special nodes is capable of generating partial certificates using their shares of the certificate signing key. In [5], Kong et al. proposed another threshold cryptography scheme by distributing the RSA certificate signing key to all the nodes in the network.

Both the approaches are based on *asymmetric* cryptosystem, which imposes a high processing overhead. In this paper, we consider a symmetric key based approach and focus on distributed pair-wise key generation. In a pairwise key scheme, each node pair shares a unique symmetric key. There are a number of applications for these types of keys. SRP for DSR [7] uses pair-wise keys for authentication. Also, any secure communication between two nodes in the absence of a public key system would require pair-wise symmetric keys between nodes.

Before discussing any further, we would like to state our assumptions and the problem. We base our system on the following assumptions:

-   A trusted server is present which initializes a set of nodes before deployment. This server is not present after the nodes have been deployed. Any un-initialized node would need to get its keying material from the network.
-   An incoming node has the computational power to generate a temporary public key-private key pair.
-   A node that joins the network and tries to obtain keying material from its neighborhood has a mechanism to prove its authenticity to the nodes it requests the shares from.

With the above assumptions in place, we state the problem as: Given an operational ad hoc network with a set of nodes initialized (by a central authority) with the keying material, a node that wishes to join the network needs to securely obtain its own keying material without the help of the central authority.

Several solutions to the problem have been proposed in literature. In [4], the authors present a probabilistic key pre-distribution technique. This idea has been extended in [2] where the authors propose a q-composite key pre-distribution.

We propose a distributed mechanism to share keying material between n nodes such that any t nodes can get together and provide another incoming node with its keying material. At the same time, an adversary listening to all the ongoing conversation and having compromised less than t members would not be able to obtain any pairwise key.

Our scheme is based on the concept of bivariate polynomials, first outlined in [1]. We extend this scheme to a distributed scenario by modifying Shamir's threshold scheme [8], so that incoming nodes can be initialized by the network into getting the keying material from the network. To the best of our knowledge, no attempt has been made to share these polynomials in a secure and distributed manner. The proposed scheme has a number of attractive properties. First, in our scheme, any incoming node would be able to get its key shares from the network and need not rely on a central server. Second, the scheme is resilient to the compromise of *t-1* nodes. Third, a node which joins the network needs to communicate only with its immediate neighborhood in order to get all its keying material. Fourth, physical capture of a node would give away only the captured node's keying material without compromising the network. Finally, our scheme is simple and does not require complex protocols to be implemented.

The rest of the paper is organized as follows. In Section 2 we introduce some background knowledge for our scheme. In Section 3 we present the proposed distributed pair-wise key generation scheme in detail. We give simulation results and discuss possible extensions to the work in Section 4. Finally, Section 5 concludes.

2.   BACKGROUND

In this section, we first take a brief look at the bivariate polynomial scheme introduced in [1], and also at the concept of threshold secret sharing introduced in [8].

2.1 Bivariate polynomial-based key pre-distribution

Consider a bivariate polynomials *f(x,y)* of degree *t*, defined as

$$f(x,y) = \sum_{i,j=0}^{t-1} a_{ij} x^i y^j \tag{1}$$

where the coefficients a_{ij} are randomly chosen over a finite field GF(q). The bivariate polynomial has a symmetric property such that

$$f(x,y) = f(y,x) \tag{2}$$

An initial server first proceeds to initialize a set of nodes by giving each node *m* the polynomial $g_m(y) = f(m,y)$, which is the polynomial obtained by evaluating $f(x,y)$ at $x = m$. That is, a deployed node would know

$$g_j = \sum_{i=0}^{t-1} a_{ij}.(m)^i \qquad (0 \le j \le t-1) \tag{3}$$

where *m* is identity of the node being deployed, and g_j is coefficient of y^j in the polynomial $f(m,y)$.

Thus, in order for a node with ID *m* to calculate the pairwise key with a node with ID n, node m simply finds out the value of $f(m,y)$ at $y=n$. Similarly, node n in turn evaluates the polynomial at $y=m$. Due to the symmetric property of the bivariate polynomial $f(x,y)$, they manage to establish a pairwise secret key known only to them.

The above procedure can be represented using a matrix notation as

$$K_{mn} = K_{nm} = \mathbf{X}^T \mathbf{A} \mathbf{Y} = \mathbf{Y}^T \mathbf{A} \mathbf{X} \qquad (4)$$

where,

$$\mathbf{X} = [m^0, m^1, m^2, m^3, \cdots m^{t-1}]^T$$
$$\mathbf{A} = \text{Coefficient Matrix for } f(x,y)$$
$$\mathbf{Y} = [n^0, n^1, n^2, n^3, \cdots n^{t-1}]^T$$
$$K_{mn} = K_{nm} = \text{Shared Key between node } m \text{ and } n$$

The deployed node m obtains the information $\mathbf{G} = \mathbf{X}^T\mathbf{A} = [g_0, g_1, \ldots g_{t-1}]$. Note that the elements of matrix $\mathbf{A}$ are not known to anyone except the initial server.

2.2 Threshold secret sharing

Secret sharing allows a secret to be shared among a group of users (also called shareholders) in such a way that no single user can deduce the secret from his share alone. One classical (t, n) secret sharing algorithm was proposed by Adi Shamir [8] in 1979, which is based on polynomial interpolation. In the scheme, the secret is distributed to n shareholders, and any t out of the n shareholders can reconstruct the secret, but any collection of less than t partial shares can not get any information about the secret. We use the scheme to share polynomials in such a manner that the coefficients of the polynomial would always remain secret. Any combination of t nodes would only derive the value of the polynomial at a certain point.

3. PROPOSED DISTRIBUTED KEY GENERATION SCHEME

In our scheme, we distribute the shares of the matrix A among n initial nodes such that:
- Any combination of t nodes would be able to derive the keying material for an incoming node. (Note that this does not amount to the nodes getting to know the coefficients of the matrix A. Instead, the incoming node would only be able to derive $\mathbf{X}^T\mathbf{A}$, as indicated in (4).
- Any combination of less than t nodes would not be able to derive any portion of the keying material for an incoming node.
- The central server initializes only a set of n nodes. A node which has not been initialized by the central server and which wishes to join the

network can do so without having to contact the central server provided at least t nodes send in keying material to this node.

To achieve the above, we modify Shamir's t-threshold scheme as outlined below. Let the newly arrived requesting node have an id of α, and the responding nodes have their ids as β_1, β_2, and so on.

We now present the scheme mathematically. In our discussion, we would also use an example to illustrate the steps. In the considered example, the node that wishes to join the network has an ID of 2, the responding nodes have IDs 1, 3 and 5 and the threshold value is 3 (i.e., t=3).

We first take a look at the shares of the key generation material that would be given to each of the initial set of nodes before deployment. The shares should have the property that no set of nodes less than or equal to t should be able to generate either the coefficient matrix A, or the vector $\mathbf{X^T A}$ for any other node.

Each node that is deployed would be initialized with a matrix A_i where A_i is of the following form:

$$\begin{bmatrix} s_{00} & s_{01} & .. & s_{0(t-1)} \\ s_{10} & s_{11} & .. & s_{1(t-1)} \\ .. & .. & .. & .. \\ s_{(t-1)0} & s_{(t-1)1} & & s_{(t-1)(t-1)} \end{bmatrix}$$

Each element s_{ij} is given by:

$$s_{ij} = a_{ij} + \sum_{m=1}^{t_1} b_{ijm}(\beta^m) \tag{5}$$

Here b_{ijm} are random numbers generated by the central server. These numbers are not known to anyone except the central server (Note that this central server is only present before deployment and would have no role to play after the network is in operation). t_1 is the threshold for sharing the matrix **A**. β is the ID of the node.

In our example, the above quantities for node 3 are given by:

$$s_{00} = a_{00} + 3.b_{001} + b_{002}.3^2$$
$$s_{01} = a_{01} + 3.b_{011} + b_{012}.3^2$$
$$.....$$
$$s_{21} = a_{21} + 3.b_{211} + b_{212}.3^2$$
$$s_{22} = a_{22} + 3.b_{221} + b_{222}.3^2$$

Also, note that t_1 and t are two separate quantities. While t_1 denotes the threshold for sharing each element of the matrix **A**, t denotes the number of terms in the vector $\mathbf{X^T A}$. For our purposes, we take $t_1 = t$, because both t and t_1 essentially represent the maximum number of members that can be compromised in a network. Although t and t_1 represent two different thresholds, we require them to be the same to have a consistent threshold value for the network.

As soon as a new node with ID α (node '2' in our example) joins the network, it sends out a temporary public key, P_k to its immediate neighborhood. The format of the message sent is

$$\{REQ\_SHARE,\ \alpha,\ P_k,\ TTL\ \}$$

Here the field TTL specifies the number of hops that the message would be broadcasted to. We start with TTL = 1. If the required number of replies is obtained within a time given by T_{rep}, the process is stopped. Otherwise a new request is sent out with TTL = 2 and so on.

Any node (1, 3 and 5 in our example) in the immediate neighborhood that receives a REQ_SHARE message responds in the following manner:
The node first computes

$$S_{ij} = s_{ij}(\alpha)^i \qquad\qquad (0 \le i, j \le t-1) \qquad (6)$$

The node now computes

$$H_{\beta j} = \sum_{i=0}^{t-1} S_{ij} \qquad\qquad (0 \le j \le t) \qquad (7)$$

and sends $E_{Pk}(H_{\beta 0}, H_{\beta 1}, H_{\beta 2}, \dots, H_{\beta(t-1)}, \beta)$ to the node α. Here E_{Pk} implies encryption using the public key P_k. For example, the quantity H_{31} sent by node 3 to node 2 would be:

$$(a_{00} + 3 \cdot b_{001} + 3^2 \cdot b_{002})\ 2^0 + (a_{10} + 3 \cdot b_{101} + 3^2 \cdot b_{102})\ 2^1 + (a_{20} + 3 \cdot b_{201} + 3^2 \cdot b_{202}) \cdot 2^2$$

Let V_j denote the quantity

$$V_j = \sum_{w=0}^{t-1} a_{wj} \alpha^w \qquad\qquad (8)$$

The row vector $\mathbf{V} = [V_0, V_1, \dots, V_{t-1}]$ thus denotes the quantity $\mathbf{X}^T\mathbf{A}$ by substituting $m = \alpha$ in Eq. (4). This is all that the node α would require in order to find out any pairwise key with any other node.
In our example, the values V_i for node 2 is given by:

$$V_0 = a_{00} 2^0 + a_{10} 2^1 + a_{20} 2^2$$
$$V_1 = a_{01} 2^0 + a_{11} 2^1 + a_{21} 2^2$$
$$V_2 = a_{02} 2^0 + a_{12} 2^1 + a_{22} 2^2$$

and the vector V is $[V_0, V_1, V_2]$.

We now show how the node α computes V after it has obtained the shares from at least t nodes. Let $V_{i\beta}$ be the share of V_i obtained from node β. Thus, rewriting Eq. (7),

$$H_{\beta j} = \sum_{w=0}^{t} a_{wj} \alpha^w + \sum_{w=0}^{t-1}(\sum_{m=1}^{t-1} b_{wjm} \beta^m) \alpha^w \qquad (9)$$

as can be derived from Eqs. (5)-(7).

Equation (9) has two terms. The first term is the quantity V_j. The second term constitutes a set of $t-1$ terms . Thus, t such equations would enable node α to calculate the value of each term. Node α would keep only the first term as the value of V_i and discard all other values.

Using the vector $\mathbf{V} = [V_0, V_1, ...V_{t-1}]$, node α can now find out its pairwise key with any other node. (Note that solving for the vector $\mathbf{V}$ is only as complicated as finding t secrets in Shamir's threshold scheme). In the considered example, the above shares for V_0, V_1 and V_2 obtained by node 2 are shown in Table 1.

	Node 1	Node 3	Node 5
Shares for V_0	$(a_{00} + 2 \cdot a_{10} + 2^2 \cdot a_{20}) +$ $1 \cdot (b_{001} + 2 \cdot b_{101} + 2^2 \cdot b_{201}) +$ $1^2 \cdot (b_{002} + 2 \cdot b_{102} + 2^2 \cdot b_{202})$	$(a_{00} + 2 \cdot a_{10} + 2^2 \cdot a_{20}) +$ $3 \cdot (b_{001} + 2 \cdot b_{101} + 2^2 \cdot b_{201}) +$ $3^2 \cdot (b_{002} + 2 \cdot b_{102} + 2^2 \cdot b_{202})$	$(a_{00} + 2 \cdot a_{10} + 2^2 \cdot a_{20}) +$ $5 \cdot (b_{001} + 2 \cdot b_{101} + 2^2 \cdot b_{201}) +$ $5^2 \cdot (b_{002} + 2 \cdot b_{102} + 2^2 \cdot b_{202})$
Shares for V_1	$(a_{01} + 2.a_{11} + 2^2.a_{21}) +$ $1 \cdot (b_{011} + 2 \cdot b_{111} + 2^2 \cdot b_{211}) +$ $1^2 \cdot (b_{012} + 2.b_{112} + 2^2.b_{212})$	$(a_{01} + 2.a_{11} + 2^2.a_{21}) +$ $3.(b_{011} + 2 \cdot b_{111} + 2^2 \cdot b_{211}) +$ $3^2.(b_{012} + 2 \cdot b_{112} + 2^2 \cdot b_{212})$	$(a_{01} + 2 \cdot a_{11} + 2^2 \cdot a_{21}) +$ $5 \cdot (b_{011} + 2 \cdot b_{111} + 2^2 \cdot b_{211}) +$ $5^2 \cdot (b_{012} + 2 \cdot b_{112} + 2^2 \cdot b_{212})$
Shares for V_2	$(a_{01} + 2.a_{11} + 2^2.a_{21}) +$ $1 \cdot (b_{011} + 2 \cdot b_{111} + 2^2 \cdot b_{211}) +$ $1^2 \cdot (b_{012} + 2 \cdot b_{112} + 2^2 \cdot b_{212})$	$(a_{01} + 2 \cdot a_{11} + 2^2 \cdot a_{21}) +$ $3 \cdot (b_{011} + 2 \cdot b_{111} + 2^2 \cdot b_{211}) +$ $3^2 \cdot (b_{012} + 2 \cdot b_{112} + 2^2 \cdot b_{212})$	$(a_{01} + 2.a_{11} + 2^2.a_{21}) +$ $5 \cdot (b_{011} + 2 \cdot b_{111} + 2^2 \cdot b_{211}) +$ $5^2 \cdot (b_{012} + 2 \cdot b_{112} + 2^2 \cdot b_{212})$

Table 1. Example scenario with threshold t=3; the requesting node with ID 2 and the responding nodes with IDs 1, 3 and 5. The values in the table indicate the shares for V_0, V_1 and V_2 as sent by nodes 1, 3 and 5 to node 2. Using these values node 2 would be able to compute the values for V_0, V_1 and V_2.

In this manner, neither an eavesdropping node, nor the incoming node would have any knowledge about the matrix $\mathbf{A}$ since the coefficients of $\mathbf{A}$ are never revealed. Also, capture of a node simply compromises that node's pairwise keys and provides no information about the pairwise keys between any other two nodes. We thus have a fully distributed key generation scheme resilient to $t-1$ nodes getting compromised.

4. PERFORMANCE EVALUATION

In this section we compare the performance of our scheme with that of other key distribution schemes which employ similar messaging systems. Based on our observations, we also suggest a variation to the key exchange mechanism to optimize on the message overhead and latency. We first look at the number and length of messages that need to be sent to a node when it joins the network.

Considering a finite field of length q, length of each share is $log_2(q)$. Number of shares sent by a node is t. Thus, the length of the message sent by a responding node is $t*log_2q$. The number of such messages is at least t. We thus have a total of at least $t^2 log_2q$ bits reaching the requesting node.

We simulated the working of our scheme to measure the performance for various network densities and threshold values. The simulations were carried out in ns-2 with the number of nodes kept at 40 over an area of

1000x1000m. The communication range was 250m. The threshold values were varied from 3 to 11.

We compared our scheme with the Threshold RSA (TRSA) scheme suggested in [5] and the ID-based scheme suggested in [3] to compare the performance in terms of the delay and message overhead. The basic idea behind the TRSA scheme is to generate authorized certificates in a distributed manner. A node joining the network requests for its share of the certificates from its neighbors and builds a certificate for its public key. The node now sends its public key along with the certificate to all the nodes in the network. In the ID based scheme, the node uses its ID as its public key and obtains shares for its private key from the network. For any communication, a session key is now established between a pair of nodes.

Both the above schemes are similar to our scheme as far as the message overheads are concerned. In all the three schemes, a node joining the network requests for its keying material from its neighbors (immediate or multihop). The behavior of our scheme is different from that of the ID-based scheme or TRSA since in our scheme, the size of the packets depends on the value of the threshold. In the other two schemes, the message size is independent of the threshold values. Thus, for small values of t, our scheme performs better and for large values, the performance of TRSA and ID-based schemes are better.

Also, our scheme introduces minimal post-key-exchange overhead on the network since no key announcements or session key establishment is needed prior to communication. This means that the performance of our system would be much better when averaged over the lifetime of the network.

We would like to note that although both the TRSA and the ID-based schemes deal with public key cryptosystems, we use them as metrics for comparison as these schemes give a good measure of the message overhead involved in obtaining keying material from the network. Also, for any communication to take place, a session key has to be established.

Our simulation goals were as follows:
- To observe the effects of the threshold values on the latency in obtaining the key shares
- To observe the effects of the threshold values on the network overhead.
- To see how our scheme performs under mobile conditions

Compare our scheme with TRSA and the ID-based scheme while considering the above parameters. We first looked at the latencies involved in getting all the required shares for the keys. The speed of the nodes is kept fixed at 30 m/s and the value of T_{rep} for our scheme was kept fixed at 1s. Figure 1 shows the obtained results. We call our scheme DBK for Distributed Bivariate Keying.

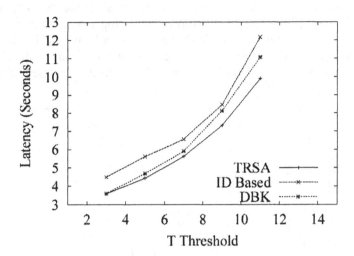

Figure 1: Latencies involved in obtaining the entire keying material from the network.

As can be seen, the latencies increase with an increase in the threshold values. An interesting point to be noted here is that as the threshold increases, the latency values for DBK approaches that of the TRSA scheme. This can be attributed to the fact that in DBK, the number of bytes being transmitted increases as the threshold values increase.

We next varied the threshold values and tried to obtain an estimate of the message overhead involved in obtaining the keying material from the network. Figure 2 plots the number of replies obtained by the requesting node. Again here the performance of DBK is midway between the performance of TRSA and the ID–based scheme. As the packet sizes increase, the latencies involved in the network also go up. Thus, with everything else remaining the same, with an increase in the packet size, packets take longer to reach their destination. If this delay becomes more than T_{rep}, a retransmission of the request packet happens and as a result, eventually more replies are obtained.

Figure 2 also shows the plots for DBK with $T_{rep} = 2$ and with $T_{rep} = 4$. The number of replies obtained for $T_{rep} = 4$ are much less than those obtained for $T_{rep} = 2$. This is evident from the fact that a larger wait would allow the node to get more replies before deciding to send out more requests at a higher latency cost.

We then see the effects of mobility on the latency. As can be seen in Figure 3, latency values drastically go down when the speeds are low. With an increase in speed, the latencies tend to become constant and stabilize to a value between 7 and 8. The reason for this is that with low mobility, nodes are able to obtain results faster from the two hop neighbors and do not send out additional requests for key shares. However, since the mode of sending requests is an n-hop broadcast (with n increasing from 1 onwards), the requesting node manages to obtain the replies faster by physically moving to new locations and new neighbors. As can be seen in Figure 3, the obtained values for DBK are lesser than TRSA and more than the ID-based scheme.

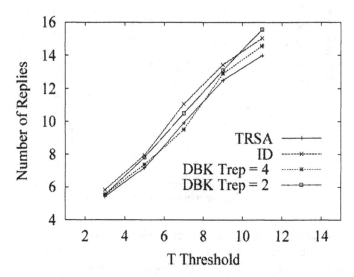

Figure 2: Total number of replies obtained by a node
after sending out the request.

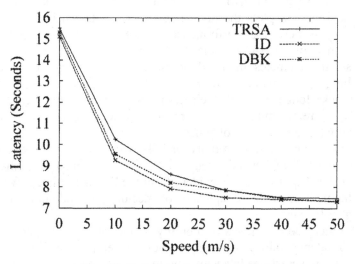

Figure 3: Variation in the latencies with the average
speed of the mobile nodes.

As shown by the simulations, lower the value of T_{rep}, higher are the number of replies obtained and thus higher is the message overhead. However, increasing the value of T_{rep} introduces higher latencies into the system. We also observe that with higher mobility, latencies go down drastically.

- We end this section by noting the following:
- The rate of increase of the TTL values should depend on the mobility of the nodes. Higher the mobility, lower should be the rate of increase.
- The initialization of the TTL field should depend on the network density and the threshold value. For sparse networks and/or high threshold, TTL should be set to a high initial value to avoid sending multiple requests. For dense networks and/or low threshold values, TTL can be initialized to a low initial value as the replies can be obtained from the immediate neighborhood.
- The T_{rep} values should be fine tuned to meet specific network densities according to the required latencies and message overhead.
- We have not investigated the above points in detail in this work and they constitute our future work.

5. CONCLUSION

In this work we have demonstrated a distributed symmetric key exchange mechanism by sharing polynomials at fixed points using Shamir's *t*-threshold scheme. Using this we have shown how a distributed scheme can provide an incoming node with the keying material. Our scheme is secure to less than *t+1* nodes getting compromised. We have shown through simulations that the message overhead and the latencies involved in the key exchange process is well within bounds of other similar protocols. Also, since no key announcements are needed and no session keys have to be established, our scheme has a much better runtime performance.

Our idea can also be extended to provide a hierarchical key generation mechanism based on the level of trust that a node wishes to provide to another node. Although we have not developed this idea, it can provide grounds for future work.

ACKNOWLEDGEMENTS

This work has been partially supported by the Ohio Board of Regents Doctoral Enhancement Funds and the National Science Foundation, under grant CCR-0113361.

REFERENCES

1. C. Blundo, A. De Santis, A. Herzberg, S. Kutten, U. Vaccaro, M. Yung, Perfectly-secure key distribution for dynamic conferences, *Advancesin Cryptology – CRYPTO* '92, LNCS 740 1993.
2. H. Chan, A. Perrig, D. Song, Random key predistribution schemes for sensor networks, *Proceedings of the IEEE Symposium on Research in Security and Privacy*, 2003.
3. H. Deng, A. Mukherjee, and D. P. Agrawal, Threshold and Identity-based Key Management and Authentication for Wireless Ad Hoc Networks, *IEEE International Conferences on Information Technology (ITCC'04)*, April 5-7, 2004.
4. L. Eschenauer and V.D. Gligor, A key-management scheme for distributed sensor networks, *Proceedings of he 9th ACM Conference on Computer and Communications Security*, November 2002.
5. J. Kong, P. Zerfos, H. Luo, S. Lu and L. Zhang, Providing Robust and Ubiquitous Security Support for Mobile Ad-Hoc Networks, *Proceedings of the IEEE 9th International Conference on Network Protocols (ICNP'01)*, 2001.
6. A. L. Murphy, G.-C. Roman, G. Varghese, An Exercise in Formal Reasoning about Mobile Communications, Proceedings of the Ninth International Workshop on Software Specifications and Design, IEEE Computer Society Technical Council on Software Engineering, IEEE Computer Society, Ise-Shima, Japan, pp.25-33, April 1998.
7. P. Papadimitratos and Z. Haas, Secure Routing for Mobile Ad Hoc Networks, Proceedings of the SCS Communication Networks and Distributed Systems Modeling and Simulation Conference, January 2002.
8. A. Shamir, How to share a secret, *Communications of the ACM*, Vol. 22, pp. 612-613, November 1979.
9. L. Zhou and Zygmunt J. Haas, Securing ad hoc networks, *IEEE Network*, 13(6):24–30, 1999.

COLLABORATION ENFORCEMENT AND ADAPTIVE DATA REDIRECTION IN MOBILE AD HOC NETWORKS USING ONLY FIRST-HAND EXPERIENCE

Ning Jiang, Kien A. Hua, Mounir A. Tantaoui
{*njiang, kienhua, tantaoui*}*@cs.ucf.edu*
School of Computer Science
University of Central Florida
Orlando, FL 32816-2362

Abstract: **In *Mobile Ad Hoc Networks* (MANETs), all participating hosts are obligated to route and forward data for others to guarantee the availability of network applications and services. Most of the contemporary collaboration enforcement techniques employ reputation mechanisms for nodes to avoid and penalize malicious participants. Reputation information is updated based on complicated trust relationships among hosts and other techniques to thwart false accusation of benign nodes. The aforementioned strategy suffers from low scalability and is likely to be exploited by adversaries. In this paper, we propose a novel approach to address the aforementioned problems. With the proposed technique, no reputation information is propagated in the network and malicious nodes cannot cause false penalty to benign hosts. Misbehaving nodes are penalized and circumvented by benign nodes within their localities based on first-hand experiences. This approach significantly simplifies the collaboration enforcement process, incurs very low overhead, and is robust against various evasive behaviors. Simulations based on various system configurations demonstrate that overall network performance is greatly enhanced.**

Key words: Mobile Ad Hoc Network, Collaboration enforcement, Reputation, First-hand experience

1.   INTRODUCTION

Mobile Ad hoc NETworks (MANETs) has attracted great research interest in recent years. A Mobile Ad Hoc Network is a self-organizing multi-hop wireless network where all hosts (often called nodes) participate in the routing and data forwarding process. The dependence on nodes to relay data packets for others makes mobile ad hoc networks extremely susceptible to various malicious and selfish behaviors. This point is largely overlooked during the early stage of MANET research. Many works simply assume nodes are inherently cooperative and benign. However, experiences from the wired world manifest that the reverse is usually true; and many works [3] [10] [9] [8] [12] [19] have pointed out that the impact of malicious and selfish users must be carefully investigated. The goal of this research is to address the cooperation problem and related security issues in wireless ad hoc networks. As a rule of thumb, it is more desirable to include security mechanisms in the design phase rather than continually patching the system for security breaches.

As pointed out in [2] [1], there can be both selfish and malicious nodes in a mobile ad hoc network. Selfish nodes are most concerned about their energy consumption and intentionally drop packets to save power. The purpose of malicious nodes, on the other hand, is to attack the network using various intrusive techniques. In general, nodes in an ad hoc network can exhibit Byzantine behaviors. That is, they can drop, modify, or misroute data packets. As a result, the availability and robustness of the network are severely compromised. A common solution to combat such problems is for each node to maintain a reputation list of other nodes, as proposed in [1] [3][13][14]. In these techniques, misbehaving nodes are detected and a rating algorithm is employed to avoid and penalize them. These schemes are not scalable and suffer from high overhead since they require synchronization of reputation information throughout the network, and manipulation of complicated trust relationships among hosts to thwart false accusation of benign nodes. In this paper, we propose a novel approach to strengthening collaboration in MANETs. With this scheme, no reputation information needs to be propagated in the network and malicious nodes cannot cause false penalty to benign hosts. Misbehaving nodes are penalized and circumvented by benign nodes within their localities based on first-hand experiences. This approach significantly simplifies the collaboration enforcement process, incurs very low overhead, and is robust against various evasive behaviors. Simulations based on various system configurations demonstrate that overall network performance is greatly enhanced.

The remainder of this paper is organized as follows. We discuss related works in Section 2. In Section 3, we introduce the selfish/malicious node detection mechanism, and also present the proactive rerouting techniques. Experimental results are given in Section 4. Finally, we conclude the paper in Section 5.

2. RELATED WORK

The current state of the art in enforcing collaboration in mobile ad hoc networks can be categorized into two groups, namely incentive motivation approaches and misbehavior penalty approaches.

The incentive motivation techniques are discussed in [5] [6] [7] [20]. These techniques either rely on tamper proof security modules or assume a central control service. The practicability and performance remain unclear.

Our research, on the other hand, falls in the second category. The main idea is to detect and penalize malicious and selfish behaviors. In [18], the authors use intrusion detection techniques to locate misbehaving nodes. A watchdog and a path rater approach is proposed in [13] to detect and circumvent selfish nodes. The main drawback of this approach is that it does not punish malicious nodes. This problem is addressed in [2] [3] [4]. The approach, called CONFIDANT, introduces a reputation system whereby each node keeps a list of the reputations of others. Malicious and selfish nodes are detected and reputation information is propagated to "friend" nodes, which update their reputation lists based on certain trust relationships. During route discovery, nodes try to avoid routes that contain nodes with bad reputations. Meanwhile, no data forwarding service is provided for low reputation nodes as a punishment. Another reputation-based technique, called CORE, is proposed in [14]. In [1], the authors attack the problem of defending application data transmission against Byzantine errors. In their approach, each node maintains a weight list of other nodes. Malicious nodes are located by an on-demand detection process and their weights are increased consequently. A routing protocol is designed to select the least-weight path between two nodes. This approach is also based on per-node reputation lists. In addition, the detection process requires that each intermediate node transmit an acknowledgement packet to the source node.

In general, most of existing detection and reaction techniques are based on global reputation mechanism and suffer from the following drawbacks. First, global reputation schemes have low scalability. Significant overhead is needed to propagate reputation information for all the benign nodes to avoid and punish "bad" citizens. Likewise, considerable efforts need to be made for malicious or falsely accused nodes to rejoin the network. Second, global reputation schemes offer incentives to various attacks. Most prominently, malicious users can "poison" the reputation lists by disseminating incorrect reputation information. Such packets can be spoofed with other nodes' addresses to hide the identity of the attacker or to pretend to be a "friend" of the receiver. In [1], digital signatures and message authentication codes [22] are employed to defeat packet spoofing. However, if a host is possessed (or physically captured) by a malicious user, cryptographic information of the particular node can be extracted and reputation poison attacks can still be mounted.

We propose a technique to address all the aforementioned problems by using only first-hand experience at each individual node instead of relying on globally propagated reputation. This strategy is both effective and efficient.

3.   THE EXPERIENCE-BASED APPROACH

In this section, we introduce the proposed experience-based techniques. Our approach is based on the following fundamental characteristics of MANETs:

- Each packet transmitted by a node A to a destination node more than one hop away must go through one of A's neighboring nodes.
- A's neighboring nodes can overhear its packet transmission.

Given a selfish node M, its un-collaborative behavior can be captured by most, if not all, of its neighboring nodes. Each of these nodes will then penalize M by rejecting all its packets. As a result, M will not be able to send any data to nodes more than one hop away. For a benign node B, if B is relaying packets for a source node S and is aware that the next hop node H is a selfish node, B can redirect the packets to avoid H. Note that the rerouting operation requires collaboration from B for S. We also present techniques to enforce such collaboration.

3.1   Node Configurations

The proposed technique is based on nodes with the following configuration. First, nodes are equipped with omni-directional antennas and wireless interface cards that can be switched to promiscuous mode to "hear" data transmission in their proximities. Second, we base our discussion on the Dynamic Source Routing protocol [11], as it is one of the most frequently used routing protocols in the literature. However, the technique can be extended to accommodate other reactive routing protocols. Overview of DSR is omitted in the interest of space. Third, 802.11 [21] is employed at the MAC layer. Finally, nodes have knowledge of their one-hop neighboring nodes. This can be achieved by either employing a HELLO protocol or by overhearing packets transmitted within the locality.

3.2   Selfish and Malicious Behaviors Considered

A selfish node can avoid the responsibility of forwarding data in two ways. First, by not participating in route discovery, a node will never show up in any routing control packets and will thus be completely released from forwarding data packets. Second, a selfish host can cooperate in route discovery, but subsequently discards data packets to save energy. We focus on the second type of selfishness in this paper as it is pointed out in [16] that such misbehavior has more negative impact on overall network throughput.

3.3   Detection and Punishment of Selfishness and Malice in Data Forwarding

In the proposed technique, each node maintains a list of its neighboring nodes and tracks their actions. Nodes make no assumption of other hosts beyond their direct observable regions.

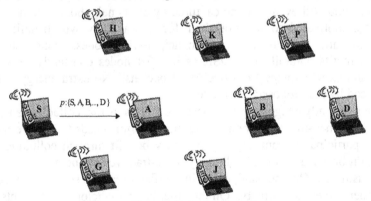

Figure 1. Detection example

We discuss the detection mechanism through an example depicted in Figure 1. It shows a node S transmitting data to a node D using a route $\{S, A, B, D\}$. Node A is a selfish node that does not forward the data packets to save energy. Assume nodes H and G are neighboring to both S and A, and Nodes K and J are neighboring nodes of both A and B. Each node allocates some memory buffer to store packets transmitted by its neighboring nodes. Let us consider node S first. After S transmits a data packet to A, it waits for a certain time interval and validates whether A has properly forwarded the packet by checking its memory buffer. Given the current validation time t, S

maintains a set Ω of all the packets it has transmitted over a time window defined by $[t - W_{UPPER}, t - W_{LOWER}]$. If the cardinality of Ω is greater than a threshold T_{SUM}, S computes the packet drop ratio for node A on Ω. If this ratio is beyond a given threshold $T_{SELFISH}$, then S tallies A as a selfish node; otherwise, S deems A as benign. The proposed detection procedure distinguishes link breakage and temporary network congestion from deliberate packet discarding, and effectively reduces false classifications. Essentially, selfish intention is sustained if and only if a node has been observed to drop a significant number of packets over a long enough timeframe. We now consider nodes G and H. They, as neighboring nodes of S, overhear all data packets sent by S and can learn about the next hop (A in this example) of each data packet p by extracting the source route option field of p's IP header. As G and H are both neighboring to A and since S is a benign node, G and H will further detect whether A relays the packet using the same technique used in S. In this example, both G and H will eventually identify A as a selfish node based on their own observations. On the other hand, although K and J are also neighbors of node A, they will not be able to detect A's misbehavior since they have no access to the packets sent by the previous hop to A (S in this example). We refer to this scenario as "*asymmetric sensing.*" In practice, A is likely to receive packets from all directions and will eventually be captured by all its neighbors. We note that users are motivated to monitor their locality as they will benefit from identifying and circumventing selfish neighboring nodes. This detection mechanism fits naturally into DSR as in DSR nodes constantly sense the media and extract routes from overheard packets. No extra energy cost is introduced by the proposed technique.

One major advantage of the proposed technique is that colluding is not an issue. In money-incentive models, significant effort needs to be invested to prevent participants from gaining monetary benefit through colluding. In reputation-based schemes, colluding is attractive to both selfish and malicious users. On one hand, colluding selfish users can successfully cover each other and escape penalty. On the other hand, malicious participants can collaboratively cause various undesirable effects to benign users. Since the proposed technique is based on direct experience at individual nodes, not through "rumor" or "propagated information," colluding is not possible in this new environment.

Punishment is enforced as follows. Consider a node H, which identifies node A as a selfish or malicious node, it will reject data packets originated by A for a period of time as a penalty. More specifically, H's decision on whether to forward a data packet p for A is based on the difference between the time H receives p and the latest recorded time when A was identified as a misbehaving node. If this difference falls within a threshold defined as *penalty interval* τ, H will reject the packet. Consequently, the penalty will not end as long as *A* continues to misbehave, and the actual penalty time is proportional to the length of *A*'s misbehavior.

3.4   Dynamic Redirection

In global reputation mechanism, two scenarios will cause a source node to reroute data packets over a particular node. First, when a node detects a selfish or malicious node, it informs other nodes (including the source node of the session) through *reputation packets* so that they can choose a "clean" route to circumvent the selfish node. Second, Route Error (RERR) packets are transmitted to the source node when broken links are encountered[1]. In both cases, source nodes are responsible for rerouting the data. In the proposed technique, we allow neither of the above packets to be propagated. An obvious question is: who should reroute the data packets to bypass both irresponsible nodes and broken links?

Our solution is that each node shares the responsibility of rerouting packets. Again, we use Figure 1 to illustrate the idea. We assume that node *S* is sending data to node *D* through a path {*S*, *A*, *B*, *D*}. Suppose the link between node *A* and node *B* is a *malfunction link* (i.e. either broken or node *B* is selfish). Without loss of generality, we assume that node *B* is a selfish node. After relaying a certain number of packets, node *A* will realize that *B* is a selfish node. We refer to node *A* as a *proxy* of source node *S*[2]. In our approach, *A* first purges all paths containing node *B* as an intermediate node from its route cache. Next, when *A* receives subsequent data packets from *S*, it broadcasts a Route Redirect (RRDIR) packet, indicating node *B* as a *bypassing target* and then reroutes the packets by obtaining an alternative clean route to node *D* from its route cache. If such a route does not exist in its cache, *A* will buffer the data packets and instantiate a route discovery process to locate a path to *D*. In Figure 1, *A* will discover a new route {*K*, *P*, *D*}, revise the embedded route of each data packet and relay them to the destination. In this case, the actual route data packets traverse from *S* to *D* is {*S*, *A*, *K*, *P*, *D*}. It is possible for several proxy nodes to adaptively reroute data packets to avoid multiple selfish nodes along the chosen route. If *A* cannot find a route to *D* after a certain number of retries, it informs *S* through a RERR packet.

The proper functioning of the proposed selfish and malicious node circumvention scheme relies on the collaboration of proxy nodes. Unfortunately, proxy nodes can act maliciously to either avoid the reroute

---

[1] Selfish nodes can falsely claim broken links to be excluded from packet transmission sessions.

[2] The proxy of a source node can be the source node itself when its next hop is selfish.

task or mount denial of service attacks. Continue the above example. When node A receives a data packet from S, it has the following choices.

- Node A can mount a denial of service attack to S by deliberately forwarding packets to B even though it is aware that B is a selfish node. Nodes K and J will detect such attack as follows. First, both nodes will identify node B as a misbehaving node and they will assume that A has reached the same conclusion. Next, as A makes no effort to bypass B, both K and J will mark A as a malicious node and starts to penalize it.

- A does not reroute the packet and simply reports a RERR back to the source. In this case, all its neighboring nodes (S, G, H, J, and K) hear the RERR packets whereas none of them is aware of any route discovery attempt made by A. Thus, all of them will deem A as a selfish node.

- A broadcasts a RRDIR packet and then starts a route discovery process. Nevertheless, A reports a RERR to the source regardless of whether it receives RREP packets from the destination. The countermeasure we design involves utilizing some context information. After A sends a RREQ packet to look for a route to D, all its neighboring nodes will wait for the RREP packet to come back. Suppose node K relays the replying RREP packet to A and assume node H also hears the packet. Both H and K will expect to see node A transmit data to node D. However, as A sends a RERR packet, both nodes will recognize A as misbehaving. Furthermore, other neighboring nodes (S, G, and J) will deduct certain number of points for node A (say, equivalent to one third of those deducted for packet dropping). In other words, failure to reroute data packets is deemed as low-weight misbehavior. The purpose of this design is to discourage un-collaborative behavior. Benign nodes always collaborate and will not suffer from such deduction.

- A broadcasts a RRDIR packet and reroutes data through a fabricated path. This attack has very limited effect in that benign nodes along the faked route will reroute the data packets and node A still has to relay data.

Another concern is that malicious nodes might exploit the reroute mechanism to disrupt data transmission. For instance, in Figure 1, suppose A is a malicious node. When it receives a data packet that it should forward to a benign node B, it redirects the packet to a different (fabricated) route, hoping that other nodes along the redirected route will drop the packet. With our redirection mechanism, A has to broadcast a RRDIR packet. Otherwise its neighboring nodes (S, H, and G) will identify it as a malicious node. In the RRDIR packet, A has to declare the correct next hop (B in this case) that it intends to bypass. Otherwise, it will be captured by S, H, and G. After receiving A's RRDIR packet, node B will realize A's attempt to deviate packets from a valid route and penalize A. Nodes K and J will also penalize A as they both recognize B as a benign node through their own experiences.

Moreover, nodes that reroute packets for an excessive number of sessions within a certain time period will be considered as malicious and penalized by their neighbors.

4.   EXPERIMENTAL STUDY

We implemented four schemes, namely the reference scheme, the defenseless scheme, the reputation-based scheme and the proposed experience-based scheme, for performance evaluation. In the reference scheme, all the nodes act collaboratively and relay data for each other. In the defenseless scheme, a certain fraction of nodes are selfish as they forward routing packets, but discard any data packet not destined at them. No detection or prevention mechanism is implemented so that the network is totally "defenseless". Next, we implemented a reputation-based system. In this scheme, each node maintains global reputation of other nodes. Nodes update reputation of others as follows. First, nodes monitor and form their opinion about the reputation of neighboring nodes. Nodes always trust their first-hand experiences with other nodes and ignore any reputation information against their own belief. Next, when a node detects a selfish node, it informs the source node of the communication session through a *reputation packet*. Finally, each node periodically broadcasts reputation of other nodes in its locality. We implemented three types of nodes in this scheme, namely benign node, selfish node, and cheating node. A benign node always truthfully broadcasts the reputation information it has observed first hand, and honestly forwards the reputation packets. A selfish node does not participate in data packet forwarding but cooperates in disseminating reputation information (i.e. it generates and relays reputation packets and never lies about other nodes). A cheating node generates genuine reputation packets and relays both data and reputation packets for others. During reputation broadcast, however, it always lies about the reputation of its neighboring nodes. For all other nodes it is aware of, the cheating node simply reports them as selfish.

We performed all the experiments based on GlomoSim [17], a packet-level simulation package for wireless ad hoc networks. Our experiments were based on a mobile ad hoc network with 50 nodes within a 700x700-square-meter 2-dimensional space. The simulation duration for each run was 10 minutes. The random waypoint model was used to model host mobility. We experimented with 5 and 10 selfish nodes, accounting for 10% and 20% of total number of nodes, respectively. Selfish nodes are randomly generated for all the simulation schemes. We tested the reputation-based system with 0 and 5 randomly selected cheating nodes. Each configuration was executed under 5 different random seeds and the average values of the metric variables were calculated. Constant Bit Rate (CBR) applications were used in this study. For each simulation run, we randomly generated a total of 10 CBR client/server sessions. In particular, we generated three *selfish sessions* (i.e. sessions originated by selfish nodes) and seven *benign sessions* (i.e. sessions started by benign nodes).

In the experiments, we evaluated the proposed scheme based on the goodput of benign sessions and selfish sessions. A good collaboration enforcement technique should ensure a high benign session goodput as well as suppressing selfish session goodput to discourage misbehaviors.

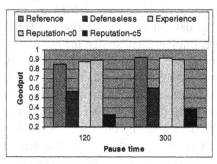
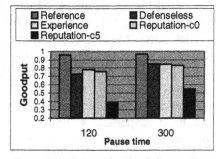

Figure 2. Benign session goodput (m=10)   *Figure 3*. Selfish session goodput (m=10)

Figure 2 and Figure 3 illustrate the goodput of the experimented schemes when there are ten selfish nodes. In both figures, we refer to the proposed scheme as "Experience," and the reputation-based scheme as "*Reputation-cX*", where X indicates the number of cheating nodes. From Figure 2 we observe that by employing the proposed scheme, significantly more data are successfully delivered to the destination nodes since proxy nodes proactively detect and reroute data around misbehaving nodes. The proposed technique lifted the goodput from around 0.6 in a defenseless network to higher than 0.85, an improvement of more than 40%. From Figure 3, we observe that the goodput experienced by selfish users is lower than what collaborative users enjoy. When the pause time is 300 second the goodput of benign sessions is approximately 0.92 (Figure 2) as opposed to 0.81 in the case of selfish sessions (Figure 3). Finally, in all the experiments, the reputation-based scheme suffered from significant performance loss (more than 50%) when only a few cheating nodes were present. We thus conclude that experience-based scheme is more suitable for MANETs due to its resilience to performance degradation caused by reputation poisoning behaviors.

5. CONCLUDING REMARKS

In mobile ad hoc networks, there is no fixed infrastructure readily available to relay packets. Instead, nodes are obligated to cooperate in routing and forwarding packets. However, it might be advantageous for some nodes not to collaborate for reasons such as saving power and launching denial of service attacks. Therefore, enforcing collaboration is essential in mobile ad hoc networks.

In most existing techniques, collaboration enforcement is achieved by a detect-and-react mechanism. In which, each node maintains global reputation of others in order to avoid and penalize misbehaving nodes. Propagation of reputation information is accomplished through complicated trust relationships. Such techniques incur scalability problems and are vulnerable to various reputation poisoning attacks.

In this paper, we proposed a novel approach to enforcing collaboration and security in mobile ad hoc networks. In our technique, nodes keep local reputation of their neighboring nodes through direct observation. No reputation advertisement is initiated or accepted. Nodes dynamically redirect data packets to avoid recognized adversaries. The redirect operation is also guarded against various evasive attempts. The advantages of this approach are many. First, since it does not rely on propagated reputation information, there is no need to maintain complex trust relationships. Second, since the misbehavior detection mechanism is based on first-hand experience at individual nodes, denial of service attacks are much more difficult to achieve. Colluding among nodes to secretly carry out fraudulent acts is not possible either.

We conducted various experiments to investigate the effectiveness and efficiency of the proposed technique. Simulation results, based on GlomoSim, indicate that this technique is very effective in improving network performance. It also works well in disciplining defecting hosts. More importantly, the success of the proposed technique does not rely on reputation exchange and is thus both scalable and robust.

REFERENCES

[1] B. Awerbuch, D. Holmer, C. Nita-Rotaru and H. Rubens. An On-Demand Secure Routing Protocol Resilient to Byzantine Failures. ACM Workshop on Wireless Security (WiSe) 2002.

[2] S. Buchegger and J. L. Boudec, IBM Research Report: The Selfish Node: Increasing Routing Security in Mobile Ad Hoc Networks. RR 3354, 2001.

[3] S. Buchegger and J. L. Boudec. Performance Analysis of the CONFIDANT Protocol: Cooperation of Nodes – Fairness In Dynamic Ad Hoc Networks. In Proceedings of

IEEE/ACM Symposium on Mobile Ad Hoc Networking and Computing (MobiHOC), Lausanne, CH, June 2002.

[4] . S. Buchegger, H. L. Boudec, Coping with False Accusations in Misbehavior Reputation Systems for Mobile Ad-hoc Networks. EPFL Technical Report IC/2003/31.

[5] L. Buttyan and J. Hubaux. Enforcing Service Availability in Mobile Ad Hoc WANs. In Proceedings of IEEE/ACM Workshop on Mobile Ad Hoc Networking and Computing (MobiHOC), Boston, MA, USA, August 2000.

[6] L. Buttyan and J. Hubaux. Stimulating Cooperation in Self-Organizing Mobile Ad Hoc Networks. Technical Report DSC/2001/046, EPFL-DI-ICA, August 2001.

[7] Stephan Eidenbenz and Luzi Anderegg, Ad hoc-VCG: A Truthful and Cost-Efficient Routing Protocol for Mobile Ad Hoc Networks with Selfish Agents. In Proceedings of the Ninth Annual International Conference on Mobile Computing and Networking (MobiCom 2003), September 2003.

[8] Y. Hu, A. Perrig, and D. B. Johnson. Ariadne: A Secure On-Demand Routing Protocol for Ad Hoc Networks. Technical Report TR01-383, Department of Computer Science, Rice University, December 2001.

[9] Y. Hu, D. B. Johnson and A. Perrig. SEAD: Secure Efficient Distance Vector Routing for Mobile Wireless AdHoc Networks. In Proceedings of the 4th IEEE Workshop on Mobile Computing Systems and Applications (WMCSA 2002). IEEE, Calicoon, NY, June 2002.

[10] J. Hubaux, L. Buttyan, and S. Capkun. The Quest for Security in Mobile Ad Hoc Networks. In Proceedings of the ACM Symposium on Mobile Ad Hoc Networking and Computing (MobiHoc) 2001.

[11] D. Johnson and D. A. Maltz. Dynamic Source Routing in Ad Hoc Wireless Networks. In T. Imielinski and H. F. Korth, editors, Mobile Computing, pages 153--181. Kluwer Academic Publishers, Dordrecht, The Netherlands, 1996.

[12] V. Kärpijoki, Security in Ad Hoc Networks. In Proceedings of the Helsinki University of Technology, Seminar on Network Security, 2000.

[13] S. Marti, T.J. Giuli, K. Lai, and M. Baker. Mitigating Routing Misbehavior in Mobile Ad Hoc Networks. In Proceedings of MOBICOM 2000, pages 255-265, 2000.

[14] P. Michiardi and R. Molva. CORE: A Collaborative Reputation Mechanism to Enforce Node Cooperation in Mobile Ad Hoc Networks. Sixth IFIP Conference on Security Communications and Multimedia (CMS 2002), Portoroz, Slovenia, 2002.

[15] P. Michiardi and R. Molva. Prevention of Denial of Service Attacks and Selfishness in Mobile Ad Hoc Networks. Research Report N° RR-02-63. January 2002.

[16] P. Michiardi, R. Molva. Simulation-based Analysis of Security Exposures in Mobile Ad Hoc Networks. European Wireless Conference, 2002.

[17] X. Zeng, R. Bagrodia, and M. Gerla. GloMoSim: A library for parallel simulation of large-scale wireless networks. Proceedings of the 12th Workshop on Parallel and Distributed Simulations (PADS '98), May 26-29, in Banff, Alberta,Canada, 1998.

[18] Y. Zhang and W. Lee. Intrusion Detection in Wireless Ad Hoc Networks. In Proceedings of MOBICOM 2000, pages 275-283, 2000.

[19] L. Zhou and Z. Haas. Securing Ad Hoc Networks. In IEEE Network magazine, special issue on networking security, Vol. 13, No. 6, November/December, pages 24-30, 1999.

[20] Sheng Zhong, Jiang Chen, Yang Richard Yang, Sprite: A Simple, Cheat-Proof, Credit-Based System for Mobile Ad-Hoc networks. IEEE INFOCOM 2003.

[21] ANSI/IEEE   Standard   802.11,   1999   Edition.   1999. http://standards.ieee.org/catalog/olis/lanman.html.

[22] Information Technology Laboratory, National Institute of Standards and Technology. The Keyed-Hash Message Authentication Code (HMAC).

A SIMPLE PRIVACY EXTENSION FOR MOBILE IPV6

Claude Castelluccia,[1] Francis Dupont,[2] and Gabriel Montenegro[3]

[1]*INRIA Rhone-Alpes*
655 Avenue de l'Europe
38334 Saint Ismier Cedex, France
Claude.Castelluccia@inria.fr

[2]*GET/ENST Bretagne*
CS 17607, 35576 Cesson-Sevigne Cedex, France
Francis.Dupont@enst-bretagne.fr

[3]*Sun Labs, Europe*
180 Avenue de l'Europe
38334 Saint Ismier Cedex, France
gab@sun.com

Abstract   In Mobile IPv6, each packet sent and received by a mobile node contains its home address. As a result, it is very easy for an eavesdropper or for a correspondent node to track the movement and usage of a mobile node. This paper proposes a simple and practical solution to this problem. The main idea is to replace the home address in the packets by a temporary mobile identifier (TMI), that is cryptographically generated and therefore random. As a result, packets cannot be linked to a mobile node anymore and traffic analysis is more difficult. With our solution, an eavesdropper can still identify the IP addresses of two communicating nodes but is not able to identify their identities (i.e., their home addresses). Furthermore since a mobile node uses a new identifier for each communication, an eavesdropper cannot link the different communications of a given mobile node together. We show that HMIPv6 can also benefit from the proposed privacy extension.

Keywords: Mobile IPv6, CGA, Privacy.

1.   Introduction

Mobile IPv6 specifies a protocol which allows nodes to remain reachable while moving around in the Internet. Each mobile node is always identified by its home address, regardless of its current point of attachment to the Internet.

In Mobile IPv6, a mobile node has two IP addresses: (1) A home address that is an address in the network the mobile node belongs to (i.e., the address in its home network). (2) A care-of address that is a temporary address in the visited network. The home address is constant but the care-of address changes as the mobile node changes links.

One privacy problem of Mobile IPv6 is that the home address of a mobile node is included in all the packets (data and signaling) that it sends and receives. As a result any eavesdropper in the network can identify packets that belong to a particular mobile node (and use them to perform some kind of traffic analysis) and track its movements (i.e., its successive care-of addresses) and usage.

The main security threat against Mobile IPv6 is the remote redirection attack, i.e., binding updates using a fake care-of address. Therefore it is critical to verify that signaling messages are properly authenticated and authorized.

In this paper, we propose a solution to prevent such tracking while still enabling route optimization. In particular, with our proposal, a mobile node can hide its identity, i.e., its home address, from any eavesdropper in the network while still being able to move. Furthermore if a mobile node initiates a communication, it can also hide its identity from its correspondent node. We only look at privacy issues in Mobile IPv6 and assume that a mobile node's identity is not revealed by other mechanisms such as network access control, IPsec setup [Kaufman, 2004], or by the applications (i.e., applications must not use any IP address in their payloads.)

Our solution is practical. It requires only few simple modifications of the Mobile IPv6 specification, it is easily deployable and it does not compromise security or affect performance.

The paper is structured as follows: Section 2 defines the problem we are addressing in this paper. Section 3 presents and analyzes some existing solutions. Section 4 details our proposal. Section 5 explains how our scheme can be combined with HMIPv6 to further improve privacy. Finally, Section 6 concludes our paper.

2.   Problem Statement

Mobile IPv6 [Johnson et al., 2004] allows nodes to move within the Internet topology while maintaining reachability and on-going connections between mobile and correspondent nodes. In Mobile IPv6, a mobile node has two IP addresses: (1) A home address that is an address in the network the mobile node belongs to (i.e. the address in its home network). (2) A care-of address that is a temporary address in the visited network. The home address is constant but the care-of address changes as the mobile node moves. The Mobile IPv6 protocol works as follows: When a mobile node moves into a new network it gets a new care-of address. It then registers the binding between its home address and its current care-of address with its home agent. A home agent is a router in the home network that is used as a redirection point. When a node wants to communicate with a mobile node, it sends the packets to the mobile node's home address. The home agent then intercepts the packets and forwards them to the mobile node current care-of address. At this point, the mobile node may decide to use the route optimization procedure. In this case, the mobile node sends a signaling message (Binding Update) to its correspondent node that contains its current care-of address. The correspondent node can then send the packets directly to the mobile node.

In Mobile IPv6, the home address of a mobile node is included in cleartext in packets it sends and receives. In fact, packets sent by a mobile node includes a home address option that contains its home address. Packets sent by a correspondent node to a given mobile node contains a routing header that includes the mobile node's home address. Furthermore when a mobile node moves to a new subnet, it sends a binding update to its correspondent nodes that contains its home address and its new Care-of Address.

As a result, any eavesdropper within the network can easily identify packets that belong to a particular home address. The eavesdropper can then identify the network the mobile node belongs to and often infer its identity. The home address can be used to perform traffic analysis and track the mobile node's movements and usage.

The goal of our work is propose a practical solution to this problem i.e. a solution that does not require to significantly modify the Mobile IPv6 specification, that is easily deployable and that does not affect performance.

Home Address Option

The home address destination option is used in a packet sent by a mobile node while away from home, to inform the recipient of that packet of the mobile node's home address. For packets sent by a mobile node while away from home, the mobile node generally uses one of its care-of addresses as the source address in the packet's IPv6 header. By including a home address option in the packet, the correspondent node receiving the packet is able to substitute the mobile node's home address for this care-of address when processing the packet, thus making the use of the care-of address transparent to the correspondent node.

The home address option must be placed as follows:

- After the routing header, if that header is present

- Before the fragment header, if that header is present

- Before the AH Header or ESP Header, if either one of those headers is present

Routing header

Before sending any packet, the sending node should examine its binding cache for an entry for the destination address to which the packet is being sent. If the sending node has a binding cache entry for this address, the sending node should use a routing header to route the packet to this mobile node (the destination node) by way of the care-of address in the binding recorded in that binding cache entry. The destination address in the packet's IPv6 header is set to the mobile node's care-of address copied from the binding cache entry.

3.   Some possible solutions

Several existing solutions are available, all with their limitations:

1 *IPv6 Privacy Extension*: a solution could be to use the privacy extension described in [Narten and Draves, 2001] to configure the home address and the care-of addresses. While this solution represents a marked improvement over the standard address configuration methods [Thomson and Narten, 1998], and should be used for the home and care-of addresses, we contend that this is not sufficient.

[Narten and Draves, 2001] causes nodes to generate global-scope addresses from interface identifiers that change over time, even in cases where the interface contains an embedded IEEE identifier. As a result when [Narten and Draves, 2001] is used to generate the home address, this address will change periodically but the network prefix (the 64 highest bits) will remain unchanged. This network prefix can still reveal much information about the mobile node's identity to an eavesdropper. This mechanism described in [Narten and Draves, 2001] must be used for the home address and care-of addresses in Mobile IPv6 but one should not rely on it to get full privacy protection.

2 *Home Address option encryption*: another solution could be to encrypt the home address option. This solution is not satisfactory because (1) it would require to modify IPsec implementation (the care-of address should then be used as traffic selector and therefore would need to be updated at each movement of the mobile node) and (2) it would make filtering difficult (currently some firewall implementations may examine the home address option for filtering purposes). Furthermore, this solution does not solve the problem of incoming packets that contain a routing header revealing the home address.

3 *IPsec bi-directional Tunnel (mobile VPN)*: a solution could be to open a bi-directional IPsec tunnel between the mobile node and its home agent [Montenegro, 2001, Arkko et al., 2004]. This solution has the following disadvantages: (1) Addition of extra bandwidth (packets need to be encapsulated) and processing overhead, (2) the routing is suboptimal: to keep Mobile IPv6 efficiency the routing optimization must remain possible.

4. Our Proposal

In our scheme a mobile node uses the privacy extension described in [Narten and Draves, 2001] to configure its home address and care-of addresses. A mobile node must use an interface identifier for its home address that is different from the one used for its care-of addresses. It should also use a new interface identifier when configuring a new care-of address. As a result, it would be more difficult for an eavesdropper to infer the mobile node's identity and track its movement.

We also assign to each mobile node a TMI (Temporary Mobile Identifier) that is a 128-bit long random number. This TMI is used by the mobile node's home agent and correspondent nodes to securely identify the mobile node.

This TMI might be used by the correspondent node as the mobile address in the traffic selectors of the IPsec security association and might also be used by firewalls to perform filtering.

4.1 Temporary Mobile Identifier (TMI)

The TMI of a mobile node must be globally unique. The consequences of two mobile nodes using the same TMI is similar than the consequences of two mobile nodes using the same home address with standard mobile IPv6.

A dedicated prefix (we assume a 16 bit prefix, previously known as a Top-Level Aggregation (TLA) identifier [Hinden et al., 1998]) would be allocated for exclusive use as the TMI space. As a result, the first 16 bits are fixed, but 112 bits are enough to keep the TMI collision probability very close to zero. Defining a specific TLA has several benefits. For example, (1) Any mobile node can be automatically authorized to use any address in this TLA, and, (2) the allocated TLA can be marked as unroutable (i.e., a wrong packet to a TMI destination will be dropped by the first router, not the first default free router). In general, misuses of TMIs become very easy to detect.

A TMI has a role similar to that of a home address in standard MIPv6. As a result, it is also subject to the redirection attack of Mobile IPv6. In Mobile IPv6, a node that communicates with a mobile node keeps a record that binds the mobile node's home address and its current care-of address. When the mobile node moves to another subnet, it sends a binding update that specifies its new care-of address. Upon receiving this signaling message, the correspondent node updates the mobile node's record with the new care-of address. However to avoid traffic redirection attacks, the mobile node has to prove ownership of the home address contained in the binding update. Otherwise any malicious host could redirect a target home address to one of its addresses and hijack the communication.

To solve this problem, IPv6 Cryptographically Generated Addresses (CGA) have been designed [Montenegro and Castelluccia, 2004, O'Shea and Roe, 2001, Aura, 2003]. CGA are IPv6 addresses where the interface identifier is generated by hashing the address owner's public key. The address owner can then use the corresponding private key to assert authority over its address by signing messages sent. This uses public key cryptography but does not require any additional security infrastructure.

For the same reason, we propose to use TMI that are Crypto-based Identifiers (CBID) [Montenegro and Castelluccia, 2004]. CBIDs have a strong cryptographic binding with their public components (of their private-public key pairs). Because of this, once a correspondent node obtains information about one of these identifiers, it has a strong cryptographic assurance about which entity created it. Not only that, it knows that this identifier is owned and used exclusively by one node: its peer in the current exchange. Hence it can safely heed its redirects when it says that the mobile node is now available at some different care-of address (and later at another). A mobile node generates its CBID as follows:

- It generates a temporary RSA key pair (PK, SK), where PK is the public key and SK the secret key.

- It computes $TMI = SHA1_{112}(PK|imprint)$, where $imprint$ is a 128-bit random value and $SHA1_{112}$ is the $SHA1$ hash function whose output is truncated to 112 bits.

A mobile node can use its CBID for the inline protection of binding updates as follow: it includes in its binding update its public key, PK, the $imprint$ value and signs the whole message with SK. Upon reception of the binding update, the correspondent node can verify that the binding update was issued by the owner of the TMI (and not by an impersonator) by verifying that (1) the TMI was generated from PK and $imprint$ and (2) the signature is valid (i.e., the sender knows SK).

There are essentially two ways an adversary can impersonate a mobile node: (1) He can try to find a RSA key pair and $imprint$ that result to the same TMI than the target node. Since the size of a TMI is 112 bits, the adversary has to try, on average, 2^{111} parameters sets. If the attacker can perform 1 billion hashes per second this would take him $8 * 10^{25}$ years. Note that our scheme is more secure than current Mobile IPv6 schemes that rely on CGA addresses generated from a 59-bit long hash function [Aura, 2003]. (2) He can try to retrieve the private key SK associated with the mobile node's public key PK. A size of the modulus n of at least 1024 bits is commonly assumed to provide a good security level.

The TMI of a mobile user must be changed periodically (every few minutes, hours or days) in order to avoid TMI leakage as explained in [Narten and Draves, 2001]. This can easily be performed with the CBIDs by keeping the same PK/SK pair but changing the random value *imprint* periodically.

4.2 Protocol description

Two scenarios have to be considered:

1 *Mobile Client: the mobile node initiates the communication*

When the mobile node initiates the communication and it is moving, we argue that the mobile node does not need to reveal its home address at all. In this case, neither the correspondent node nor any eavesdropper should be able to identify the mobile node home address and thereof its identity.

In our proposal, a mobile node that initiates a communication uses standard Mobile IPv6 with the TMI as Home address. Packets sent and received by a mobile node will contain its TMI instead of its home address. As a result, the mobile identity is hidden from correspondent nodes and from potential eavesdroppers in the network.

Note that in this case correspondent nodes must never route directly to the "home address" (because this "home address" is a non-routable TMI), but should use the care-of address instead.

Since the TMI is a random value unrelated to the home address, neither a correspondent node nor any eavesdropper can link a TMI to a mobile node. Furthermore we suggest that the mobile node change its TMI periodically and use a different one per *session* or per *connection* to make linkability more difficult.

Mobile IPv6 uses a procedure called *Return Routability test* to authorize the establishment of the binding between a home address and a care-of address. This procedure enables the correspondent node to verify that the mobile node is really reachable at its claimed care-of address as well as at its home address. This is done by testing whether packets addressed to the two claimed addresses are routed to the mobile node. The mobile node can pass the test only when it is able to supply proof that it received certain data which the correspondent node sends to those addresses. Note that this procedure requires that the correspondent node know the mobile home address. Therefore our scheme is incompatible with the return routability procedure since a correspondent does not have to know the mobile node's home address.

2 *Mobile server: the correspondent node initiates the communication*

When the correspondent node initiates the communication, it knows by definition the home address of the mobile node. In this case it is meaningless to hide the home address from it.

However the mobile node might still want to hide its mobility, i.e., its care-of address, to a particular correspondent node. In this case, it must not send any binding update to this correspondent node and use bi-directional tunneling. As a result, packets sent to the mobile node are addressed to its home address and encapsulated by the home agent to its current care-of address. The decision to send or not to send a binding update to a correspondent node is a policy issue that is out of the scope of this paper. Any eavesdropper between the home agent and the mobile node is able to identify and track mobile movements by looking at inner packets. Therefore we suggest to encrypt all packets that are sent between the mobile node and its home agent [Arkko et al., 2004].

If the mobile node decides to use route optimization (and therefore reveal its care-of address to its correspondent node), it must then send a binding update to its correspondent node. This binding update contains the TMI in the home address option and the actual home address is encoded in a newly defined binding update sub-option. Of course to preserve privacy the binding update must be encrypted (the security association should be indexed with the TMI and not the home address). The correspondent node uses the binding update to bind the TMI with the home address and the care-of address.

Subsequent packets between the mobile node and the correspondent node will contain the TMI in the home address option and in the routing header extension instead of the actual home address. As a result an eavesdropper won't be able to identify the packets belonging to a particular node.

The mobility signaling (i.e., the binding update/binding acknowledgment exchange) may be protected by IPsec. For instance in the first scenario, the mobile client could establish an IPsec security association pair for mobility messages using its TMI as its address in its traffic selector, its care-of address for running IKE over, its RSA public key for signing and putting the *imprint* value in the IDi payload of type ID_KEY_ID or in a new type of CERT payload. The local policy on the correspondent node can recognize this special case and apply a specific authorization, for example accepting only ESP protection of mobility signaling messages. As in IKEv2 [Kaufman, 2004] the authentication and the negotiation of the first IPsec security association are done in the same exchange, the support of this kind of policy could be easily provided.

5. Privacy with Hierarchical Mobile IPv6

Hierarchical Mobile IPv6 (HMIPv6) is an optimization of Mobile IPv6 that is designed to reduce the amount of signaling required and to improve handoff speed for mobile connections [Soliman et al., 2004]. With HMIPv6, a mobile node gets two care-of addresses: a local one, the local care-of address (LCoA), and a global one, the regional care-of address (RCoA). It then registers the binding between its LCoA and its RCoA with a local server, the *Mobility Anchor Point (MAP)* and the binding between its RCoA and Home Address with its Home Agent and its correspondent nodes. As a result, the correspondent nodes or the home agent only know the global address but don't know where the mobile really is within the domain. This is clearly an improvement over Mobile IPv6 in term of privacy. Note that in HMIPv6, the Mobile Anchor Point (MAP) does not know the home address (i.e., the identity) of the mobile node. The MAP only knows the binding between the mobile's node regional and local care-of addresses. One may argue that a MAP could snoop the mobile host's packets to discover its home address. This is true but however this is still an improvement over Mobile IPv6.

When combining the privacy extension presented in this paper with HMIPv6, a mobile node uses the privacy extension to register with its home agent, its correspondent nodes and the local MAP. We can achieve full privacy protection because the mobile node's identity is hidden from its correspondent nodes and the local MAP. Its local care-of address is hidden from its home agent and correspondent nodes. No node knows the mobile node's identity (home address) and its care-of address together. Furthermore the MAP cannot find out the mobile node identity by snooping its packets (because the home address is not included in packets anymore). We argue that the combination of HMIPv6 with the privacy extension of this paper provides a level of privacy to a mobile node that is superior to that which a VPN provides (bi-directional tunnel between the mobile node and its home agent) but without the cost of a VPN.

Indeed when using HMIPv6 with the proposed privacy extension, we can:

- Hide the location (care-of address) of a mobile node from its Home Agent (this is not provided by a VPN),

- Hide the location (care-of address) of a mobile node from its correspondent nodes (provided by a VPN),

- Hide the identity of a mobile node from its correspondent nodes when the mobile is the initiator (not provided by a VPN),

- Prevent any eavesdropper in the network from identifying the packets that belong to a particular mobile and to track its location.

6.   Conclusions

In Mobile IPv6, each packet sent and received by a mobile node contains its home address. As a result, it is very easy for an eavesdropper or for a correspondent node to track the movement and usage of a mobile node. This paper proposes a new, simple and practical solution to this problem. The main idea is to replace the home address in the packets by temporary crypto-based identifiers (CBIDs). As a result, packets cannot be linked to a mobile node anymore and traffic analysis is more difficult. With our solution, an eavesdropper can still identify the IP addresses of two communicating nodes but is not able to identify their identities (i.e., their home addresses). Furthermore since a mobile node uses a new identifier for each communication, an eavesdropper cannot link the different communications of a given mobile node together. We show that HMIPv6 can also benefit from the proposed privacy extension.

References

[Arkko et al., 2004] Arkko, J., Devarapalli, V., and Dupont, F. (2004). *Using IPsec to Protect Mobile IPv6 Signaling between Mobile Nodes and Home Agents*. IETF, RFC3676.

[Aura, 2003] Aura, T. (2003). Cryptographically generated addresses (CGA). In *6th Information Security Conference (ISC'03*, volume 2851, pages 29–43, Bristol, UK. LNCS.

[Fasbender et al., 1996] Fasbender, A., Kesdogan, D., and Kubitz, O. (1996). Analysis of security and privacy in mobile ip.

[Hinden et al., 1998] Hinden, R., O'Dell, M., and Deering, S. (1998). *An IPv6 Aggregatable Global Unicast Address Format*. IETF, RFC2364.

[Johnson et al., 2004] Johnson, D., Perkins, C., and Arkko, J. (2004). *Mobile IP for IPv6*. IETF, RFC 3775.

[Kaufman, 2004] Kaufman, C., E. (2004). *Internet Key Exchange IKEv2 Protocol*. IETF, draft-ietf-ipsec-ikev2-14.txt.

[Montenegro, 2001] Montenegro, G. (2001). *Reverse Tunneling for Mobile IP, revised*. IETF, RFC3024.

[Montenegro and Castelluccia, 2004] Montenegro, G. and Castelluccia, C. (2004). Crypto-Based Identifiers (cbids): Concepts and applications. *ACM TISSEC*, 7(1).

[Narten and Draves, 2001] Narten, T. and Draves, R. (2001). *Privacy Extensions for Stateless Address Autoconfiguration in IPv6*. IETF, RFC3041.

[O'Shea and Roe, 2001] O'Shea, G. and Roe, M. (2001). "Child-proof Authentication for MIPv6 (CAM). *ACM Computer Communications Review*.

[Reed et al., 1998] Reed, M. G., Syverson, P. F., and Goldschlag, D. M. (1998). Anonymous connections and onion routing. *IEEE Journal on Selected Areas in Communications*, 16(4).

[Soliman et al., 2004] Soliman, H., Castelluccia, C., El-Malki, K., and Bellier, L. (2004). *Hierarchical MIPv6 mobility management*. IETF, draft-ietf-mipshop-hmipv6-01.txt, work in progress.

[Thomson and Narten, 1998] Thomson, S. and Narten, T. (1998). *IPv6 Address Autoconfiguration*. IETF, RFC2462.

A TRUST-BASED ROUTING PROTOCOL FOR AD HOC NETWORKS

Xiaoyun Xue and Jean Leneutre
INFRES department, ENST Paris - CNRS LTCI-UMR 5141, 46 rue Barrault 75634 Paris Cedex 13, France Tel: +33 1 45 81 71 98 Fax: +33 1 45 81 71 58
Xiaoyun.Xue, Jean.Leneutre@enst.fr

Jalel Ben-Othman
Prism laboratory, UVSQ - CNRS UMR 8144, 45 avenue des Etats-Unis 78035 Versailles Cedex, France
Jalel.Benothman@prism.uvsq.fr

Abstract   Ad hoc networks are particularly vulnerable as compare to traditional networks mainly due to their lack of infrastructure. A malicious node can easily disrupt both the routing discovery phase and the data forwarding phase of a routing protocol if it is not secured enough. This paper proposes a new secure reactive routing protocol named TRP (Trust-based Routing Protocol) that relies on a distributed trust model managing trust levels. The model provides an estimation of trust level to each route to help a source node to chose the most secure one. Our security mechanism is protected and does not affect significantly the network performance.

Keywords: Ad hoc networks, trust model, secure routing, blackmail attack

1.   Introduction

MANET (Mobile Ad hoc NETworks) are mobile wireless networks without fixed infrastructure. In an ad hoc network, each node is at the same time a router and a terminal, and is free to change its position with any speed and at any time. Many applications are possible: battlefields, conferences, urgency services...

Although the security requirements are different from one application to another, they are not negligible in most cases. Unfortunately, ad hoc networks are particularly vulnerable due mainly to their lack of infrastructure. Other reasons could be: high mobility, wireless links, limited bandwidths, lack of boundaries, short lifetime batteries and weak capacity of equipments.

Current ad hoc network security research works can be classified into two principal categories according to their main problematics: the distribution and management of keys and the authentication scheme, and the security of various routing protocols.

Traditional authentication and key management schemes are not applicable to ad hoc networks because they usually depend on some central server to establish trust relationships among nodes. In the contrary, an ad hoc environment requires a more distributed and robust solution. There are mainly three types of solutions proposed: emulation of a distributed certificate authority [Zhou and Haas, 1999] based on threshold cryptography; use of "trust chains" as done in PGP [J.P.Hubaux et al., 2001]; generation of one or more symmetric keys shared by the whole or a subset of the network [Asokan and Ginzboorg, 2000].

Most ad hoc routing protocols have been initially designed to deal with frequently changing topology, none of them have considered security issues in their design, suppose that it will be addressed later or by another layer's security mechanisms (for example 802.11's security mechanism or SSL/TLS). However a security mechanism at another layer is not sufficient because a mistake in routing choice can already welcome attacks, and security considerations must be integrated into routing protocols at the very first time.

A large part of newly proposed secure ad hoc routing protocols are applied to two reactive protocols: DSR and AODV, especially DSR. The original DSR is not secured at all, but since DSR controls every hop on a route and can work in a multi-routes mode, it is often used as a base of a secure routing protocol, just as what we have done in this work.

An ad hoc routing protocol has to be secured in both the routing discovery (and maintenance) phase and the data forwarding phase. In the first one, incorrect topology information should be forbidden and in the second, nodes who do not correctly forward data, intentionally or not, should be identified and excluded from the network.

This paper proposes a new secure reactive routing protocol named TRP based on DSR and a distributed trust model.

The rest of the paper is organized as follows: the section 2 discusses related work, the section 3 describes the proposition, the section 4 is dedicated to performance evaluation, residual vulnerabilities are discussed in the section 5 and finally our conclusion and perspectives are presented in the section 6.

2.   Related work

Recent works trying to mitigate problems in the routing discovery phase usually integrate cryptographic mechanisms into routing protocols to guarantee the authentication and the integrity of routing control messages. The SRP (Secure Routing Protocol) protocol proposed in [Papadimitratos and J.Haas,

2002] uses only light-weight cryptographic operations limited to MAC calculations and is able to avoid a large variety of attacks.

For the security in the data forwarding phase, the protocol SMT (Secure Message Transmission) [P.Papadimitratos and Z.J.Haas, 2003] takes advantage of the presence of multiple routes through the distribution of data on several routes and the addition of a feedback mechanism to control further retransmissions. SMT is designed as a complement of SRP and it can effectively shield attacks even with up to 50% attackers. However, it induces overloads, and it is not able to isolate attackers.

Otherwise, by introducing a distributed supervision system using promiscuous mode to each node, it is possible to detect undesired behaviors made by malicious or failing nodes. To achieve this, every node memorizes its own observations [Marti et al., 2000], or as well as negative recommendations in form of alarm messages [Buchegger and Boudec, 2002, Lakshmi, 2001] or positive confirmations made by traffics [Michiardi and Molva, 2002]. The choice of routes is then carried out from trust values. However, the approach in [Marti et al., 2000] do not accept second-hand reputations, therefore its training time will be longer. And since [Buchegger and Boudec, 2002, Lakshmi, 2001] accept negative reputations, they are vulnerable to blackmail attacks (attacks where a malicious node send false accusations to tarnish reputations of honest nodes). Furthermore, the problem of authentication in supervision systems, an essential issue, is only partially addressed in [Buchegger and Boudec, 2002] and [Lakshmi, 2001].

Most of current works encounter the same difficulty which is to propose a robust but light-weight security mechanism. Our proposition TRP adopts a mechanism close to SRP to provide security to the routing discovery phase, and a light-weight supervision mechanism together with a trust model are integrated to secure the data forwarding phase. The exchanges of recommendations are protected by the same MAC used in SRP. In addition, a particular care has been taken in the definition of the reputation model used: it is based on the work presented in [Beth et al., 1994].

3. TRP protocol

The following assumptions are applicable to the rest of the paper: each node has at least a sufficient storage capacity for the supervision of a restricted part of the traffics forwarded by itself; each node has a minimum calculation capacity to carry out simple arithmetic calculations; transmission ranges of nodes are identical; each node has a unique identifier (ID), and it is possible to authenticate nodes; and lastly, there is at most one attacker on a route, and in the case there are more than one attacker on a route, they are not neighbors.

Routing discovery phase

TRP is largely inspired by SRP with regard to the security in the routing discovery phase. We suppose also that a sender node S trusts its destination node D (a Security Association with mainly a shared secret key $K_{S,D}$ exists between S and D).

Like SRP, the proposed security mechanism adds an additional header into routing control messages. The sender S initiates a routing discovery process by adding two integers, a sequence number Q_{seq} and a random number Q_{id}, to a normal DSR RREQ. A MAC using key $K_{S,D}$ is also added at the end. During the broadcast of a RREQ, intermediate nodes add their identities into the request, and continue to relay the request until it reaches its destination. The receiver D verifies the MAC, and sends back a RREP including the found route, Q_{seq}, Q_{id} and a new MAC calculated over the RREP. S verifies the MAC and then the route included can be written into S's cache.

This mechanism is able to resist to a great part of attacks (see [Papadimitratos and J.Haas, 2002] for a detailed description), anyway, it remains vulnerable:

- since RERRs are not authenticated, malicious nodes can invalidate correct routes by sending invented RERRs with spoofing; a fast moving attacker could thus invalidate a lot of routes;

- by using the promiscuous mode, a malicious node can refuse to add its own identity into a RREQ, so that the route will seem shorter; otherwise, if an attacker is a neighbor of a destination node, wrong routes can be created by spoofing IP addresses and finally, a loop can be inserted into a RREQ;

- wormhole attacks are not treated, selfish nodes and cooperating attackers neither.

The first attack can be avoided by adding an authentication mechanism to RERRs. The attacks in the second item could be detected by using the supervision mechanism presented later in the paper. The attacks in the last item are more sophisticated and are not addressed in TRP.

TRP uses above mechanisms, together with an additional header added to RREQs and RREPs to exchange trust informations between nodes, as described below.

Data forwarding phase

A mini supervision mechanism is integrated into each node. According to events that a node observed, trust values for each neighbors could be evaluated dynamically. The goal of the mechanism is to let the source node concludes a trust degree for each route so that the most secure one will be chosen.

The trust values are calculated using the following reputation (trust) model:

Trust evaluation. Our reputation model is a variant of the trust model initially introduced in [B.K.R.Yahalom and T.Beth, 1994] and developed in [Beth et al., 1994] for its valuation part (it is modified to adapt it to the ad hoc context). The original model gives us the possibility to take into account various classes of trust relations: an entity is not absolutely trusted, but with regard to one or more specified tasks (for example, key generation, nondisclosure of secrets, or in our case, the routing function). Moreover, it allows valuation of trust relationships: from the numbers of positive and negative experiences an entity has assigned to another entity (with regard to a giving function), the former computes a trust level associated to the latter. Furthurmore, it is also a distributed trust model and it is possible to derive trust relationships from recommendations using transitivity.

Thus, three types of trust relationships are considered in our model:

- *direct trust relationship between neighbors*: that is valuated by positive and negative direct observation experiences;

- *indirect trust relationship between two nodes*: that is derived from direct trust relationships using transitivity;

- *trust relationship between a node and a route*: that is computed using direct and indirect trust values.

All these trust values are taken in $\{-1\} \cup [0, 1[$, and when the value -1 is associated to a node, it means this node is considered as malicious (or failing).

An initiator of a RREQ will obtain in each returned RREP a series of direct trust values given by the nodes on the route. With these direct values, the sender of the RREQ is then able to evaluate the indirect trusts for the nodes on the route for which it has no direct trust values. Finally, a trust value of the route will be computed in order to avoid encountering malicious nodes during the data forwarding phase.

Direct trusts. All direct trust values are initialized to 0 by default. But since the model is totally local, nodes are free to initiate some trust values to some wanted values when there exists some pre-established trust relationships.

The evolution of the trust value of the node n_i on the node n_j is given by the following formula:

$$C_{n_i,n_j}^D(t) = \begin{cases} -1 & \text{if } p_{n_i,n_j}(t) < 0 \\ 1 - \alpha^{p_{n_i,n_j}(t)} & \text{otherwise} \end{cases}$$

here $\alpha \in (0,1)$, (higher is α, slower a direct trust goes from 0 to 1 with time, vice versa) and $p_{n_i,n_j}(t)$ depends on the number of positive experiences $p_{n_i,n_j}^+(t)$ (the number of good behaviors) and the number of negative experiences $p_{n_i,n_j}^-(t)$ (the number of bad behaviors) of n_j observed by n_i until time t. The value of $p_{n_i,n_j}(t)$ is defined as:

$$p_{n_i,n_j}(t) = p_{n_i,n_j}^+(t) - \beta * p_{n_i,n_j}^-(t)$$

where β ($\beta > 1$) is a parameter which allows the modulation of the importance of negative experiences (greater is β, larger is the influence of negative experiments). β is introduced so that a certain number of faults may be tolerated. Both α and β should be relatively high to keep the efficiency of the model. According to the form of the formula, we can see that:

- if a node always behaves well, its trust value will rapidly increase to 1;

- if a node is moderately malicious or failing, its trust value will be stable;

- if the node is malicious or quite failing, then it will immediately become untrusted.

Indirect trust. Since a source node has not inevitably a neighborhood relationship with all nodes on a route, it is sometimes needed to derive indirect trust values using recommendations, when a RREP is returned to a source. For example, we consider a route made up of $k+1$ nodes, where n_i is the ith node and n_0 and n_k are respectively the source node and the destination node, the indirect trust values are defined as:

$$C_{n_0,n_{k-1}}^I = \begin{cases} -1 & \text{if } C_{n_0,n_k}^D \, or \, C_{n_k,n_{k-1}}^D = -1 \\ 1 - (1 - C_{n_k,n_{k-1}}^D)^{C_{n_0,n_k}^D} & \text{otherwise} \end{cases}$$

and for $1 \leq i \leq k-2$

$$C_{n_0,n_i}^I = \begin{cases} -1 & \text{if } C_{n_0,n_{i+1}}^I \, or \, C_{n_{i+1},n_i}^D = -1 \\ 1 - (1 - C_{n_{i+1},n_i}^D)^{C_{n_0,n_{i+1}}^I} & \text{otherwise} \end{cases}$$

For our propose, we only need to consider the indirect trust relations starting from n_0 but our definition could easily be extended.

The indirect trust values are defined so that:

- if one of the direct trust values used in the recommendation chain indicates that one of the nodes on the route is a potential attacker (the corresponding direct trust value is -1), then the afterward derived indirect trust values are no more relevant (will equal to -1);

- if $C^D_{n_{i+1},n_i}$ is based on a experience level p, then $C^I_{n_0,n_i}$ is based on the experience level $p * C^I_{n_0,n_{i+1}}$, because $C^I_{n_0,n_i} = 1 - (1 - (1 - \alpha^p))^{C^I_{n_0,n_{i+1}}} = 1 - \alpha^{p*C^I_{n_0,n_{i+1}}}$.

An optimization could be to check beforehand if there is a node n_l between n_i and n_k with which n_0 has a direct trust value. If so, the $C^I_{n_0,n_i}$ $(1 \leq i < l)$ can be calculated starting from $C^D_{n_0,n_l}$ instead of $C^D_{n_0,n_k}$.

All indirect trust values are deleted after the derivation of the trust value of the route so that influences of malicious recommendations are limited.

Route trust. According to the principle that the security level of a whole system corresponds to the security level of its weakest component, our trust level of a route corresponds to the source's lowest trust level on its intermediate nodes. For example, the trust value of a route $n_0, ..., n_k$ equals:

$$C^R_{n_0,...,n_k} = min^k_{i=1}(C_{n_0,n_i})$$

where

$$C_{n_0,n_i} = \begin{cases} C^D_{n_0,n_i} & \text{if } p^+_{n_0,n_i} + p^-_{n_0,n_i} \geq 1 \\ C^I_{n_0,n_i} & \text{otherwise} \end{cases}$$

The main difference between [Beth et al., 1994] and our model is that we tolerate some faults (we added the -1 value to keep a rating for untrustful nodes), and for our particular objective, we introduced trust values for routes.

TRP implementation details.

Supervision system. In order not to overload nodes, we chose a restricted supervision mode: a node does not supervise all traffics in its neighborhood, but only supervises traffics passed by itself. We called this mode "supervision on routes" comparing to the mode "supervision in the neighborhood" in all other works.

Each node maintains a trust information table memorizing the number of good behaviors p^+, the number of bad behaviors p^- and the direct trust value ("rating") of each neighbor observed. In the data forwarding phase, for each data packet sent or forwarded, the sender or the forwarder n_i save a copy of the packet in its buffer, and then supervises the action of the next node n_{i+1} (when n_{i+1} is not the destination). Depending on whether or not n_{i+1} correctly forwards the data within a limited time period, n_i increments the value of $p^+_{n_i,n_{i+1}}$ or $p^-_{n_i,n_{i+1}}$. The rating will also be updated if necessary.

We suppose of course that the authentication of neighbors is performed.

Modifications to RREQs and RREPs.   The header of SRP is extended to include a new table memorizing the trust values reported by intermediate nodes. Every intermediate node adds its direct trust value on the former node into the request.

All trust values collected in a RREQ should be send back to the source node within a RREP, and the MAC in the RREP covers equally their integrity. Furthermore, when a RREP is received by an intermediate node n_i, $C_{n_i,n_{i-1}}^D$ is verified by n_i to ensure that $C_{n_i,n_{i-1}}^D$ has not been modified by one or more nodes among $n_{i+1},...,n_k$. A packet with a modified trust value should be dropped, so that a attacker has no possibility to modify any value not reported by itself.

As an option, every intermediate node n_i checks also if $C_{n_i,n_{i+1}}^D$ equals to -1, and if it is the case, it stops to forward the RREP. This option has been implemented in an optimized version of TRP called TRP$^+$.

When a RREP is received by its source, the latter checks the MAC. If the MAC examination succeeds, the direct trust values included in the RREP will be used to calculate the necessary indirect trust values and a route trust value will be computed.

The only way for a malicious node n_i to introduce a wrong trust value is to give a wrong $C_{n_i,n_{i-1}}^D$ to a RREQ. If $C_{n_i,n_{i-1}}^D$ is too low, the route will not be chosen so that the attacker will not be on an active route; if $C_{n_i,n_{i-1}}^D$ is too high, because there is a great probability that trust levels C_{n_0,n_i}^D and C_{n_{i+1},n_i}^D are low or negative and also because the trust value of the route only depends on the minimum value of all the trust values, the probability that the trust level of the route increases is low. Even in an extreme situation, where all intermediate nodes are malicious, the destination which is trusted by the source can still give a bad reputation to n_{k-1}, so that the route will not be chosen.

Route management.   Since each route has a specific trust value, routes accepted by a node should be stored separately into its route cache, and all routes will expire after a timeout (which is important for refreshing trust levels of routes). Routes with a trust value equal to -1 are not stored into caches. Moreover, if all nodes on a given route are completely included in another route resulting from a same RREQ, the second route will be discarded.

Route choice.   Each data packet must obtain a route before its sending. Two strategies are possible: always choose the route with the highest trust value; or set a threshold as the lowest acceptable trust level, and chose the shortest route among all routes having a trust level greater than the threshold. The first strategy emphasizes the security requirement, while the second is a compromise between the security and the network performance.

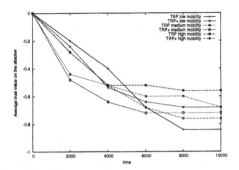

Figure 1. Average trust value on the attacker

4. Performance evaluation

Some simulations have been carried out under NS-2 using OpenSSL's library for cryptographic functions. TRP is implemented on the DSR module.

The simulation network contains 24 normal nodes and a attacker. The attacker modifies a packet when forwarding it, and it will never send back (initiate or forward) a RERR whether there is a broken link or not.

Three mobility scenarios have been tested:

- low mobility - 100s as pause time and 2m/s as maximum speed;

- medium mobility - 20s as pause time and 5m/s as maximum speed;

- high mobility - 5s as pause time and 20m/s as maximum speed.

We use FTP as application, with 22 random CBR sources and with a packet rate of 2 packets/s. The simulation time is 10,000s and the simulation range is 700m*700m. $\alpha = 0.75$, $\beta = 10$ and the promiscuous buffer size is 30.

The simulations are realized on TRP as well as on TRP$^+$.

At first, the figure 1 shows that the average direct trust value on the attacker decreases with time, whatever is the mobility scenario. TRP$^+$ is generally better than TRP, because in TRP$^+$ more nodes could have a forwarding relationship with the attacker.

The figures 2, 3 and 4 show us that the number of successful attacks could be stabilized after a time with TRP, and this time will still be shorter with TRP$^+$. Furthermore, as we could foresee, stronger is the mobility, better are the results: a frequently changed network topology can help nodes to detect the attacker.

We also observed that the average route length increased not more than 3% in all simulations.

In term of communication overhead, no new message is added but only the size of RREQs and RREPs are slightly increased due to the addition of

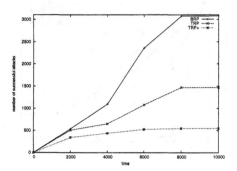

Figure 2.   Losses comparison: low mobility

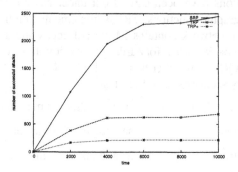

Figure 3.   Losses comparison: medium mobility

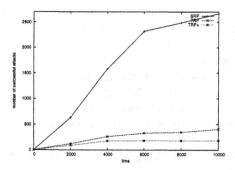

Figure 4.   Losses comparison: high mobility

the new trust header. However, the routing overload increases considerably comparing to DSR because of the deletion of many DSR optimizations and the refreshments of caches: since routes expire after a lifetime, we have to more often initiate RREQs to refresh routes.

5. Residual vulnerability

Firstly, we need a training phase before TRP can really be efficient; secondly, we have not considered selfish nodes: in our solution, a node can simply give a "-1" as a trust value to save its energy; thirdly, cooperate attackers are not treated; and lastly, spoofing of IP addresses should be stopped by using some authentication schemes.

6. Conclusion and future work

We presented in this article a new solution relying on a trust model which allows us to secure a reactive ad hoc routing protocol. The proposition can counter most of attacks during both the routing discovery phase and the data forwarding phase and it seems more suited to ad hoc networks with a long lifetime and a frequently changing topology.

Contrary to other approaches adopting a similar reputation system [Marti et al., 2000, Buchegger and Boudec, 2002, Michiardi and Molva, 2002], our solution protects its reputation exchanges to be not vulnerable to blackmail attacks. Compare with a solution which transfers data on multiple routes [P.Papadimitratos and Z.J.Haas, 2003], our solution has disadvantages such as needing a training phase before becoming operational and needing authentications of intermediate nodes, but it has the advantages of not requiring Ack messages and isolating attackers.

As an immediate work, we plan to:

- optimize the trust calculation in particular by calibrating parameters α and β;
- carry out more intensive simulations, especially by considering more attackers, and by adopting less systematic attackers' behaviors;
- extend the supervision to the neighborhood in order to reduce the training phase.

In the future, we will study if our solution can be adapted to other reactive protocols, or even to proactive protocols. And lastly, we hope to tackle the complex problem of key management and authentication scheme.

References

[Asokan and Ginzboorg, 2000] Asokan, N. and Ginzboorg, Philip (2000). Key-agreement in ad-hoc networks. *Computer Communications*, 23(17):1627–1637.

[Beth et al., 1994] Beth, T., Borcherding, M., and Klein, B. (1994). Valuation of trust in open networks. In *Proc. 3rd European Symposium on Research in Computer Security – ESORICS '94*, pages 3–18.

[B.K.R.Yahalom and T.Beth, 1994] B.K.R.Yahalom and T.Beth (1994). Trust relationships in secure systems - a distributed authentication perspective. *Computer Systems*, 7(1):45–73.

[Buchegger and Boudec, 2002] Buchegger, Sonja and Boudec, Jean-Yves Le (2002). Cooperative Routing in Mobile Ad-hoc Networks: Current Efforts Against Malice and Selfishness. In *Lecture Notes on Informatics, Mobile Internet Workshop, Informatik 2002*, Dortmund, Germany. Springer.

[J.P.Hubaux et al., 2001] J.P.Hubaux, L.Buttyan, and S.Capkun (2001). The quest for security in mobile ad hoc networks. In *Proceedings of ACM Symposium on Mobile Ad Hoc Networking and Computing (MobiHOC 2001)*, Long Beach, CA, USA.

[Lakshmi, 2001] Lakshmi, Venkatraman (2001). Secured routing protocol for ad hoc networks. Master thesis, University of Cincinnati, Computer Science.

[Marti et al., 2000] Marti, S, Giuli, T, Lai, K, and Baker, M (2000). Mitigating routing misbehavior in mobile ad hoc networks. *Proceedings of MOBICOM*.

[Michiardi and Molva, 2002] Michiardi, Pietro and Molva, Refik (2002). Core: A COllaborative REputation mechanism to enforce node cooperation in mobile ad hoc networks. In *Communication and Multimedia Security 2002 Conference*.

[Papadimitratos and J.Haas, 2002] Papadimitratos, Panagiotis and J.Haas, Zygmunt (2002). Secure routing for mobile ad hoc networks. *Proceedings of the SCS Communication Networks and Distributed Systems Modeling and Simulation Conference*.

[P.Papadimitratos and Z.J.Haas, 2003] P.Papadimitratos and Z.J.Haas (2003). Secure data transmission in mobile ad hoc networks. In *Proceedings of the 2003 ACM workshop on Wireless security table of contents*, pages 41–50.

[Zhou and Haas, 1999] Zhou, Lidong and Haas, Zygmunt J. (1999). Securing ad hoc networks. *IEEE Network*, 13(6):24–30.

SHORT-TERM FAIRNESS
OF 802.11 NETWORKS
WITH SEVERAL HOSTS

Gilles Berger-Sabbatel, Andrzej Duda, Martin Heusse,
and Franck Rousseau
LSR-IMAG Laboratory
Grenoble, France
{gberger, duda, heusse, rousseau}@imag.fr

Abstract   Previously, we have analyzed the short-term fairness of the 802.11 DCF (Distributed Coordination Function) access method in the case of a network with two hosts. In this paper we extend the analysis to an increased number of hosts. We use two fairness indices. The first one is the number of inter-transmissions that other hosts may perform between two transmissions of a given host. The second index is based on the sliding window method that considers the patterns of transmissions and computes the average Jain fairness index in a window of an increasing size. Computed over traces gathered during measurement sessions the indices show that the fairness of 802.11 is pretty good even on the short term time scale. We also evaluate the impact of the short-term fairness on performance by providing the measurements of the delay.

Keywords: Wireless LANs, Fairness of 802.11, CSMA/CA, Network measurements

1.   Introduction

Fairness is related to the ability of a MAC (Medium Access Control) layer to equitably share a common channel among N contending hosts. A MAC layer can be considered as *long-term fair* if the probability of successful access to the channel observed on a long term converges to $1/N$ for N competing hosts. A stronger property is *short-term fairness*: over short time periods, the access to the channel should also be fair. A MAC layer may be long-term fair, but short-term unfair: one host may capture the channel over short time intervals. Short-term fairness is extremely important for obtaining low latency: if a MAC layer presents short-term fairness, each host can expect to access the channel during short intervals, which in turn results in short delays.

In another paper [4] we have analyzed the fairness of the 802.11 DCF
(Distributed Coordination Function) access method in the case of a
network with two hosts. We have shown that contrary to the common wisdom, 802.11 does not exhibit short-term unfairness. The belief
in the short-term unfairness of 802.11 comes from an analysis of the
Wavelan CSMA/CA (Carrier Sense Multiple Access/Collision Avoidance) medium access protocol [6]. Its results have been extrapolated to
802.11 without realizing that the access method of 802.11 has changed
with respect to that of the Wavelan cards: in the 802.11 standard [1]
the exponential backoff is only applied after a collision and not when a
host finds the channel busy.

This paper extends the analysis of the fairness in 802.11 networks
to an increased number of hosts. Having more than two hosts in a
cell changes considerably the access conditions to the radio channel—
fairness may be degraded because of an increasing number of collisions.
In this case the exponential backoff applied after a collision may have a
negative impact on the fairness: a host doubles its maximum congestion
window after a collision and has a higher probability of choosing a large
interval. During this period other hosts may benefit from transmission
opportunity.

We report on the measurements of two fairness indices in a 802.11
cell. The first one, proposed and extensively studied previously [4], is
the *number of inter-transmissions* that other hosts may perform between
two transmissions of a given host. The second index is based on a largely
used method for evaluating fairness: the *sliding window method* that
considers the patterns of transmissions and computes the *average Jain
fairness index* in a window of an increasing size. Computed over traces
gathered during measurement sessions the indices show that the fairness
of 802.11 is pretty good even on the short term time scale.

The rest of the paper is organized as follows. We start with the
review of the existing work on fairness in wireless local area networks
(Section 2). Then, we define the notion of fairness (Section 3) and
present the results of measurements (Section 4). Finally, we present
some conclusions (Section 5).

2.   Related work

The fairness of 802.11 when all hosts have equal opportunity of using a shared common channel has been largely analyzed in the literature. Koksal et al. analyzed the short-term unfairness of the Wavelan
CSMA/CA medium access protocol [6]. They proposed two approaches
for evaluating fairness: one based on the sliding window method with the

Jain fairness index and the Kullback-Leibler distance, and the other one that uses renewal reward theory based on Markov chain modeling. The authors used Slotted ALOHA as an example of an access method with better fairness, but with a higher collision probability. This paper clearly identifies the short-term unfairness problem of an access method in which hosts perform exponential backoff whenever the channel is sensed busy.

Since this paper, many authors have stated that 802.11 suffers from short-term unfairness and cited it as the paper that proves the short-term unfairness of 802.11 [3][9][7]. However, they have not realized that the access method of 802.11 has changed with respect to that of the Wavelan cards: in 802.11 standard [1] exponential backoff is only applied after a collision. This misleading common wisdom has emerged from the confusion of these two different access methods.

The confusion of the access methods in Wavelan and 802.11 dies hard: recently, some authors have described the 802.11 access method as based on the same principle as in the Wavelan cards, i.e. exponential backoff applied when the channel is sensed busy [5].

The belief in the short-term unfairness of 802.11 also comes from some properties of wired CSMA networks, for example several authors have studied the *Ethernet capture effect*, in which a station transmits consecutive packets exclusively for a prolonged period despite of other stations contending for access [8]. However, the 802.11 DCF method eliminates this problem by adopting a CSMA/CA strategy instead of CSMA/CD.

3.   Fairness

Our goal is to study the intrinsic fairness properties of the 802.11 DCF access method, so we concentrate on the homogeneous case in which all hosts benefit from similar transmission conditions: no host is disadvantaged by its signal quality, traffic pattern, or spatial position. This means that we do not take into account the problem of hidden or exposed terminals and we do not consider the RTS/CTS extension. In particular, we do not deal with the problems of unfairness due to different spatial host positions [5, 2]. Once we got insight into the intrinsic fairness of the 802.11 MAC layer, we can investigate the influence of other factors such as different spatial positions or traffic patterns.

In general, the fairness of a MAC layer can be defined in a similar way to Fair Queueing: assume N hosts and let $W_i(t_1, t_2), i \in \{0, 1, 2, ...N\}$, be the amount of bandwidth allocated to host i in time interval $[t_1, t_2]$. The fair allocation requires that $W_i(t_1, t_2) = W_j(t_1, t_2), i, j \in \{0, 1, 2, ...N\}$, regardless of how small the interval $[t_1, t_2]$ is.

We consider the case of greedy hosts (they always have a frame to send) that send frames of equal size. In this case, it is sufficient to only take into account the number of transmissions: the fair allocation needs to guarantee that over any time interval, each host transmits the same number of frames.

To evaluate fairness we will use two methods. The first one uses the *number of inter-transmissions* that other hosts may perform between two transmissions of a given host and the second one computes the average Jain fairness index in a window of an increasing size.

3.1 Number of inter-transmissions

Consider the case of $N = 2$: two hosts A and B share a common channel. To characterize fairness we take the point of view of host B and investigate K, the number of inter-transmissions that host A may perform between two transmissions of host B:

- $K = 0$ means that after a successful transmission of B, the next transmission will be done once again by B,

- $K = 1$ means that A will transmit once and then the next transmission will be done by host B,

- $K = 2$ means that A will transmit twice and then the next transmission will be done by host B, and so on.

Consider the following example pattern of transmissions: BBAAABABAAB— random variable K takes the following values: 0, 3, 1, 2.

In a deterministic channel sharing system such as TDMA, the distribution of K will simply be $P(K = k) = 0$ for $k = 0$ and $P(K = k) = 1$ for $k = 1$, meaning that both hosts perfectly alternate transmissions. The mean number of inter-transmissions in TDMA is $E(K) = 1$.

An example of a randomized access protocol that presents good fairness properties is Slotted ALOHA—it has been previously used for fairness comparisons [6]. Time in Slotted ALOHA is divided into slots, each access is independent from the previous one and when a collision occurs, a transmitting host waits a random number of slots distributed geometrically. If we ignore collisions, Slotted ALOHA with two hosts can be modeled as a simple Markov chain with two states. In this case the number of inter-transmissions is geometrically distributed with the parameter $1/2$ (this expression only holds for two hosts):

$$P(K = k) = \frac{1}{2^{k+1}}, \quad k \in \{0, 1, 2, ...\} \tag{1}$$

Note that for Slotted ALOHA $P(K = 0) = 1/2$, so that each host has the equal probability of accessing the channel and the mean number of inter-transmissions is $E(K) = 1$, which is the same as for the TDMA.

We can generalize the number of inter-transmissions to a larger number of hosts: we choose one host and observe how many times other hosts transmit frames before another transmission by the chosen host. Consider for example the following sequence of transmissions by five hosts: BAACEDCAB. The number of inter-transmissions observed from the point of view of B is 7.

Observing each outcome of random variable K gives us information on short-term fairness, whereas its distribution and moments convey indication about both short-term and long-term fairness. We can notice that large values of K mean lower fairness, because other hosts may capture the channel for several successive transmissions. Similarly, too small values of K also indicate lower fairness, as the chosen host captures the channel in an unfair way.

More precisely, the distribution of inter-transmissions $P(K = k)$ enables us to quantify fairness by means of:

- *capture probability:* $P(K = 0)$ characterizes the chances of a host to capture the channel. If $P(K = 0) = 1/N$, then all hosts have equal probability of accessing the channel.

- *mean number of inter-transmissions:* $E(K) = 0$ means that one host monopolizes the channel and $E(K) = N - 1$ means that on the average each host performs one transmission at a time, the situation that can be considered as fair ($N - 1$ is the number of inter-transmissions in TDMA). Values $E(K) < N - 1$ indicate a shorter tail of the distribution and better fairness, whereas $E(K) > N - 1$ indicates increased unfairness.

- *100qth percentile:* characterizes the tail of the distribution, it is the smallest l for which $\sum_{k=0}^{l} P(K = k) < q$, $0 < q < 1$. For instance the 95th percentile tells us that in 95% of cases the number of inter-transmissions will be less than the 95th percentile. Putting it another way, in only 5% of cases a host should wait more than the 95th percentile transmissions before the next access to the channel.

We can notice that the number of inter-transmissions is directly related to delays perceived by a host competing with other hosts for the channel access: when a host experiences large values of K, it also suffers from large delays, because it has to wait for the channel access while other hosts transmit several frames.

3.2   Sliding window method with the Jain fairness index

The sliding window method considers the patterns of transmissions and computes the average Jain fairness index in a window of an increasing size [6]. It is defined as follows: let γ_i be the fraction of transmissions performed by host i during window w; the fairness index is the following:

$$F_J(w) = \frac{(\sum_{i=1}^{N} \gamma_i)^2}{N \sum_{i=1}^{N} \gamma_i^2}. \tag{2}$$

Perfect fairness is achieved for $F_J(w) = 1$ and perfect unfairness for $F_J(w) = 1/N$.

The definition of window w also should take into account N, the number of competing hosts. We propose to normalize the window size with respect to the number of hosts and compute the Jain index for the window sizes which are multiples of N, because only in this case computing the Jain index makes sense. We call m such that $w = m \times N, m = 0, 1, 2, ...$, a *normalized window size*. The Jain index will be computed as $F_J(m)$.

Both indices have the nice property of being able to capture the short-term as well as the long-term fairness.

4.   Experimental results

To evaluate the fairness in a 802.11 cell with several hosts, we have set up an experimental platform to measure the fairness indices and delays. We use notebooks running Linux RedHat 8.0 (kernel 2.4.20) with 802.11b cards based on the same chipset (Lucent Orinoco and Compaq WL 110). The results concerning fairness and delays also apply to 802.11a and 802.11g, because they use the same DCF access method as 802.11b with some parameter values modified.

The wireless cards work in the infrastructure mode—an access point is connected to the wired part of the network. Notebooks are greedy in our experiments—they try to send fixed sized UDP packets to the access point as quickly as possible and they always have a packet to send. We gather traces on the destination host: the arrival instant and the source host. Then we compute the fairness indices and delay statistics from the traces.

There are two ways of organizing measurement sessions:

- *Synchronized hosts.* All hosts start at the same instant so that their congestion window counters are reset. To do this, we use a wired 100 Mb/s Ethernet interface in addition to a 802.11b wireless card. Sending a multicast packet on the wired network allows us to synchronize the hosts at the beginning of a measurement session.

- *Non synchronized hosts.* Hosts start at different instants so when a chosen host performs a successful transmission, other hosts may have lower values of their congestion windows because they use residual time intervals. These potentially shorter intervals mean that they may perform more inter-transmissions than in the previous case. The resulting distribution is different and corresponds to the worst case with respect to the fairness.

The case of synchronized hosts is easier to analyze theoretically, because all hosts begin to operate in a known initial state. However, it does not correspond to realistic working conditions in which hosts operate independently in a non synchronized manner. For the case of a 802.11 cell with two hosts we have done measurements with synchronized hosts to validate the theoretical analysis of the number of inter-transmissions [4]. Such a set up is cumbersome for a larger number of hosts so for this paper we have organized measurement sessions with non synchronized hosts. The results may present worse fairness, but if they are satisfactory, the measurements of synchronized hosts should result in even better fairness.

4.1   Number of inter-transmissions

We briefly recall the results of the analysis for the case of $N = 2$ (the analysis assumes the case of synchronized hosts).

Figure 1 shows the analytical, simulated, and measured distributions of the number of inter-transmissions in 802.11b with two competing hosts. We can see that all three distributions are close to each other, the approximate analytical distribution slightly overestimating the other values for $K = 1$ and underestimating for $K > 2$. The figure also compares the distribution of the number of inter-transmissions in 802.11b with Slotted ALOHA. Note that $P(K = 0) = 1/2$ which confirms that the access probability is equal for both hosts.

Table 1 presents $E(K)$, the mean number of inter-transmissions for two hosts. It is lower than 1 showing that the fairness of 802.11 is better than that of Slotted ALOHA because the tail of its distribution is shorter.

Figure 2 shows the measured distribution of the number of inter-transmissions for several competing hosts. In the measurement session hosts are not synchronized, so that the measured distribution is slightly different from Figure 1: $P(K = 0) < 1/N$, because the host we observe has less chances to capture the channel than the other hosts—after a

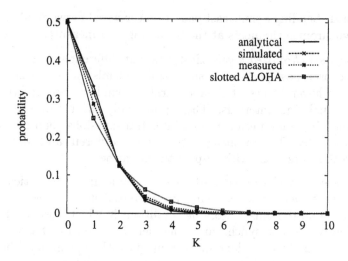

Figure 1. Distribution of the number of inter-transmissions in 802.11b and Slotted ALOHA, $N = 2$, synchronized hosts

Table 1. Mean number of inter-transmissions, $N = 2$

Case	$E(K)$
analytical Slotted ALOHA	1.0
analytical 802.11	0.718
simulated 802.11	0.768
measured 802.11	0.853

successful transmission by the chosen host, the other hosts benefit from smaller congestion window values corresponding to residual time intervals.

We can see from this figure that the distribution decays quickly in function of K. As for an increasing number of hosts, the number of collisions increases, one may expect that fairness degrades. However, we observe that the collisions do not have strong negative impact on the fairness: for $N = 5$ most of the values of K remain lower than 10.

The statistics presented in Table 2 confirm the good fairness properties of the access method: the average number of packets between two transmissions of a given host is equal to the number of hosts with which the host competes for the channel. Our measurements also show that the 95th percentile is fairly low, which confirms the fairness of the access

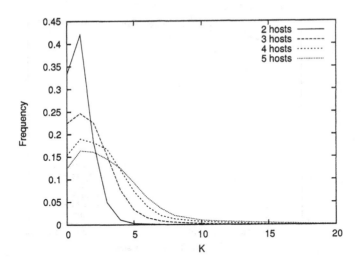

Figure 2. Measured distribution of the number of inter-transmissions in 802.11b, non synchronized hosts.

Table 2. Statistics of the measured distribution

Number of hosts N	mean	standard deviation	95th percentile
2	1.00	0.97	3
3	2.001	2.3	6
4	3.05	4	10
5	4.001	4.5	11

method. More precisely, when 5 (resp. 2) hosts contend for the channel, a host that has just transmitted a frame has probability of 5% to transmit a frame after the next 11 (resp. 3) successful transmissions.

4.2 Sliding window method with Jain fairness index

We have also computed the Jain fairness index over sliding windows of size $w = m \times N, m = 0, 1, 2, ...$ The index of 1 represents perfect fairness and we use the threshold value of 0.95 to characterize how quickly it is achieved.

Figure 3 shows the Jain fairness index measured in a 802.11 cell with several hosts in function of the normalized window size. It can be seen

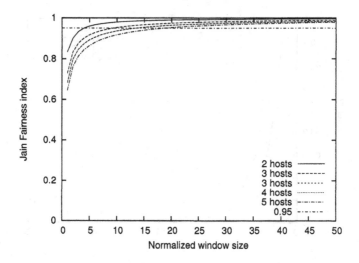

Figure 3.   Jain fairness index of 802.11 for normalized window size

that the threshold value of 0.95 is quickly attained for small values of
the normalized window. Note that for an increasing number of hosts,
the Jain index begins at a lower value, e.g. for $N = 5$ it starts with the
value of 0.2.

Table 3.   Window size to achieve 0.95 fairness index

Number of hosts N	2	3	4
Wavelan	475	83	112
802.11	4	9	13

Table 3 compares the window size required to achieve the threshold
of 0.95 for the Wavelan cards and 802.11. To obtain the normalized
window size, we have divided the results reported by Koksal et al. [6]
by the number of hosts (only the values of the Jain fairness index for
$N = 2, 3$, and 4 hosts have been given in this paper).

4.3   Delay

To evaluate the impact of the short-term fairness on the delay we
have measured delays perceived by one host when competing with other
hosts. A delay is the interval between the ends of the last successful
transmissions by a given hosts. It is composed of the waiting time while
other hosts are transmitting and the actual transmission time of a frame.

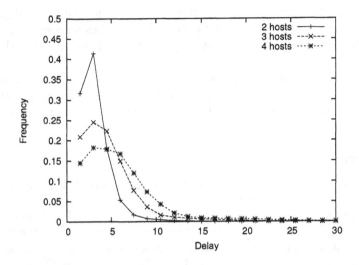

Figure 4. Measured distribution of the delays experienced by one host in milliseconds

Figure 4 presents the measured distribution of delays for an increasing number of hosts $N = 2, 3, 4$. We can see that the delay remains fairly low and the shape of the delay distribution is similar to that of the number of inter-transmissions (cf. Figure 2). This confirms the strong relationship between this fairness index and transmission latency.

5. Conclusion

In this paper we have analyzed the short-term fairness of the 802.11 DCF access method. The results of this paper complete our previous analysis of the 802.11 network with two hosts. They show that contrary to the common wisdom, 802.11 does not exhibit short-term unfairness. This fact has an important influence on transmission latency. As 802.11 presents short-term fairness, each host can access the channel during short intervals, which in turn results in short delays. The results also show that an increased number of collisions in a cell with several hosts does not have strong negative impact on the fairness.

Such a good behavior comes from the fact that hosts in 802.11 use their residual congestion intervals—when a host chooses a long interval, it will wait during one or several turns, but then it will eventually succeed, because its congestion interval becomes smaller and smaller.

Although our measurements were only done for several hosts ($N \leq 5$), the results are also meaningful for a cell with much more hosts that generate real-life traffic: in our experiments the traffic sources were greedy so the load corresponds to a network with many hosts generating traffic from intermittent sources statistically multiplexed over the shared radio channel.

We work on extending our analysis to take into account other factors such as different spatial positions or traffic patterns. We also investigate improvements to the 802.11 DCF access method able to increase short term fairness and lower the delay.

References

[1] ANSI/IEEE. 802.11: Wireless LAN Medium Access Control (MAC) and Physical Layer (PHY) Specifications, 2000.

[2] V. Barghavan, A. Demers, S. Shenker, and L. Zhang. MACAW: a Media Access Protocol for Wireless LAN's. In *Proceedings of ACM SIGCOMM*, August 1994.

[3] C. L. Barrett, M. V. Marathe, D. C. Engelhart, and A. Sivasubramaniam. Analyzing the Short-Term Fairness of IEEE 802.11 in Wireless Multi-Hop Radio Networks. In *10th IEEE International Symposium on Modeling, Analysis, and Simulation of Computer and Telecommunications Systems (MASCOTS'02)*, Fort Worth, Texas, October 2002.

[4] G. Berger-Sabbatel, A. Duda, O. Gaudoin, M. Heusse, and F. Rousseau. Fairness and its Impact on Delay in 802.11 Networks. In *Proceedings of IEEE GLOBECOM 2004*, Dallas, USA, 2004.

[5] Z. Fang, B. Bensaou, and Y. Wang. Performance Evaluation of a Fair Backoff Algorithm for IEEE 802.11 DFWMAC. In *Proceeding of MOBIHOC'02*, Lausanne, Switzerland, June 2002.

[6] C.E. Koksal, H. Kassab, and H. Balakrishnan. An Analysis of Short-Term Fairness in Wireless Media Access Protocols. In *Proceedings of ACM SIGMETRICS*, 2000. extended version available at http://nms.lcs.mit.edu/papers.

[7] Y. Kwon, Y. Fang, and H. Latchman. A Novel MAC Protocol with Fast Collision Resolution for Wireless LANs. In *Proceedings of IEEE INFOCOM 2003*, San Francisco, USA, March-April 2003.

[8] K.K. Ramakrishnan and H. Yang. The Ethernet Capture Effect: Analysis and Solution. In *IEEE 19th Conference on Local Computer Networks*, 1994.

[9] N. H. Vaidya, P. Bahl, and S. Gupta. Distributed Fair Scheduling in a Wireless LAN. In *Sixth Annual International Conference on Mobile Computing and Networking*, Boston, August 2000.

RAAR: A RELAY-BASED ADAPTIVE AUTO RATE PROTOCOL FOR MULTI-RATE AND MULTI-RANGE INFRASTRUCTURE WIRELESS LANS*

JainShing Liu,[1] and Chun-Hung Richard Lin [2]

[1] *Providence University, R.O.C.,* [2] *National Sun Yat-Sen University, R.O.C.*

Abstract   Auto rate adaptation mechanisms have been proposed to improve the throughput in wireless local area networks with IEEE 802.11a/b/g standards that can support multiple data rate at the physical layer. However, even with the capability of transmitting multi-packets with multi-rate IEEE 802.11 PHY, a mobile host (MH) near the fringe of the Access-Point's (AP's) transmission range still needs to adopt a low-level modulation to cope with the lower signal-to-noise ratio (S-NR), Thus, it can not obtain a data rate as high as that of a host near AP in most cases. According to the characteristics of modulation schemes, the highest data rate between a pair of mobile hosts will be inversely proportional with the transmission distance. Considering these factors, we here demonstrate a Relay-Based Adaptive Auto Rate (RAAR) protocol that can find a suitable relay node for data transmission between transmitter and receiver, and can dynamically adjust its modulation scheme to achieve the maximal throughput of a node according to the transmission distance and the channel condition. The basic concept is that the best modulation schemes are adaptively used by a wireless station to transmit an uplink data frame, according to the path loss condition between the station itself and a relay node, and that between the relay node and AP, thus delivering data at a higher overall data rate. Evaluation results show that this scheme provides significant throughput improvement for nodes located at the fringe of the AP's transmission range, thus remarkably improving overall system performance.

Keywords: Relay-based MAC protocol, Multi-Rate and Multi-Range, Wireless LANs

1.   Introduction

In both wired and wireless networks, adaptive transmission techniques are often used to enhance the transmission performance. These techniques generally include varying the transmission power, packet length, coding rate/scheme, and modulation technology over the time-varying channel. Related research in

*This work was supported by the National Science Council, Republic of China, under grant NSC92-2213-E-126-004.

cellular networks indicates that throughput would be improved by permitting a mobile station near the center of a cell to use a high-level modulation scheme, while those near the fringe of a cell should use a low-level modulation to cope with the lower signal-to-noise ratio (SNR).

In wireless local area networks (WLANs), the same issue is also considered. Providing that the new high-speed IEEE 802.11a MAC/PHY is adopted, MHs can vary their modulation schemes to transmit packets with different data rates ranging from 6 to 54 $Mbps$. Given such capabilities, all MHs should insist on using the highest-level modulation scheme to obtain the maximal channel utilization. However, as in cellular networks, such insistence is not reasonable because the data rate is inversely proportional to the transmission distance between a pair of MHs, and a high-level modulation scheme requires a higher SNR to obtain the same bit error rate (BER) when compared with that of a lower-level modulation scheme. Thus, without increasing the transmission power to decrease the lifetime of a node, the maximal data rate can be obtained only when the transmitter and receiver are close enough, and a lower data rate should be adopted if the distance is larger. For reference, the expected data rates of IEEE 802.11a (802.11b, and 802.11g) at different ranges measured in [1] are shown in Fig. 1.

Given that, we consider the infrastructure WLAN (IWLAN) defined in IEEE 802.11. In such networks, an AP can provide MHs the access to the Distribution System (DS) and relay packets for all MHs in its transmission range. At the same time, a MH, if equipped with a multi-rate wireless interface such as IEEE 802.11a, can deliver packets to an AP with a suitable data rate based on its distance to AP. These constitute a so-called Multi-rate, Multi-range IWLAN (MMI-WLAN).

In the MMI-WLAN, several auto rate adaptation mechanisms have been proposed to exploit the capability of multi-rate IEEE 802.11 physical layer. Among these, a so-called Opportunistic Auto Rate (OAR) rate adaptation scheme [2] opportunistically grants channel access for multiple packet transmissions in proportion to the ratio of the achievable data rate over the base rate. Similar to OAR, the authors of [3] provide an Adaptive Auto Rate (AAR) protocol that can also transmit multiple packets when the channel conditions are good. For this, AAR carries additional information in each ACK or CTS to indicate the transmission rate of the next data packet, and thus can adapt very quickly to the change of channel quality.

The capability of transmitting multiple packets in a basic slot-time let these methods have the potential to solve the anomaly problem [4] if MHs are located near AP. However, if MHs are located far from AP, these methods still require a low-level modulation to cope with the lower signal-to-noise ratio (SNR), as indicated in the very beginning. This can be an even more serious problem in the MMI-WLAN because, in such networks, if two neighboring MHs desire to communicate to each other, their packets have to be relayed by an AP no matter how close they are. In this environment, it becomes a challenge that how a MH can use a data rate as high as possible to obtain the maximal throughput, without increasing transmitting power and the overall interference in the network.

One possible solution is to deliver data packets through relay nodes. However, even recently, few studies have focused on the issue of directly using IEEE 802.11 MAC to realize a relay mechanism so that the system throughput can be improved [5]. This task is not trivial because unlike TDMA, IEEE 802.11 Distributed Coordination Function (DCF) MAC protocol has the 4-way handshake of RTS/CTS/DATA/ACK, and requires that the roles of sender and receiver should be interchanged several times between a pair of communication nodes.

To provide higher throughputs for MHs not close to AP in the MMI-WLAN, in this work we propose a Relay-Based Adaptive Auto Rate protocol (RAAR) as an enhanced protocol for the MMI-WLAN. RAAR slightly modifies the IEEE 802.11 MAC protocol by introducing a new message exchange procedure for a relay node between a pair of communication nodes. The core idea of RAAR is that after the 4-way handshake of a pair of communication nodes, the relay node should not compete for the channel again, wasting the valuable bandwidth, because the channel is already reserved by the original communication nodes. Our analysis and simulation show that the throughputs of MHs at the fringe of the AP's transmission range can be significantly improved if a suitable relay node can be obtained.

2. Relay-Based Adaptive Auto Rate Control protocol (RAAR)

In this section, the proposed RAAR is introduced in terms of protocol principles and major operations. Before this, the system model of MMI-WLAN under consideration is summarized as follows. Given M different modulation schemes, the MMI-WLAN can be logically segmented into M concentric circles surrounding AP as shown in Fig. 2. This network can be further divided into M disjoint regions: the innermost circle (R_1) and a number of $M - 1$ 'doughnut' like regions which are numbered as R_2, R_3, ..., R_M from inner to outer. The data rate that can be obtained in R_i is denoted as TR_i. For simplicity, we consider only upstream traffic in the MMI-WLAN, i.e., traffic from MHs to AP. This is because, in contrast to MHs having a battery as the usual power supply, AP is fully power-supported, and thus it can choose the data rate more flexibly for transmitting data to MHs. However, this model can be also applied to a general WLAN if the transmission condition over wireless channel is symmetrical.

Transmission of Multiple Back-to-Back Packets

The first aim in RAAR is to keep the channel for an extended number of packets once the channel is measured to be of sufficient quality to allow transmission at rates higher than the base rate. For this, RAAR allows the same time to be granted to a sender as if the sender is transmitting at the base rate. For example, if the IEEE 802.11a interface is adopted in the MMI-WLAN, the base rate is 6 Mbps and a node, MH_i, is located in the innermost region, R_1, wherein 54 Mbps data rate is achievable with 64-QAM modulation scheme. This node is granted a channel access time to send $\lfloor 54/6 \rfloor$ or $\lceil 54/6 \rceil$ (= 9 in both cases) packets to AP conservatively or aggressively. In the former case, the unused time quantum can be released while the latter, the original time quantum should be extended to complete the last packet. This method provides temporal fairness for all MHs in each region. That is, if $\phi_i(t_1, t_2)$ is defined as the service time that a flow in R_i receives during (t_1, t_2), the measure of fairness of RAAR with equal weights for different regions, R_i and R_j, is considered as

$$|\phi_i(t_1, t_2) - \phi_j(t_1, t_2)| \tag{1}$$

This ensures that RAAR provides near base-rate time shares for all MHs in different regions. As indicated in [2], the temporal fairness is more suitable for multi-rate networks than the throughput fairness [6, 7] since normalizing flow throughputs in the latter would cancel the throughput gains available from a multi-rate physical layer.

Relay Node Selection

The second aim in RAAR is for nodes near the fringe of AP's transmission range to transmit packets at a higher rate. For this, RAAR introduces a suitable relay node between the sender and AP to segment the long distance between the communication ends into two shorter pieces. According to the characteristics of modulation schemes, the highest data rate between a pair of mobile hosts will be inversely proportional with the transmission distance. Consequently, even with the same noise level, the reduced path length, and hence the reduced path loss, lead to the sender being able to transmit its data packets through the relay node to AP with a higher rate.

To this end, whenever a MH wants to communicate with an other node, AP should find a suitable relay node, and guide the sender to deliver its data via this node. This can be done because, in the MMI-WLAN, AP has the capability of collecting all MHs' location information, as well as their modulation schemes now adopted [8]. AP can collect such information from periodical status reports or routing information exchanges between MHs and the AP. When a suitable relay node is found for a MH, such information can be delivered from the AP to this node and the relay node just found with the same mechanism of periodical reports or routing exchanges. Consequently, if a mobile host, say MH_i, can find a relay node suggested by AP, it will not deliver its data to AP directly. Instead, MH_i will try transmitting its data via the relay node, say MH_j, to obtain a higher data rate. Otherwise, the transmission will be direct.

This is exemplified in Fig. 3, where the transmission ranges with corresponding data rates are drawn based on the measure data given in Fig. 1. In this figure, MH_i at the fringe of MMI-WLAN is the sender while the destination may be AP, or some other node in the network (not shown). In Fig. 3(b), MH_i, uses MH_j as its relay node to AP, which provides a higher overall data rate. Whereas, in Fig. 3(a), MH_i directly transmits data to AP at only 6 Mbps data rate without the aid of a relay node.

Control Message Flow

Fig. 4 illustrates the RAAR time-line for transmission of data packets. In principle, this is the IEEE 802.11 fragmentation mechanism extended for the incorporation of relay nodes. As the IEEE 802.11 standard, each frame in RAAR contains information that defines the duration of the next transmission. The duration information from the RTS/CTS frame is used to update the NAV to indicate that the channel is busy until the end of ACK0. Both DATA/FRAG0 and ACK0 also contain information to update the NAV to indicate a busy channel until the end of ACK1. This continues until the last fragment which carries the duration ending at the last ACK. The ACK for the last fragment has the duration field set to zero.

As shown above, the control flow indeed follows the same principle of the IEEE 802.11 fragmentation mechanism, and eliminates unnecessary RTS/CTS exchanges to improve the network throughput. That is, in contrast to the direct implementation of IEEE 802.11 4-way handshake, we let the relay node, MH_j, forward DATA/FRAGs from the sender, MH_i, to AP without the exchange of RTS/CTS in advance. This can improve the network throughput because at the very beginning, MH_i and AP have reserved the channel using RTS/CTS transmitted at the base rate. Consequently, the relay node, MH_j, has no need to reserve the channel for MH_i and AP again, which obviously would waste bandwidth for the same transmission.

Traffic and Channel Conditions. There are some issues which arise while using the RAAR protocol. These issues mainly revolve around the behavior of the sender, MH_i. In general, MH_i should calculate the overall transmission time based on its own data rate, TR_s, and the relay node's data rate, TR_r, suggested by AP, to determine whether or not this transmission should go through the relay node. If so, MH_i will estimate the distance between itself and the relay node using information provided by AP, and select a modulation scheme for error free transmission to the relay node.

However, such estimation may be incorrect due to mobility or channel condition change, and consequently, with the modulation scheme just chosen, the data frame could be received incorrectly by AP. In this case, when expecting a new arrival from the relay node after RTS/CTS exchange, AP estimates its ACK time for this frame as well, and if the time expires without receipt of the expected frame, AP can send a NACK to explicitly inform MH_i of this failure. As a result, after obtaining the negative acknowledgement, MH_i sends the following frames to AP directly, with data rate TR_s, and the protocol reverts back to a degraded version of this method, as discussed in the next section.

3. Throughputs of IEEE 802.11 MAC, RAAR and D-RAAR

In the MMI-WLAN, multiple nodes located in multiple regions, using the common wireless channel of IEEE 802.11 PHY are usual cases. In this environment, we first examine the performance anomaly problem of IEEE 802.11 MAC indicated in [4]. Then, we show how this problem can be solved with a degraded version of RAAR. Finally, we demonstrate the capability of RAAR that can further increase the system throughput beyond the solution of D-RAAR. These analytical results are further verified with the simulation results of GloMoSim [9], which is a scalable simulation environment for wireless and wired networks based on PARSEC [10].

IEEE 802.11 performance anomaly

Consider that in a M-region MMI-WLAN, N_j mobile hosts in each region j are ready to transmit an uplink frame with l-bytes UDP data payload using PHY mode m_j with transmit power P_{tx}. In this environment, with the IEEE 802.11 MAC, the total time expended by a MH in transmitting a l-bytes datagram will be

$$T'_{I,overall}(l,j) = T_{tx}(l,m_j) + T_{ov} + T'_{CW} \tag{2}$$

where $T_{tx}(l,m_j)$ denotes the data transmission time for a node in R_j using modulation scheme m_j, T_{ov} represents the constant overhead, $T_{DIFS} + 3 \cdot T_{SIFS} + T_{RTS} + T_{CTS} + T_{ACK} + 4 \cdot T_{PLCP}$, and T'_{CW} denotes the contention window time, $\sum_{j=1}^{K} \left[(1 - P_{coll}) \cdot P_{coll}^{j-1} \cdot \frac{CW(j)}{2} \right] + P_{coll}^{K} \cdot \frac{CW(K)}{2}$. In this window time, $CW(j)$ is $(2^{j-1} \cdot (aCWmin+1) - 1) \cdot aSlotTime$ when $1 \leq j \leq 6$ and becomes $aCWmax \cdot aSlotTime$ when $j \geq 6$[1], K denotes the maximum number of retransmission, and P_{coll} presents the collision probability in the network, and is now obtained with simulations of GloMoSim.

---

[1]$aCWnin$ is the minimum contention window size, $aCWmax$ is the maximum contention window size, and aSlotime is the slot time.

With these, we can analyze a node's throughput by considering that the channel utilization of a node in R_j is the ratio between the time to send one packet and the time during which all other mobile hosts transmit once with possible collisions. Because the channel access probability of CSMA/CA is equal for all mobile hosts, the channel utilization factor of a node in R_j can be calculated as

$$U_I^j(l) = \frac{T'_{I,overall}(l,j)}{\sum_{i=1}^{M} N_i \cdot T'_{I,overall}(l,i)} \tag{3}$$

where N_i denotes the number of nodes in R_j. Given that and l-bytes UDP data payload, the node throughput in R_j can be obtained with

$$
\begin{aligned}
\psi_I^j(l) &= U_I^j(l) \cdot \frac{l}{TR_j \cdot T'_{I,overall}(l,j)} \cdot TR_j \\
&= \frac{l \cdot T'_{I,overall}(l,j)}{\left(\sum_{i=1}^{M} N_i \cdot T'_{I,overall}(l,i)\right) \cdot T'_{I,overall}(l,j)} \\
&= \frac{l}{\sum_{i=1}^{M} N_i \cdot T'_{I,overall}(l,i)} \tag{4}
\end{aligned}
$$

It can be readily seen in the above that no particular terms with regard to the R_j are involved in the final formula, which implies that a node located in an inner region and using a higher rate to transmit l-bytes data obtains the same throughput as a node located in an outer region and using a lower rate for the same data. Contrary to all expectations of the multiple-rate PHY, this result exhibits the fact that a higher data rate does not bring the throughput higher than the others if it competes with a lower rate. This can be regarded as a performance anomaly problem, and its reason is clearly shown in Eqs. (3) and (4): when only one packet can be transmitted with whatever rate, the equal long term channel access probability leads to the same throughput for all nodes.

Solving The Performance Anomaly Problem of IEEE 802.11 MAC with Degraded RAAR

The basic approach to solving the performance anomaly problem in IEEE 802.11 MAC is to use the same base data rate to send multiple back-to-back data packets whenever the channel is held. This is the so called base-rate time share given in Eq. (1) at the very beginning, and it is indeed the core concept of OAR [2] and AAR [3]. However, the focus of these methods is on the opportunistic performance gain when the channel quality is good for transmission, not on the solution of the anomaly problem.

In our work, D-RAAR aims at both getting the opportunistic performance gain and solving the anomaly problem together when RAAR's relay scheme can not work at that time. That is, when no relay nodes can be found or when a given relay node is missed due to mobility or channel condition change, a sender locating in R_j will use the modulation m_j to transmit multiple back-to-back packets directly to AP, if the channel condition is allowed (referring to Fig. 5). In this case, a MH can transmit the number of $G_D^{(j)}(l)$ back-to-back packets, i.e., the number of $\lfloor \frac{T_{tx}(l,1)}{T'_{tx}(l,m_j)} \rfloor$ or $\lceil \frac{T_{tx}(l,1)}{T'_{tx}(l,m_j)} \rceil$ packets to AP. In above, $T_{tx}(l,1)$ denotes the transmission time of l-bytes UDP data payload using the basic rate with the slowest modulation scheme, 1, and $T'_{tx}(l,m_j)$ denotes the transmission time for the same payload in D-RAAR, which can be obtained as $T'_{tx}(l,m_j) = T_{tx}(l,m_j) + T_{ACK} + 2 \cdot T_{SIFS}$ (referring to Fig. 5).

We now estimate the D-RRAR's throughput by analysis. Consider that a mobile host locating in R_j is ready to use the modulation scheme m_j to transmit l-bytes UDP data payload to AP directly when no relay nodes can be found at that time. Based on the base rate adopted in the MMI-WLAN, the overall transmission time can be calculated as

$$T'_{D,overall}(l,j) = G_D^{(j)}(l) \cdot T'_{tx}(l,m_j) + T_{ov} + T'_{CW} \qquad (5)$$

Under the same consideration that the long term channel access probability is equal for all nodes, in D-RAAR, the channel utilization factor of a node in R_j can be

$$U_D^j(l) = \frac{T'_{D,overall}(l,j)}{\sum_{i=1}^M N_i \cdot T'_{D,overall}(l,i)} \qquad (6)$$

Replacing $U_I^j(l)$ in Eq. (4) with $U_D^j(l)$ in above and considering the overall transmission time of D-RAAR, the resulted equation leads to the fact that in spite of the utilization factor, if multiple packets can be transmitted in a basic slot-time, the node throughput of D-RAAR can increase by several times when compared with IEEE 802.11 MAC. That is,

$$
\begin{aligned}
\psi_D^j(l) &= U_D^j(l) \cdot \frac{G_D^{(j)}(l) \cdot l}{TR_j \cdot T'_{D,overall}(l,j)} \cdot TR_j \\
&= \frac{T'_{D,overall}(l,j) \cdot G_D^{(j)}(l) \cdot l}{\left(\sum_{i=1}^M N_i \cdot T'_{D,overall}(l,i) \right) \cdot T'_{D,overall}(l,j)} \\
&= \frac{G_D^{(j)}(l) \cdot l}{\sum_{i=1}^M N_i \cdot T'_{D,overall}(l,i)}.
\end{aligned}
\qquad (7)
$$

Note that if the value of the denominator is similar to that of IEEE 802.11 MAC, $G_D^{(j)}(l)$ in the numerator in fact represents the gain factor that D-RAAR can increase on the throughput.

Throughput Analysis of RAAR

As shown above, if the performance-gain factor of D-RAAR, $G_D^{(j)}(l)$, can be larger than 1, the MMI-WLAN's throughput can benefit from only D-RAAR. However, with the degraded method, nodes in the outer regions may not enjoy such performance gain much since the rates that can be obtained there are only marginally larger than the base rate in usual cases. Accordingly, RAAR provides suitable intermediate nodes for relaying data to AP, and further increases the throughput of these nodes, as analyzed below.

At first, the overall transmission time in the multiple node environment can be obtained with

$$T'_{R,overall}(l, m, m') = G_R^{(j)}(l) \cdot T_{tx}(l, m, m') + T_{ov} + T'_{CW} \tag{8}$$

where $G_R^{(j)}(l)$ represents the number of multiple back-to-back packets that can be obtained with RAAR, given by $\lfloor \frac{T_{tx}(l,1)}{T_{tx}(l,m,m')} \rfloor$ or $\lceil \frac{T_{tx}(l,1)}{T_{tx}(l,m,m')} \rceil$, and $T_{tx}(l, m, m')$ denotes the data transmission time when the relay node is involved. Referring to Fig. 4, this time can be obtained with $T_{tx}(l, m, m') = 3 \cdot T_{SIFS} + 3 \cdot T_{PCLP} + T_{tx}(l, m) + T_{tx}(l, m') + T_{ACK}$, where $T_{tx}(l, m)$ and $T_{tx}(l, m')$ denote the transmission time of this packet from the sender to the relay node and that from the relay node to the AP, respectively.

Given this time, in RAAR, the channel utilization factor of a node in R_j can be obtained with

$$U_R^j(l) = \frac{T'_{R,overall}(l, m, m')}{\sum_{i=1}^{M} N_i \cdot T'_{overall}(l, m, m')} \tag{9}$$

Accordingly, when considering $G_R^{(j)}(l)$ packets delivered in the basic slot-time, $T_{tx}(l, 1)$, the node throughput of RAAR in R_j can be estimated by

$$
\begin{aligned}
\psi_R^j(l) &= U_R^j(l) \cdot \frac{G_R^{(j)}(l) \cdot l}{TR_j \cdot T'_{R,overall}(l, m, m')} \cdot TR_j \\
&= \frac{T'_{R,overall}(l, m, m') \cdot G_R^{(j)}(l) \cdot l}{\left(\sum_{i=1}^{M} N_i \cdot T'_{R,overall}(l, m, m') \right) \cdot T'_{R,overall}(l, m, m')} \\
&= \frac{G_R^{(j)}(l) \cdot l}{\sum_{i=1}^{M} N_i \cdot T'_{R,overall}(l, m, m')} \tag{10}
\end{aligned}
$$

Throughput Comparison of IEEE 802.11 MAC, D-RAAR, and RAAR

In this subsection, the analysis indicated in the previous section is verified and its results are further confirmed with simulations of GloMoSim. With both analysis and simulation, we exhibit the possible performance benefit of D-RAAR and that of RAAR.

The environment under consideration is shown in Fig. 6, wherein five nodes are located in each region, and in total eight regions are taken into account in the MMI-WLAN according to the IEEE 802.11a measure data shown in Fig. 1. For simplicity, no nodes change their regions in the run time and a suitable relay node can be found when a sender want to delivery its data to AP. With this scenario, the methods under consideration are analyzed and simulated, and their results are given in Fig. 7.

Fig. 7(a) shows the node throughputs in R_8 with RAAR, and those in R_1 to R_7 with IEEE 802.11 MAC, for different sizes of UDP data payload. It can be seen from both analysis and simulation that IEEE 802.11 MAC results in the same throughput in spite of a node's location, which is the so-called 802.11 performance anomaly problem indicated previously, while RAAR significantly improves the node throughput in the outermost region, as expected. Fig. (7)b shows the node throughput obtained with D-RAAR for different sizes of UDP data payload. Obviously, D-RAAR gives nodes in inner regions higher throughput than nodes in outer regions, fully utilizing the capability of multi-rate IEEE 802.11a PHY. Finally, Fig. (7)c shows the node throughput obtained with RAAR in the outermost region, R_8. For comparison, the theoretical results of all regions and simulation results of R_8 with D-RAAR are also given. It is readily observable that the node throughput of R_8 is significantly improved by RAAR. This confirms our argument that, even though D-RAAR can transmit multiple back-to-back packets when nodes near AP, in the outermost region, R_8, such a degraded method can provide at most one packet with the lowest data rate, 6 Mbps. In such situation, RAAR shows its capability most remarkably.

Given the results in Fig. (7)c, the question arises that, in addition to the outermost region, how many other regions can benefit from RAAR. To answer this question, we extend the experiment in Fig. (7)c to collect not only D-RAAR's results, but those of RAAR for each region and each UDP payload size. These experimental results are now shown in Fig. 8. As shown in Fig. (8)a, the performance-gain-factor ratio, $G_R^{(j)}(l)/G_D^{(j)}(l)$, is larger than or equal to one in R_8 to R_5 for most data payload sizes, which implies the possible throughput benefits existing in these regions. This is confirmed in Fig. (8)b, which shows the corresponding ratios between the RAAR's throughputs and the D-RAAR's throughputs. For example, the value of 3 for payload size of 2000 bytes indicates that the RAAR's throughput can be $3\times$ that of D-RAAR, in the outermost region. Apart from this most remarkable example, other values larger than 1 can be also seen from R_8 to R_5 for most payload sizes.

These results suggest that the node throughput can benefit from RAAR when nodes are located in an outer region of the MMI-WLAN. In particular, the outermost four regions are preferred when using the measure data given in [1]. With this evaluation scheme, we estimate which regions can benefit from RAAR, and except for IEEE 802.11a, the same evaluation scheme is also considered to be applicable to other multi-rate 802.11 PHY. This would be an interesting possibility for future research.

4. Conclusion

In this paper, we introduce the Relay-Based Adaptive Auto Rate protocol, which can find a suitable relay node for data transmission between transmitter and receiver. RAAR can dynamically adjust a node's modulation scheme to achieve its maximal throughput regarding the transmission distance and the channel condition. When mobility and channel condition changes cause the relay node's information to be out of date, RAAR reverts back to a degraded version of this method, namely D-RAAR, to transmit multiple back-to-back packets directly from the sender to AP without the aid of a relay node. Experiment results indicate that with this scheme, significant improvement on throughput can be achieved for nodes located at the fringe of the AP's transmission.

References

[1] Broadcom Corporation. "The new mainstream wireless LAN standard," White Paper, 18 March 2003.

[2] B. Sadeghi, V. Kanodia, A. Sabharwal, and E. Knightly. "Opportunistic media access for multirate ad hoc networks," Proceedings of MobiCom 2002.

[3] Chunchung Richard Lin, and Yuan-Hao Johusom Chang. "AAR: An adaptive rate control protocol for mobile ad hoc networks," Proceedings of ICON 2003.

[4] Martin Heusse, Franck Rousseau, Gilles Berger-Sabbatel, and Andrzej Duda. "Performance anomaly of 802.11b," Proceedings of INFOCOM 2003.

[5] Arup Acharya, Archan Misra, and Yorktown Heights. "A label-switching packet forwarding architecture for multi-hop wireless LANs," IBM Research Report, RC22512(W0206-141), June 2002.

[6] S. Lu, V. Bhargharan, and R. Srikant. "Fair Scheduling in wireless packet networks," IEEE/ACM Transactions on Networking, 7(4):473-389, Aug. 2003.

[7] T. Ng, I. Stoica, and H. Zhang. "Packet fair queueing algorithms for wireless networks with location dependent errors," Proceedings of INFOCOM'98, MAY 1998.

[8] S. T. Sheu, Y, H. Lee and M. H. Chen. "Providing multiple data rates in infrastructure wireless networks," Proceedings of GLOBECOM'01, 2001.

[9] Global Mobile Information Systems Simulation Library. http://pcl.cs.ucla.edu/projects/glomosim.

[10] PARSEC parallel simulation language. http://pcl.cs.ucla.edu/projects/parsec.

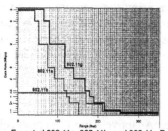

Fig. 1.   Expected 802.11a, 802.11b, and 802.11g Data Rates at Varying Distance from Access Point

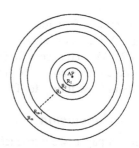

Fig. 2.   Network architecture of the MMI-IWLAN

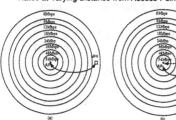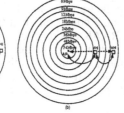

Fig. 3.   Environment for the single node's throughput comparison
(a) direct transmission
(b) transmission with relay node

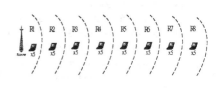

Fig. 4.   The control message flow of RAAR

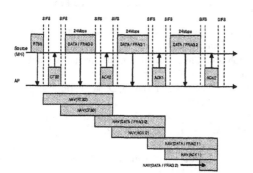

Fig. 5.   The control message flow of D-RAAR

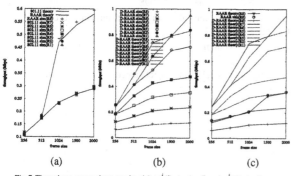

Fig. 6.   The environment with multiple nodes deployed in the MMI-WLAN

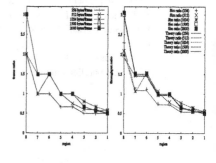

(a)   (b)   (c)

Fig. 7. Throughput comparison results : (a) $\psi_j^i(l)$, $j = 1,...,7$ and $\psi_R^i(l)$, $j = 8$
(b) $\psi_D^i(l)$, $j = 1,...,8$ and (c) $\psi_D^i(l)$, $j = 1,...,8$ and $\psi_R^i(l)$, $j = 8$

Fig. 8. Performance ratio : (a) performance - gain factor ratio, $G_R^i(l)/G_D^i(l)$, and (b) throughput ratio, $\psi_R^i(l)/\psi_D^i(l)$

A NON-TOKEN-BASED-DISTRIBUTED MUTUAL EXCLUSION ALGORITHM FOR SINGLE-HOP MOBILE AD HOC NETWORKS

Romain Mellier and Jean-Frederic Myoupo
LaRIA, CNRS FRE 2733, Université de Picardie Jules Verne,
5, rue du Moulin Neuf 80000 Amiens, France
{mellier, myoupo}@laria.u-picardie.fr

Vlady Ravelomanana
LIPN, CNRS, UMR 7030, Université Paris 13,
Av. J-B Clément 93430 Villetaneuse, France
vlady.ravelomanana@lipn.univ-paris13.fr

Abstract   This paper presents a simple mutual exclusion algorithm for ad hoc mobile networks. Our algorithm does not use the token circulation technique. A station which requests a Critical Section (CS) competes in order to be alone to use the unique channel dedicated to this CS. To reach this goal, we derive a Markov process which guarantees that each station will enter the CS. More precisely, we show that, in presence of collision detection, n/ln2 broadcast rounds are necessary in the average case to satisfy n (n unknown) stations wishing to enter the same CS.

Keywords: Mutual Exclusion , Ad Hoc Mobile Networks, Wireless Networks.

1.   Introduction

Nowadays, the research in Mobile Ad hoc NETworks (MANET for short) is attractive and desirable, due to the development of wireless networks and personal communication ([Bose et al., 1999], [Chlebus et al., 2002], [Garg and Wilkes, 1996], [Hayashi et al., 2000], [Lakshdisi et al., 2001], [Lin and Stojmenovic, 2001], [Malpani et al., 2000], [Malpani et al., 2001], [Myoupo, 2003], [Myoupo et al., 2003], [Vaidya et al., 2001]) . Mobile ad hoc networks are formed by a collection of mobile wireless nodes. Communication links form and disappear as nodes come into and go out of each other communication range. Such networks have many practical applications, including home networking, search-and-rescue, and military operations. We assume that critical sections are disseminated in the network : some stations can be dedicated or specialized to collect or to give some information concerning the characteristics of the network. We can consider a rescue ad hoc network for example. One of its station can be dedicated to centralize the material damages collected by all stations. Another one can be specialized in the collection of medical information and so on. A *unique channel* is assigned to each dedicated station. It is used by the other stations to send or to receive the needed information. So the access to a dedicated station can only be done in mutual exclusion way. More precisely, a station which desires to send or to receive an information from a specialized station, through a channel k_0, may compete in order to be alone to broadcast in k_0. Therefore, the access to one of these dedicated or specialized stations is considered as a *Critical Section (CS)* in the mobile network.

1.1   Related Works

Intuitively, one can believe that a mutual exclusion algorithm for ad hoc networks can be obtained by a simple adaptation of a mutual exclusion algorithm for wired networks. But it is not obvious due to the permanent change of the topology of the network in a MANET. Token-based mutual exclusion algorithms provide access to CS through the maintenance of a single token that cannot simultaneously be present at more that one station in the network. Only the station holding the token can enter the CS. From this token consideration, we can quote first the work of Raymond ([Raymond, 1989]) in which the requests are sent over a static spanning tree of the network, toward the token holder. Chang et al ([Chang et al., 1990]) extend the previous algorithm by inducing a token oriented Directed Acyclic Graph (DAG) for the search of the token holder. Next, Dhamdhere and Kulkarni ([Dhamdhere and Kulkarni, 1994]) showed that the algorithm of Chang et al ([Chang et al., 1990]) can suffer from deadlock and brought a solution to this problem by assigning a dynamic changing sequence number to each node yielding a total ordering of stations in the network. Recently J.E. Walter et al ([Vaidya et al., 2001]) proposed a token-based mutual exclusion algorithm, for a MANET, which combines the ideas from ([Dhamdhere and Kulkarni, 1994]) adapted to a mobile environment. They use the partial reversal technique introduced in [Bertsekas and Gafni, 1981] to maintain a token oriented DAG with a dynamic destination. Their approach induces a logical DAG on the network, dynamically modifying the logical structure to adapt to the changing physical topology in the ad hoc environment.

1.2   Our contribution

This paper presents a distributed mutual exclusion protocol obtained from a Markov process. Our approach does not use a token circulation technique. It guarantees that each candidate for a CS will be satisfied. We give the average number of broadcast rounds necessary for n stations, n unknown, to enter the same CS. Our idea is to construct a splitting process which yields a random binary tree. With the help of the probabilistic divide-and-conquer technique, we derive a single hop protocol which requires $n/\ln 2$ broadcast rounds in average case. Where *ln* is the logarithmic function in basis e and n is the number of stations wishing to enter the same CS.

The rest of this work is organized as follows : the environment considered in this work is presented in section 2. In section 3 a simple single-hop protocol is presented, and its performance is obtained with the help of average case analysis. Section 4 contains brief comments on experimental results. Concluding remarks end the paper

2.   Basic definitions

An Ad Hoc Network is a set S of N radio transceivers or stations which can transmit and/or receive messages through a set C of k channels (a MANET(N, k) for short). The time is assumed to be slotted and all stations have a local clock that keeps synchronous time. In any time slot, a station can tune into one channel and/ or broadcast on at most one channel. A broadcast operation involves a data packet whose length is such that the broadcast operation can be completed within one time slot. So, in the MANET with *collision detection* (CD for short), the status of an n-station MANET channel is :

- **NULL** : if no station broadcasts on the channel in the current slot,

- **SINGLE** : if exactly one station broadcasts on the channel and

- **COLLISION** : if two or more stations broadcast on the channel in the current time slot.

Also, all communications are performed at time slot boundaries, i.e. the duration of broadcast operations is assumed to be sufficiently short.

1 Let us consider N stations which communicate by message passing over a wireless network through k distinct communication channels. Each station runs an application process and a mutual exclusion process (to get a resource) that communicate with each other to ensure that the station cycles between its REMAINDER section (not interested in the CS), its WAITING section (waiting for access to CS), and its CRITICAL section. **Only the station which broadcast yields a single status of the channel executes the CS.** When leaving the CS, it broadcasts in the unique channel to inform the other stations that they can compete to enter the CS.

2 We suppose that n is the number of candidates for entering the same critical section. It is clear that $n \leq N$

3 The system is a **single-hop** network.

We assume that critical sections are disseminated in the network : some stations can be dedicated or specialized to collect or to give some information concerning the characteristics of the network. For example we can consider a rescue ad hoc network . One of its stations can be dedicated to centralize the material damages collected by other stations. Another one can be specialized in the collection of medical information and so on. *A unique channel* is assigned to each dedicated station. It is used by the other stations to send or receive the needed information. So the access to a dedicated station can only be done in a mutual exclusion way. Therefore these specialized stations are considered as *resources* by the others. More precisely, a station which wants to send or to receive an information from a *resource* station through a channel, say k_0, may compete to be alone to broadcast in k_0. Therefore, communications with all these dedicated or specialized stations are *Critical Sections* (CS).

3.   A single-hop mutual exclusion algorithm

Here, S is the set of stations which require access to the same critical section, say CS_0. First, we give a procedure that can split randomly the set S of stations into two non-empty subsets, say S_1 , S_2. Each station is assumed to have the computing power of small laptop such that they can generate random bits and store few data. The time is slotted. The following procedure is an implementation of *Bernoulli* process where the protocol *Single-Hop Ex-Mut* below randomly partitions a given set S (for example the initial set of stations), into two subsets S_1 and S_2. The process is repeated until there is two non-null parts. Each station has a counter initialized to zero and runs the following protocol in order to enter in CS_0.

3.1   Processing an example

We begin this section by presenting an example of MANET(5, 1) in figure 1 will help to understand the basic idea our approach. In this figure, we suppose that the protocol works first with the most left branch until its leaves are reached. Then, it goes backward working again in the most left branch with nodes containing more than one station, and so on. This process is recursively done until the leaves of all branches are considered.

In figure 1, the number of stations that have chosen the same bit are indicated in the nodes. The stations which have chosen bit 1 are on nodes of the right sub-tree. Those which chosen bit 0 are in the nodes of the left sub-tree. The numbers in bold around the node means a step of our example. Before talking our example, we first give a brief description of our algorithm. The set of stations is recursively partitioned into two subsets S_1 and S_2. Each station owns a counter. All counters are initialized to 0.

1 A station has the right to broadcast in the channel only if it is in S_1 and if its counter shows 0.

2 A station in S_2 can broadcast only if the status of channel is either Single or Collision and if its counter shows 0 and if S_1 is empty.

3 If S_1 is empty and S_2 is not empty then the stations of S_2 which counter show 0 move in S_1.

4 After the random choice of a bit from set $\{0, 1\}$, if S_1 (if collision status) and S_2 (if collision status) are not empty, then the stations in S_2 increase their counters by 1.

5 If there are two consecutive single status of the channel, then each station in S_2 decreases its counter by 1.

6 If the status of the channel is single, then the station which has broadcasted executes its CS.

We are now in a position to process our example :

- **Step 1** : Initially the five stations broadcast in the unique channel, obviously with collision. Then, each of them chooses a bit in $\{0, 1\}$ at random. Three of them choose bit 1 (set S_1) and two bit 0 (set S_2). Each station of S_2 increments its counter by one.

- **Step 2** : The three stations in S_1 broadcast in the unique channel and obviously with collision again. Each of them, again chooses a bit in $\{0, 1\}$ at random. Two of them choose 1 (S_1) and one chooses bit 0 (S_2). The unique station in S_2 executes its CS.

- **Step 3** : None of the two stations in S_1 chooses bit 1. Hence $S_1 = \emptyset$.

- **Step 4** : Since $S_1 = \emptyset$ then $S_1 \leftarrow S_2$.

- **Step 5** : One station of S_1 chooses 1 and the other bit 0. Therefore the status of the channel is single. Then only station in S_1 executes its CS.

- **Step 6** : The unique station in S_2 which counter shows 0 moves in S_1 and broadcasts with single status . Then it executes its CS.

- **Step 7** : We then have two consecutive single status. Therefore each station in S_2 decreases its counter by one. Consequently the counters of the two stations which went in S_2 in step 1 show 0 each. So, they move in S_1. One of them chooses 1 and the other bit 0.

- **Step 8** : The unique station in S_1 executes in its CS.

- **Step 9** : Since $S_1 = \emptyset$, $S_1 \leftarrow S_2$. The unique and last station in S_1 finally also executes in its CS.

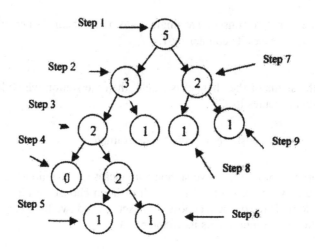

Figure 1.   Mutual Exclusion on a MANET(5, 1)

3.2   The Algorithm

Now that we have cleared the ideas which guide our motivation, we now turn to give the procedure that each station must run in order to perform the mutual exclusion. More precisely a station runs the algorithm which follows :

Procedure Single-Hop-Ex-Mut (INPUT : S, OUTPUT : S_1, S_2)

1. $S_1 \leftarrow S$; $S_2 \leftarrow \emptyset$;
2. **WHILE two consecutive status of the channel are not NULL (i.e.** $S_1 \neq \emptyset$ **AND** $S_2 \neq \emptyset$) **DO**
3. Each station listens to the unique channel while the protocol is running
4. **REPEAT**
      "WAITING Section"
5. Each station in S_1 broadcasts in the unique channel assigned to CS_0.
6. IF the status of the channel is SINGLE then
      BEGIN
            the unique station in S_1 executes CS_0
            the stations of S_2 which counters show zero move in S_1
      END
7. IF the status of the channel is COLLISION then
      BEGIN
            Each station in S_2 broadcasts in the unique channel.
            IF the status of the channel is SINGLE then the unique station in
                  S_2 executes CS_0
            IF the status of the channel is COLLISION then each station in
                  S_2 increments its counter by one
      END

8. IF the status of the channel is NULL then
    BEGIN
        IF $S_2 \neq \emptyset$ then all stations of S_2, which counters
            show zero move in S_1
    END
9. Each station in S_1 chooses a bit in the set $\{0, 1\}$ at random.
10. Those which have chosen 1 stay in S_1 and those which have chosen
    0 move in S_2.
UNTIL the unique channel has two consecutive SINGLE status.
The stations of S_2 decrease their counters by one. Those which counters
show zero move in S_1.
    "REMAINDER section"
11. The station which is in the CS_0 broadcasts a signal on the unique
channel to mean that it is leaving the CS_0.
The effect of this message is to inform the stations in S_1 that they can
now request the CS_0.
END WHILE
END Single-Hop Ex-Mut

Consider a subgroup of m stations. The probability of failure of the splitting
process applied on this subset is then given by : $Pr[failure] = 2*\frac{1}{2^m}$. So with
probability equals to $1 - \frac{1}{2^m-1}$, the procedure *Single-hop Ex-Mut* subdivides
the initial set of stations into two non-empty subsets of stations. At the end of
the *repeat-until* loop, each station can determine if whether

1 it is the only one that selected the bit 1, in this case the status of the
   channel is **SINGLE** or

2 there are at least two stations that selected bit 1, in this case the status of
   the corresponding channel is **COLLISION**. A station which broadcast
   yields a single status has the right to enter its CS. Since the power of
   stations allow us to store the status of the channel, only one broadcast
   per turn of the "repeat-until" loop in the *Single-hop Ex-Mut* procedure
   is needed to perform the protocol. In other words, our protocol randomly generates a binary-tree-like structure (see figure 1). Concretely,
   our principal problem is then to compute the number of passes through
   all he internal nodes (nodes of degree > 1), including repetitions which
   are exactly the number of broadcast rounds.

3 Note that when the channel is of single status, all stations of S_2 wait till
   they receive a signal from a station leaving CS_0.

3.3   The use of a counter in each station

The management of the counter is one of the key points of our algorithm.
The consistency of our approach is guaranteed by the following considerations

1 each station monitors the channel, and so it knows the status of the channel at every broadcast round.

2 root of sub-tree : in the WAITING section, at level 7 of the procedure two consecutive COLLISION status (one yielded by the stations of S_1 and the second by the stations of S_2) show that after the random choice of bits 0 or 1, we have $S_1 \neq \emptyset$ and $S_2 \neq \emptyset$. Therefore the node considered has two non empty sub-trees. The stations of the right sub-tree will enter the CS_0 after those of the left sub-tree. Incrementing their counters indicate the backward processing of our algorithm.

3 decreasing the counter : according to 2. above, decreasing a counter means that all stations of the left sub-tree have been the CS_0. It is taken into account in Level 11 of the procedure.

4 It can happen that no station chooses bit 1. Then $S_1 \leftarrow S_2$ with the condition that S_2 must not be empty (see level 8 of the procedure).

3.4   Evaluation of the number of broadcast rounds necessary for n stations to enter the same CS

In this paragraph, Our goal is to introduce basic methods that are useful to analyse performance of such protocols. As we said earlier, our approach generates a binary tree and it is shown in [Flajolet and Sedgewick, 1996] that such process always ends. However we next show that the average number of broadcast rounds is given by an absolute convergent Fourier series. Then proving that the process of splitting always ends will be done.

Theorem 3.1. *Let us consider a MANET(N, 1) with CD. Suppose n stations, $n \leq N$, request the same CS. The protocol Single-Hop Ex-Mut terminates in approximately $\frac{n}{\ln 2}$ broadcast rounds on the average.*

Proof. The goal of this proof is to show that the process of splitting in the Single-Hop Ex-Mut always ends as stated earlier. It is easy to see that protocol Single-Hop Ex-Mut generates a random complete binary tree and it terminates when reaching all the nodes of degree 1. At each step of the protocol, the probability of splitting a given set of size m can be depicted as in the figure 2. Our idea is similar to the one in [Myoupo et al., 2003]. In the figure 2, each edge is weighted with $\binom{m}{p}/2^m$ which is the probability that a set of m stations will split into exactly 2 non-empty subsets of respective sizes p and $m - p$. So, the probabilistic model is here a *Markov chain* (see [Feller, 1957]) where reaching a node of the tree of degree 1 corresponds, now to an absorbing state (self-loop of the state with a probability equals to 1). If we denote by α_i the *average waiting time* to reach an absorbing state (partition recursively n stations until getting n parts), the computing is classical (cf. [Feller, 1957]) by means of linear formulae

$$\alpha_i = 1 + \sum_{r+s=i, r \geq 0, s \geq 0} \Pr[i \xrightarrow{splits} r, s] \alpha_{r,s} \qquad (1)$$

where $\Pr[i \xrightarrow{splits} r, s]$ is the probability of a transition from the "state" i to the "state" (r, s). Since in our case we have to split *sequentially*, in their turn, the two non-empty subsets (containing respectively r and s stations), i.e. to terminate all subdivisions must be done one by one.

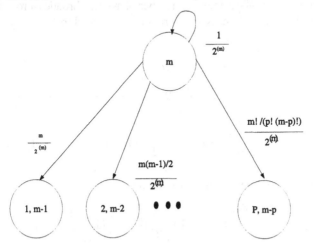

Figure 2.   Splitting randomly a given number m into two parts

So, we have simply

$$\alpha_{r,s} = \alpha_r + \alpha_s \tag{2}$$

Here, $\alpha_0 = \alpha_1 = 0$ and α_n, $n \geq 2$ verifies

$$\alpha_n = 1 + \frac{1}{2^{n-1}}\alpha_n + \sum_{p=0}^{n=1} \frac{\binom{n}{p}}{2^{n-1}}\alpha_p \tag{3}$$

Now, let us introduce the exponential generating function (see for example [Comtet, 1974] for more details)

$$\alpha(x) = \sum_{n=2}^{\infty} \alpha_n \frac{x^n}{n!} \tag{4}$$

In formula (4), the average number of broadcast rounds, α_n is given as the coefficient of $\frac{x^n}{n!}$ of the power series $\alpha(x)$. Recall that if $P(x)$ is a polynomial, the notation $[x^n]P(x)$ gives the coefficient of $P(x)$. This notation applies to our function $\alpha(x)$ and the average number of broadcast rounds is then

$$\alpha_n = n![x^n]\alpha(x) \tag{5}$$

Replacing α_n in (4) by its expression from the formula (3), we obtain the following linear functionnal equation :

$$\alpha(x) = e^x - x - 1 + 2e^{\frac{x}{2}}\alpha(\frac{x}{2}) \tag{6}$$

First terms are given below

$$\alpha(x) = 2x^2 + \frac{10}{3}x^3 + \frac{100}{21}x^4 + \frac{652}{105}x^5 + \frac{24922}{3255}x^6 + \frac{9874}{1085}x^7 + \cdots \quad (7)$$

Thus, the coefficient $\frac{652}{105}$ is then the number of average broadcast rounds needed to initialize 5 stations. Successive iterations of (6) lead to

$$
\begin{aligned}
\alpha(x) &= e^x - x - 1 + 2e^{\frac{x}{2}}\alpha(\frac{x}{2}) \\
&= e^x - x - 1 + 2e^{\frac{x}{2}}(e^{\frac{x}{2}} - \frac{x}{2} - 1) + 4e^{\frac{3x}{4}}\alpha(\frac{x}{4}) \\
&\cdots \\
&= \sum_{j=0}^{\infty} 2^j exp((1 - \frac{1}{2^j})x)(exp\frac{x}{2^j} - \frac{x}{2^j} - 1)
\end{aligned}
$$

When expanding the exponentials, we have

$$\alpha_n = \sum_{j=0}^{\infty} 2^j (1 - (1 - \frac{1}{2^j})^n - \frac{n}{2^j}(1 - \frac{1}{2^j})^{n-1}) \quad (8)$$

As, we have

$$(1 - \epsilon)^n = e^{-n\epsilon + O(n\epsilon^2)} \quad (9)$$

splitting the right hand side of (8), one can legitimate the use of (9) and we have

$$\alpha_n \approx \frac{n}{\ln 2} \quad (10)$$

Note that *Mellin transforms* methods, see for instance [Knuth, 1973] and [Flajolet and Sedgewick, 1996] for more details, are well suited for asymptotic estimates of coefficients of linear functionnal equations like (6). Here, it gives an additional fluctuating and periodic term $P(\frac{\ln n}{\ln 2})$ where P is an absolute convergent Fourier series of variations $\leq 10^{-5}$. In [Knuth, 1973], Knuth derived explicit expressions of the fluctuating term. Note also that, $1/\ln 2 \approx 1.44$. So there are 44% "waste of broadcast rounds".   □

4.   Experimental results

This section presents graphics on the evolution in time of the variation of the number of stations which request the CS. The irregularities of the slopes of these curves are due to the random choice of it 1 or 0.

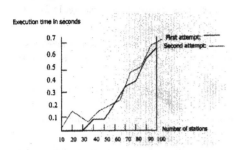

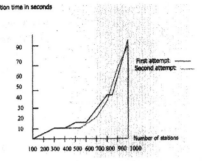

Figure 3. *Mutual exclusion for 100 stations : two attempts*

Figure 4. *Mutual exclusion for 1000 stations : two attempts*

5. Concluding remarks

In this paper we considered a MANET (N, k) from which we derived a distributed mutual exclusion protocol from a Markov process. Our approach does not use a token circulation technique. However it guarantees that each candidate for a CS will be satisfied. The performance of our protocol in terms of broadcast round is evaluated. More precisely, it requires $n/\ln 2$ broadcast rounds in average. Where ln is the logarithmic function in basis e and n is the number of stations wishing to enter the same CS. An interesting challenge is to derive a k-mutual exclusion protocol in a MANET from our approach. In our approach, we assume that confidential data items are encoded. The reader may argue that the use of token as in [19] guarantees more confidentiality of data items to be broadcasted. It is not always true, because it is well known that hijacking data items broadcasted in an air channel is easy to realize. Therefore even with the use of token the confidential data items must be encoded. Finally, a station can execute its CS as many times as its needs : after it has left its CS, it waits until it hears two consecutive NULL status of the channel. Then it runs the *Single-Hop-Ex-Mut*.

References

[Bertsekas and Gafni, 1981] Bertsekas, D. and Gafni, E. (1981). Distributed algorithms for generating loop-free routes in networks with frequently changing topology. In *IEEE Transactions on Communications, C-29(1) : 11-18.*

[Bose et al., 1999] Bose, P., Morin, P., Stojmenovic, I., and Urrutia, J. (1999). Routing with guaranteed delivery in ad hoc wireless networks. In *Proc. of 3rd ACM Int. Workshop on Discrete Algorithms and Methods for Mobile Computing and Communications DIAL M99, Seattle, August 20, 1999, 48-55.*

[Chang et al., 1990] Chang, Y., Liu, M., and Singhal, M. (1990). A fault tolerant algorithm for distributed mutual exclusion. In *Proc. of 9th IEEE Symp. On Reliable Distr. Systems, pp. 146-154.*

[Chlebus et al., 2002] Chlebus, B.S., Gasienec, L., Gibbons, A., Pelc, A., and Rytter, W. (2002). Deterministic broadcasting in ad hoc networks. *Distributed computing*, 15:27–38.

[Comtet, 1974] Comtet, L. (1974). *Advanced combinatorics*. Reidel, Dodrecht.

[Dhamdhere and Kulkarni, 1994] Dhamdhere, D.M. and Kulkarni, S.S. (1994). A token based k-resilient mutual exclusion algorithm for distributed systems. *Inform. Process. Letters*, (50):151–157.

[Feller, 1957] Feller, W. (1957). *An introduction to probability theory and its applications*. John Wiley, New-York.

[Flajolet and Sedgewick, 1996] Flajolet, P. and Sedgewick, R. (1996). *Introduction to the analysis of algorithms*. Addison-Wesley.

[Garg and Wilkes, 1996] Garg, V.K. and Wilkes, J.E. (1996). *Wireless and personnal communication systems*. Prentice-Hall PTR, Upper Saddle River, NJ.

[Hayashi et al., 2000] Hayashi, T., Nakano, K., and Olariu, S. (2000). Randomized initialization protocols for ad hoc networks. *IEEE Trans. Parallel Distr. Syst.*, 11:749–759.

[Knuth, 1973] Knuth, D.E. (1973). *The art of computer programming, vol 3 : Sorting and Searching*. Addison-Wesley.

[Lakshdisi et al., 2001] Lakshdisi, M., Lin, X., and Stojmenovic, I. (2001). Location based localized alternate, disjoint, multi-path and component routing algorithms for wireless networks. In *ACM Symposium on Mobile Ad Hoc Networking And Computing MobiHoc, Long Beach, California, USA, October 4-5, 2001, 287-290*.

[Lin and Stojmenovic, 2001] Lin, X. and Stojmenovic, I. (2001). Loop-free hybrid single-path/flooding routing algorithms with guaranteed delivery for wireless networks. *IEEE Transactions on Parallel and Distributed Systems*, 12(10):1023–1032.

[Malpani et al., 2000] Malpani, N., Vaidya, N., and Welch, J.L. (2000). Leader election algorithms for mobile ad hoc networks. In *DIALM for Mobility 2000, pp. 96-103*.

[Malpani et al., 2001] Malpani, N., Vaidya, N., and Welch, J.L. (2001). Distributed token circulation in mobile ad hoc networks. In *Proceedings of the 9th International Conference on Network Protocols (ICNP)*.

[Myoupo, 2003] Myoupo, J.F. (2003). Dynamic initialization protocols for mobile ad hoc networks. In *11th IEEE Intern. Conf. on Networks (ICON2003), Sydney, Australia, pp. 149-154*.

[Myoupo et al., 2003] Myoupo, J.F., Ravelomanana, V., and Thimonier, L. (2003). Average case analysis based-protocols to initialize packet radio networks. *Wireless Communication and Mobile Computing*, 3:539–548.

[Raymond, 1989] Raymond, K. (1989). A tree-based algorithm for distributed mutual exclusion. *ACM Trans. Comput. Systems*, 7:61–77.

[Vaidya et al., 2001] Vaidya, N.H., Walter, J.E., and Welch, J.L. (2001). A mutual exclusion algorithm for ad hoc mobile networks. *Wireless Networks*, 7(6):585–600.

THE RECEIVER'S DILEMMA

John P. Mullen[1], Timothy Matis[1], Smriti Rangan[2]

[1] *Center for Stochastic Modeling, Industrial Engineering Department, New Mexico State University;* [2] *Center for Stochastic Modeling, Klipsch School of Electrical and Computer Engineering, New Mexico State University*

Abstract　　In Mobile Ad Hoc Networks (MANETs), each node has the capacity to act as a router. The performance of the MANET relies on how well the nodes perform this function. In simulations, the receipt of a Route Reply (RREP) packet is evidence that the associated link is reliable, but in the real world of wireless links, it is not. Thus a node often has to decide whether the RREP indicates a reliable link or is due to an outlier in the distribution of received power. If the node accepts an atypical RREP when the link is not reliable, the subsequent attempt to communicate will fail. On the other hand, if it rejects a representative RREP, it will fail to use a reliable link. This paper examines this selection problem using a stochastic model of link behavior and explores some techniques that may be used to deal with the situation.

Keywords:　MANET routing; wireless propagation; multipath fading; stochastic models; decision rules; power averaging.

1.　　Introduction

In a *Mobile Ad Hoc Network* (MANET), each node has the ability to act as a router, permitting adaptable multihop communications. Although simulations show that MANET protocols can support nets containing hundreds of nodes, practical considerations limit most implementations to five or fewer mobile nodes [1–4]. A significant cause of this disparity is the stochastic nature of wireless links. This paper describes this nature, demonstrates how it affects MANETs, and summarizes some promising directions for future study.

In this paper, Section 2 describes the basic problem, presents two test scenarios, and develops the basic evaluation measures. Section 3 outlines several strategies to deal with the problem and estimates their effectiveness. Section 4 presents a simulation of the impact of combining two particular strategies: unicast route replies and multiple attempts

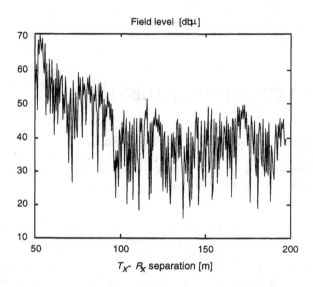

Figure 1.   Field Measurements of Signal Levels from [5]

at the link level. Finally, Section 5 summarizes, draws conclusions, and outlines future work.

2.   A Fundamental MANET Problem

As illustrated in Figure 1, wireless signals are subject to fine grained, high magnitude fluctuations which may cause two detrimental events:

1 A packet may be dropped over a reliable link, and

2 A packet may be received over an unreliable link.

This phenomena certainly impacts stub and cellular networks. However, in MANETS, it creates much more severe problems because , in addition to Event 1 causing data packets to be lost, the occurrence of either event may impact the stability of the MANET's routing mechanism.

Event 1 is liable to cause a node to conclude that a current route is no longer usable, when it actually is. The resulting unnecessary route search significantly reduces performance. A more serious difficulty arises when a node receives a *route reply* (RREP) over an unreliable link due to Event 2. This may cause the node to conclude that the link is reliable and to include it in the selected route. When the unreliable link fails, a new search becomes necessary, which degrades performance further. Not only that, the new search creates another opportunity for Event 2 to occur. This has proven to be a serious problem in the field [1, 2, 4].

In most MANET simulations and analysis, it is assumed that receipt of a RREP over a link is proof that it is reliable. However, in the real world, receipt of a RREP could indicate either a reliable route or that Event 2 has occurred. Thus, when a MANET node receives a RREP from another node, it is faced with a dilemma. If it chooses to use a link which is, in fact, unreliable, the subsequent attempt to transfer information will fail. On the other hand, if it rejects a link that is actually reliable, an opportunity to transfer information will be lost.

Because whether a node accepts or rejects a particular RREP, it may make a serious error, the performance of a MANET will depend greatly on how well each node guesses which links are reliable and which are not. Failing to account for this dilemma in protocol design will generally lead to poor performance in the "real world." Therefore, the manner in which this guessing occurs is an important MANET design consideration. This paper explores this dilemma as well as several methods to deal with it.

A Simple Stochastic Link Model

There are a number of models which can predict $\bar{p}(d)$, the average received signal strength as a function of distance [5–7]. In this paper, it is assumed that power decreases exponentially with distance. Thus,

$$\bar{p}(d) = p_0 \left(d_0/d\right)^c, \tag{1}$$

where d_0 is a reference distance, p_0 is the average power measured at d_0, and c is the rate of exponential decay. Assuming there are no other nodes transmitting in the vicinity, interference would be minimal and the probability of reception would be mainly a function of received power. In addition, letting $p_0 = p_{\min}$, the the minimum amount of power for reliable reception, means that d_0 is the nominal range. Finally, it is assumed that $c = 3$, a typical value [6, 7].

Given any model to predict $\bar{p}(d)$, the fine grain variations in received power can be represented as a stochastic process. This variability is chiefly due to effects of fading and shadowing, together with some other effects not completely understood [5–7]. Because multipath fading can cause signals to be stronger, as well as weaker, than expected, it plays a primary role in the occurrence of Event 2. There are several common multipath fading models. In this paper, the Rayleigh model is used. Although simple, this model is realistic [1, 5–7] and will serve to illustrate the fundamental difficulties induced by multipath fading.

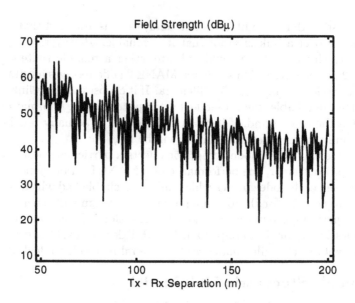

Figure 2.   Synthetic Trace Generated by Eq. (5)

If Rayleigh fading is in effect, the probability that instantaneous received power (p_i) is at least p would be:

$$\Pr\{p_i(d) \geq p\} = \int_p^\infty [1/\overline{p}(d)] \exp\left[-s/\overline{p}(d)\right] ds$$

$$= \exp\left[-p/\overline{p}(d)\right]. \tag{2}$$

Substituting Eq. (1) into (2) leads to:

$$\Pr\{p_i(d) \geq p\} = \exp\left[-(d/d_0)^c (p/p_{\min})\right]. \tag{3}$$

Finally, the probability of reception is:

$$\Pr\{p_i(d) \geq p_{\min}\} = \exp\left[-(d/d_0)^c\right]. \tag{4}$$

To demonstrate the ability of this model to predict field measurements, a simulation was run in Scenario 1.

Scenario 1.   This is a simple test in which a transmitter is initially 5m away from a receiver and then moves at a constant rate of 0.5m/s until it is 100m away. This scenario demonstrates the nature of reception over a range of separation distances. It is similar to tests in [1] and [5].

The inverse transform of Eq. (3)

$$p_i(d; d_0, p_{\min}, c) \;=\; -p_{\min}(d_0/d)^c \ln r. \tag{5}$$

was used to generate random received power levels. Here r is a random number uniformly distributed on $(0,1)$ and $p_i(\cdot)$ is a random instance of received power. In this test, the transmitter sent one 1024 bit UDP packet each second. Power was adjusted to give a mean response similar to that in [5]. Figure 2 shows the simulated received power levels that resulted. The variation in these values is very similar to that in Figure 1.

Evaluation Model

The impact of each strategy discussed in the following section will be estimated in the context of Scenario 2.

Scenario 2.   The layout for this scenario, shown in Figure 3, was inspired by a field test conducted in [2], except that in this scenario none of the nodes move. The task is for Node A to find a route to Node B. The nominal range is 50m. Also there are no other nodes in the vicinity. Assume the routing protocol seeks routes on demand and will select that route with the fewest number of hops. Note that if the originating node had perfect knowledge, it would choose the AXB two-hop route. This scenario, therefore, focuses on the impact of uncertainty.

As each strategy is introduced, the probability of selecting the better route and the probability of successfully transmitting five data packets are estimated analytically. Let the probability of a control packet being received over either link AX or XB be p_1 and that for Link AB be p_2. The three possible ways in which Node A could receive a RREP are listed in Table 1. This table also states their probabilities in terms of p_1 and p_2. From Eq. (4),

$$p_1 = \exp(-0.6^3) \approx 0.806 \quad \text{and} \quad p_2 = \exp(-1.2^3) \approx 0.178. \tag{6}$$

This leads to the values listed in Table 1.

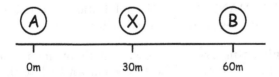

Figure 3.   Relative Position of Nodes A, B, and X in Scenario 2

Table 1. Possible ways of A receiving a RREP for a Path to B

Possibility	RREQ Path	RREP Path	Probability Expression	Value
a	AXB	BXA	$p_a = p_1^4$	0.421
b	AB	AB	$p_b = p_2^2$	0.032
c	AXB	AB	$p_c = p_1^2 p_2$	0.115
b or c	AXB or AB	AB	$p_{c \cup b} = \left(p_1^2 + p_2 - p_1^2 p_2\right) p_2$	0.126

In the cases that follow, for Case n, let $P_{\text{SAXB}n}$ denote the probability of selecting the more reliable AXB route and $P_{\text{s}n}$ be the probability that a five-packet message will be transmitted successfully.

Case 1: Baseline.   It is assumed that Node A will accept a RREP from either B or X, regardless of the route the RREQ followed. Hence,

$$P_{\text{AXB1}} = \frac{p_a \left(1 - p_{b \cup c}\right)}{1 - \left(1 - p_a\right)\left(1 - p_{b \cup c}\right)} \approx 0.744.$$

Thus, the poorer route will be selected about 25% of the time. Under the simplifying assumption that $p_{\min}$ is independent of packet size, and ignoring the various delays associated with route searches,

$$P_{S1} = P_{\text{AXB1}}\, p_1^{10} + (1 - P_{\text{AXB1}})p_2^5 \approx 0.086.$$

Although this baseline model is not very good, it serves to illustrate the basic computations. The following section discusses possible improvements on this simple protocol.

3.   Some Strategies to Deal with Fading

This section discusses five general approaches that can help reduce the severity of this problem, organized by OSI level.

The MANET Protocol

Two possible MANET protocol modifications are unicasting RREP packets and specifying a minimum reliability.

Case 2: Unicasting rrep Packets.   A simple improvement to the basic protocol is to require that RREPs be unicast back to the node from which the RREQ came. This simple change has been implemented in a number of protocols [8, 9]. Requiring that the RREP be unicast back to the source eliminates the third possibility in the baseline scenario. Thus

$$P_{\text{AXB2}} = \frac{p_a \left(1 - p_b\right)}{1 - \left(1 - p_a\right)\left(1 - p_b\right)} \approx 0.928.$$

Although this improves P_{AXB2} appreciably, it only increases P_{S3} to about 0.107. However, combining this strategy with other strategies leads to further improvement. This will be explored further in Section 4.

Case 3: Minimum Reliability Criterion. Another approach is to include a minimum reliability value in the RREQ. In this scheme, nodes continually estimate the probability that they will be able to communicate with their one-hop neighbors. If Node B knows that its link to A is not sufficiently reliable, it would not reply to a RREQ from A, even if it receives one. A simple rule could be that unless two packets from A are received in a row, do not reply, but other rules are possible. The strategy of having nodes curtail their RREPs reduces the consequences of Event 3 and the number of RREPs on the channel. The primary disadvantage is that this requires some pre-existing traffic to work.

It is difficult to state how this would impact the probability of success, since it depends on the method chosen to estimate reliability. However, assuming the simple two-in-a-row rule, the probability of two successes in a row AB link is $p_2^2 \approx 0.032$. For the AXB path, the probability of two successes is $p_1^4 \approx 0.649$. So, $P_{\text{AXB3}} \approx 0.927$ and $P_{S3} \approx 0.466$.

The Link Level

A common strategy in wireless nets is to retry packets at the link level up to some limit (R) times before reporting a failure to the higher levels.

Case 4: Up to R Retries. If a packet will be sent up to R times, there will have to be $R + 1$ failures in a row before the MANET layer is advised of the failure. In each attempt, there could be a failure in the packet transmission or in the receipt of the ACK. Thus, the probability of success on a single link is:

$$P(R) = 1 - (1 - p^2)^{(R+1)}, \tag{7}$$

where p is the probability of success for a single attempt. Note that this assumes that the transmitter will wait for an ACK after the last attempt. That is, the case in which $R = 0$ is not the same as in Case 1.

Key probabilities are listed in Table 2 in Scenario 2 for several values of R. Note that while $P_1(R)$ increases with R, the probability of Event 2 also rises. As a consequence, $P_{\text{SAXB4}}(R)$ decreases. In addition, P_{S4}

Table 2. Effect of R on Success in Scenario 2

R	$P_1(R)$	$P_2(R)$	$p_a(R)$	$P_{b \cup c}(R)$	$P_{SAXB3}(R)$	$P_{S3}(R)$
0	0.649	0.032	0.178	0.014	0.927	0.012
2	0.957	0.092	0.838	0.085	0.901	0.579
4	0.995	0.148	0.979	0.147	0.851	0.806
6	0.999	0.201	0.997	0.201	0.799	0.794

seems to peak around $R = 4$. However, there are other detrimental effects, such as an increase in the number of packets attempted and mean packet delay, which influence the optimal choice of R. The impact of these factors will be explored in further in Section 4.

The Physical Layer

The layer with the greatest knowledge of what is happening is, of course, the physical layer.

Case 5: Power Averaging. In [1], the authors employed an exponentially-smoothed average of received power to estimate the reliability of the link. In such a scheme, $\widehat{P}_i$, the i-th estimate of average received power, is defined to be:

$$\widehat{P}_i = \alpha p_i + (1 - \alpha)\widehat{P}_{i-1} \qquad (8)$$

where p_i is the $i - th$ observed power level and $0 < \alpha < 1$ is the exponential smoothing parameter. This average is equivalent to a weighted average of all observations to date, with greater emphasis on the most recent [10, p. 594], reducing the impact of node movement on the estimate. The expectation of $\widehat{P}$ is $E[P]$ and its variance is $\alpha/(2 - \alpha)V[P]$.

Let $\alpha = 0.5$, and the minimum threshold for a usable link be $\widehat{P} > F p_{min}$ where F is a factor chosen to improve the chances of reception on a selected link. Here, let $F = 1.05$. Approximating $\widehat{P}$ by a normal random variable with mean $E[\widehat{P}(d)] \approx F p_{min} (d_0/d)^c$ and standard deviation equal to $0.58\ E[\widehat{P}(d)]$, yields a probability of about 0.945 of accepting the AX or XB links and about 0.134 of selecting AXB. Combining this with the probabilities of receiving RREPs, leads to $P_{AXB5} \approx 0.989$ and $P_{S5} \approx 0.114$. The effectiveness depends on a number of technical factors, such as bandwidth and rate of node movement [1].

Case 6: hello Messages. A major problem with reliability estimation and power averaging is that one must have traffic present. Most wireless networks employ HELLO messages to establish the degree of local connectivity. Unfortunately, due to differences in data rate and packet size, one cannot directly infer reliability for message packets from that of HELLO packets [2]. However, this is not the case for power averaging. Thus, one can estimate receive power (or SNR) with any packet. The main difference is that if each HELLO packet generates a single estimate of p_i, then data and other longer packets should generate more estimates.

HELLO packets have another advantage for power averaging. Because HELLO packets are transmitted on a schedule, the receiver can detect missing HELLO packets and use that information to generate more accurate estimates of the average received power. Alternatively, one can simply count the number of HELLO messages received with $p_i > p_{min}$ or some other critical value that would indicate proper reception of data packets. This would reduce the need to estimate power levels, reducing the computational load.

This approach requires each node to estimate and store link reliability values, even if that link is not needed. However, HELLO messages are very short and this need only be done in the one-hop neighborhood. The information is propagated to others by the simple means of not replying to RREQs from nodes sharing unreliable links. Since HELLO packets are normally send only when other traffic is not present, they would only be useful at very light loads in Scenario 2. However, they would be more useful in larger nets.

4. Simulation Analysis

There are some effects that were not considered in the Section 3. In this section, the combination of unicast RREPs and retries at the link level are explored by means of simulations. These simulations are performed in OPNET[1] using a modified wireless lan model in which the exponential decay factor is 3 and Raleigh fading is employed. Details of this modification are in [11].

Figure 4 illustrates the impact of retries for $R = 0$, 2, and 6 in Scenario 1 with a generation rate of 10 packets/s. The dotted line is the performance predicted by by considering only $\overline{p}(d)$. Note that while increasing R can increase reliability on $(0, d_0)$ to nearly that predicted by the non-fading model, large values of R also increase the probability of Event 2. However, this analysis is at a very light load and does not consider such things as the impact of the repeated transmissions of failed packets.

Figure 5 shows the relative throughput for a range of offered loads in Scenario 2 using unicast and several values of R on a one Mbit channel. The MANET protocol is AODV. Here the relative offered load is the ratio of the number of bits in data packets received to that of data

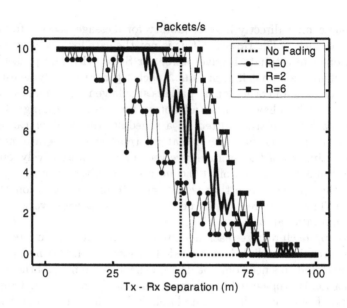

Figure 4.   Simulated Effect of Range and the Number of Retries on Throughput

packets sent. The simulation consisted of two replicates, each consisting of a 500 second run for each value of offered load and R. The replicates were consistent with each other, roughly indicating sufficient run length. The values Figure 5 are the averages from the two replicates.

The results indicate that while having some number of retries improves performance, having too many is also detrimental. This is likely because of the increased utilization of the channel due to the multiple transmissions. It appears that for this particular situation, Either $R = 1$ or $R = 2$ is indicated. The smaller value yields lower overall throughput, but more consistency, while $R = 2$ brings higher performance at the cost of less stability. It also appears that an offered load of more than 10% of channel capacity will sharply decrease relative throughput and that the best one can expect in this configuration is between 70% and 80% of the packets to get through. This agrees with results in [2].

5.   Summary and Conclusions

When searching for routes, a MANET node can make two sorts of errors: 1) rejecting a reliable link on the basis of an unusually weak signal and 2) accepting an unreliable link on the basis of an unusually strong one. Although either error has significant consequences, because it may lead to the need for a new route search, the second is more

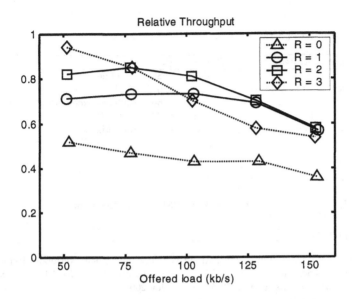

Figure 5.   Effect of the Number of Retries on Throughput

serious than the first. Because the power of a received packet is a random variable, it is not possible for a node to anticipate, on the basis of a single packet, how reliable a route is likely to be. It is essential that some sort of repeated measures be incorporated to achieve acceptable throughput. The probability of selecting a better path over a shorter, but less reliable, one under several rules is summarized in Table 3. While some seem to work better than others, each introduces its own problems and requires some thought about parameter settings. Additionally, it is likely that the ultimate solutions to this problem will involve combinations of two or more basic approaches. For example, in [3] the authors describe an integrated approach that uses the number of retransmissions as the route selection metric, rather than shortest-path. This seems promising, if for no other reason than they have a working MANET containing 29 nodes.

There are many possibilities to consider. Doubtless, the unreliability problem will be mollified to some degree by technological advances at the physical and link levels, but there will still be a residual problem at the MANET level. In the end, however, the most important thing may be to develop more analytical and computer simulation models that will permit the consideration of the impact of this high magnitude, fine grained stochastic link behavior in MANET design.

Table 3.   Impact of Strategies

Case	Strategy	P_{SAXB}	P_{S}	Notes
1	None	0.744	0.086	Baseline
2	Unicast RREPs	0.928	0.107	Typical?
3	$R_{\min}$	0.927	0.466	Reduces excess RREPs, too.
4	Link retries	0.901	0.579	Using $R = 2$.
5	Power Averaging	0.989	0.114	Has side effects.

Notes

1. OPNET is a registered trademark of OPNET Technologies, Inc.

References

[1] Kwan-Wu Chin, John Judge, Aidan Williams, and Roger Kermode. Implementation experience with MANET routing protocols. *ACM SIGCOMM Computer Communications Review*, 32(5):49–59, Nov 2002.

[2] Ian D. Chakeres and Elizabeth M. Belding-Royer. The utility of Hello messages for determining link connectivity. *Wireless Personal Multimedia Communications*, 2:504–508, October 27 2003.

[3] Douglas S. J. De Couto, Daniel Aguayo, John Bicket, and Robert Morris. A high throughput path metric for multihop wireless routing. In *MobiCom 03*, San Diego, CA, USA, September 14-19, 2003.

[4] S. Desilva and S. Das. Experimental evaluation of a wireless ad hoc network. In *Proceedings of the 9th Int. Conf. on Computer Communications and Networks (IC3N)*, Las Vegas, NV, October 2000.

[5] Aleksandar Neskovic, Natasa Neskovic, and George Paunovic. Modern approaches in modeling of mobile radio systems propagation environment. *IEEE Communications Surveys*, pages 2–12, October 2000.

[6] Jean-Paul Linnartz. *Narrowband Land-Mobile Radio Networks*. The Artech House Mobile Communications Library. Artech House, Boston, 1993.

[7] Theodore S. Rappaport. *Wireless Communications: Principles and Practice*. Prentice-Hall, Inc., Upper Saddle River, NJ, second edition, 2002.

[8] C. Perkins, E. M. Belding-Royer, and I. D. Chakeres. Ad hoc on-demand distance vector (AODV) routing. http://moment.cs.ucsb.edu/pub/draft-perkins-manet-aodvbis-00.txt, 2003. Work in progress.

[9] D. B. Johnson, D. A. Maltz, and Yih-Chun Hu. The dynamic source routing protocol for mobile ad hoc networks (DSR). http://ietf.org/internet-drafts/draft-ietf-manet-dsr-09.txt, 2003. Work in progress.

[10] George E. P. Box, William G. Hunter, and J. Stuart Hunter. *Statistics for Experimenters*. John Wiley & Sons, New York, NY, 1978.

[11] John P. Mullen. Efficient models of fine-grain variations in signal strength. In *OPNETWORK 2004*, Washington, DC, August 30 - September 3, 2004. OPNET Technologies.

THEORETICAL CAPACITY OF MULTI-HOP WIRELESS AD HOC NETWORKS*

Yue Fang
RWIN-Lab, Department of Electrical & Computer Engineering
Northeastern University, Boston, Massachusetts, USA
yfang@ece.neu.edu

A. Bruce McDonald
RWIN-Lab, Department of Electrical & Computer Engineering
Northeastern University, Boston, Massachusetts, USA
mcdonald@ece.neu.edu

Abstract   The capacity of multi-hop wireless ad hoc networks is presented extending the cross-layer model for link capacity in [1]. Two semantics for network capacity are proposed and discussed. The effect of node location on capacity is analyzed within the context of the boundary condition and its impact on evaluation of network capacity. The main focus is on the bottleneck capacity referred to as the "maximum instantaneous network capacity" (MIC). The metric is intended to characterize the true information capacity of the network as a whole—reflecting all possible destinations. The optimization problem is shown to be NP-complete and heuristic algorithms are applied to bound the solution from above. The asymptotic results compare favorably to the well-known results of the highly abstract model in [2].

1.   Introduction

Theoretical models capable of accurate ad hoc network characterization have become increasingly important — robust, efficient and scalable network services depend on understanding the dynamic processes and limitations inherent from these systems. Interest in applications including wireless sensor networks and ubiquitous inter-net access underscores the practical importance of understanding fundamental properties associated with ad hoc network systems (ANS). Capacity bounds and other invariant performance characteristics are crucial elements required for the development of future ANS designed to support real-time and other performance bound applications. A primary obstacle, however, is that ANS exhibit dynamic interactions between entities at different protocol "layers". Hence, these interactions must be understood and integrated into performance analysis and network design.

*This work is supported in part by the National Science Foundation under the Faculty Early Career Development <CAREER> Program: NSF Award ID 0347698

The objective of this paper is to present the sequel to the authors' cross-layer analysis of channel capacity in multi-hop ANS. Specifically, a comprehensive investigation focusing on two related approaches to the problem of ad hoc network capacity is presented. The extensibility of the channel model is demonstrated as the previously developed concepts of "the deferral set" and "the equivalent competitor" constitute the basis for an elegant and novel node-oriented model for performance analysis given ideal channel and quasi-static assumptions [1]. The results have interesting theoretical and practical significance with respect to future ad hoc system design—most significant is the parametric model that adapts readily to different MAC protocols and provides direct insight to network design and control problems. This is fundamentally different than previous results that are either too abstract or tightly coupled with specific MAC and routing algorithms. Moreover, under similar constrains the results agree with the well-known bounds in [2], however, provide more practical information and more optimistic results when the constrains are relaxed. Future work will generalize the problem as fundamental limit that is MAC invariant, and present a thorough sensitivity analysis comparing the results to those in [2].

Existing literature focuses primarily on single-hop scenarios or on fully connected networks. Important contributions exist, however, the failure to capture the essence of complex cross-layer interactions and the impact induced by multi-hop environments represent significant shortcomings. With respect specifically to the *network capacity* problem much of the current literature focuses on simulation results or specific routing protocols. Recently, however, important theoretical results have been reported for multi-hop networks: In [2], the capacity of ad hoc networks is presented with results based on a randomly selected source-destination pair. A significant shortcoming of this model, however, stems from its lack of parametric network characterization. Although the results illustrate multi-hop performance bounds they are derived from highly abstract models. As such, they lack practical insight that can be applied to design more effective networks, which is the objective of cross-layer design. Moreover, the results promote an overly pessimistic vision. Specifically, by failing to account for the benefits of temporal and spatial diversity, coupled with the a non-parametric approach a pre-maturely negative tone emerged in the research community. In contrast, although the assumptions in [4] lack practical relevance, the authors illustrate how diversity, namely, that provided by node mobility can improve the network capacity.

One of the difficulties encountered in the present work is the obtuse nature of the *network capacity* problem itself. In contrast to *channel capacity* the definition of network capacity lacks a universal semantic. The ambiguity, however, is used to advantage, namely, by engaging in multiple interpretations more insight is provided. In this paper, network capacity is interpreted in two ways, namely, as (1) "maximum instantaneous capacity" (MIC) and (2) "network saturation capacity" (NSC). The MIC is the maximum amount of data flow in the network at any instant given ideal routing and scheduling; whereas the NSC is the sum of the capacity of all the channels in the network assuming that nodes and traffic are uniformly distributed independent of routing and scheduling algorithms. Both metrics are important for performance analysis, the first is an upper-bound, whereas, the second reflects an achievable flow rate under back-logged conditions.

The remainder of this paper is organized as follows: Section-2 presents analysis of the "network saturation capacity"including characterization and analysis of the "boundary conditions". Section-3 presents alternative approaches for estimating "maximum instantaneous capacity" discussing the merits and shortcomings of each approach. Conclusions are presented in Section-4 elaborating on the contributions and limitations of the results.

2. Analysis of Network Saturation Capacity

Without deeper inspection one may mistakenly assume that given an ANS with uniformly distributed nodes, all the "links" will have same capacity, thus, the network saturation capacity is the product of the channel capacity and the number of links in the network. Unfortunately, it is not this simple. Node location, for example, has a direct impact on network capacity. In order estimate network capacity with sufficient precision it is necessary to study the relation between the capacity and node location; one difficulty arises due to the "boundary conditions".

2.1 Boundary Conditions

Given a network with N nodes, each having n_{avg} neighbors there are $N_l = N \times n_{avg}/2$ "links". The derivation for arbitrary channel capacity, S_{chan}, is given in [1]. Several examples of NSC (from *ns2* simulation using parameters from [1]) are given in Table-1.The table compares simulation results with the $S_{chan} \times N_l$ formulation. In all cases the NSC obtained from simulation are greater than $S_{chan} \times N_l$. The error increases with increasing node density near the boundary (for a fixed network radius). The reason is that the nodes close to the boundary of the network have fewer neighbors, hence, less channel contention. Consequently, in general links close to the boundary have greater available capacity than those in the center of the network. The following definition is required to formalize the problem:

n_{avg}	N	NSC (Mb/s)		n_{avg}	N	NSC (Mb/s)	
		$N_l \times S_{chan}$	Simulation			$N_l \times S_{chan}$	Simulation
3	49	1.75	1.779	4	64	1.56	1.636
6	81	1.06	1.52	8	81	0.957	1.17
11	75	0.417	0.86	12	81	0.405	0.82

Table 1. NSC: simulation vs. estimation for fixed network radius and transmission range

Definition 2.1 Let X_i be a random variable that measures the distance from node $i \in G$ to the network boundary. Assume without loss of generality that transmission is omni-directional and the fixed value r is an accurate estimate of the nominal transmission range given an ergodic, homogeneous network. The boundary zone is defined as the doughnut shaped region occupied by all nodes $i \in G | X_i < 2r$.

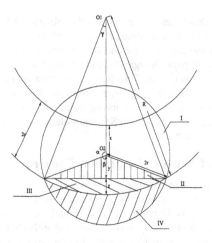

Figure 1.   Boundary zone and analysis of boundary condition

The boundary condition quantifies the overestimation of access contention within the boundary zone. Without loss of generality a circular boundary is assumed for the geometric analysis. Figure-1 illustrates the "boundary zone". The arc at the lower part of the figure is the network boundary, while the dashed circle represents the "deferral set" zone [1] of node O_2. Observe, however, that for nodes in the boundary zone the physical area covered by the actual deferral set must exclude the shaded area — Area$_{IV}$, which lies outside the network, hence, contains no active nodes. To find the conditions for which the error is negligible assume that the area is occupied by "phantom nodes" that do not produce traffic. The ratio of "phantom nodes" ($N_{phantom}$) to N should be small. The question is how small for a desired precision? First, it is necessary to find the area of the "phantom zone":

$$
\begin{aligned}
\text{Area}_I &= \pi(2r)^2 \frac{2\alpha}{2\pi} \\
&= 4r^2 \arccos\left(\frac{8r^2 + x^2 - 4Rr - 4rx + 2Rx}{4r(R - 2r + x)}\right) \\
\text{Area}_{II} &= \frac{1}{2}(2 \cdot 2r \sin \beta)(2r \cos \beta) \\
&= 2r^2 |\sin(2\alpha)| = 4r^2 \sin \alpha |\cos \alpha| \\
\text{Area}_{III} &= \text{Area of Fan} - \text{Area of Triangle} \\
&= \gamma R^2 - 2r \sin \beta (R - z) \\
\text{Area}_{IV} &= \pi(2r)^2 - \text{Area}_I - \text{Area}_{II} - \text{Area}_{III} \\
&= 4\pi r^2 - 4r^2\alpha - 2r^2|\sin(2\alpha)| - \gamma R^2 - 2r \sin \beta (R - z),
\end{aligned}
$$

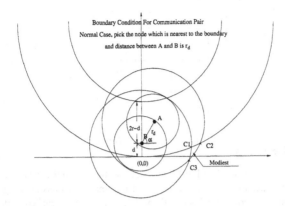

Figure 2. Estimation of phantom zone associated with communicating nodes

where α, β, γ, y and z are shown in Figure-1 and can be determined geometrically. Figure-2 illustrates the boundary condition with respect to a pair of actively communicating nodes (A,B). The "phantom zone" is the union of Area$_{IV}$ with the parameterized triangular region bounded by C_1, C_2 and C_3, which is the portion of the zone covered by only one of the nodes (referred to as "Modiest" in the equations). Construct an x,y coordinate system with the x axis tangent to the network boundary and the y axis perpendicular to the axis at the x coordinate of the node nearest to the network boundary (node B in the figure). The coordinates of C_1 C_2 C_3 can be obtained geometrically and the area is calculated as follows:

$$
\begin{aligned}
\text{Area}_{\text{phantom}}(d, \alpha, r_d) &= \text{Area}_{IV} + \text{Modiest}(d, \alpha, r_d) \\
&= \text{Area}_{IV} + \int_{C3(x)}^{C1(x)} \int_{d+r_d \sin \alpha - \sqrt{4r^2 - (x - r_d \cos \alpha)^2}}^{d - \sqrt{4r^2 - x^2}} dy dx + \\
&\quad \int_{C1(x)}^{C2(x)} \int_{d+r_d \sin \alpha - \sqrt{4r^2 - (x - r_d \cos \alpha)^2}}^{R - \sqrt{R^2 - x^2}} dy dx
\end{aligned}
$$

The overestimation of the "equivalent competitors" [1] affecting boundary zone nodes is given by the total number of phantom nodes, $N_{phantom}$:

$$
\begin{aligned}
\overline{N}_{\text{phantom}} &= \rho \left(\int_0^{2r} \frac{1}{2r} (\text{Area}_{\text{phantom}}(d, \alpha, r_d)) \right) \\
&= \rho \left(\int_0^{2r} \frac{1}{2r} (\text{Area}_{IV}(d) + (\int_0^\pi \frac{1}{\pi} d\alpha \int_0^r \frac{1}{r} \text{Modiest}(d, \alpha, r_d) dr_d)) \, dd \right)
\end{aligned}
$$

The limits of integration reflect the coordinate system and are chosen with respect to the location of the node closest to the network boundary. Thus, the other node must be closer to the center: α varies from 0 to π to cover all positions of node A in this region; the distance r_d between nodes A and B is varied from 0 to r reflecting the uniform distribution of nodes.

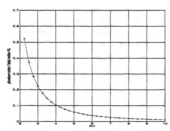

Figure 3. $N_{phantom}/N$ vs R/r

2.2   Discussion

Numerical analysis depicted in Figure-3 shows that when the network diameter (R) is on the order of 10 or more times the nominal transmission range (r), the effect of "boundary condition" is less than 1%. This can be regarded as negligible, thus, the NSC can be approximated by the product of S_{chan} and N_l. The analysis illustrates how the location of the nodes affects channel capacity, and in turn the NSC. Hence, for moderate to large size networks the boundary effect can be ignored without affecting the accuracy of the NSC.

There are additional parameters that may affect network capacity. The most important of these include: spatial and temporal variation of the distribution of nodes, traffic characteristics, the wireless channel and node mobility. A significant advantage of the present model is that the analysis of these factors is facilitated through probabilistic interpretation of the "equivalent competitor" [1] and enumeration of the effects of the dynamics of the aforementioned parameters. Sensitivity analysis will be included in the extended version of this article.

3.   Analysis of Maximum Instantaneous Capacity

The analysis of MIC reflects the bottleneck achievable throughput between *any set of sources and destinations*. Given ideal transmission scheduling it represents a lower-bound on maximum simultaneous flows between *all* node pairs — given the "ideal scenario" every link must either be transmitting, receiving or in deferral due to the "coupling" effect [3]. Thus, there are a fixed number of links that may be activated simultaneously. The idea is to find a sequence of simultaneously active links that cover the connected network; at each step the number of active links is maximized—the minimum size set represents the desired bottleneck. The shortest covering sequence minimizes the delay as well. The MIC is approximated in two steps. First, find the maximum concurrent active links, second, find the bottleneck of the concurrent active links — the capacity in this case is the MIC.

3.1   Maximum Number of Simultaneously Active Links

The first step in solving for MIC is finding the maximum feasible number of concurrent active links. After showing that this problem is NP complete, a suboptimal solution is found using a greedy algorithm.

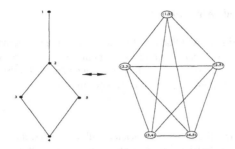

Figure 4. Graph transformation for showing NP-completeness

NP-completeness: Given a network G(V,E) and the definition of deferral link set $D_{Link}(i,j)$ [2] [1] associated with link (i,j), consider the following definitions, which are required for the problem formulation:

Definition 3.1 An aggregate deferral link set, D_n, is defined as a set of deferral link sets associated with a set of simultaneously active links (i,j). The size of D_n is equal to the number of deferral link sets in D_n and is denoted $S(D_n)$.

$$D = \bigcup_{(i,j)} D_{Link}(i,j)$$

Definition 3.2 Given network G(V,E), an aggregate deferral link set is defined as a deferral partition of G: $D_n = \mathcal{DP}_n(G(V,E))$, if and only if it includes all the edges (links) in G:

$$\bigcup_{(i,j)} D_{Link}(i,j) = E(G)$$

Problem formulation:

> Given : $G(V,E)$
> Maximize : $S(D_n)$
> Over : all $n|D_n = \mathcal{DP}_n(G(V,E))$
> Such that :

$$\forall D_{Link}(i,j) \in D, and\ (\forall D_{Link}(m,n) \in D_i) \neq D_{Link}(i,j), (i,j) \ni D_{Link}(m,n)$$

This problem can be reduced in polynomial time to the maximum independent set problem from graph theory [3], which has been shown to be NP-complete [5] [6]. Thus, the problem of finding the maximum number of simultaneously active links in a network is equivalent to finding the maximum number of independent sets in in the transformed graph G'(V',E'). The rules for transforming G(V,E) to G'(V',E') are given as follows (an example is depicted in Figure-4):

- $\forall e \in E$, create a corresponding node n' in graph G' such that $\bigcup n' = V'$;

- $\forall e \in E$, let $E_2(e) = \{e_2|e_2$ is less than or equal to two hops away from $e\}$, create a corresponding link e' connects nodes generated by e and every $e_2 \in E_2(e)$ such that $\bigcup e' = E'$.

3.1.1   The Greedy Algorithm

Heuristic algorithms exist that are efficient and capable of finding optimal solutions to the maximum independent set problem under a well-defined set of conditions. Sufficiently interesting results, however, are attainable for the present capacity problem using a simple iterative greedy algorithm as follows:

Algorithm 3.1

Step 1:   A unique deferral set is associated with each adjacent pair of nodes. A given deferral set is "feasible" if and only if the pair of nodes are not otherwise deferred. List all the feasible deferral sets in ascending order by the number of links in each set.

Step 2:   Pick the first deferral set in the list; the corresponding link is assumed to active since its link deferral set has the fewest links. Hence, transmission or flow on the link causes the minimum possible access contention.

Step 3:   Update the set of feasible link deferral sets: Sets associated with links within two hops of any active link must be removed from the feasible set because flow on these links will interfere with already active transmissions.

Step 4:   Update the size of remaining feasible deferral sets: Care must be taken not to double count any links. Any link that has already been deferred by an active transmission must be removed from any other deferral sets.

Step 5:   If more than one deferral set has the same size, the tie is broken by activating the link incident to the pair of nodes with the strongest and most stable signal, or, alternatively, the minimum LOS (line-of-sight) distance.

Step 6:   Repeat steps 1-5 until the set of feasible deferral sets is empty.

An example of the results from execution of Algorithm-3.1 is depicted in Figure-5. Based on execution of the algorithm there will be 5 simultaneously active links in the network. At any instant when all five links are active *all remaining links in the network* must defer any attempt to access the transmission medium. Numerical examples of the maximum number of simultaneously active links (l_{max}) for different configurations based on algorithm-3.1 are given in Table-2.

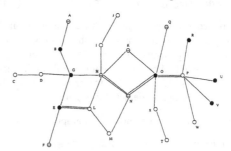

Figure 5.   Simultaneously active links using Algorithm-3.1

N	n_{avg}	l_{max}	N	n_{avg}	l_{max}	N	n_{avg}	l_{max}	N	n_{avg}	l_{max}
432	3	118	530	4	119	732	6	139	952	8	163

Table 2. Number of simultaneously active links l_{max} using the greedy algorithm

3.1.2   The Random Link Selection Algorithm

A non-determinist algorithm for estimating l_{max} uses random selection: The algorithm is initialized by placing all the links on the feasible list. At each iteration a link is randomly selected from the feasible list and all the links in its link deferral set are removed from the feasible list. The algorithm iterates until the list is empty.

Random link selection is faster than the greedy heuristic. Moreover, given a uniformly distributed topology of heterogeneous nodes in steady-state the results are of the same order. Given this scenario the number of links is roughly uniform over all the deferral sets other than in the boundary zone. Statistically significant simulation result for random link selection are summarized in Table-3 below:

N	n_{avg}	l_{max}			N	n_{avg}	l_{max}		
		mean	std	95% c.i.			mean	std	95% c.i.
432	3	99	2.7806	1.1125	530	4	87	2.4815	0.9928
732	6	108	1.8173	0.7271	952	8	115	1.7425	0.6971

Table 3. Simulation result for random link selection (20 independent replications)

3.2   The Bottleneck Aggregate Link Set

The first step in finding the maximum instantaneous capacity is to find the maximum possible number of simultaneously active links. This intermediate result is an upper bound that reflects the maximum instantaneous flow. However, in discussing the motivation for this analysis the desired metric was to reflect the *bottleneck* flow with respect to an arbitrary set of communicating nodes. Hence, MIC must represent the maximum lower bound for the flow of data among arbitrary entities in the network. The optimal solution requires multiple iterations of the independent set problem, hence, it is NP-complete. The following algorithm consists of a polynomial bounded number of iterations of the greedy algorithm:

Algorithm 3.2

Step 1: Repeat a modified version of Algorithm-3.1 for each link; use the corresponding link to select the initial deferral set and construct a candidate set.
Step 2: Sort the sets by the number of simultaneously active links.
Step 3: Select the largest aggregate deferral set from the candidate list. If all the active links are in the covered link list remove it from the candidate list.
Step 4: Repeat Step-3 until a set is selected with at least one link not in the covered link set.
Step 5: Move the set from the candidate list to the selected list and add any new links to the covered link set.
Step 6: Repeat Steps 2-5 until all the links are in the covered link set.

At each iteration the algorithm selects the aggregate deferral set from the candidate list that maximizes the number of simultaneously active links so long as it contains *at least* one active link that has not previously been covered. Thus, the algorithm terminates and provides reachability between all nodes in the connected network. The final iteration represents the bottleneck. Due to the uniformity of the network and the application of the sub-optimal greedy algorithm the upper and lower bounds do not differ significantly. Table-4 shows the simulation result of algorithm-3.1 and 3.2.

N	n_{avg}	iterations	lower bound	N	n_{avg}	iterations	lower bound
432	3	116	111	530	4	241	110
732	6	591	128	952	8	1899	152

Table 4. Number of required iterations and corresponding lower bound of algorithm-3.1 and 3.2

Algorithm-3.1 and 3.2 provide approximate solutions to the "Maximum instantaneous capacity" problem. However, under ideal conditions it may be possible to achieve better lower bounds, and, thus, show that under worst case analysis it is possible to exceed the results in [2]. Moreover, the solution is not sufficiently efficient. Selected links tend to be "re-activated" in numerous aggregate deferral sets, whereas, other links may be activated only once during the search for the solution. For the purpose of comparison in terms of efficiency and precision a random algorithm is again utilized. Table-5 shows the results. The table shows that the greedy algorithm achieves a tighter lower-bound, however, random selection requires significantly less computation.

N	n_{avg}	iterations needed			lower bound		
		mean	std	ci (95%)	mean	std	ci (95%)
432	3	45	8.474065	1.729761	94	1.295897	0.264524
530	4	82	13.15728	2.685719	82	0.999094	0.203939
732	6	178	21.71956	4.433486	101	1.007220	0.205598
952	8	341	56.42641	11.51799	107	1.559798	0.318393

Table 5. Number of required iterations and corresponding lower bound: random algorithm

3.3   Discussion

The significance of this work is based on the following observations: (1) In contrast to previous work the semantic of network capacity itself is analyzed in order to provide a clearer understanding and basis for comparison, (2) the results, while sub-optimal, and based on worst-case analysis improve on the most often cited results from [2], (3) the analysis is central of a broad cross-layer framework, hence, it has practical significance with respect to network design, and (4) the underlying models for channel capacity are node-based and parametric, thus they are extensible in terms of access protocols, generalization and application to real control problems as opposed to being based entirely on abstract models.

In light of the widely accepted limitations and their impact on the broader research community it is critical to examine this work with respect to [2] which studies the capacity of wireless networks. The main results from their analysis relevant to this work is summarized in what follows:

- Throughput is defined as the time average of the number of bits per second that can be transmitted by every node to its destination.
- Capacity is expressed as: $\lambda(N) = \Theta(\frac{W}{\sqrt{NlogN}})$ where W is the channel bandwidth and N is the number of nodes in the network. Note that capacity has end-to-end significance with respect to a specific source-destination pair.
- Given ideal routing and scheduling their result is shown to improve to $\Theta(\frac{W}{\sqrt{N}})$

The network configuration and parameters are the same in both analyses. However, the semantic with respect capacity differs, namely, the MIC analysis is a general bottleneck considering the entire network, whereas, in [2] an upper bound us determined on a per-node basis. This form of result can be misleading as any consumer will experience diminishing returns given a fixed network of resources. The two semantics for capacity in this paper are more useful as they reflect the aggregate scaling effect of the network versus an individual consumer. Furthermore, in [2] the asymptotic results are bounded from above and below. The results in this paper reflect "worst-case" analysis and are bounded only from below. Hence, it can be reasoned that the optimal results are even better.

In order to make a meaningful comparison of the results it is necessary to use information about the present analysis to (1) consider the worst case throughput for a single source destination pair, and, (2) look at asymptotic bounds. From [1] the average area covered by level-1 interference set [4] is approximately $(\pi + 0.98)r^2$, thus the number of non-overlapping level-1 interference sets is given by:

$$\frac{\text{Area}_{network}}{\text{Area}_1} = \frac{\pi}{\pi + 0.98} \frac{N}{n_{avg} + 1}$$

Based on comparison of results from the above estimation and simulation using random link selection it can be shown that the number of non-overlapping level-1 interference sets is roughly equivalent to the number of maximum simultaneously active links. Corresponding asymptotic bounds can expressed as follows:

$$O(\frac{N}{(n_{avg} + 1)}) = \Omega(\frac{N}{(n_{avg} + 1)}) \rightarrow \Theta(\frac{N}{(n_{avg} + 1)})$$

Considering only unicast communications there are at most N/2 source-destination pairs in the network. The average hop count is approximately $\sqrt{N/n_{avg}}$. Let C represent a constant that corrects for access capacity W be the channel bandwidth. The resulting capacity per-node is given by (where $C' = 2C$):

$$\lambda(N) = \frac{C \cdot \frac{N}{n_{avg}+1} \cdot W}{\frac{N}{2} \cdot \sqrt{N/n_{avg}}} = \frac{C'W}{\sqrt{Nn_{avg}}} = O(\frac{W}{\sqrt{N}})$$

4.   Conclusions

In this paper the authors' previous work [3] [1] has been extended. Building
on cross-layer models for link capacity this paper introduces two new metrics for
the capacity of wireless ad hoc networks under ideal conditions: "network saturation capacity" and "maximum instantaneous network capacity" are defined and
compared. The instantaneous capacity problem reflects the true flow capacity of
the network between any nodes—it is the bottleneck capacity, as such it reflects a
lower bound on total throughput for all possible destinations. This property dif-
ferentiates the metric from related work, for example, the asymptotic throughput
analyzed in [2]. Determination of the bottleneck capacity is shown to be an NP-
complete problem; two heuristic algorithms are presented for finding approximate
solutions. Using the heuristic results and taking limiting values the results that
reflect worst-case analysis when bounded from below are shown to agree with the
results reported in [2]. The agreement mutually validates the two models, however,
it also suggests that the previous work is pessimistic and does not provide insight
regarding how to more effectively leverage available network capacity.

Notes

1. "Deferral set" and "equivalent competitor" in the context characterize the multi-hop wireless network
environment and reflect the competition faced by a specific communication, in short, "deferral set" is the union of
the nodes which is less than and equal to the two hops away from the active nodes, only transmission originated
from those nodes may affect the ongoing transmission, while "equivalent competitors" is the weighted number of
nodes in "deferral set", refer to [1] for more details.

2. $D_{link}(i,j)$ is the set of links which are less than or equal to two hops away from link (i,j), while the size
of $D_{link}(i,j)$ is the number of links in it.

3. An independent set is the largest subset of vertices of V such that no pairs of vertices defines an edge of E
for a given graph G(V,E).

4. level-1 interference set is the set of direct neighbors of the communication pair

References

[1] Yue Fang and A. Bruce McDonald, "Theoretical Channel Capacity in Multi-hop Ad Hoc Net-
    works," Proceedings of the 13th IEEE workshop on Local and Metropolitan Area Networks,
    April 25th-28th, 2004

[2] P. Gupta and P.R. Kumar, "The Capacity of Wireless Networks," IEEE Transactions on Infor-
    mation Theory, vol. 46, no. 2, pp. 388-404, March, 2000

[3] Yue Fang and A.Bruce Mcdonald, "Cross-Layer Performance Effects of Path Coupling in Wire-
    less Ad Hoc Networks: Power and Throughput Implications of IEEE 802.11 MAC," Proceed-
    ings of 21st IEEE International Performance, Computing, and Communications Conference
    (IPCCC 2002), April 3-5, pp. 281-29, 2002, Phoenix, Arizona

[4] Matthias Grossglauser and David N. C. Tse, "Mobility Increases the Capacity of Ad-hoc Wire-
    less Networks," INFOCOM, pp. 1360-1369, 2001

[5] James Abello, Sergiy Butenko, Panos Paradalos and Mauricio G.C Resende, "Finding Inde-
    pendent sets in a graph using continuous multi-variable polynomial formulations," Journal of
    Global Optimization, 2001, no. 21, pp. 111-137

[6] Robert Endre Tarjan and Anthony E Trojanowski, "Finding a Maximum Independent Set,"
    SIAM Journal of Computing, September, 1977, vol. 6, no. 3, pp. 537-546,

HOW TO DISCOVER OPTIMAL ROUTES IN WIRELESS MULTIHOP NETWORKS

Michael Gerharz, Christian de Waal, and Peter Martini
University of Bonn, Roemerstr. 164, 53117 Bonn, Germany
{gerharz, dewaal, martini}@cs.uni-bonn.de

Abstract   In this paper, we introduce a distributed algorithm that is able to discover optimal routes in mobile wireless multihop networks using reactive routing protocols. The algorithm is based on Dijkstra's shortest-path algorithm and maps the quality of a path to a delay of the corresponding route request to allow high-quality paths to surpass low-quality paths. With a proper selection of the delay mapping, this approach yields a low overhead and interoperable integration of maximisable routing metrics into existing protocols like AODV and DSR while keeping the route setup delay at a moderate level.

Keywords: Ad hoc networks, Routing, Quality-of-Service, Dijkstra

1.   Introduction

Reactive routing protocols provide a ressource efficient solution to the routing challenge in highly dynamic network topologies by discovering a route only when it is actually needed. The source node floods across the network a route request (RREQ) which is unicast back as a route reply from the destination to the source along the discovered route.

In the recent past, much effort has been spent on integrating all kinds of different routing metrics into reactive routing protocols. Motivations come e.g. from QoS-considerations (e.g. maximising the route reliability or the bottleneck capacity) as well as power aware protocols (e.g. minimising the sender-receiver distance in terms of a distance metrics such as energy consumption or number of weak links).

By default, common reactive routing protocols like AODV and DSR [Perkins, 2001] do not support sophisticated metrics, because they process only the first arriving RREQ. While DSR provides limited support by replying to multiple

This work was supported in part by NOKIA and the German Federal Ministry of Education & Research (BMBF) as part of the IPonAir project.

RREQs, this approach is still not able to discover optimal routes, but merely selects the best of several short-delay paths. There are several approaches to incorporate optimal route discovery into reactive routing protocols. However, these approaches focus on special metrics or suffer from high overhead.

In this paper, we will present a generic algorithm applicable to a wide range of diverse routing metrics with very low overhead. The main idea is to use a distributed version of Dijkstra's shortest path algorithm. The key is to schedule the transmission of RREQs in an order that is equivalent to the treatment of the stations in Dijkstra's algorithm.

There are several requirements to an algorithm supposed to discover optimal routes according to some routing metric. Firstly, the *overhead* of the route discovery should be as low as possible compared to conventional reactive routing protocols. Two competing goals are to minimise the number of messages sent over the medium and to minimise the route setup delay. Our algorithm, just like conventional reactive routing protocols, requires exactly one broadcast per station while adding a slight delay to distinguish paths of different quality.

Secondly, the route discovery process should be *interoperable* to the plain routing protocol. This allows for gradual deployment and enables different devices to stress different requirements on the discovered route (whether this is reasonable depends on the scenario). Our approach is fully interoperable with AODV and DSR, but the caching strategy of the latter protocol has to be chosen with care in order not to base routing decisions on outdated information.

The rest of this paper is structured as follows. Section 2 provides preliminaries for the discussions to follow. In section 3, we present related work. Section 4 describes our generic approach to optimal routing and outlines several design choices. After discussing implementational aspects in section 5, we draw conclusions and outline perspectives for further research in section 6.

2. Shortest Path Algorithms & Routing Metrics

A thorough discussion of maximisable routing metrics may be found in [Gouda and Schneider, 2003]. The authors define a routing metric as a 5-tuple $(M, W, \text{MET}, p_0, \prec)$ where M is the set of all possible metric values, W is the set of possible edge weights, $\text{MET} : M \times W \rightarrow M$ is a metric function which calculates metric values cumulatively, $p_0 = \max(M)$ is used as the initial path metric of a route discovery, and $\prec$ is a less-than total order relation over M so that the routing metric selects the paths of maximal metric values. Sometimes, we will use $\text{MET}(c_1, \ldots, c_i)$ as an abbreviation for $\text{MET}(\text{MET}(\ldots(\text{MET}(p_0, c_1), \ldots), c_i)$.

According to this we define a *distance metric* as the 5-tuple $(\mathbb{R}, W, +, 0, >)$. A *reliability metric* is given by $(M = \{x \in \mathbb{Q} | 0 \leq x \leq 1\}, W = M, \cdot, 1, <)$. A *flow metric* is defined as $(M \subset \mathbb{N}, W = M, \min, \max(M), <)$. Defining a maximal weight for the source station has technical reasons as required in section 4. Interestingly, this requirement may be relaxed for a practical implementation (cf. section 5). Anyway, the maximally recordable weight will be limited by a finite value in any real implementation.

Numerous examples for these routing metrics exist in the literature. Power consumption is a frequently used distance metric, flow metrics are commonly used when minimum bandwidth requirements have to be met, and the expected packet loss rate is a typical reliability metric. Further examples for all of these metrics can be found in [Gerharz et al., 2003].

In the following, we will provide a short overview of Dijkstra's single source shortest paths algorithm [Dijkstra, 1959], because it forms the basis for our distributed approach. Consider a network G with a weight function w and a source node s. The shortest paths to all other nodes in the network are basically calculated by the following procedure (cf. [Cormen et al., 1990]):

> DIJKSTRA(G, w, s):
> INITIALIZE(G, s)
> $Q \leftarrow V[G]$
> **while** $Q \neq \emptyset$
> $u \leftarrow$ EXTRACT-MAX(Q)
> **for** each vertex $v \in Adj[u]$
> **do** RELAX(u, v, w)

The set Q contains all nodes for which the optimal path is not yet known. EXTRACT-MAX selects the node u whose currently best known path is maximal of all nodes in Q. For this node u the maximal metric value is already found. We will denote this maximal value as OPT(u).

The central task of the DIJKSTRA algorithm is the RELAX procedure. It takes three parameters: the two endpoints of a link u, v as well as the weight function w and calculates the metric value $d[v]$:

> RELAX(u, v, w) :
> **if** $d[v] \prec$ MET$(d[u], w(u, v))$
> **Then** $d[v] \leftarrow$ MET$(d[u], w(u, v))$
> $nexthop(v) \leftarrow u$

The important property of Dijkstra's algorithm in our context is that the RELAX procedure is called exactly once for every edge. This property permits a distributed computation of the algorithm if the distributed calls to RELAX follow an equivalent order as in the centralised case.

3. Existing Distributed Algorithms for Optimal Routing Ad Hoc Networks

A lot of previous work exists on the discovery of optimal routes with reactive routing protocols which basically splits into two groups. The first group tries to keep the route setup delay at a minimum at the price of an increased routing load while the other group favours the opposite.

In DSR, a rudimentary support for optimal routing is provided by the destination which replies to all incoming RREQs. DSR itself uses this approach to discover shortest paths. But, it has also been adopted by other publications, sometimes with slight modifications. In [Tickoo et al., 2003] e.g., the destination does not send multiple route replies, but rather delays the reply in order to answer to the best RREQ just once. This reduces the routing overhead but on the other hand increases the route setup time.

While being very simple, this approach is unable to find the actual optimal route, because the destination is provided with only a subset of all paths. Therefore, extensions were proposed (e.g. [Gupta and Das, 2002], [Bergamo et al., 2004]) to have intermediate stations forward multiple RREQs instead of just the first. While this approach may be able to find the optimal path with low latency, it consumes a huge amount of capacity, because the inherently harmful broadcast storm problem [Tseng et al., 2002] will even be augmented. In contrast, our approach will even lessen the broadcast storm problem.

A different approach is proposed in [Cho and Kim, 2002] and [Chakeres and Belding-Royer, 2003] which is based on the standard AODV discovery scheme. But in contrast to the basic scheme, RREQs are delayed depending on a local state maintained in every station. By this means, the probability for the station to be on the selected route is influenced. Although this approach is specified with weights being assigned to nodes rather than edges, a generic mapping to edge weights is possible due to the fact that only the first arriving RREQ is processed. With that transformation, this approach is merely a special case of ours.

A related approach, also operating on a specific distance metric (power consumption) is described in [Aslam et al., 2003] with algorithm 5, this time using edge weights and assuming a global clock. A station receiving a RREQ delays the forwarding of this RREQ according to the accumulated distance.

In this paper, we will generalise this concept to arbitrary maximisable routing metrics without requiring a global synchronisation of all stations.

4.   A Distributed Version of Dijkstra's Shortest Path Algorithm

In this section, we assume that neither the stations nor the medium introduce any further latency other than the one enforced by the algorithm. Furthermore, we assume that the clocks of all stations are synchronised and that without loss of generality the clock starts at 0 for every route discovery. This synchronisation requirement will be relaxed in later sections.

4.1   Key Concepts & Basic Algorithm

The key idea is to make the EXTRACT-MAX procedure implicit by scheduling the broadcast of RREQs distributedly in an equivalent order as the nodes are extracted from the set Q and distribute the computation of RELAX to those nodes receiving the RREQ. In other words, the broadcast time $\mathrm{BT}(u)$ of a RREQ at node u has to fulfil the following condition:

$$\forall u, u' \in G : \mathrm{OPT}(u) \prec \mathrm{OPT}(u') \Rightarrow \mathrm{BT}(u) > \mathrm{BT}(u') \tag{1}$$

To achieve this, we assign to every path in the network a total RREQ-delay corresponding to the path's cost. Formally, we define a function $D : M \to \mathbb{R}_0^+$ which is strictly monotonically decreasing. (Note that this is equivalent to finding a mapping of the routing metric to a distance metric.) Having this mapping, a RREQ which is received along a path of value p, is scheduled to be forwarded at global time $D(p)$. Should a better RREQ arrive before $D(p)$ has elapsed, the transmission has to be re-scheduled to the earlier period. Worse RREQs will be discarded. Ties are broken arbitrarily. Formally, we define $\mathrm{BT}(u)$ as the minimal delay of all paths to u. In conjunction with D's monotonicity, it immediately follows that $\mathrm{BT}(u) = D(\mathrm{OPT}(u))$.

This leads to a generic formulation of a distributed version of Dijkstra's algorithm. The set Q is only maintained implicitly and not centrally administered. By broadcasting a RREQ, a node is extracted from Q. Subsequently, the relaxation of an edge is distributed to the broadcasting station's neighbours and triggered by the reception of the RREQ. The RELAX procedure also needs three parameters, however in an accumulated form: the id of the previous hop, the cumulated pathcost p transmitted in the RREQ, and the linkcost c of the last hop:

> RELAX(u, p, c):
> **if** $met \prec \mathrm{MET}(p, c)$
> **Then** $met \leftarrow \mathrm{MET}(p, c)$
> $nexthop \leftarrow u$
> RESCHEDULE-BT$(D(met))$

Under the assumption that no additional delay is introduced by the medium or the stations, this algorithm is able to discover optimal routes which follows immediately from Dijkstra's optimality, because the stations broadcast their RREQ in an equivalent order as the nodes would be extracted from Q. The effect of increasing medium and station latency is out of scope of this paper. However, note that although additional delays lead to suboptimal route assignments this is not necessarily a drawback in practice. By limiting the detour of an optimal route compared to the shortest path, the capacity of the network as well as the energy resources are potentially spared.

As already mentioned in section 3, to discover optimal routes with reactive protocols, a tradeoff has to be found between overhead and route setup delay. Our algorithm does not introduce any overhead in terms of packet transmissions. In terms of byte-overhead, RREQs need to be extended with a *met* field which on the one hand is quite negligible in size and on the other may be spared completely if certain conditions are met (cf. section 5.2). Additionally, note that delaying some of the RREQs stretches the route discovery broadcast storm in time and thereby reduces the peak load on the network.

4.2   Mapping Metric Values to RREQ-Delays

In this section, we will take a look at some example delay mappings, namely linear and logarithmic transformations.

Linear Transformation.   In general, a linear transformation of metric values to delays will look like this:

$$D(p) = mp + o \tag{2}$$

where $m > 0$, if $\prec$ actually is a greater-than operator and $m < 0$ otherwise. In general, $o \neq 0$ for $m < 0$ or $\min(M) > 0$. Reasonably, we require o to be chosen such that $D(\max(M)) = 0$ in order to guarantee that optimal paths do not experience any delay at all.

For distance metrics, this leads to delays proportional to the distance (note that this is the special case of algorithm 5 in [Aslam et al., 2003], cf. sec. 3):

$$D(p) = kp \tag{3}$$

where $k \in \mathbb{N}, k > 0$ which we will assume throughout the rest of the paper. For reliability metrics, we have $\prec \equiv <$ and thus $m < 0$. Consequently, with $m = -k$ we define $o = k$ in order to guarantee $D(1) = 0$ which leads to delays proportional to the fragility of the path:

$$D(p) = -kp + k = k(1 - p) \tag{4}$$

Similarly, for flow metrics we define $m = -k$ and $o = k\max(M)$ to get delays proportional to the unused or preoccupied resources:

$$D(p) = k(\max(M) - p) \tag{5}$$

We observe that the maximal delay a RREQ may experience is bounded by k for reliability metrics and by $k\max(M)$ for flow metrics while it is unbounded for distance metrics.

This means, with distance metrics a RREQ may in principle travel through the network arbitrarily long which seems undesirable at first sight. Firstly, RREQs should arrive in a timely manner, because old RREQs will possibly carry outdated information. Furthermore, late RREQs may increase the route setup delay if a better route is not available. However, it may be doubted that this has a great impact in practice. The use of a performance metric in scenarios where a good performance may not be expected in the first place, may be doubted at all. It should be expected that usually a better path is available whose delay is accordingly short.

Logarithmic Transformation.   An alternative approach is to use a logarithmic delay transformation which provides underproportional growth of delay for high-quality paths and overproportional growth of delay for lower-quality ones. As an example, we will look at reliability metrics and define:

$$
\begin{aligned}
D(p) &= k \log p^{-1} \quad , p \neq 0 \\
D(p) &= \infty \qquad\quad , p = 0
\end{aligned}
\tag{6}
$$

In principle, a logarithmic transformation is also possible for other routing metrics, but care has to be taken to keep the delay positive.

As with a linear transformation, this mapping guarantees a zero delay for 100% reliable paths. But different to linear transformations, 0% reliable paths will be totally discarded (which is a reasonable thing to do). Furthermore, the delay is not bounded but approaches infinity for reliabilities close to zero.

Depending on the number of alternative paths, two pragmatic solutions exist to this problem: unreliable paths below a certain threshold may be totally discarded or delayed by a constant upper limit. The latter approach disregards differences in the reliability of a path and leads to sub-optimal routes while the former approach in effect reduces the connectivity of the network. Which solution is preferable depends on the scenario.

Fig. 1 provides a comparison of linear and logarithmic transformations for reliability metrics. The log-transformation is $D(p) = \frac{3}{10} \log_{10} p^{-1}$ while the linear transformation is $D(p) = \frac{1}{3}p$. Thereby, paths with a reliability of at least 10% will be discovered within 300ms (plus medium and station latency).

Although many more special mappings may be chosen, we refrain from discussing details here.

5.   Implementational Aspects

Until now, we have assumed to have the clocks of all stations globally synchronised which is clearly undesirable in real implementations. This may be avoided by computing delays incrementally.

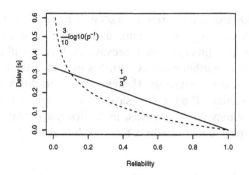

Figure 1.   Comparison of linear and logarithmic transformation of reliability metrics

In section 5.1, we describe a straightforward approach that makes use of the path metric value propagated in the RREQ. If this value is implemented as an optional field which is not modified by stations not supporting the extension, this will allow an interoperable integration of routing metrics into existing routing protocols, which is generally desirable. Section 5.2 will show that the message format may even be left unchanged when utilising special delay mappings which allows a fully backwards compatible implementation of routing metrics.

While this approach allows for gradual deployment of novel routing metrics and permits different devices and applications to focus on different performance aspects, having only parts of the stations support a routing metric certainly yields suboptimal routes. Be aware that a partial approach will for some metrics lead to wrong and sometimes even counterproductive decisions (if in particular those stations with high quality links support this and in particular those with bad quality links do not).

5.1   Differential Delay Mapping

In this section, we assume that the pathcost is propagated in the RREQ. With p_{i-1} we denote the metric value contained in the RREQ transmitted on the $i-th$ hop of a path. c_i denotes the linkcost of that link. Additionally, we do not require a global synchronisati on but assume that the deviation of the stations clocks remains in sensible bounds. We formally define:

DEFINITION 1 Differential Delay Mapping
  A differential delay mapping *is a function* $d : M \times W \rightarrow \mathbb{R}_0^+$ *such that for a delay mapping* $D : M \rightarrow \mathbb{R}_0^+$:

$$\forall p \in M, c \in W : D(p) + d(p,c) = D(\mathrm{MET}(p,c)) \tag{7}$$

The application of this definition to previously introduced delay mappings provides some interesting insights. For linear delay mappings e.g. we get:

$$\begin{aligned} d(p_{n-1}, c_n) &= (mp_n + o) - (mp_{n-1} + o) \\ &= m(p_n - p_{n-1}) \end{aligned} \tag{8}$$

We observe that the delay calculation comes with very low computational overhead. The procedure requires only two arithmetic operations, because p_n has to be calculated anyway to measure the quality of the path and p_{n-1} is extracted from the RREQ.

Additionally, we notice that d is independent of o. Note that although the local delay is proportional to the absolute difference of the pathcosts, this does not generally imply that d is proportional to the linkcost. This is however true for linear differential mappings of distance metrics which deserves a closer look (recall that $D(p) = kp$, $p_n = \sum_{i_1}^{n} c_i$):

$$\begin{aligned} d(p_{n-1}, c_n) &= k(p_n - p_{n-1}) \tag{9} \\ &= k\left(\sum_{i=1}^{n} c_i - \sum_{i=1}^{n-1} c_i\right) \\ &= kc_n \tag{10} \end{aligned}$$

Obviously, the local delay that a RREQ experiences in a station is independent of the path cost and depends only on the local linkcost which means that it is actually redundant to include the pathcost in the RREQ. This leads us to the notion of local mappings, defined in the following section.

5.2 Local Delay Mapping

In the previous sections, the delay mapping has been calculated from the pathcost. But for distance metrics, it has been shown that this is actually redundant. In this section, we will see that also for other metrics it is possible to make the calculation of the pathcost implicit and to compute the delays of RREQs directly from the linkcosts. Formally, we define:

DEFINITION 2 Local Delay Mapping

A local delay mapping *is a function* $d : W \to \mathbb{R}_0^+$ *such that* $(n, m \in \mathbb{N})$:

$$\mathrm{MET}(c_1, \ldots, c_n) \prec \mathrm{MET}(c_1', \ldots, c_m') \Rightarrow \sum_{i=1}^{n} d(c_i) > \sum_{i=1}^{m} d(c_i') \tag{11}$$

Figure 2 Flow metrics in general are not local

Note that a local delay mapping as defined here is not a special case of a differential delay mapping. Two paths with the same pathcost may arrive at different delays with local mappings whereas they will be guaranteed to experience the same delay with differential mappings by definition.

Reliability metrics may be implemented with a local delay mapping using a logarithmic mapping of reliabilities to delays which we will derive from the differential mapping of a logarithmic transformation (cf. equation 6):

$$
\begin{aligned}
d(p_{n-1}, c_n) &= \log p_n^{-1} - \log p_{n-1}^{-1} \\
&= \log\left(\prod_{i=1}^{n} c_i^{-1}\right) - \log\left(\prod_{i=1}^{n-1} c_i^{-1}\right) \\
&= \sum_{i=1}^{n} \log\left(c_i^{-1}\right) - \sum_{i=1}^{n-1} \log\left(c_i^{-1}\right) \\
&= \log c_n^{-1}
\end{aligned}
$$

Since the delay is independent of the pathcost, a local delay mapping for reliability metrics exists via $d(c) = \log c^{-1}$.

For flow metrics, we state the following theorem:

THEOREM 1 *A local delay mapping for flow metrics exists if and only if the size of the network is bounded by a constant N which is known in advance or $|W| < 3$.*

PROOF We will first prove that flow metrics are not local if neither of the two conditions is met by providing a counter-example:

The delay mapping shall impose a lower delay on any path with lower cost, regardless of how much longer this path is. It is easy to see that this is not generally possible: Consider a network of size $n + 1, n \in \mathbb{N}$ with circular shape as depicted in figure 2. Let c_i be the linkcost of link $(i, i + 1)$ and c_n be the cost of link $(n, 0)$. Furthermore, let $c_n = \min(W)$ and $c_n < c_i < \max(W), 0 \le i < n$ (recall that $|W| \ge 3$). Then, by definition:

$$
\sum_{i=0}^{n-1} d(c_i) < d(c_n) \tag{12}
$$

However:

$$\sum_{i=0}^{n-1} d(c_i) > (n-1)\min_{i<n}(d(c_i)) \tag{13}$$

Since $\forall i < n : c_i < \max(W)$, we have $\min_{i<n}(d(c_i)) > 0$. Furthermore, $\min_{i<n}(d(c_i))$ and c_n are constant. This leads to a contradiction to eq. 12, if n is large enough. In general:

$$\exists n \in \mathbb{N} : d(c_n) < (n-1)\min_{i<n}(d(c_i)) < \sum_{i=0}^{n-1} d(c_i) \tag{14}$$

On the other hand, if the network size is bounded by a constant $N \in \mathbb{N}$ which is known in advance, a local delay mapping for flow metrics exists:

$$d(c_n) = kN^{-c_n} \tag{15}$$

Consider a network of size N. Consider a link $(0, N-1)$ of cost c. In the worst case, a path $0, \ldots, N-1$ of length $N-1$ exists that just contains links of linkcost c' only marginally better than c, i.e. $c' = c + 1$. Then:

$$
\begin{aligned}
(N-1) \cdot d(c') &< d(c) \\
\Leftrightarrow N - 1 &< d(c)d(c')^{-1} \\
\Leftrightarrow N - 1 &< N^{-c}N^{c+1} \\
\Leftrightarrow N - 1 &< N
\end{aligned}
$$

Furthermore, if $|W| = 1$, a valid local delay mapping is trivially defined by $d(c) = 0$. If $|W| = 2$, a valid local delay mapping is defined by $d(c_{min}) = k$ and $d(c_{max}) = 0$. *qed.*

At first sight, it might seem straightforward to simply choose N large enough to meet any imaginable realistic scenario. However, this would require us to increase k as well in order to be able to distinguish also small differences in path quality. But, a large choice of k may result in very large delays even for high quality paths.

6. Conclusions & Further Work

In this paper, we have presented a generic approach to discover optimal routes with reactive routing protocols. The key idea is to delay the forwarding of RREQs according to the pathcost of the discovered path which was used to develop a distributed version of Dijkstra's shortest path algorithm.

This approach does not increase the routing load in terms of packet transmissions and in fact even reduces the peak load during a broadcast storm. On the other hand, the route setup delay is increased. Thus, the delay of RREQs has to be chosen carefully in order to find a good tradeoff between a minimal route setup time and a reliable distinction of high-quality from lower-quality paths. Finally, we have presented a local version of our algorithm which is fully backwards compatible to existing reactive protocols.

Future work will focus on several aspects. First of all, the impact of medium and station latency on a sensible choice of the RREQ delay has to be analysed. Possible improvements may be achieved by using the plain reactive routing protocol to quickly discover *some* route and refine this route selection by additionally running our proposed algorithm. A similar approach would be to limit the maximal RREQ-delay to a relatively short period, but as a compensation forward multiple RREQs if one arriving late yields a significantly better path.

References

[Aslam et al., 2003] Aslam, J., Li, Q., and Rus, D. (2003). A lifetime-optimizing approach to routing messages in ad-hoc networks. In Cheng, X., Huang, X., and Du, D.-Z., editors, *Ad Hoc Wireless Networking*, pages 1–43. Kluwer Academic Publishers.

[Bergamo et al., 2004] Bergamo, P., Giovanardi, A., Travasoni, A., Maniezzo, D., Mazzini, G., and Zorzi, M. (2004). Distributed power control for energy efficient routing in ad hoc networks. *Wireless Networks*, 10(1):29–42.

[Chakeres and Belding-Royer, 2003] Chakeres, I. D. and Belding-Royer, E. M. (2003). Resource biased path selection in heterogeneous mobile networks. Technical Report 2003-18, UCSB, Santa Barbara.

[Cho and Kim, 2002] Cho, W. and Kim, S.-L. (2002). A fully distributed routing algorithm for maximizing lifetime of a wireless ad hoc network. In *Proc. 4th IEEE Conf. on Mobile and Wireless Communication (MWCN 2002)*.

[Cormen et al., 1990] Cormen, T. H., Leiserson, C. E., and Rivest, R. L. (1990). *Introdcution to Algorithms*, chapter 25 Single-Source Shortest Paths, pages 514–527. The MIT Press.

[Dijkstra, 1959] Dijkstra, E. W. (1959). A note on two problems in connexion with graphs. *Numerische Mathematik*, 1:269–271.

[Gerharz et al., 2003] Gerharz, M., de Waal, C., Martini, P., and James, P. (2003). Strategies for finding stable paths in mobile wireless ad hoc networks. In *Proc. 28th IEEE Conf. on Local Computer Networks (LCN'03)*, Bonn, Germany.

[Gouda and Schneider, 2003] Gouda, M. G. and Schneider, M. (2003). Maximizable routing metrics. *IEEE/ACM Transactions on Networking*, 11(4):663–675.

[Gupta and Das, 2002] Gupta, N. and Das, S. R. (2002). Energy-aware on-demand routing for mobile ad hoc networks. In *Proc. 4th Intl. Workshop on Distributed Computing (IWDC 2002)*, pages 164–173.

[Perkins, 2001] Perkins, C., editor (2001). *Ad Hoc Networking*. Addison-Wesley.

[Tickoo et al., 2003] Tickoo, O., Raghunath, S., and Kalyanaraman, S. (2003). Route fragility: A novel metric for route selection in mobile ad hoc networks. In *Proc. 11th IEEE Intl. Conf. on Networks (ICON 2003)*.

[Tseng et al., 2002] Tseng, Y.-C., Ni, S.-Y., Chen, Y.-S., and Sheu, J.-P. (2002). The broadcast storm problem in a mobile ad hoc network. *Wireless Networks*, 8(2/3):153–167.

ASYMPTOTIC PHEROMONE BEHAVIOR IN SWARM INTELLIGENT MANETS

An Analytical Analysis of Routing Behavior

Martin Roth and Stephen Wicker
*Wireless Intelligent Systems Laboratory**
School of Electrical and Computer Engineering
Cornell University
Ithaca, New York 14850
USA

roth@ece.cornell.edu

wicker@ece.cornell.edu

Abstract   An analytical justification is proposed for the design and global routing performance of three pheromone update methods proposed for use in Termite, a swarm intelligent routing algorithm for mobile wireless ad-hoc networks. A simple model is used in order to determine the average amount of pheromone present on a link, as well as some basic aspects of the pheromone dynamics. This includes a tendency towards a one-zero pheromone distribution favoring the better link. The pheromone update methods are investigated with the perspective that link pheromone is more an estimate of link utility than simply a routing heuristic. This allows the routing solution to be rephrased from a biological analogy to a more traditional best-metric routing terminology. A signal estimation perspective is suggested.

1.   Introduction

Recent applications of biologically inspired algorithms to routing in mobile wireless ad-hoc networks (MANETs) have shown increased performance over traditional approaches in many critical metrics [1] [2] [3]. It remains unclear as to exactly why they work as well as they do, and how to best take advantage of their positive and negative feedback mechanisms.

This paper presents a simple analytical model of Termite, a swarm intelligent MANET routing algorithm [1]. The purpose of this model is to discover

---

*http://wisl.ece.cornell.edu/

how individual parameters are related to each other and how they affect global metrics, such as the reliability of message delivery and adaptability to changes in the network environment. The critical element under study is pheromone. Because Termite is based on a model of social insect behavior, much of the biological terminology remains. Pheromone is a measure of the metric that the network is optimizing for; it is a measure of link or route utility. The model will first be used to characterize the behavior of pheromone on a single communications link. This will establish an intuition for determining the dynamics of pheromone in a system of two links.

1.1   Previous Work

Routing algorithms are often difficult to formalize into mathematics; they are instead tested using extensive simulation [14]. The interaction between parameters is unclear, as are their global effects, or even how the values should be determined for optimal performance.

Swarm intelligent routing algorithms lend themselves to mathematical analysis. Their routing update and decision procedures are mathematical functions themselves. One of the earliest works on swarm intelligent routing is by Schoonderwoerd et al. on the Ant Based Control (ABC) algorithm [9]. This algorithm is for a wired circuit switched network, such as a telephony network. ABC was later modeled analytically in [6]. This work demonstrates and proves the behavior of pheromone and its effect on global system performance in ABC.

Substantially more work has been done to characterize the behavior of such systems by means of simulation. Many were inspired by the Ant Colony Optimization (ACO) algorithm [10]. A number of routing algorithms for packet switched networks are also available based on this framework [11]. These ideas have also been adapted to the field of mobile wireless ad-hoc networks [1] [3] [5]. Finally, a great deal of work has been done with the original biologically inspired models. This work spans several fields, including experimental biology, theoretical biology, and the various disciplines of engineering which apply the models derived by the former. Some summaries may be found in [12] and [13].

1.2   Structure of Paper

Section II gives a brief introduction to the MANET routing problem, as well as a review of the Termite routing algorithm. An analytical model of a MANET is presented in Section III. This model captures some of the most critical aspects of the network, and allows for easy integration with the routing equations used by Termite. Section IV illustrates in analytical detail the average behavior of pheromone on a link, both alone and coupled with other links. Comparisons are made between three pheromone update methods. Section V analyzes the results of previous section and explains earlier experimental results. Section VI concludes the paper with a note about future work and final remarks.

2. Termite Routing for MANETs

This section will give a short introduction to the routing problem in mobile wireless ad-hoc networks. A brief review of the Termite swarm intelligent routing algorithm follows. The section will conclude with some problems exhibited by the algorithm, which will motivate the goals of the analysis.

2.1 A Short Introduction to Ad-Hoc Networks

A mobile wireless ad-hoc network is a collection of mobile computers able to communicate wirelessly with others who are within radio range. Each computer, or node, is able to forward messages for others such that any pair of nodes in the network are able to communicate with each other. Using this forwarding technique, a member of the network with only minimal communications capability is able to take advantage of the resources offered by the sum. No installed infrastructure is necessary for the network to operate.

Due to node mobility, the topology of the network changes often. Some of the outstanding problems include network scalability, the speed with which a routing algorithm is able to adapt to a new topology, the amount of control traffic needed to maintain network connectivity, and data packet delivery reliability [14].

2.2 Termite

Termite is a distributed routing algorithm for mobile wireless ad-hoc networks [1]. It is designed using the swarm intelligence framework in order to achieve better adaptivity, lower control overhead, and lower per-node computation. The algorithm is inspired by the hill building behavior of termites.

The Termite algorithm is explained in detail in [1] and updated in [2], however it may be described simply as follows. Each node in the network has a specific pheromone scent. As packets move through the network on links between nodes, they are biased to move in the direction of destination pheromone gradients. Packets follow this gradient while laying pheromone for their source on the links that they traverse. The amount of pheromone deposited by a packet on a link is equal to the utility of its traversed path. Using this method of pheromone updating, consistent pheromone trails are built through the network. Changes in the network environment, such as topological or path quality changes, are accounted for by allowing pheromone to decay over time. This requires paths to be continuously reinforced by new traffic; new information about the network is added to links. Each node records the amount of pheromone that exists for each destination on each of its links. This creates a routing table similar to those found in traditional link-state routing algorithms.

Pheromone Update. The pheromone update equation is a function which updates the pheromone of the packet source at a node upon its arrival. The update function shown here is the traditional approach and is known as the γ Pheromone Filter. Pheromone on all links decays simultaneously upon packet arrival, and proportionally to packet interarrival times. This is known as continuous pheromone decay and was originally presented in [3]. This idea is extended in this work by also decaying pheromone when it is checked to send a packet. In general, pheromone is decayed whenever it is observed. This allows the system to truly keep track of the real time deterioration of the reliability of available information about the network. Each packet maintains the total utility of the path it has traversed (updated at each node visited), γ, and deposits this amount of pheromone on each link. The γ Pheromone Filter pheromone update equation is,

$$\forall i \; P_{i,s}^n = P_{i,s}^n \cdot e^{-(t-t_{s,obs}^n)\tau}$$
$$P_{r,s}^n = P_{r,s}^n + \gamma \tag{1}$$

where $P_{i,s}^n$ is the amount of pheromone from source node s, on the link from neighbor node i, at node n. The previous hop of the packet is node r. The variable γ is the amount of pheromone carried by the packet, which will vary from packet to packet depending on its path. The time $t_{s,obs}^n$ is the last instant that the pheromone from source node s at node n was observed, either due to packet sending or packet receiving. The pheromone decay rate is τ.

Forwarding Equation. The forwarding equation is used to determine the probability of using a link based on the amount of pheromone on it.

$$p_{i,d}^n = \frac{(P_{i,d}^n + K)^F}{\sum_{j=1}^{N_n}(P_{j,d}^n + K)^F} \tag{2}$$

where $p_{i,d}$ is the probability of using neighbor node i in order to get to destination node d, at node n. N_n is the number of neighbors of node n. $K \geq 0$ is the pheromone threshold and $F \geq 0$ is the pheromone sensitivity.

Termite Summary. Termite has been shown to perform well, especially in regions of high node mobility [2] [1]. However, packets often take substantially longer than necessary paths. In part, this is required in order to maintain current pheromone through the network, but this behavior can also generate significant resource inefficiencies.

3. The Model

A model of an ad-hoc network is presented which will be used to evaluate the behavior of the pheromone update methods. The network is modeled as two communicating nodes with two independent paths available between them. These paths abstract all other connections between the two nodes, including additional nodes, mobility issues, or communications effects. The physical structure is shown in Figure 1, and is the same as that used in [6].

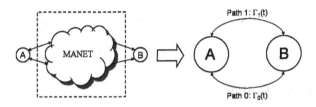

Figure 1.   Diagram of the MANET Model

Each node sends packets to the other with iid exponentially distributed interarrival times. The average rate at which node A sends to B is λ_A, and λ_B in the opposite direction. Each node is also able to decay the pheromone on its links independently. The decay rates at each node are τ_A and τ_B.

Each path i has a utility characterized by a non-negative random process $\Gamma_i(t)$ with mean $\mu_i(t)$. The pheromone contained in a packet arriving on a link, γ, is a sample of that process. Γ is non-stationary since link utilities change over time due to mobility and other effects. Γ is considered to be stationary in this work for ease of analysis. Since each packet passes through the network indepedently of all other packets, there is no correlation between successive samples of the link utility process. The forwarding equation independently considers each packet.

4. Pheromone Update Analysis

This section analyzes the amount of pheromone found on a link. The results will explain the performance of the γ Pheromone Filter (γPherFilt), Joint Decay IIR Filter (IIR2), and pDijkstra pheromone update methods, which are reprinted in Figure 2 from [2]. Namely, why γPherFilt achieves such high goodput while being slow to adapt at high mobility, why pDijkstra adapts better, and why IIR2 is able to achieve the highest performance with regards to the Delivery Efficiency metric.

Goodput is the fraction of successfully delivered data packets, while the latter metric also includes the ratio of the achieved path metric to the best possible available at the time.

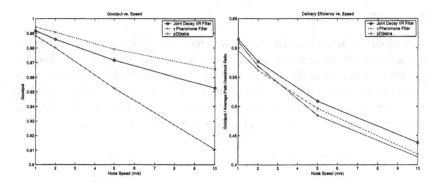

Figure 2. Performance vs. Pheromone Update Method

4.1   Single Link Pheromone

A system of one link is first considered in order to give insight on the more interesting system of two links. A formula is established for the time average pheromone deposited on a link, given that the packet arrival rate is poisson distributed with mean λ [packets/second]. The decay rate, τ [1/second] remains constant. The poisson packet arrival rate assumption implies that the packet interarrival times are exponentially distributed with mean, λ^{-1} [seconds/packet]. The average value of the received pheromone is $E\Gamma = \mu$. The amount of pheromone on the link before packets begin arriving is P_0.

By applying the pheromone update equation n consecutive times, an expression is derived for the amount on a link given that n packets have arrived, $P(n)$. The packet interarrival time of the nth packet is independently and identically distributed, t_n.

$$
\begin{aligned}
P(n) &= (((P_0 e^{-t_1 \tau} + \mu) \cdot e^{-t_2 \tau} + \mu) \cdot \ldots) \cdot e^{-t_n \tau} + \mu \\
&= P_0 e^{-\left(\sum_{i=1}^{n} t_i\right)\tau} + \mu \left[\sum_{i=2}^{n} e^{-\left(\sum_{j=i}^{n} t_j\right)\tau}\right]
\end{aligned}
\tag{3}
$$

The expectation of $P(n)$ with respect to packet interarrival time is found according to standard methods.

$$
EP(n) = \frac{P_0 \lambda^n}{(\lambda + \tau)^n} + \sum_{i=0}^{n-1} \frac{\mu \lambda^i}{(\lambda + \tau)^i}
\tag{4}
$$

In order to simplify Equation 4, substitute $\beta = \frac{\lambda}{\lambda + \tau}$, and further reduce the expression.

$$
EP(n) = P_0 \beta^n + \mu \left(\frac{1 - \beta^n}{1 - \beta}\right)
\tag{5}
$$

To arrive at an expression for the expected amount of pheromone on a link over time, note that the number of packet arrivals, n, within a given time, t, is distributed according to the poisson distribution with parameter, λt.

$$
\begin{aligned}
EP(t) &= \sum_{n=0}^{\infty} [poisson(\lambda t, n)] [EP(n)] \\
&= \sum_{n=0}^{\infty} \left[e^{-\lambda t} \frac{(\lambda t)^n}{n!} \right] \left[P_0 \beta^n + \mu \left(\frac{1 - \beta^n}{1 - \beta} \right) \right] \\
&= \frac{\mu}{1 - \beta} + e^{-\frac{\lambda \tau t}{\lambda + \tau}} \left(P_0 - \frac{\mu}{1 - \beta} \right)
\end{aligned}
\tag{6}
$$

The long term behavior of the link pheromone is defined as its mean.

$$
\begin{aligned}
\lim_{t \to \infty} EP(t) &= \frac{\mu}{1 - \beta} \\
&= \frac{\mu(\lambda + \tau)}{\tau} \\
&\overset{\text{def}}{=} EP
\end{aligned}
\tag{7}
$$

A similar analysis shows the variance of the link pheromone,

$$
VAR(P) = \frac{\mu^2 \lambda}{2\tau}
\tag{8}
$$

Scale Invariance. The ratio of λ and τ is a scale invariant parameter in this system. Primarily characterized by the expected decay factor in Equation 4, the expected pheromone on a link may be held constant as long as $\frac{\lambda}{\tau}$ remains the same. The scale invariant parameter is proportional to β.

4.2   Two Link Pheromone

The following analysis shows the average value of pheromone on each link in a two link system. The two link system is that described by the model in Section III. The γ Pheromone Filter, Joint Decay IIR Filter, and pDijkstra update methods are reviewed. Each generates different pheromone dynamics and maintains varying amounts of link pheromone in equilibrium. Each method falls into a one-zero pheromone distribution in the correct parameter space, which echoes previous results from an analysis of the Ant Based Control routing algorithm [6].

A system of equations is presented which recursively compute the mean pheromone at each node on each link, $P_{0,B}^A$, $P_{1,B}^A$, $P_{0,A}^B$, and $P_{1,A}^B$. Pheromone changes either when it is checked in order to send a packet, or when it is updated due to packet arrival. The total rate at which pheromone is observed is

$\lambda = \lambda_A + \lambda_B$. The *Pheromone Check* proceedure only decays the pheromone and accounts for the fraction of the instances when a packet must be sent. During *Packet Arrival*, which accounts for the fraction of instances that a packet arrives, the pheromone is not only decayed, but also incremented if the packet arrives on the correct link.

The average amount by which pheromone decays between observations has already been implicitly derived in Equation 4. Suppose random variable Y is the interarrival time between packets and is distributed exponentially with mean λ^{-1}, as described in the model definition. Random variable X is defined such that $X = e^{-Y\tau}$, which describes the fraction of pheromone decayed in between packet arrivals. Its probability distribution function is, $f_X(x) = \frac{\lambda}{\tau}x^{(\frac{\lambda}{\tau}-1)}$ where $0 \le x \le 1$. $EX = \frac{\lambda}{\lambda+\tau} = \beta$.

The γ Pheromone Filter. The γ Pheromone Filter has been defined previously. The example below develops the average link pheromone equation for $P_{0,B}^A$ based on the previous description of pheromone influences.

$$
\begin{aligned}
P_{0,B}^A &= \; Pheromone\ Check \\
&+ \; Update\ Not\ On\ This\ Link + Update\ On\ This\ Link \\
P_{0,B}^A &= \; \frac{\lambda_A}{\lambda}\left[\left(\frac{\lambda}{\lambda+\tau_A}\right)P_{0,B}^A\right] \\
&+ \frac{\lambda_B}{\lambda}\left\{p_{1,A}^B\left[\left(\frac{\lambda}{\lambda+\tau_A}\right)P_{0,B}^A\right] + p_{0,A}^B\left[\left(\frac{\lambda}{\lambda+\tau_A}\right)P_{0,B}^A + \mu_0\right]\right\} \\
&= \; \left(\frac{\lambda}{\lambda+\tau_A}\right)P_{0,B}^A + \left(\frac{\lambda_B}{\lambda}\right)p_{0,A}^B\mu_0 \qquad (9)
\end{aligned}
$$

where $p_{0,A}^B$ is the link probability described in Equation 2. The evolution of pheromone on a link at one node as packets are received is illustrated in Figure 3.

Note that when $K = 0$, the asymptotic probability mass function between the two links follows a one-zero distribution; the algorithm uses the better link exclusively. The reason for this is that when a packet arrives at a node, its link is positively reinforced, while all other links at that node are negatively reinforced; pheromone decays on all links but is only replaced on one. Because packets are also biased towards pheromone, this produces a strong positive feedback which eventually transfers all traffic to the link with the highest utility. With no pheromone threshold there is no incentive to use a lesser link. Consequently the probability of using it disappears. Under these conditions, the dominant link follows the behavior of the single link. Only when $\mu_0 = \mu_1$ do the links have equal probability.

With $K > 0$, traffic is allowed to be forwarded over all links, regardless of their utility. Currently bad links may be tested on occasion for a change

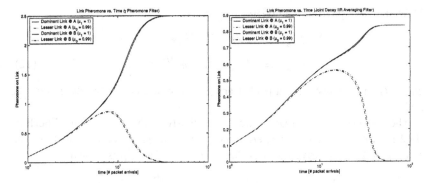

Figure 3.   Pheromone vs. Packets Received ($\beta = \frac{\lambda=2}{2+\tau=1} = \frac{2}{3}, K = 0, F = 5, R = 0.5$)

in utility. This allows for a measure of adaptivity, however K must be set appropriately in order to allow for links to be tested *often enough*. Too few tests of other links will not overcome the positive feedback towards to the dominant link.

The mean link pheromone on the dominant link in the case of a one-zero pheromone distribution can be found by setting the probability of using the dominant link to one and solving for the remaining pheromone. Solving this case for $\tilde{P}^A_{0,B}$,

$$\tilde{P}^A_{0,B} = \frac{\lambda_B (\lambda + \tau_A)}{\lambda \tau_A} \mu_0 \qquad (10)$$

Joint Decay IIR Filter (IIR2).   The Joint Decay IIR Filter implements a simple one-tap infinite impulse response (IIR) averaging filter with joint pheromone decay. The inspiration for such an approach comes from the *Single Link Pheromone* analysis. There, the expected link pheromone may be represented by the sum shown by Equation 11. This is a reexpression of Equation 4. Indices represent packet arrivals; $\gamma(n)$ is the amount of pheromone contained in the nth packet, and $P(n)$ is the expected pheromone after its arrival. Assume $P_0 = 0$ and $B(n) = \beta^n = \left(\frac{\lambda}{\lambda+\tau}\right)^n, n \geq 0$.

$$\begin{aligned} P(n) &= P_0\beta^n + \sum_{i=0}^{n-1} \gamma(n-i)\beta^i \qquad (11) \\ &= \gamma(n) * B(n) \end{aligned}$$

In this interpretation, the series $B(n)$ is the average impulse response of the γ Pheromone Filter in the single link case. This is a simple unnormalized one tap IIR averaging filter. IIR2 is derived from γPherFilt by normalizing the received pheromone to create a version of the well known one-tap IIR averaging filter.

Thus, the IIR2 pheromone update equation is,

$$\forall i \; P_{i,s}^n \;=\; P_{i,s}^n \cdot e^{-(t-t_{s,obs}^n)\tau}$$
$$P_{r,s}^n \;=\; P_{r,s}^n + \left[1 - e^{-(t-t_{i,s}^n)\tau}\right]\gamma \qquad (12)$$

where $t_{i,s}^n$ is the last time a packet arrived from source s from neighbor i at node n.

The analysis of IIR2 may be treated similarly to the analysis of γPherFilt. IIR2 requires that arriving pheromone be normalized according to the time since a packet last arrived on that link. The result for $P_{0,B}^A$ is shown.

$$P_{0,B}^A = \left(\frac{\lambda}{\lambda + \tau_A}\right) P_{0,B}^A + \left(\frac{\lambda_B}{\lambda}\right)\left[1 - \left(\frac{p_{0,A}^B \lambda_B}{p_{0,A}^B \lambda_B + \tau_A}\right)\right] p_{0,A}^B \mu_0 \qquad (13)$$

Time pheromone evolutions is also shown in Figure 3. Mean link pheromone in the one-zero distribution case is shown,

$$\tilde{P}_{0,B}^A = \frac{\lambda_B (\lambda + \tau_A)}{\lambda (\lambda_B + \tau_A)} \mu_0 \qquad (14)$$

pDijkstra.   The simulations of [2] show that the previous pheromone update methods provide the best results with a high pheromone sensitivity, F. The results in Figure 3 above show that a high sensitivity will quickly force the links pheromone towards a one-zero distribution and thus to the exclusive use of the best link (save for any threshold). The determination and selection of a best neighbor link to a destination is reminiscent of typical link-state routing such as Dijkstra's algorithm [15]. pDijkstra is developed where the pheromone update equation is,

$$\forall i \quad P_{i,s}^n = P_{i,s}^n \cdot e^{-(t-t_{s,obs}^n)\tau}$$
$$if \quad P_{r,s}^n < \gamma, \; P_{r,s}^n = \gamma \qquad (15)$$

Similarly to Dijkstra's algorithm, if a link with higher utility is found, then the link pheromone is updated. All pheromone decays concurrently as in the previous methods.

The pDijkstra pheromone update method cannot be analyzed in a similar way because it is nonlinear. For the purposes of the analysis presented here it suffices to note that the link pheromone is upper bounded by the utility of the link.

$$\tilde{P}_{0,B}^A \le \mu_0 \qquad (16)$$

5.   Analysis

As shown in the simulation results of Figure 2, γPherFilt has the best goodput performance but poor ability to track changes in the network. pDijkstra is able to do better in delivery efficiency, while IIR2 scores best in this regard. These results can be explained based on the analysis in the previous section.

γPherFilt maintains the largest equilibrium pheromone level on the best link. Since pheromone takes time to decay, this allows it to route on a particular link longer in case the underlying link metric change; large amounts of pheromone implies a large link forwarding probability. Due to this hysteresis effect on pheromone decay, suboptimal links are used longer, which prevents the algorithm from adapting quickly. In essence, γPherFilt tends to use a known good route, and thus achieves a high goodput, but it is unwilling to change that route in the face of varying route and link metrics.

IIR2 and pDijkstra see less goodput but higher adaptivity compared to γPherFilt, both for similar reasons. They maintain less pheromone on the links and because of this are able to adapt to changes faster. Less pheromone requires less time to decay. Note that $\tilde{P}_{IIR2} < \tilde{P}_{\gamma PherFilt}$ due to an additional term of $\lambda\lambda_B$ in the demoninator of the former. Differences in pheromone between links is relatively less than with γPherFilt, thus the link probabilities are also less decisive during transition periods while the algorithm is choosing a new link. This leads to wandering packets which eventually timeout in the network and result in lower goodput.

It is also important to note the overall superior performance of IIR2 was derived from a simple application of linear filtering. The pheromone delivered on a packet is considered a measure of the packet's path, and it is averaged to create an expectation of the utility of that path to the destination.

6. Conclusion

This paper has presented an extension of an analytical model of a mobile wireless ad-hoc network. This model was used to investigate the properties of the Termite swarm intelligent MANET routing algorithm. The mean pheromone on a single link and in a system of two links were determined. These results were compared for three different pheromone update methods. Relationships between parameters were explored and scale invariance was found. It is shown that the amount of pheromone put on a link influences performance, not only in goodput, but also in adaptability. The analysis also revealed a linear filtering perspective in which link utility is directly estimated with the pheromone rather than simply using it as a routing heuristic.

Future work in this area should more completely characterize the dynamics of the pheromone update methods shown here. While there exists a good intuition for their behavior, a true formalization is necessary. An understanding of exactly what parameters allow a zero-one distribution, how long it takes to arrive at said distribution, and what conditions allow the system to evolve out of the distribution if link costs change would be helpful.

References

[1] M. Roth, S. Wicker, *Termite: Emergent Ad-Hoc Networking*, The Second Mediterranean Workshop on Ad-Hoc Networks, 2003.

[2] M. Roth, S. Wicker, *Performance Evaluation of Pheromone Update Methods in Termite*, in preparation.

[3] M. Gunes, M. Kahmer, I. Bouazizi, *Ant Routing Algorithm (ARA) for Mobile Multi-Hop Ad-Hoc Networks - New Features and Results*, The Second Mediterranean Workshop on Ad-Hoc Networks, 2003.

[4] M. Gunes, U. Sorges, I. Bouazizi, *ARA - The Ant-Colony Based Routing Algorithm for MANETs*, Proceedings of the ICPP Workshop on Ad Hoc Networks (IWAHN 2002), IEEE Computer Society Press, 2002, 79-85.

[5] M. Heissenbuttel, T. Braun, *Ants-Based Routing in Large Scale Mobile Ad-Hoc Networks*, Kommunikation in Verteilten Systemen (KiVS), 2003.

[6] D. Subramanian, P. Druschel, J. Chen, *Ants and Reinforcement Learning: A Case Study in Routing in Dynamic Networks*, Proceedings of the International Joint Conference on Artificial Intelligence, 1997.

[7] G. Di Caro, M. Dorigo, *Mobile Agents for Adaptive Routing*, Technical Report, IRIDIA/97-12, Universit Libre de Bruxelles, Belgium, 1997.

[8] B. Baran, R. Sosa, *A New Approach for AntNet Routing*, Proceedings of the Ninth International Conference on Computer Communications and Networks, 2000.

[9] R. Schoonderwoerd, O. Holland, J. Bruten, L. Rothkrantz, *Ant-Based Load Balancing In Telecommunications Networks*, Adaptive Behavior, 1996.

[10] M. Dorigo, G. Di Caro, L. M. Gambardella, *Ant Algorithms for Discrete Optimization*, Artificial Life, Vol. 5, No. 2, 1999.

[11] K. M. Sim, W. H. Sun, *Ant Colony Optimization for Routing and Load-Balancing: Survey and New Directions*, IEEE Transactions on Systems, Man, and Cybernetics - Part A: Systems and Humans, Vol. 33, No. 5, September 2003.

[12] E. Bonabeau, M. Dorigo, G. Theraulaz, *Swarm Intelligence: From Natural to Artificial Systems*, Oxford University Press, 1999.

[13] M. Resnick, *Turtles, Termites, and Traffic Jams: Explorations in Massively Parallel Microworlds*, Bradford Books, 1997.

[14] J. Broch, D. A. Maltz, D. B. Johnson, Y. Hu, J. Jetcheva, *A Performance Comparison of Multi-Hop Wireless Ad Hoc Network Routing Protocols*, Proceedings of the Fourth Annual ACM/IEEE International Conference on Mobile Computing and Networking, 1998.

[15] E. Dijkstra, *A note on two problems in connection with graphs*, Numerische Mathematik, Vol. 1, 269-271, 1959.

RANDOMIZED ROUTING ALGORITHMS IN MOBILE AD HOC NETWORKS

T. Fevens, I.T. Haque, and L. Narayanan
Department of Computer Science and Software Engineering,
Concordia University,
Montreal, Quebec, Canada, H3G 1M8
{fevens,it_haque,lata}@cs.concordia.ca

Abstract   Position-based routing is a well-known paradigm for routing in mobile ad hoc networks. We give several new randomized position-based strategies for routing in mobile ad hoc networks. Our algorithms combine the greedy heuristic of minimizing the distance remaining to the destination and the directional heuristic of staying close to the direction of the destination with the use of randomization to retain some flexibility in the chosen routes. Our experiments show that randomization gives a substantial improvement in the delivery rate over the deterministic greedy and directional routing algorithms. For some of the algorithms we propose, this improvement comes at the expense of only a small deterioration in the stretch factor of the routes.

Keywords: Wireless networks, mobile ad hoc networks, routing, position based routing, stretch factor, delivery rate.

Introduction

A mobile ad hoc network (MANET) is a collection of autonomous mobile devices that can communicate with each other without having any fixed infrastructure. Each node in the network has an omni-directional antenna and can communicate using wireless broadcasts with all nodes within its transmission range. Thus a MANET can be represented by a unit disk graph, where two nodes are connected if and only if their Euclidean distance is at most the transmission range [Barriere et al., 2001]. Since nodes may not directly communicate with all other nodes because of the limited transmission range, multi-hop communication is needed in the network. The nature of MANETs include issues such as dynamic topology changes, absence of infrastructure, autonomous heterogeneous nodes, and resource constraints that contribute to making the problem of routing in these networks a tremendous challenge.

In the last few years, a plethora of routing protocols for MANETs has been proposed in the literature. As yet there is no consensus and no standards have been adopted. The proposed protocols can be divided into two main categories: *proactive* and *reactive* protocols. Proactive or table-driven protocols [Perkins and Bhagwat, 1994] are based on Internet distance-vector and link-state protocols, and maintain consistent and updated routing information about the entire network by exchanging information periodically. Randomized versions of some of these proactive protocols have been proposed, such as R-DSDV [Boukerche et al., 2001; Choi and Das, 2002; Boukerche and Das, 2003]. Reactive or on-demand routing protocols [Perkins and Royer, 1999; Johnson et al., 2002] discover routes only when data needs to be sent or the topology is changed. Reactive protocols typically use less bandwidth in terms of control packets to discover topology information, but even so, packets to discover new routes must sometimes be flooded through the network, which consumes a huge amount of bandwidth [Chlamtac et al., 2003].

One way of limiting flooding is by using information about the position of nodes in the network. In position-based routing protocols, a node forwards packets based on the location (coordinates in the plane) of itself, its neighbors, and the destination [Giordano et al., 2003]. The position of the nodes can be obtained using GPS, for example, if the nodes are outdoors. There are numerous ways of using position information in making routing decisions. For instance, in DREAM [Basagni et al., 1998] and LAR [Ko and Vaidya, 1998], information about the position of the destination is used to limit the extent of flooding. Nodes whose position makes it unlikely for them to be on a shortest path to the destination will simply not forward packets. In Terminode and Grid [Liao et al., 2001], position-based routing is used to cover long distances and non-position based algorithms are used for shorter distances. In another class of algorithms, which has been termed *progress-based* algorithms in [Giordano et al., 2003], the algorithm forwards the packet in every step to exactly one of its neighbors, which is chosen according to a specified heuristic. Finally, position information can be used to extract a planar sub-graph such that routing can be performed on the perimeter of this sub-graph as in [Bose et al., 1999] and [Karp and Kung, 2000]. The advantage of this last approach is that delivery of packets can always be guaranteed.

In this paper we shall focus on the progress-based routing algorithms. In *greedy routing* [Finn, 1987; Stojmenovic and Lin, 2001], a node forwards the packet to its neighbor which is closest to the destination. *Compass* or *directional* routing [Kranakis et al., 1999] moves the packet to a neighboring node such that the angle formed between the current node, next node, and destination is minimized. Clearly the next node selected by the two heuristics is not always the same (see Figure 1 for an example). Both of these algorithms are known to fail to deliver the packet in certain situations. For examples of such situations, see [Bose and Morin, 1999; Karp and Kung, 2000].

In this paper, we propose several variations of the greedy and directional heuristics. To improve on the delivery rate of these algorithms, we use randomization. Our heuristics combine in various ways the two goals of covering as much distance as possible to the destination (as in greedy routing) and staying as close as possible to the direction of the destination (as in directional routing), while using randomization to allow flexibility with respect to the actual path followed. We evaluate all our heuristics in terms of delivery rate and stretch factor (ratio of the number of hops of the path given by the algorithm to the shortest path in the network). For purposes of comparison, we also study the performance of the greedy and compass routing strategies, as well as RCOMPASS. RCOMPASS chooses the next node uniformly at random from the two nodes that satisfy the directional heuristic on each side of the line from the current node to the destination (see Figure 1 for an example; a precise definition is given in Section 1.1). Our results show that randomization leads to a definite improvement in the delivery rate. Conversely, the best stretch factors is achieved by the deterministic algorithms. However, some of the randomized strategies do very well in terms of both measures of performance. In particular, one of our algorithms, WEIGHTEDRCOMPASS has the best delivery rate of all the algorithms while having one of the best stretch factors.

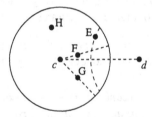

Figure 1. GREEDY chooses E, and COMPASS chooses F as next node, while RCOMPASS chooses uniformly at random from F and G to find the next node.

The rest of the paper is organized as follows. The next section gives relevant definitions including our routing strategies. Section 1.2 gives the empirical results of our simulations and provides an interpretation of the behavior of the algorithms. We conclude with a discussion of the results and future directions of this research in Section 1.3.

1.1 Definitions of Routing Algorithms

We assume that the set of wireless nodes is represented as a set S of n points in a two-dimensional plane. Two nodes are connected by a link if the Euclidean distance between them is at most r, where r represents the transmission range of the nodes. The resulting graph $UDG(S)$ is called a unit disk graph. For node u, we denote the set of its neighbors by $N(u)$. Given a unit disk graph $UDG(S)$ corresponding to a set of points S, and a pair (s, d) where $s, d \in S$, the problem of online position-based routing is to construct a path in $UDG(S)$ from s to d, where in each step, the decision of which node to go to next is based only on the current node c, $N(c)$, and d. Here, s is termed the source and d the destination. A position-based routing algorithm is *randomized* if the next neighboring node is chosen randomly out of the neighbors of the current node [Bose and Morin, 1999]. The routing algorithm may or may not succeed in finding a path from s to d. The performance measures we are interested in are the *delivery rate*, that is, the percentage of times that the algorithm succeeds, and the *stretch factor*, the average ratio of the length of the path returned by the algorithm to the length of the shortest path in the graph. Here the length of the path is taken to mean the number of hops in the path; while other papers consider stretch factor based on Euclidean distance, we do not consider it here. Finally, nodes are assumed to be static for the duration of the packet transmission.

Given a node u, we denote the disk centered at node u with radius ℓ by $disk(u, \ell)$. Given an angle θ such that $0 \leq \theta \leq 2\pi$, we define $Sector(c, d, \theta)$ to be the sector given by angle θ in $disk(c, r)$ that is bisected by the line segment $\overline{cd}$. Further, given an α such that $0 \leq \alpha \leq 1$, we define *Periphery*, *Core*, and *Wing* as follows. Also, Figure 2 illustrates the given definitions.

- $Periphery(c, d, \theta, \alpha) = Sector(c, d, \theta) \cap disk(d, R)$ where R is chosen such that α is the ratio of the area of $Periphery(c, d, \theta, \alpha)$ to the area of $disk(c, r) \cap disk(d, R)$.

- $Core(c, d, \theta, \alpha) = Sector(c, d, \theta) - Periphery(c, d, \theta, \alpha)$.

- $Wing(c, d, \theta, \alpha) = (disk(c, r) \cap disk(d, R)) - Periphery(c, d, \theta)$.

1.1.1   Randomized Algorithms

In what follows, we always assume that the current node is c, the next node is x, and the destination node is d. The algorithms below differ in how to choose x from among the set $N(c)$.

FARINSECTOR: The next node x is chosen uniformly at random from the first non-empty set in the following sequence: $Periphery(c, d, \theta, \alpha)$, $Wing(c, d, \theta, \alpha)$, $Core(c, d, \theta, \alpha)$, $disk(c, r)$.

GREEDYINSECTOR: If $Sector(c, d, \theta)$ is not empty then x is chosen to be the neighbor of c with minimum distance from d, in $Sector(c, d, \theta)$. Otherwise, x is chosen uniformly at random from the set $N(c)$.

RANDOMINSECTOR: The next node x is chosen uniformly at random from

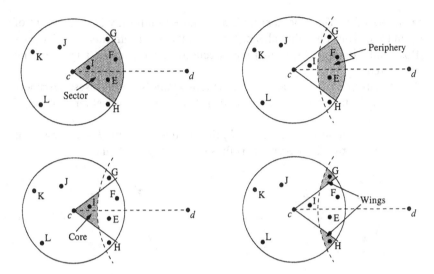

Figure 2. Illustrations of the definitions of *Sector, Periphery, Core,* and *Wing* (represented by the shaded regions). Note that $Sector(c, d, \theta) = Core(c, d, \theta, \alpha) \cup Periphery(c, d, \theta, \alpha)$. In this example, FARINSECTOR selects x uniformly at random out of nodes E and F, the only nodes in $Periphery(c, d, \theta, \alpha)$, while GREEDYINSECTOR picks $x = F$ since F is closest to d out of E, F, and I, the nodes in $Sector(c, d, \theta)$.

the first non-empty set in the following sequence: $Sector(c, d, \theta_1)$, $Sector(c, d, \theta_2)$, $disk(c, r)$, where $0 \leq \theta_1 \leq \theta_2 \leq 2\pi$.

RCOMPASS: Let $\angle xuy$ denote the angle formed by x, u, and y measured counterclockwise. Let n_1 be the neighbor of c above the line $\overline{cd}$ such that $\angle dcn_1 = \theta_1$ is the smallest among all such neighbors. Similarly, n_2 is the neighbor of c below $\overline{cd}$ such that $\angle n_2cd = \theta_2$ is the smallest among all such neighbors. The next node x is chosen uniformly at random from n_1 and n_2. This algorithm differs from the algorithm *Random Compass* proposed in [Bose and Morin, 1999] in the context of triangulations in a small way. In *Random Compass*, one neighbor chosen, ccw(c), is the neighbor of u that minimizes $\angle dc\{ccw(c)\}$, and the second neighbor, cw(c), is the neighbor of c that minimizes $\angle \{cw(c)\}cd$, Therefore, both neighbors could lie on the same side of the $\overline{cd}$ line, whereas in RCOMPASS, if there are no neighbors on one side of the $\overline{cd}$ line, only one neighbor is considered.

WEIGHTEDRCOMPASS: Let n_1 be the neighbor of be defined as in RCOMPASS. The next node x is chosen from n_1 and n_2 with probability $\theta_2/(\theta_1 + \theta_2)$ and $\theta_1/(\theta_1 + \theta_2)$, respectively.

BEST2COMPASS: Let n_1 and n_2 be the neighbors of c such that $\angle dcn_1$ (or $\angle n_1cd$) and $\angle n_2cd$ (or $\angle dcn_2$) are the two smallest such angles among all neighbors of c. Then the next node x is chosen uniformly at random out of n_1 and n_2. This algorithm also differs from the algorithm *Random Compass* [Bose and Morin, 1999] in that the directions of the smallest angles are not considered.

BEST2GREEDY: Let n_1 and n_2 be the closest and second closest neighbors of c to the destination d. The next node is chosen uniformly at random out of these two nodes.

All the algorithms above take $O(k)$ time to find x, where d is the degree of c. The behavior of the algorithms is illustrated in Figures 2 and 3.

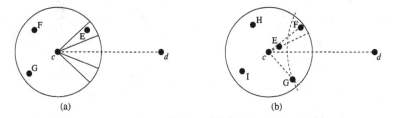

(a)   (b)

Figure 3.   In (a) RANDOMINSECTOR selects E as the next node since $Sector(c, d, \theta_1)$ is empty and E is the only node in $Sector(c, d, \theta_2)$. In (b) WEIGHTEDRCOMPASS sets $n_1 = E$ and $n_2 = G$, BEST2GREEDY sets $n_1 = F$ and $n_2 = G$ whereas BEST2COMPASS sets $n_1 = E$ and $n_2 = F$.

1.2   Empirical results

We have implemented, in C++, all the algorithms discussed in Section 1.1. With the exceptions of GREEDY and COMPASS, all of the algorithms considered here are randomized. To evaluate the performance of these randomized algorithms we will consider their packet delivery rates and stretch factors. We first describe our simulation environment, including the choice of algorithm-specific parameters, and then describe and interpret our results, comparing our algorithms with previous work, as well as with each other.

1.2.1   Simulation Environment

In the simulation experiments, a set S of n points (where $n \in \{75, 100, 125, 150\}$) is randomly generated on a square of 100m by 100m. For the transmission range of nodes, we use 15m or 18m (experiments showed that with lower transmission radii, the graph was too often disconnected, and with higher transmission radii, the generated graphs were so dense that the delivery rate of all algorithms approached 100%). After generating $UDG(S)$, a source and destination node are randomly chosen. If there is no path from s to d in $UDG(S)$, the graph is discarded, otherwise, all routing algorithms are applied in turn. Clearly, an algorithm succeeds if a path to the destination is discovered. The deterministic algorithms are deemed to fail if they enter a loop, while the randomized algorithms are considered to fail when the number of hops in the path computed so far exceeds the number of nodes in the graph. To compute the packet delivery rate, this process is repeated with 100 random graphs and the percentage of successful deliveries determined. To compute an average packet delivery rate, the packet delivery rate is determined 100 times and an average taken. Additionally, over the 100×100 runs, the average hop stretch factor is computed.

Several of the randomized routing algorithms use experimentally optimized parameters. In particular, FARINSECTOR depends on the parameters θ and α. A smaller value of θ clearly means a smaller number of eligible neighbors of c. Similarly, a smaller value of α means the area of the periphery is smaller compared to the area of the sector, which changes the number of eligible neighbors that are closer to the destination. We use $\theta = \pi/3$ and $\alpha = 0.6$ after comparing the performance of the algorithm with θ varying from $\pi/3$ to π and α from 0.1 to 0.9. For GREEDYINSECTOR, $\theta = \pi/3$ was found experimentally to give the best performance. Also, RANDOMINSECTOR depends on the size of the nested sectors, and the experimentally determined optimum values are $\pi/6$ and $\pi/3$, respectively.

1.2.2 Discussion of Results

Detailed simulation results for all the routing algorithms, along with the associated standard deviations, are given in Tables 1 and 2 for the case when the transmission radius is 15 [1]. In particular, we are interested in the performance of our proposed randomized routing algorithms with the previously published routing algorithms GREEDY, COMPASS and RCOMPASS.

BEST2GREEDY and BEST2COMPASS are straightforward randomizations of the greedy and compass strategies, where the next node is chosen randomly from the top two candidates according to the respective heuristics. BEST2-GREEDY is the worst of the randomized strategies in terms of delivery rate, but the best in terms of the stretch factor for both values of transmission radius. BEST2COMPASS has the second-best stretch factor and the second-worst delivery rate.

The sector-based algorithms improve significantly on the delivery rate of GREEDY, COMPASS and the above two strategies. The key idea behind the sector-based algorithms is to restrict the extent to which we stray away from the direction of the destination while keeping some flexibility regarding exactly which neighbor to forward to next. The three algorithms differ in the

---

[1]The trend of the results is the same when the transmission radius is 18 so the details are omitted here.

Table 1. Average packet delivery rate and standard deviation, σ, in terms of percentages, for transmission radius $r = 15$m.

| | $n = 75$ | | $n = 100$ | | $n = 125$ | | $n = 150$ | |
Algorithms	Aver.	σ	Aver.	σ	Aver.	σ	Aver.	σ
GREEDY	70.67	4.61	75.92	3.66	85.70	3.47	92.37	2.53
COMPASS	72.10	4.42	77.14	4.07	87.83	3.16	93.59	2.37
RCOMPASS	86.93	3.07	92.61	2.66	97.97	1.37	99.51	0.64
FARINSECTOR	80.60	3.92	85.51	3.44	94.14	2.37	97.60	1.58
GREEDYINSECTOR	80.73	3.93	86.16	3.50	94.33	2.13	97.82	1.68
RANDOMINSECTOR	80.51	3.87	85.90	3.63	94.25	2.32	97.80	1.61
BEST2GREEDY	78.97	4.08	82.64	3.81	90.67	2.97	95.13	1.98
BEST2COMPASS	80.94	3.84	85.18	3.42	91.92	2.70	96.51	1.77
WEIGHTEDRCOMPASS	87.53	3.08	91.97	2.71	97.09	1.65	99.10	0.95

Table 2. Average stretch factor and standard deviation, σ, for transmission radius $r = 15$m.

| | $n = 75$ | | $n = 100$ | | $n = 125$ | | $n = 150$ | |
Algorithms	Aver.	σ	Aver.	σ	Aver.	σ	Aver.	σ
GREEDY	1.01	0.05	1.02	0.06	1.02	0.07	1.02	0.07
COMPASS	1.06	0.11	1.08	0.13	1.10	0.14	1.10	0.14
RCOMPASS	2.72	2.47	2.69	2.47	2.34	2.27	1.95	1.75
FARINSECTOR	2.03	1.98	2.02	2.18	1.83	1.99	1.62	1.75
GREEDYINSECTOR	2.21	2.17	2.28	2.45	2.03	2.25	1.75	2.05
RANDOMINSECTOR	2.31	2.22	2.39	2.49	2.18	2.39	1.89	2.08
BEST2GREEDY	1.79	1.26	1.60	1.07	1.41	0.83	1.28	0.59
BEST2COMPASS	1.86	1.43	1.68	1.18	1.48	0.92	1.35	0.65
WEIGHTEDRCOMPASS	1.73	1.44	1.74	1.58	1.55	1.35	1.42	1.22

choice of the neighbor within the sector[2]. GREEDYINSECTOR chooses the neighbor closest to the destination from within the sector, while FARINSEC-TOR chooses a node randomly from among the nodes closest to the destination ($Periphery(c, d, \theta, \alpha)$), and RANDOMINSECTOR chooses randomly from among nodes closest to the direction of the destination. FARINSECTOR also allows for the choice of next neighbor from $Wing(c, d, \theta, \alpha)$, if the sector is empty, on the grounds that it may be worthwhile straying outside the sector provided that we cover a lot of distance to the destination. While the delivery rates of all sector-based algorithms are almost identical for both values of transmission radius, FARINSECTOR appears to have a slight edge over the

---

[2]For simplicity, we use "the sector" to refer to $Sector(c, d, \theta)$ in this paragraph.

other two in terms of stretch factor. However FARINSECTOR also is slightly more complicated to implement than the other two.

The RCOMPASS algorithm proposed in [Bose and Morin, 1999] is the best in terms of delivery rate but is significantly worse than all other randomized algorithms in terms of the stretch factor. Recall that RCOMPASS chooses the next node with equal probability among the two nodes making the smallest angles in clockwise and counterclockwise directions from the the $\overline{cd}$ line. This means that if there is a problem with the route on one side of the line, there is a good chance of avoiding it, which gives a good delivery rate. However, by potentially moving far away from the direction of the destination, we may end up increasing the path length. In contrast, WEIGHTEDRCOMPASS still allows for moving on both sides of the $\overline{cd}$ line, but by weighting the probability of the choice of each neighbor x on the angle $\angle xcd$, it reduces the chances of moving too far away from the right direction and therefore taking too long a path. As a result, WEIGHTEDRCOMPASS achieves the same delivery rate as RCOMPASS but significantly improves on its stretch factor.

In summary, the experimental results show that all the randomized algorithms perform better than the deterministic algorithms in terms of average packet delivery rate. In particular, WEIGHTEDRCOMPASS and RCOMPASS outperform the other randomized algorithms. However, in terms of average hop count stretch factors, the deterministic algorithms outperform all the randomized algorithms, with GREEDY having the lowest stretch factors. The fact that the randomized algorithms are able to continue and find alternative and possibly longer routes even when encountering the same node again in a path accounts for both the higher delivery rates and the higher stretch factors as compared to the deterministic algorithms. However, it is interesting to note that our best algorithm WEIGHTEDRCOMPASS has not only the best delivery rate but also one of the best values of stretch factor. In fact, when computing the stretch factor over only those instances where GREEDY or COMPASS succeed as well, the stretch factor of WEIGHTEDCOMPASS is even better than the values shown in Table 2.

1.3 Summary

In this paper, we proposed six new randomized position-based strategies for routing in mobile ad hoc networks. We compared the performance of our algorithms with the previously proposed GREEDY, COMPASS, and RCOMPASS algorithms. Our simulation results demonstrate that randomization yields a definite improvement over the deterministic algorithms in terms of the delivery rate. Conversely, the best stretch factors are achieved by the deterministic algorithms. One of our new algorithms WEIGHTEDRCOMPASS achieves the best delivery rate and has one of the best stretch factors.

We are currently working on improving further the the the performance of WEI-GHTEDRCOMPASS by varying the choice of the qualifying candidates for the next node as well as the weighting function. We have recently completed a study of a class of such randomized algorithms [Fevens et al., 2004]. We are also interested in the performance of WEIGHTEDRCOMPASS on other types of graphs, such as planar graphs and triangulations. Using weighted probabilities to choose the next neighbor in a sector-based algorithm is another interesting avenue for research. Finally we are interested in characterizing the kinds of networks for which our randomized algorithms deliver the packet with probability 1.

Acknowledgments

The authors wish to thank the anonymous referees for their helpful comments. The research of T. Fevens and L. Narayanan is supported in part by NSERC.

References

Barriere, L., Fraignaud, P., Narayanan, L., and Opatrny, J. (2001). Robust position-based routing in wireless ad hoc networks with irregular transmission ranges. In *Proc. of 5th ACM Int. Workshop on Discrete Algorithms and Methods for Mobile Computing and Communications*.

Basagni, S., Chlamtac, I., Syrotiuk, V.R., and Woodward, B.A. (1998). A distance routing effect algorithm for mobility (DREAM). In *4th ACM/IEEE Conference on Mobile Computing and Networking (Mobicom '98)*, pages 76–84.

Bose, P. and Morin, P. (1999). Online routing in triangulations. In *10th Annual International Symposium on Algorithms and Computation (ISAAC '99)*, pages 113–122.

Bose, P., Morin, P., Stojmenovic, I., and J.Urrutia (1999). Routing with guaranteed delivery in ad hoc wireless networks. In *3rd Workshop on Discrete Algorithms and Methods for Mobile Computing and Communication (DIALM)*.

Boukerche, A. and Das, S.K. (2003). Congestion control performance of R-DSDV protocol in multihop wireless ad hoc networks. *ACM/Kluwer Journal on Wireless Networks*, 9(3):261–270.

Boukerche, A., Das, S.K., and Fabbri, A. (2001). Analysis of randomized congestion control scheme with DSDV routing in ad-hoc wireless networks. *Journal of Parallel and Distributed Computing*, 61(7):967–995.

Chlamtac, I., Conti, M., and Liu, J. (2003). Mobile ad hoc networking: Imperatives and challenges. *Ad Hoc Network Journal*, 1(1):13–64.

Choi, W. and Das, S.K. (2002). Performance of randomized destination-sequence distance vector (R-DSDV) protocol for congestion control in ad hoc wireless network routing. In *Proc. of Applied Telecommunications Symposium - Wireless Track (ATS)*.

Fevens, T., Haque, I.T., and Narayanan, L. (2004). A class of randomized routing algorithms in mobile ad hoc networks. Submitted for publication.

Finn, G.G. (1987). Routing and addressing problems in large metropolitan-scale internetworks. Technical Report ISU/RR-87-180, USC ISI, Marina del Ray, CA.

Giordano, S., Stojmenovic, I., and Blazevic, Lj. (2003). Position based routing algorithms for ad hoc networks: A taxonomy. In Cheng, X., Huang, X., and Du, D.Z., editors, *Ad Hoc Wireless Networking*. Kluwer.

Johnson, D., Maltz, D., Hu, Y-C., and Jetcheva, J. (2002). The dynamic source routing protocol for mobile ad hoc networks. Technical Report Internet Draft, draft-ietf-manet-dsr-07.txt (work in progress), IETF.

Karp, B. and Kung, H. (2000). GPSR: greedy perimeter stateless routing for wireless networks. In *Proc. of 6th ACM Conference on Mobile Computing and Networking (Mobicom '00)*.

Ko, Y.B. and Vaidya, N.H. (1998). Location-aided routing (LAR) in mobile ad hoc networks. In *4th ACM/IEEE Conference on Mobile Computing and Networking (Mobicom '98)*, pages 66–75.

Kranakis, E., Singh, H., and Urrutia, J. (1999). Compass routing on geometric networks. In *Canadian Conference on Computational Geometry (CCCG '99)*, pages 51–54.

Liao, W.H., Tseng, Y.C., and Sheu, J.P. (2001). GRID: A fully location-aware routing protocols for mobile ad hoc networks. *Telecomm. Systems*, 18(1):37–60.

Perkins, C.E. and Bhagwat, P. (1994). Highly dynamic destination-sequenced distance-vector routing (DSDV) for mobile computers. In *ACM SIGCOMM '94 Conference on Communications Architectures, Protocols and Applications*, pages 234–244.

Perkins, C.E. and Royer, E. (1999). Ad hoc on-demand distance vector routing. In *2nd IEEE Workshop on Mobile Computing Systems and Applications*, pages 90–100, New Orleans, LA.

Stojmenovic, I. and Lin, X. (2001). Loop-free hybrid single-path/flooding routing algorithms with guaranteed delivery for wireless networks. *IEEE Transactions on Parallel and Distributed Systems*, 12(10):1023–1032.

RBR: REFINEMENT-BASED ROUTE MAINTENANCE PROTOCOL IN WIRELESS AD HOC NETWORKS

JainShing Liu,[1] and Chun-Hung Richard Lin [2]

[1] *Providence University, R.O.C.,* [2] *National Sun Yat-Sen University, R.O.C.*

Abstract   In this paper, we propose a so-called refinement-based routing protocol that uses dynamic route redirection to provide proactive route selection and maintenance to on-demand routing algorithms so that the benefits of both types of routing algorithms can be combined and their drawbacks minimized. Experimental results demonstrate that adding the refinement-based routing protocol to AODV significantly reduces the number of broken paths and the end-to-end packet latency when compared with the pure on-demand routing protocol, AODV.

Keywords: route maintenance, ad hoc wireless network

1.   Introduction

As the popularity of mobile computing increases, cooperative communication using wireless devices is attracting interest. A Mobile Ad-Hoc Network (MANET) is a collection of such mobile devices, denoted as nodes, without the required intervention of any centralized access point or existing infrastructure. Each node in the network is equipped with a wireless transmitter and a receiver, and can act as both a host and a router forwarding packets to other nodes.

An important issue for achieving efficient resource utilization in the network is how to update route information depending on a change of network topology and connectivity. Since mobility in MANET causes frequent, unpredictable and drastic changes to the topology, it is especially important for nodes to be able to adapt to such changes and find an efficient route between communicating source and destination. To provide routes in such dynamic environments, many routing protocols have been proposed over the last few years. These protocols can be broadly classified onto three categories, namely, proactive, reactive, and hybrid.

On the one hand, pro-active routing protocols are proposed as a means to dynamically adjust an on-going route when network topology changes, or equivalently to provide different routes at different points of time so as to control the dynamics of a MANET. However, when providing such adaptability, a pro-active routing protocol uses a large portion of network capacity to keep the routing information current, which is an inefficient use of bandwidth to flood control messages to the network.

On the other hand, on-demand routing protocols only maintain the active paths to those destinations to which data must be sent and thus significantly reduce routing overheads. These have recently attracted more attention than the pro-active protocols. However, on-demand routing protocols accommodate route changes only when an active route is disconnected. They cannot adapt to the change of network topology even if another route with less hops becomes available by the movement of intermediate nodes, unless any link in-between is disconnected. To alleviate this problem, several preemptive-based routing protocols were proposed [1][2][3][4]. Among these, Preemptive Routing [1] is in principle, a preventive mechanism, which in advance initializes a route discovery when links of a path are likely to be broken, and then hands off the dangerous path to the shortest new path just found. Such an approach reduces a handoff delay experienced by the data packets if the path-break predicted actually happens. However, it is no help in reducing routing overheads because, whenever a path is suspected to be broken, this method will flood the route-request packets to the network. This may even increase such overheads.

Router Handoff [2] deals with the path-broken problem by finding an alternate node in the vicinity of a potential link break, and then handing off the route to this node. However, the approach involves no optimizations for an active path; that is, a router handoff cannot shorten a path since such an approach only replaces an intermediate node with another which connects the replaced node's upstream node and its downstream node, with stronger links.

OR2 [3] is an adaptive path tuning scheme for mobile ad hoc networks. This method reduces the hop count of an active route while data packets are sent without link disconnection. However, it targets the "one-hop path shortening" between nodes that are adjacent with a single intermediate node, and provides no "N-hop path shortening".

Recently, a so called $P_r AODV$ protocol [4] is proposed as an extension to AODV [5]. This method invokes a path rediscovery routine from the source according to the received information about the path which is contained in the route reply packet or according to a warning message received at the source. Moreover, instead of monitoring the signal power of the receiving packet, it monitors the transfer time for the hello packets to predict the link break. However, when a source node receives warning messages and decides to rediscover the path, it still needs to flood the route-request packets to the whole network, increasing the possible routing overhead as that noted for the preemptive routing [1].

In contrast to the above research, "RBR" provides path optimizations that simultaneously reduce the routing overheads for a broken path and the end-to-end delays of an active route. To achieve the first goal of reducing routing overheads, when a path breaks, RBR does not flood the route-request packets into the network; instead, it repairs the broken path using only local broadcast messages. Regarding the second goal of decreasing delays, RBR progressively shortens a path by introducing a suitable redirector to the path with fewer hop counts. We have incorporated the two mechanisms into the AODV routing protocol. Experimental results demonstrate that adding refinement-based routing to AODV significantly reduces the number of broken paths and the end-to-end packet latency, with only small extra control overheads when compared with the pure on-demand routing protocol, AODV.

This paper is organized as follows. In Sections 2 and 3, two major mechanisms involved in RBR are introduced. Then, performance evaluation results are given in Section 4. Finally we draw up our conclusions in Section 5.

2. Passive Probe Route Redirection

This section describes the details of the proposed Passive Probe Route Redirection P-PR_2, the first component of the proposed Refinement-Based Routing Protocol. First, we introduce the concept of "vicinity" of two nodes and that of the "redirector" of a path in our RBR. Then, we explain the design of P-PR_2 on the basis of overheard data, recorded in a so-called Overhear Table without the aid of source route information.

Vicinity

To argue for the concept of "nearness" of two nodes and that of "redirector" of a path more formally for our RBR, we introduce the notions of vicinity and bypassed hop count based on the observation of the distance relationship between two nodes. For this purpose, let us define the following notations:

- $D_{(AB)}$: The distance estimated from B to A.

- $P_{(A)}$: The vicinity of A.

- $R_{uf}(A)$: The upper-stream adjacent node of A in relation to flow f.

- $R_{df}(A)$: The down-stream adjacent node of A in relation to flow f.

- $HC_f(K)$: The bypassed hop count of a Redirector K in relation to flow f.

In this context, we assume $D_{(AB)} = D_{(BA)}$. Based on this, B is said to be in the vicinity of A, or $B \in P_{(A)}$, implying that $A \in P_{(B)}$ as well. Furthermore, suppose that a flow, f, now traverses A, B, C and D in this order, and X

and Y are two nodes not in f. Using the above definitions, the relationship between the four nodes involved in f can be represented as $A = R_{uf}(B) = R_{uf}(R_{uf}(C))(= R_{uf}^2(C)) = R_{uf}(R_{uf}(R_{uf}(D)))(= R_{uf}^3(D))$.

As shown in Fig. 1, when C moves away from B, and $C \in P_{(B)}$ is no longer true, there is a possibility that X or Y can act as a redirector to repair the broken path. In this scenario, X may be $R_{df}(B)$ and $R_{uf}(C)$ while Y may be $R_{df}(B)$ and $R_{uf}(D)$. Comparing the two candidates, X and Y, it can be seen that X can only connect to the direct down-stream node of B, i.e., $C = R_{df}(B)$ while Y can connect to the more descendent down-stream node of B, i.e., $D = R_{df}^2(B)$. This can be represented more precisely as $HC_f(X) = 1$ and $HC_f(Y) = 2$. Consequently, if X and Y compete to be a redirector for the broken link, $\overline{BC}$, simultaneously, and B can choose Y as the result, then the need of reinitializing a route discovery will be avoided and a shorter repaired path, when compared with that done by X, can be obtained. From this, we can develop the scheme described in the next section.

Design of P-PR_2

Passive Probe Route Redirection is, in principle, a preventive, preemptive approach to deal with path-breaks. We set two design goals to P-PR_2: repairing a broken path, and minimizing the number of additional control packets. On the one hand, the first goal is obvious in the context of the aforementioned problem. On the other hand, as regards the second goal, we aim at a scheme that can result in only a small extra routing overhead for repairing a broken path. In particular, we do not allow a path to be broken, which wastes huge bandwidth when flooding routing packets to find a new path, whenever there exists at least one redirector, in the vicinity of a broken link which can repair it. This is an important consideration for an ad hoc network since nodes in the network need to reduce their power consumption whenever possible. We design a scheme in which control packets are transmitted only when a node determines that a path should be changed based on when the alarm is given that there is a link-break. We call the scheme P-PR_2 and it is the first component of RBR. In what follows, we first introduce the Overhear Table used in P-PR_2, and then discuss our solutions for the so-called Route Redirection Reply Storm Problem.

Overhear Table.   In P-PR_2, each node makes use of its *Overhear Table*. This table contains information about the status of each neighbor overheard in promiscuous receive mode. In particular, without the aid of source route information, an overhear table under AODV can be implemented with the following fields:

- The identity of the overheard node.

- The overhear time. It aids the flush work triggered when the overheard data is out of date.

- The path source and the path destination: two fields used to identify a flow.

- The time to live (TTL). This represents the sequence of overheard nodes in the same route.

- The estimated distance: a value obtained with received power and the transmitted power announced by the sender. With this distance and the hop count derived from TTL, a node can decide on its preference to be a redirector.

With this table, let us now explain the fundamental messages passed among the nodes in a MANET, which involves three control messages (packets): P-PR_2_RRREQ (Route Redirection Request), P-PR_2_RRREP (Route Redirection Reply), and P-PR_2_RRACK (Route Redirection Acknowledgement).

At first, when movement of an intermediate node or the destination causes a link to break, or when heavy traffic in an area makes a link useless, a node that uses this link to its next-hop, retransmitting RTS more than m_{rt} times, changes its state from normal to probe and locally broadcasts a P-PR_2_RRREQ. Upon receiving the P-PR_2_RRREQ, each neighboring node triggers its own P-PR_2 reply procedure. That is, the node that receives P-PR_2_RRREQ first checks if itself can be a redirector by searching the initiating node and the downstream nodes in the same path, in its own overhear table. If both are found, the by-passed hop counts between the sender and the found downstream nodes, HCs, will be calculated, and the distances from the node to these downstream nodes, d_rs, and the distance from the node to the initiating node, d_s, will be all retrieved. These values are then used to decide the preference of this node to be a redirector in the path, which helps to solve the Route Redirection Reply Storm Problem to be discussed in the next section. According to the preference value, the node randomly delays for a period of time, and then sends a P-PR_2_RRREP to the initiator. By receiving the P-PR_2_RRREP, the initiator decides which redirector will be its new next-hop, sending its decision with a P-PR_2_RRACK to the redirector, which makes the redirector update its routing table and completes the redirection procedure.

Preventing Route Redirection Reply Storms. The ability of nodes to reply to a P-PR_2_RRREQ on the basis of information in their Overhear Tables can result in a possible Route Redirection Reply "storm" in some cases. That is, when a node broadcasts a P-PR_2_RRREQ, each neighbor may attempt to send a P-PR_2_RRREP, thereby wasting bandwidth and possibly increasing the number of network collisions in the area.

This is illustrated by an example in Fig. 2, the same one given in Fig. 1 but with more details. As shown in this figure, when link $\overline{BC}$ is likely broken or congested, nodes X, Y, Z and W at this time will receive B's P-PR_2_RRREQ for the path from S to I, and each has some overheard nodes cached for this route. Normally, they all attempt to reply from their own Overhear Tables and all send their P-PR_2_RRREP at about the same time because they all receive the broadcast P-PR_2_RRREQ at about the same time. Such simultaneous replies from different nodes may create packet collisions among some or all of these replies and may cause local congestion in the wireless network.

When a node puts its network interface into promiscuous receive mode, it can delay sending its own P-PR_2_RRREP for a short period of time and listen to see if the initiating node begins using a redirected route first. That is, this node should delay sending its own P-PR_2_RRREP for a random period, which is now obtained as

$$T = \delta \times (M - HC + \frac{d_s}{TR} \times r_1 + \frac{d_r}{TR} \times r_2), \qquad (1)$$

where δ is a small constant delay, M is the maximum number of hop counts that can be bypassed, HC is the bypassed hop count, d_s and d_r are the distance to the initiating node, and that to the found downstream node, respectively, TR is the node's transmission range, and r_1 and r_2 are two random numbers between 0 and 1. This delay effectively randomizes the time at which each node sends its P-PR_2_RRREP. That is, in the probability sense, all nodes sending P-PR_2_RRREP messages giving larger bypassed hop counts, HCs, will send these replies earlier than all nodes sending P-PR_2_RRREP messages giving smaller HCs. In addition, a node's relative location, represented as (d_s, d_r), serves as a perturbation factor in this randomization. More precisely, a node with location near both the initiating node and the downstream node may reply sooner than the others.

Provided with the delay period, a neighboring node promiscuously receives all packets, looking for P-PR_2_RRACK from the initiator of this P-PR_2_RRREQ. If such an ACK, destined to another node, is overheard by this neighboring node during the delay period, the node may infer that the initiator of the P-PR_2_RRREQ has already received a P-PR_2_RRREP giving an equally good or better redirected route. In this case, this node cancels its own P-PR_2_RRREP for this P-PR_2_RRREQ.

In this example, Y receives the P-PR_2_RRREQ and checks its Overhear Table, finding that it can overhear three nodes, B, C, and D, in this path. Y then chooses D as its next-hop for this path since it gives a larger bypassed hop count, 2, compared with that provided by C, 1. Given this value, 2, and two distances (to the initiator B, and the chosen next-hop, D, respectively), 150 and 180, Y can compute its delay period according to equation 6, sending a P-PR_2_RRREP at the indicated time if no other replies are overheard during this period. Additionally, since the hop count given by Y is larger than those provided by X, Y is very likely to send its reply successfully if its location is not too far from the initiator and the chosen next-hop when compared with that of its competitor, X.

For reference, the control message flow summarizing the salient features of P-PR_2 is given in Fig. 3. This figure mainly depicts the control messages exchanged between the initiator, B, and the successful redirector Y in the example. In addition, the original next-hop, C, is also shown, which explicitly indicates the possibility that C can also give a response for the P-PR_2_$RRREP$ even though this does not happen in this example.

3. Active Probe Route Redirection

This section describes the details of A-PR_2, the second component of RBR. The goal of A-PR_2 is that without the aid of source routing information, by using the Overhear Table only, it can automatically shorten a path if one or more of its intermediate hops becomes unnecessary. This is particularly useful when the path is unavoidably lengthened by P-PR_2 as the redirectors found at that time have their own HC_fs equal to 1 at most. The Active Probe Route Redirection (A-PR_2) is designed in such a way that even MANET is operating with a distance vector routing algorithm.

More precisely, like P-PR_2, A-PR_2 uses the same Overhear Table, filled with overheard path information obtained under the promiscuous receive mode, to check if a node itself can be a redirector to shorten an active path without the need for reinitializing a route-discovery procedure. That is, whenever a node discovers in its Overhear Table that there exists at least two nodes, say i and j, in the same path, having their TTLs, $TTL(i)$ and $TTL(j)$, with difference equal to or greater than 3, i.e., $|TTL(i) - TTL(j)| >= 3$, or $HC_f >= 3$, the node will trigger an Active Route Redirection Request, A-PR_2_$RRREQ$, relayed to the upstream neighboring node, say i, to redirect the next-hop of the route to this node. Upon receiving the A-PR_2_$RRREQ$, node i makes a decision as to whether the request should be responded to or not based on certain criteria. If the decision is positive, an A-PR_2_$RRREP$ will be sent to the requestor (redirector).

One of the possible criteria is the number of hop counts to be shortened in the path if several A-PR_2_$RRREQ$ messages are received from different nodes at the same time. In this case, a redirector providing a shorter path, i.e., a larger HC_f, is preferred. In addition, the criteria should be considered with those redirections that just took place in the near past in mind. That is to say, if node i changes its next-hop to a node, say k, as its new next-hop for a redirection request delivered by k already, the following request from the node in question should be rejected as long as the request suggests a HC_f is not an improvement on the previous one.

Fig. 4 illustrates an example demonstrating the control flow in A-PR_2, which involves only two control messages: A-PR_2_RRREQ and A-PR_2_RRREP. Let us assume that $A = R_{uf}(B)$, $B = R_{uf}(C)$, $C = R_{uf}(D)$, and $D = R_{uf}(E)$ in relation to flow f. Moreover, suppose that C' has overheard data packets transmitted from A to B and data packets transmitted from D to E, simultaneously. Based on the overheard information, C' determines that it may be a redirector between A and D; that is, $R_{uf}(C') = A$ and $R_{df}(C') = D$ may be established. In this case, this node, C', sends an A-PR_2_RRREQ to A, and upon receipt of this request, A updates its routing table and sends an A-PR_2_RRREP back to C' if the criteria mentioned previously are satisfied. In particular, to ensure that the downstream node D can also be connected to C', C' has the option to ask D to reply to the same A-PR_2_RRREQ, which is shown with dot-lines in the figure. Finally, when A-PR_2_RRREP is received, C' can update its own routing table accordingly. Note that since this route is now changed, the refreshed routing information can be piggybacked from the initiating node toward the path endpoints to inform other nodes on the path of the fact that the path length may be changed, which is the same issue considered in P-PR_2.

4. Performance Evaluations

In this section, we report on theoretical analysis and simulation studies (with GloMoSim [6] in order to come to an understanding of the potential improvements in performance obtained by the proposed *Refinement-Based Routing*.

Theoretical Analysis

Some definitions and basic results are given first. Let the number of nodes in the network be N, the transmission range of each node be R, and the average path length of all possible traffic be $\bar{L}$. The average hop count from source to destination can be approximated as $H = \frac{\bar{L}}{R}$, and the number of hops to discover a route is $2H$ on average.

Overhead of repairing a broken link. Consider the control overheads that AODV may involve. Supposing that AODV's RREQ packets can reach all nodes in the network, the overhead of flooding such messages will be N. Further, if each of the H hops has the same broken probability and the number of routes affected by a link breakage is ψ on average, then the number of RERR packets for the link break can be estimated as $\psi \cdot \frac{1+2+,...,+H-1}{H}$.

In pure AODV, a broken link causes RERRs sent to the sources affected. Assuming all the sources initiate route rediscovery after receiving such R-

ERRs, the average number of control overhead, $Pkt^f_{AODV_p}$, and the average delay, $Del^f_{AODV_p}$, for the broken link can be obtained with $Pkt^f_{AODV_p} = Pkt^f_{RERR} + Pkt^f_{RREQ} + Pkt^f_{RREP} = \psi \cdot (\frac{1+2+,...,+H-1}{H} + N + H) \approx \frac{\psi \cdot (3H+2N)}{2}$ and $Del^f_{AODV_p} = Del^f_{RERR} + Del^f_{RREQ} + Del^f_{RREP} = \frac{1+2+,...,+H-1}{H} + H + H \approx \frac{5H}{2}$, respectively, where $Pkt^f_{RERR}(Del^f_{RERR})$, $Pkt^f_{RREQ}(Del^f_{RREQ})$, and $Pkt^f_{RREP}(Del^f_{RREP})$ are the number of RRER packets (the delay of R-RER) sent to the sources, the number of RREQ packets (the delay of RREQ) sent to find new routes to destinations, and the number of RREP packets (the delay of RREP) replied to the sources, respectively.

Equipping AODV with local repair function results in the upstream node of a broken link being able to find the new routes to the destinations affected by itself. Thus, the control overhead and the corresponding delay can be reduced as $Pkt^f_{AODV_{lr}} = Pkt^f_{RERR} + Pkt^f_{RREQ} + Pkt^{f'}_{RREP} = \psi \cdot (\frac{1+2+,...,+H-1}{H} + N + \frac{H}{2}) \approx \psi \cdot (H+N)$, and $Del^f_{AODV_{lr}} = Del^f_{RREQ} + Del^{f'}_{RREP} = \frac{H}{2} + \frac{H}{2} = H$, respectively, where $Pkt^{f'}_{RREP}$ is the number of RREP packets replied to the upstream node, and Del^f_{RREQ} and $Del^{f'}_{RREP}$ are the delay to reach the destinations and the delay to reach the upstream node, respectively.

Now consider the control overhead of RBR for repairing a broken link. Unlike the AODV's approaches, which flag an error and re-initiate a route-discovery procedure at the source or at an intermediate node where the route is broken, P-PR_2 repairs a broken link with only local broadcasts to find a suitable redirector, and thus, its control overhead and the corresponding delay are future reduced as $Pkt^f_{RBR} = Pkt^f_{P-PR_2\_RRREQ} + Pkt^f_{P-PR_2\_RRREP} + Pkt^f_{P-PR_2\_RRACK} = 1 + 1 + 1 = 3$, and $Del^f_{RBR} = Del^f_{P-PR_2\_RRREQ} + Del^f_{P-PR_2\_RRREP} + Del^f_{P-PR_2\_RRACK} = 1+1+1 = 3$, respectively, where $Del^f_{P-PR_2\_RRREQ}$, $Del^f_{P-PR_2\_RRREP}$ and $Del^f_{P-PR_2\_RRACK}$ are regarded as the same and all given with 1 unit of time for analytical simplicity.

Overhead of shortening an active route.   Consider the Automatic Route Shortening (ARS) mechanism of DSR at first. According to the Internet draft [7], in DSR, nodes promiscuously listen to packets, and if a node receives a packet found in the Flow Table but the MAC-layer (next hop) destination address of the packet is not this node, the node determines whether the packet was sent by an upstream or downstream node by examining the Hop Count field in the DSR Flow State header. If the Hop Count field is less than the expected Hop Count at this node, the node will adds an entry for the packet to its Automatic Route Shortening Table, and returns a Gratuitous Route Reply to the source of the packet.

According to this, the communication overhead and packet delay of ARS could be $Pkt^s_{DSR} = Pkt^s_{G_{RouteReply}} = \frac{H}{2}$, and $Del^s_{DSR} = Del^s_{G_{RouteReply}} = \frac{H}{2}$, where $Pkt_{G_{RouteReply}}$ represents the number of Gratuitous Route Reply unicasting from the intermediate node to the source, and $Del_{G_{RouteReply}}$ is the corresponding delay. On the other hand, RBR requires only three control messages to shorten an active path, and costs $Pkt^s_{RBR} = Pkt^s_{A-PR_2\_RRREQ} + Pkt^s_{A-PR_2\_RRREP} = 1 + 1 = 2$, and $Del^s_{RBR} = Del^s_{A-PR_2\_RRREQ} + Del^s_{A-PR_2\_RRREP} = 1 + 1 = 2$.

In above, we assume that if RBR causes path lengths changed in the run time, the routing protocol under consideration, e.g., AODV, can record this fact and piggyback these changes to the downstream nodes. Therefore, no additional control overhead should be considered beyond that of RBR itself.

Simulation

In this subsection, we simulate the RBR-enhanced method and the original AODV to obtain their possible performance differences in mobile environment. For an unbiased comparison, scenarios similar to the previous work [1] were simulated with and without the use of RBR. However, instead of only 35 nodes, we experimented using scenarios with a set of 100 nodes in an area of 2000 square meters. In addition to this, we created the following environment. The transmission range was 250 meters. The mobility adopted was the random waypoint model. In this experiment, MOBILITY-WP-MIN-SPEED was 0 meters/sec, the maximum speed, MOBILITY-WP-MAX-SPEED, of 10 meters/sec was taken as our low speed, and that of 20 meters/sec as our high speed. The MAC layer adopted was IEEE 802.11 designed to note a link-break in advance.

With above, we examined the impact of dynamic channel quality change caused by mobility to the performance of RBR. For this aim, every Constant Bit Rate (CBR) flow is randomly selected from among 35 nodes, for each speed (low, medium, and high). Each CBR source sent one 512-byte packet every second to its destination for a duration of 1000 seconds of simulation time. In total 100 simulations for each speed and each number of flows were carried out, and the performance metrics measured, that is, the end-to-end delay, jitter, throughput, control overhead, and number of path-breaks, were averaged to show any possible performance differences that may exist between the original AODV and our RBR.

Figs. 5 and 6 show the number of path break and the packet latency (end-to-end delay) for both mobility scenarios. As shown in these figures, the number of broken paths of RBR is only 0.33% (0.38%) of that of AODV in the low (high) mobility scenario. The end-to-end delay is improved by RBR up to 30% in both mobility scenarios. In addition, as shown in both figures, the higher the mobility, the higher the number of broken paths and the higher the packet latency is.

Other information of the experiment is given in Figs. 7 and 8. Fig. 7 shows that the two methods provide similar throughputs. The maximum difference between these throughputs is no greater than 1%. Fig. 8 shows the control overhead, i.e., the number of control messages sent by each method. The control messages include routing packets (AODV's RREQ, RREP, and RERR), route redirection requests (P-PR_2_RRREQ and A-PR_2_RRREQ), route redirection replies (P-PR_2_RRREP and A-PR_2_RRREP), and route redirection acknowledgements (P-PR_2_RRACK). As shown in this figure, the number of control message of RBR is only half of that of AODV in both mobility scenarios. Observing Figs. 5 and 8, it is evident that RBR results in fewer path-breaks and greater savings in control overheads.

5. Conclusion

In this paper, we investigate adding proactive route selection and maintenance to on-demand routing algorithms to combine the benefits of both types of routing algorithms while minimizing their drawbacks [1]. Evaluation results show that adding the refinement-based mechanisms to AODV significantly reduces the number of broken paths as well as the end-to-end packet latency when compared with the pure on-demand routing protocol, AODV.

References

[1] T. Goff, N.B. Abu-Ghazaleh, D. S. Phatak and R. Kahvecioglu. "Preemptive routing in ad hoc networks," Proceedings of ACM MobiCom, 2001.

[2] Srianth Perur, Abhilash P., Sridhar Iyer. "Router handoff: A preemptive route repair strategy for AODV," IEEE Intl. Conference on Personal Wireless Computing, December 2002.

[3] Masato Saito, Hiroto Aida, Yoshito Tobe, Yosuke Tamura and Hideyuki Tokuda. "OR2: A path tuning algorithm for routing in ad hoc networks," Proceedings of IEEE Conference on Local Computer Networks, pages 560-567, Nov. 2001.

[4] A. Boukerche and Liqin Zhang. "A preemptive on-demand distance vector routing protocol for mobile and wireless ad hoc networks," The 36th Annual of Simulation Symposium, pages 73-80, April 2003.

[5] Charles Perkins and Elizabeth Royer. "Ad hoc on-demand distance vector routing," Proceedings of the 2nd IEEE Workshop on Mobile Computing Systems and Applications.

[6] Global Mobile Information Systems Simulation Library. http://pcl.cs.ucla.edu/projects/glomosim.

[7] David B. Johnson, David A. Maltz and Yih-Chun Hu. "The dynamic source routing protocol for mobile ad hoc networks (DSR)," IETF MANET Working Group INTERNET-DRAFT, draft-ieft-manet-dsr-09.txt, 15 April 2003.

[1] This work was supported by the National Science Council, Republic of China, under grant NSC92-2213-E-126-004.

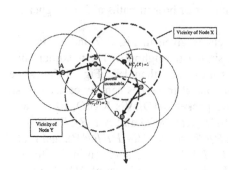

Fig. 1. Vicinity of node

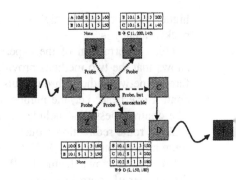

Fig. 2. Route redirection reply storm

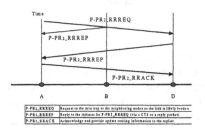

Fig. 3. P-*PR₂* control message flow

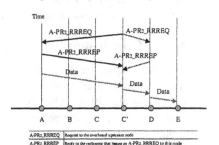

Fig. 4. A-*PR₂* control message flow

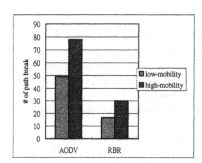

Fig. 5. The number of path break

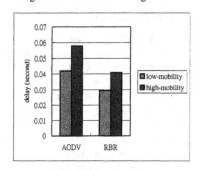

Fig. 6. The end to end delay

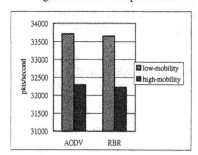

Fig. 7. The throughput

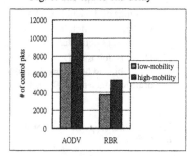

Fig. 8. The number of control message

ENABLING ENERGY DEMAND RESPONSE WITH VEHICULAR MESH NETWORKS

Howard CheHao Chang[1], Haining Du[2], Joey Anda[3]
Chen-Nee Chuah[1], Dipak Ghosal[3], and H. Michael Zhang[2]

[1] *Department of Electrical and Computer Engineering, UCDavis, Davis, CA 95616, USA*
{hcchang, chuah}@ece.ucdavis.edu

[2] *Department of Civil and Environmental Engineering, UCDavis, Davis, CA 95616, USA*
{hndu, hmzhang}@ucdavisledu

[3] *Department of Computer Science, UCDavis, Davis, CA 95616, USA*
{anda, ghosal}@cs.ucdavis.edu

Abstract   Inter-vehicle communication is becoming increasingly important in recent years. Traditional research efforts on vehicular networks have been put into safety or infotainment applications. In this article, we propose a Vehicular Mesh network (VMesh) to inter-connect disjoint sensor networks and act as a data transit network. Our VMesh networks are designed to provide a low cost, high fidelity, scalable, and fault resilient solution for data collections and disseminations. Preliminary simulation results based on a random mobility model show that the system has better performance when mobile routers are allowed to talk to peers. Mobile router density and transmission ranges dominate the simulation results. When the router density is high and/or transmission range is large, the data success rate will be high and latency will be small. There also exists a critical value of transmission ranges. There are no significant performance improvements once this critical value is reached.

1.   INTRODUCTION

US FCC has allocated a block of spectrum from 5.85 to 5.925 GHz band for inter-vehicle and roadside-to-vehicle communications that can support a wide range of applications. Such ad-hoc communication modes complement cellular communications by providing very high data transfer rates in circumstances where minimizing latency in the communication link and isolating relatively small communication zones are important. Previous work on vehicular communication has mainly focused on supporting two broad categories of applications: a) vehicular safety, such as exchanging safety relevant information or remote diagnostics using data from sensors built into vehicles and b) mobile internet access. However, there is a large untapped potential of using such vehicular networks as powerful and distributed computing platforms or as transit networks.

Vehicular networks have very different properties and design challenges compared to their counterparts such as laptops in nomad computing or sensor networks. First, vehicles travel at a much higher speed, making it challenging to sustain communications between stationary sites and moving vehicles, as well as handing-off the communication link from one site to the other as the vehicles pass between them. Second, despite the common assumption of random mobility patterns in many simulation studies on ad hoc networks, the vehicular traffic has a more well defined structure that depends on the transportation grid (highway, city roads, etc). Lastly, vehicles, as communication nodes, have abundant life and computing power as compared to sensor networks.

Our research focus is on designing vehicular mesh (VMesh) networks as reliable data transit networks. A mesh network is an ad-hoc network with no centralized authority or infrastructure. Nodes can move, be added or deleted, and the network will realign itself. The benefits of a mesh network are that it has the abilities of self-forming, self-healing, and self-balancing. As shown in Figure 1, one of the applications of using VMesh as a transit network is to establish connections between disjoint sensor networks.

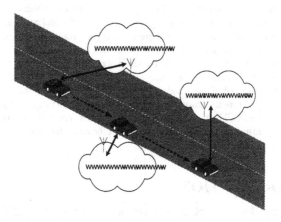

Figure 1.   Using VMesh to connect disjoint sensor networks

One of our expectations for VMesh is to enable demand response (DR) [1] for automatic utility usage retrievals and price dispatching. DR is a project initiated by California Energy Commission (CEC). The main spirit of the project is to allow utility companies to adjust utility prices based on real-time feedback from end users. With the advancement in consumer technologies, we assume that buildings are equipped with wireless sensors/transceivers that are capable of monitoring gas, water, and electricity usage, transmitting measured data, and receiving and processing price information. VMesh can be used to interconnect these sensors and the backbone wide-area network (WAN) infrastructure, e.g., Internet backbone, wireless cellular infrastructure, DSL or Cable. We will discuss some key properties that we expect VMesh to have to support DR in the next section. In Section III, we will describe VMesh architecture and some design considerations. We will summarize and introduce our proposed routing algorithm in Section IV. Lastly, we will show some preliminary results in Section V.

2. VMESH DESIGN RATIONALE FOR DEMAND RESPONSE

We envision that VMesh leverages a variety of vehicles, ranging from public transit (e.g., buses, light rail) to vendor trucks (e.g., FedEx, UPS) and police cars, to form a dynamic group of mobile routers. These vehicles, or MRs, are responsible for disseminating price information to and retrieving information from different households to support dynamic tariffs and demand-response. In this case, VMesh is designed to meet the following goals:

- **Low Deployment and Maintenance Cost:** Since the routers in this case are "mobile", one can drive these mobile routers into a station (e.g., main bus station or police headquarters) for repairs or software/hardware upgrades, instead of having to send a repair team out to various locations to fix the problems if they were stationary. Moreover, the number of mobile routers required can be minimized, e.g., using buses traveling on different routes is sufficient to cover the entire city. Hence, both the installation and maintenance costs are greatly reduced.

- **Adaptive Fidelity:** By using a combination of vehicles as mobile routers, VMesh provides a wide spectrum of flexibility in terms of the frequency of message retrieving and the granularity of demand response control loops. For example, while buses run every 0.5 - 1 hour, they do not stop at every single household. On the other hand, garbage trucks may stop at every household but they only come by once a week.

- **Scalability:** VMesh can support incremental deployment easily as the number of sensor nodes grows. In fact, VMesh benefits from the economy of scales, i.e., the cost of introducing a new mobile router goes down as the number of sensor nodes it is capable of serving increases.

- **Broadcast and Multicast Capabilities:** Broadcast and multicast capabilities are inherent in the wireless communications used between the mobile routers and sensor nodes, between the mobile routers and aggregation points to the VMesh network backbone, and between mobile routers and other vehicles equipped with wireless transceivers.

- **High Level of Redundancy:** VMesh has a high level of built-in redundancy by leveraging different vehicles that overlap in spatial coverage and temporal samplings. For example, there may be different routes for public transit through the city, but these routes often overlap in the main streets. In addition, garbage and postal trucks will visit the households on the same street at different times of the day. Therefore, the same end-user can be connected by two or three different types of vehicles. There are multiple available paths to ensure the delivery of the DR messages.

- **Failure Resiliency via Deflection Routing:** VMesh ensures the survivability of the DR messages by deploying the deflection routing technique. The key premise lies in the ability of VMesh to deflect messages until a valid path is found to the destination instead of dropping them when the original path fails due to faulty mobile routers, broken communication links, or vehicular accidents.

DR is one type of applications the VMesh network is capable of supporting. In addition to supporting demand response, VMesh also enables the deployment of other large-scale societal applications such as amber alert (e.g., broadcasting information of a kidnapper's vehicle and detecting this moving target), vehicular traffic control, and temperature/air-pollution monitoring. In the next section, we will first introduce a generic VMesh network architecture and then describe the method to support DR with this generic architecture.

3.   VMESH ARCHITECTURE

Traditional sensor networks are developed in a way that sensors are installed close enough to provide direct communications. If the deployment area is vast, one or more data collection base stations and clustering techniques are required to reduce the overhead of data transmissions. When the covering area is large, base stations may use other sensors, which have limited power, as gateways to relay messages or increase transmission power to increase transmission range if possible. However, increasing the power also increases the interference to other nodes in the network and using other sensors as gateways drains out the energy of some specific nodes more quickly. Both methods have their own drawbacks and may decrease the stability of the network. If the latency requirement of data in a network is not critical, we can alleviate this problem by using the VMesh network, which has one or more mobile routers as an essential component. As these mobile routers move by, they can collect and disperse information to static base stations. An example of using mobile routers to collect data is represented in the data mule paper [2] from Intel research lab. The difference between our work and [2] is that the data mule paper focused on sparse networks whereas we put emphasis on dense networks.

Vehicular mesh networks are ad hoc networks formed by vehicles enabled with wireless networking. As the vehicles move, the connectivity between the vehicles and other static network nodes changes. The network is dynamic and as a result, nodes may be disconnected from the network at times. To address this, nodes will store the data during the period they are disconnected from the network. An example vehicular mesh network is shown in Figure 2. The figure shows two bus routes (Route 136 and Route 108) and some key components of VMesh. The key components of the architecture are as follows:

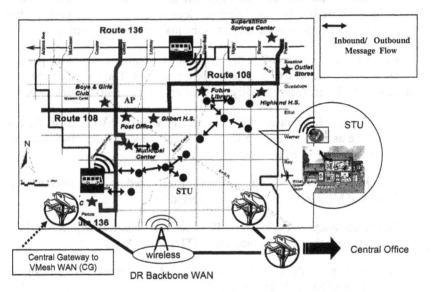

Figure 2. An illustration of proposed VMesh architecture

1 **Sensing and Transceiving Units (STUs)** are wireless enabled sensors that will transmit collected data to the central office (inbound message). They also receive new configurations such as new pricing information generated by the central office (outbound message). These are shown as solid circles in Figure 2. If STUs cannot directly connect to the VMesh backbone network, they can form a network themselves to route inbound and outbound messages.

2 **Aggregation Points (APs)** are nodes that act as gateways to the VMesh. They can aggregate inbound messages, relay the messages to the VMesh,

accept outbound messages, and route these outbound messages to one or more STUs. These gateways could be special nodes deployed at appropriate locations or specific STUs that are enabled with the gateway functionality.

3 **Mobile Routers (MRs)** are wireless enabled mobile objects that have the ability to store and forward data. For example, buses, along with other various types of vehicles equipped with storage and wireless networking, form ad hoc networks with other mobile routers and connect to the static gateways.

4 **Central Gateways to VMesh (CGs)** are gateways which connect different VMesh networks. These are located at specific locations on the paths of mobile routers, such as at the main terminal stop of buses.

The characteristics of VMesh depend on the mobility pattern of the participating mobile routers as well as neighboring vehicles. The choice of vehicles that are suitable for VMesh heavily depends on their attributes which include:

1 **Coverage**: How many STUs are directly accessed? We refer to the coverage as being fine-grain if the MRs can directly access individual or small groups of STUs directly.

2 **Schedule and Periodicity**: Is there a fixed schedule in the mobility pattern of MRs and if so, what is the period?

3 **Redundancy**: Are STUs or APs covered by multiple MRs? Are there multiple paths from the individual STUs to the APs?

4 **Cost**: What is the cost of deployment and maintenance in terms of the number of STUs, MRs, and APs?

For example, buses provide coverage to almost every major street in a dense major city such as San Francisco and their schedules coincide with peak electricity usage (e.g., buses run more frequently during work hours when energy consumption is high than at night). The mobility pattern of the vehicles depends on the type of vehicles, which include personal automobiles, public transport buses and light rails, postal vans, garbage trucks, various types of vendor trucks and vans such as UPS and FedEx, law enforcement vehicles such as police cars, and other monitoring vehicles such as those that monitor parking violations.

Various types of vehicular mesh networks can be characterized along these parameters depending on the targeted geographic area. An example of such characterization for a suburban area is shown in Table 1.

	Cars	Buses	Postal Vans	Police Cars
Coverage	Fine (1-10 m)	Coarse (10-1000 m)	Fine (1-10 m)	Coarse (10-1000 m)
Schedule	Regular	Regular	Regular	Random
Period	24 hrs	15 mins	12/24 hrs	None
Redundancy	Low	Medium	Low	High
Cost	High	Low	Low	Low

Table 1. A qualitative characterizations of various types of MRs for a VMesh in a suburban area

For the case of demand response, DR is enabled in the following manner. Periodically, the sensors transmit collected data that are routed to one or more aggregation points through a local ad-hoc network. When the mobile router travels by these aggregation points, it downloads the collected data and uploads new configurations which are distributed to the sensor nodes using local ad-hoc networks as well. The collected data can be opportunistically routed by the vehicular mesh network to the central gateways. In the worst case, the central gateway may be located at the terminal point of the route of the mobile router.

We will discuss about routing issues based on this architecture in the next section.

4. ROUTING IN VMESH

Many routing algorithms have been proposed for MANET. However, not all of them are fit well in our VMesh network. For example, Destination-Sequenced Distance-Vector Routing (DSDV) [3] needs network-wide update messages. If every household has at least one sensor, like the case in demand response, the density of sensors would be extremely high. Hence, network wide route update messages will drain out the energy of sensors quickly. Ad-hoc On-Demand Distance Vector Routing (AODV) [4] uses bi-directional links but one field study [5] has shown that this may not always be true in real life. The main disadvantage of Dynamic Source Routing (DSR) [6] is the long routing table. Because nodes in a DSR network need to store all the routing sequences toward intended destinations, the routing table will be prohibitedly long if the network is dense. For detail routing algorithm reviews, please refer to [7][8][9] [10]. In the following subsections, we will discuss routing issues from STUs to APs, APs to MRs, and MRs to MRs separately.

4.1   *Between STUs and APs*

STUs can easily obtain location informations because we can hard-code the informations into each static STUs. We use the *most forward progress within radius* (MFR) greedy algorithm as the general packet forwarding algorithm to minimize hop counts. MFR always tries to find a node within transmission range which can provide the smallest remaining distance toward destination.

In a case like DR, information will be sent out frequently by the APs. During the new information dispatching phase, new information are broadcasted over the whole network. In [11] and [12], low overhead flooding algorithms have been presented. Instead of establishing static multicast trees, we can use the algorithm in [12] to dispatch messages. A node in [12] sets up a random backoff timer based on the forwarding progress when a new broadcast message is received for the first time. If the node receives the same message from all directions (NE, NW, SE, SW) before the timer has expired, it cancels the timer. Otherwise, the node resends the message.

In our VMesh network, we may have hundreds of STUs trying to send data to an AP at the same time. This situation puts heavy loading on the MAC layer. Therefore, we will apply clustering techniques to decrease MAC layer loading and retransmissions. Conventionally, when a node wants to transmit information back to the aggregation points, it sends packets to the cluster head first. The cluster head will then forward the packets to the next cluster head through a one-hop broadcast or use other sensors as gateways and perform multihop relays. Broadcasting and clustering are two similar techniques but work in the opposite ways. Brodcasting best suits for one-to-many scenarios while clustering is many-to-one. The optimal solution for both broadcasting and clustering is to use the minimum number of nodes to cover the whole network space. Nodes in the VMesh network can record the nodes that they received messages from and then use them as the cluster heads for return paths.

In the traditional MFR, a node tries to select another node that can provide the most forwarding progress as the next hop. This characteristic will quickly drain out the energy of certain nodes that are located on critical paths. In our proposed routing method, next hops are the cluster heads that are chosen randomly according to the forwarding progresses during each data dispatching phase. Since the backoff timers are chosen randomly, the nodes that rebroadcast messages are different from time to time. This indicates that the next hops, which are actually cluster heads, are distributed in a set of nodes instead of being fixed. Because the cluster heads are chosen dynamically during each new message dispatching phase, this method is also more resilient to network disconnection caused by a single node failure. Suppose now a node who claims itself as a cluster head dies after rebroadcasting a message. The node running out of energy is excluded from the clusterhead candidate list automatically so new cluster heads are chosen in the next message dispatching phase.

4.2   *Between MRs and APs:*

In the VMesh architecture, aggregation points are static transceivers by the roadside so mobile routers can get very close to the aggregation point. Therefore, simple one hop broadcasting is sufficient for message transferring.

4.3   *Between MRs:*

The fundamental routing protocol between mobile routers is also MFR. One very important usage of the addressed VMesh network is to deliver time sensitive information such as dispatching new utility prices to end users. In this case, a utility company can provide more accurate pricing information if the company gets more feedback from users. Therefore, we plan to setup a deflection routing environment to increase the robustness of the network and to decrease the message delivery latency to allow more information being sent back to the central gateway. Deflection routing has been proposed to reroute packets during a node or link failure in the IP networks to reduce the transient link overloads and packet loss rates [13]. We propose to apply similar techniques to our VMesh network by deflecting messages toward the destination from one mobile router to nearby vehicles or backup mobile routers when the primary communication path fails, or when the primary mobile router is trapped in a traffic congestion that may cause intolerable end-to-end latency. This ensures that the messages are routed around trouble instead of being dropped.

5.   PRELIMINARY RESULTS

This paper focuses on the design of VMesh network architecture. The simulation results shown in this section are used to evaluate and study the worst case bound by using a random mobility model. We have implemented the algorithm discussed in the data mule paper [2] in C with some minor modifications under the Unix environment. In some of our simulation settings, mules can communicate with each other while this is prohibited in the original work. We use a 40 x 40 grid network. Three kinds of devices, aggregation point, sensor nodes and mobile routers are initially scattered arbitarily in the grids. The number of aggregation points is set to 1 and the numbers of sensor nodes and mobile routers are either 16 or 80 for different simulation scenarios. Given that the densities of sensor nodes and mobile routers are number of devices over total number of grids spaces, 16 and 80 stand for a density of 0.01 and 0.05 respectively. The grid network is setup to be a torus so that the mobile routers moving past the edge will appear at the opposite side of the network. Thus, the initial position of the devices will not influence the movement or simulation results. We use

random walk as our mobility model in this simulation. That is, a mobile router
will move 1 grid space into one of the adjacent spaces (with probability 0.25
for each space) for every simulation tick, while the positions for sensor nodes
and aggregation points are permanently fixed. The simulation duration is set
to 10000 ticks. A sensor node can generate one data packet per simulation tick
and transmit packets to mobile routers which can transport the packets to the
aggregation points. We assume all these devices have infinite buffer and that
there is no packet loss during transmission.

There are two different communication methods between mobile routers during simulations. In the first configuration (NOTALK), mobile routers can send
data to the aggregation point only, while disallowing MR to MR communications. In the second configuration (BROADCAST), mobile routers can broadcast
their data to all other mobile routers within the transmission range. The performance metrics we consider are data success rate (DSR) and average delay
latency. If NC is the number of non-duplicated data packets collected by the
aggregation point during time T and NG is the number of data packets generated by all the sensors during time T, the way to calculate DSR is $\frac{NC}{NG}$. When the
simulation ends, if some generated data packets are still left in the buffer of the
MRs or sensor nodes and not yet collected by the aggregation point, the DSR
will be less than 1. Average latency is the average time difference between the
time when a packet is generated and the time when it is collected. Since we
consider four different node densities, 2 communication methods, and 10 different transmission ranges(from 1 to 10), there are 80 different scenarios in
total.

From Figure 3, we can see that the DSRs increase linearly with the increase of the transmission range of mobile routers when the NOTALK method
is used. When the BROADCAST method is used, as shown in Figure 4, the
DSRs increase more rapidly and hit a saturation value when the transmission
range is 6. We can also notice that the density of mobile routers is the dominant factor that affects the performance of the system. The system performs
better when the density of mobile routers is high. In Figure 5, we compare latency values derived from BROADCAST and NOTALK when sensor density
is 0.05. Results show that latencies in NOTALK and BROADCAST decrease
with a similar slope except when the mobile router density and sensor density
are both 0.05. In this case, the latency drop much faster and also saturate when
the transmission range is 6.

Finally, we look at how the differences between BROADCAST and NO-
TALK vary with the changes of the transmission range. The BROADCAST
results in Figure 6 are normalized to NOTALK when the sensor and mobile
router densities are both 0.05. It shows that the results of DSR and latency for
BROADCAST are not far from 1, which is the mark of NOTALK, given small
transmission range. This indicates that BROADCAST does not provide obvious

advantage over NOTALK when transmission range is relatively small. As the transmission range increases, the difference between the two transmission methods will first increase then decrease. This is due to the fact that the transmission range for BROADCAST would reach a saturation value after which we can not get obvious benefit if we keep increasing the range.

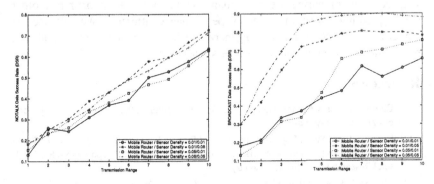

Figure 3.   NOTALK DSR vs Transmission Range

Figure 4.   BROADCAST DSR vs Transmission Range

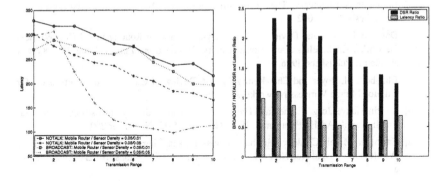

Figure 5.   BROADCAST Latency vs Transmission Range

Figure 6.   Normalized DSR and Latency when AP/Sensor Density = 0.05/0.05

6.   CONCLUSION AND FUTURE WORK

While there are many advantages of the network solution described above, for sake of brevity, we enumerate only the important ones and use the case of utility pricing dissemination and usage retrieval as an example. In this case, the network is scalable and the performance of the Vmesh network is also dependent on the design of bus routes. The vehicular mesh network can minimize the number of user-side and utility-side gateways thereby minimizing the deployment cost. Since the maintenance can be done at the bus depot, the cost will be low. The next step of this project is to implement and simulate our proposed routing algorithms (including deflection routing) in Qualnet [14]. Future work will focus on comparing the performance of the proposed algorithm to other existing routing algorithms to see the performance difference. We will also use a microscopic traffic simulator to generate realistic vehicle trace files and then re-generate the latency and data success rates.

References

[1] California Energy Commission (CEC). Demand Response. http://www.energy.ca.gov/demandresponse/.

[2] Rahul C. Shah, Sumit Roy, Sushant Jain, and Waylon Brunette. Data MULEs: Modeling a Three-tier Architecture for Sparse Sensor Networks. Technical Report IRS-TR-03-001, Intel Research Lab at Seattle, January 2003.

[3] Charles Perkins and Pravin Bhagwat. Highly Dynamic Destination-Sequenced Distance-Vector Routing (DSDV) for Mobile Computers. In *ACM SIGCOMM'94 Conference on Communications Architectures, Protocols and Applications*, pages 234–244, 1994.

[4] Charles E. Perkins and Elizabeth M. Royer. Ad-hoc On-Demand Distance Vector Routing. In *IEEE WMCSA'99*, pages 90–100, New Orleans, LA, February 1999.

[5] Benjamin A. Chambers. The Grid Roofnet: a Rooftop Ad Hoc Wireless Network. Master's thesis, MIT, 2002.

[6] David B Johnson and David A. Maltz. Dynamic Source Routing in Ad Hoc Wireless Networks. In Imielinski and Korth, editors, *Mobile Computing*, volume 353. Kluwer Academic Publishers, 1996.

[7] Elizabeth M. Royer and Chai-Keong Toh. A Review of Current Routing Protocols for Ad-Hoc Mobile Wireless Networks, April 1999.

[8] Martin Mauve, Jorg Widmer, and Hannes Hartenstein. A Survey on Position-Based Routing in Mobile Ad-Hoc Networks. IEEE Network 15 (6), November 2001.

[9] Carlos de Morais Cordeiro, Hrishikesh Gossain, and Dharma Agrawal. Multicast over Wireless Mobile Ad Hoc Networks: Present and Future Directions. IEEE Network, January 2003.

[10] Martin Mauve, Holger Fussler, Jorg Widmer, and Thomas Lang. Position-Based Multicast Routing for Mobile Ad-Hoc Networks. Technical Report TR-03-004, Department of Computer Science, University of Mannheim, 2003.

[11] Vamsi S. Paruchuri, Arjan Durresi, Durga S. Dash, and Raj Jain. Optimal Flooding Protocol for Routing in Ad-hoc Networks. Submitted to WCNC 2003.

[12] Jesus Arango, Mikael Degermark, Alon Efrat, and Stephen Pink. An Efficient Flooding Algorithm for Mobile Ad-hoc Networks. In *WiOpt 2004*.

[13] Sundar Iyer, Supratik Bhattacharyya, Nina Taft, and Christophe Diot. An Approach to Alleviate Link Overload as Observed on an IP Backbone. In *Proc. of IEEE Infocom*, April 2003.

[14] Scalable Network Technologies. QualNet. http://www.scalable-networks.com/.

CONTEXT-AWARE INTER-NETWORKING FOR WIRELESS NETWORKS

Franck Legendre, Marcelo Dias de Amorim, and Serge Fdida*
LIP6/CNRS – University of Paris VI
8, rue du Capitaine Scott – 75015 – Paris, France
{legendre,amorim,sf}@rp.lip6.fr

Abstract   Key issues of wireless ad hoc networks are auto-configuration and flexibility. Due to mobility, wireless networks are subject to frequent splits
and merges. In many situations, overlapping networks must combine to
form a new single network requiring a partial or total reconfiguration.
In this context, existing proposals assume that merging networks use
similar routing schemes. Nevertheless, there is no guarantee that different networks implement the same routing protocols. In such a scenario,
routing protocols must adapt to the changing nature of the network
characteristics. The wide spectrum under which ad hoc wireless networks are used lead to the design of multiple routing protocols (*e.g.*,
proactive, reactive and adaptive). Merging of wireless networks using
distinct routing protocol is thus unavoidable and leverages particular
difficult problems. We propose in this paper a flexible solution that
enables heterogeneous networks to efficiently cope with merging. We
designed a routing translator daemon based on the neighborhood context enabling neighbor nodes using distinct protocols to interoperate.
The neighborhood context is obtained from the Neighborhood Routing
Protocol Discovery Protocol. Several scenarios under different assumptions show that AODV, DSR and OLSR can efficiently interoperate.
We give insights on how routing must adapt and evolve in order to cope
with merging.

1.   Introduction

Wireless technologies show a convergence between standards of the
telecommunication industry and standards of the computer industry in
the name of IP. Nevertheless, no unique standard has been defined for the
routing algorithm in mobile ad hoc networks. Different classes of routing

*This work has been supported by CNRS and Euronetlab.

protocols are under standardization. Examples are the Ad hoc On-Demand Distance Vector (AODV [Perkins et al., 2003]), the Dynamic Source Routing (DSR [D.B. Johnson and Hu, 2003]), the Optimized Link State Routing (OLSR [Clausen and Jacquet, 2003]), and the Topology Dissemination Based on Reverse-Path Forwarding (TBRPF [R. Ogier and Lewis, 2004]).

Routing protocols can be classified in three categories: reactive, proactive, and adaptive. In reactive routing protocols (AODV, DSR), routes are discovered on demand, which reduces overhead by sending routing information only when needed. Proactive protocols (OLSR, TBRPF) exchange routing information in a background process. While on-demand routing protocols suit sparse communication patterns and small networks, proactive protocols have immediate route availability that is more adapted to dense communication patterns, larger networks, and applications requiring QoS guarantees. Hybrid or adaptive routing protocols like the Zone Routing Protocol (ZRP) [Haas, 1997] have been proposed combining advantages of both on-demand and proactive approaches. Even though hybrid protocols are promising solutions, they are not candidates for standardization. We can expect for a single adaptive protocol that will automatically and optimally fit all network situations.

Meanwhile, the restricted number of standardized routing protocols will enable equipments with sufficient capabilities to implement several protocols and use the appropriate one when joining a network. It will be possible for a group to switch from one routing protocol to another in order to efficiently adapt to new network conditions. But it is likely that equipment vendors will either choose a standardized protocol or develop proprietary protocols based on recent developments on adaptive routing. In both cases, these observations legitimate the urgent need for a general routing framework which allows interaction and interoperation between heterogeneous ad hoc routing protocols. This issue has already been tackled in [Lu and Bhargava, 2001], where a hierarchical architecture is setup between merging networks with group agents dedicated to inter-network communication.

We propose a decentralized approach where a new protocol, the Neighborhood Routing Protocol Discovery Protocol (NRPDP) enables nodes to discover the routing capabilities (i.e., supported routing protocols) of their neighbors. Based on these acquired information about the neighborhood, heterogeneous nodes are able to translate routing control packets of their respective routing protocols so that it can be understood by all their neighbors. This translation is realized on the fly by the Routing Translator Daemon (RTD) using a set of defined rules dependent on the context. Note that NRPDP and RTD are not intended to replace existing approaches, but to allow heterogeneous nodes to interoperate.

We will show the features of our approach through two scenarios, one where AODV interoperates with DSR and another where AODV interoperates with OLSR.

The paper is organized as follows. Section 2 introduces our network model. In Section 3, we present the advantages and requirements for a *smooth* merging. Section 4 formalizes our framework and introduces the design and mechanisms required. In Section 5, we illustrate our solution with two networks using distinct routing schemes, (*i.e.*, AODV, DSR, and OLSR). Finally, Section 6 discusses future research investigation and concludes the paper.

2. Network model: the cell approach

In this paper, we define the concept of network cells. A cell, C, is the spatial region spanned by a set of nodes, N_C, willing to participate in some collaborate activity (*e.g.*, emergency disaster relief, battlefield troops, conference attendees) or with similar interest in spatial movement (*e.g.*, public transport users). The evolution of a cell is progressive, *i.e.* new arriving nodes acquire the address allocation scheme and routing scheme from an existing cell member. The address assignment may be of any kind (conflict-free, conflict-detection, or best-effort). For example, nodes can acquire an address from their neighbors or randomly generate their addresses and verify their uniqueness using a duplicate address detection (DAD) [Nesargi and Prakash, 2002] mechanism. For routing, we assume that due to the diversity of situations, nodes implement a set of existing protocols. However, since equipments have disparate capabilities (*i.e.*, processing, memory), they do not always implement the same set of routing protocols but it is likely that they will implement the most adopted ones. We believe that a restricted list of protocols will be supported by vendor equipments, some respecting standards and others proprietary implementations. When cells C_1 and C_2 are about to merge, the overlapping region $S_{C_1,C_2} = C_1 \cap C_2$ is called a *scar-zone* and is delimited by a *scar-zone membrane* (cf. Fig. 1). Nodes located in the scar-zone are called *scar-zone* nodes, S_C, while nodes outside the scar-zone, I_C, are named *interior* nodes. Depending on the respective mobility of the two cells, the scar-zone can evolve between a minimal overlapping where the two cells are interconnected via one radio link to a complete overlapping where the spatial extent of one cell is included in the other.

When C_1 and C_2 merge, two cases may occur:

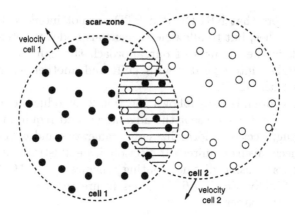

Figure 1.   Network cell model.

- **Merging cells are homogeneous**: in this case, both cells use the same routing protocol. Existing approaches address only this case. We believe this is an ideal case that does not hold all the time.

- **Merging cells are heterogeneous**: heterogeneity has been widely studied in ad hoc environments, but restricted to PHY and MAC layers. For example, several frameworks have been proposed to enable the convergence of different radio access technologies known as Software Definable Radio (SDR) [JSAC, 1999]. With IP as the universal standard, heterogeneity at upper layer levels has been superficially considered. With the various routing protocols available and under standardization, it is clear that protocol heterogeneity must be tackled in a similar manner to what is done in wired networks.

3.   Heterogeneous merging: a smooth approach

Heterogeneous merging can be addressed in two different ways. The first approach – inspired from proposals related to addressing schemes that cope with merging – would be to completely reconfigure both or only one of the cells. The second approach consists in enabling interaction between heterogeneous protocols, which does not require a global reconfiguration. In the following, we discuss the advantages and drawbacks of the first approach and from conclusions we propose our scalable framework.

3.1   The case of heterogeneous cell interoperability

In the first approach, nodes of the two cells agree on a common routing protocol. This approach is similar to what is done in addressing schemes where one of the two cells completely reconfigures its addresses in order to have a coherent and homogeneous addressing space. Even if this approach is the simplest one, we demonstrate that under some network conditions it is not achievable and raises scalability problems.

First, the reconfiguration of one of the networks requires all nodes to implement and swap to the required protocol. Unless an active approach is available, this case is hardly achievable.

Second, considering that all nodes can converge to the same routing protocol, all on-going routing control traffic, communications, and states held by each node (*i.e.*, routing tables, forwarding tables, etc.) would be lost. For example, changing from OLSR to a reactive protocol would invalidate proactive routing tables.

Third, depending on the network conditions and more specifically on the cell mobility, swapping to another routing protocol may be sub-optimal. For example, if several successive merges occur in a short period of time it would require multiple reconfigurations. The system would become instable and no more routing would be possible. Another example is if merging is only transient. In this case again, reconfiguration can be sub-optimal in some cases.

As described in the introduction, the different categories of routing protocols are adapted to particular network conditions and must be chosen accordingly. If reconfiguration is required, a lap of time is needed to discover the new network situation when merging occurs. During this short period of time it might not be optimal to swap to an inappropriate routing protocol.

For all these reasons, we plead for a smooth merging protocol. This protocol can be complementary to a possible posterior reconfiguration of two merging cells.

3.2   Addressing heterogeneous cell interoperability

We propose a completely different approach where nodes do not reconfigure themselves when merging. Instead, we benefit from the fact that routing protocols can be of the same family, which may differ in implementation but not in their general philosophy. We also take advantage of existing routing states already stored by nodes, as in the case of link-state (LS) routing protocols. In order to do this we must, depending on the context and the routing protocols involved, either translate routing control packets from one protocol to another or allow the interaction of nodes relevant to routing. While the former proposal is designed for protocols of the same family, the latter is suited for protocols that are completely different in their conception and do not enable a direct translation.

With this approach, we want to conceive a new protocol with the following design goals:

- Routing control packets (link-state updates, route requests, route replies) must not be modified to suit our needs.

- Inter-cell communication should avoid dedicated and centralized translator cluster-head or group leader.

- Routing loops and unnecessary recursive translations must be avoided.

4.   Design and mechanisms

In order to achieve inter-cell operability, scar-zone nodes – nodes located in the overlapping region – must define their neighborhood environment context (*i.e.*, routing capabilities of neighbors). Based on this context, specific translation must be performed or, if translation is not possible, nodes must execute appropriate neighborhood interaction.

Proactive and reactive protocols can be implemented as routing daemons (executed in user space), like wired routing protocols such as RIP and OSPF. We define then a general ad hoc routing layer that enables the flexible use of several routing protocols. This daemon intercepts the I/O of routing control packets and process these packets (requests, replies, updates) based on the status of the node:

- The node operates normally if it is in the interior zone.

- The node executes a context-dependent algorithm (neighborhood's context) if it is in the scar-zone.

We propose to associate this new daemon with the Neighborhood Routing Protocol Discovery Protocol (NRPDP). This protocol enables neighbor nodes to discover their respective routing capabilities (*i.e.*, the implemented routing protocols).

4.1   The NRPDP Protocol

Nodes of the scar-zone must be aware that they are possible candidates for network routing protocol translation or interaction. In order to achieve this, nodes must detect that a merging is about to occur and then determine if they are scar-zone nodes or not. Previous approaches proposed the use of Network IDentifiers (NID) to detect merging between distinct cells. A NID is a random number generally broadcasted over the entire cell or piggybacked during the address allocation process. NRPDP sends periodic `Hello` messages containing a cell's NID. This procedure enables the detection of merging and defines a scar-zone if merging cells use the same or a different routing scheme.

Nodes located in the scar-zone exchange their routing capabilities (*i.e.*, supported routing protocols) and the current protocol in use – the *mother* routing protocol. For example, in Figure 3 node X sends $\{*\text{AODV}*, \text{DSR}\}$ to Y, pointing out the routing protocols it supports and the current routing protocol in use in its cell, indicated by $\{*\text{Protocol}*\}$. Similarly, node Y sends its NRPDP `Hello` packet to X, $\{*\text{DSR}*, \text{AODV}\}$.

The NRPDP daemon interacts with the routing translator daemon by communicating the neighborhood context (cf. Fig. 2).

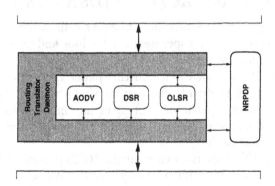

Figure 2.   NRPDP and routing translator.

4.2   The Routing Translator Daemon

Once a node is aware of its neighbors' capabilities, thanks to NRPDP, the routing translator daemon can take appropriate actions relative to routing. Depending on the context, a translator is either off or on. If the node is surrounded by a homogeneous neighborhood, the translator is off, otherwise the translator is on. In the latter case, depending on the context, the translator may take one of the following three actions: (a) it forwards the packet as it is to homogeneous neighbors, (b) it translates the packet for heterogeneous neighbors using a different routing protocol of the same category, otherwise (c) it interacts with an heterogeneous neighbor node to retrieve required information for the routing process. In the following section, we will explicitly detail these different contexts through examples.

Table 1. The Routing Translator Daemon table.

RREQ ID	Dest. IP Address	Dest. Seq. Num.	Orig. IP Addr.	Orig. Seq. Num.	Hop count	DSR request id.
m	B	0	A	seq_A	1	$\text{hash}(seq_A)$
⋮	⋮	⋮	⋮	⋮	⋮	⋮

5.   Application: AODV ↔ (DSR,OLSR)

In this section, we illustrate how merging cells using heterogeneous routing protocols can interoperate in a flexible and efficient manner. The vocabulary and acronyms used in the following refer specifically to each protocol. We do not discuss in this paper the route maintenance inherent to reactive protocols and assume that the mobility relative to the establishment of a path does not incur any link breakage. We only consider how paths are established and do not consider how packets are forwarded on these paths.

OLSR and TBRPF are two experimental RFCs proposed for standardization by the IETF MANET Working Group. We have chosen OLSR for our case study. OLSR is a LS routing protocol where nodes periodically exchange topology information with other nodes in the network. After a time needed for convergence, each node has an optimized view of the current topology. OLSR differs from pure LS in that it optimizes the flooding of updates and reduces the size of LS packets.

AODV (experimental RFC) and DSR (Internet-Draft) are the two most referenced reactive protocols. In reactive protocols, paths from a source to a destination are discovered on-demand with query packets for a destination flooded across the entire network. In AODV, intermediate nodes learn the path to the source and enter the route in their forwarding table. When the destination is reached, it uses the path traced by query packets as reverse route toward the source. DSR distinguishes from AODV by the use of source-routing, *i.e.* intermediate nodes concatenate their IDs in the route query packet header. The destination then retrieves the path and uses source-routing to respond to the source. Compared to AODV, intermediate nodes in DSR do no keep any state related to a given path.

5.1   AODV ↔ DSR

Consider the scenario shown in Fig. 3, where cell 1, C_1, runs AODV and cell 2, C_2, runs DSR. Consider also two nodes, A and B, with $A \in I_{C_1}$, $B \in I_{C_2}$ (*i.e.*, $\{A, B\} \not\subset$ scar-zone). Here, we study how paths can be established in these cells in both ways, from A to B and vice-versa. Note that, in this case, protocols belong to the same family which clearly simplifies the interoperability – only a simple translation is required.

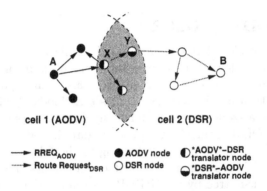

Figure 3.   Translation AODV ↔ DSR.

For the establishment of a path from A to B, $A{\to}B$, A floods a RREQ (Route Request). When a node in the scar-zone receives the request, here X, it translates this AODV RREQ into a DSR Route Request. X associates the AODV's RREQ with an entry in the table maintained by its RTD (Tab. 1). AODV's RREQ sequence number, seq_A, is associated with the newly emitted DSR Route Request sequence number result of hash(seq_A), where hash($\cdot$) is a hash function shared by all nodes. This entry indicates that a translation will be required at the reception of the DSR Route Reply.

When the DSR Route Reply is received, an AODV RREP is sent using the reverse route entry established by the initial RREQ. In order to avoid loops, the translation requires the use of the same sequence number and the precise translation of the number of hops. We cannot use the same sequence number since the sequence number fields of these protocols have different sizes. This is why we use a hash function common to all nodes. For the hop count, in the AODV header the number of hops is represented by a hop count field, while in DSR counting the number of concatenated IDs gives the hop number. This translation is done by referring to the corresponding entry in the RTD table. As well, the correct association between routing control packets prevents recursive translations. For example, X receives an AODV RREQ from Y in reply to its original DSR Route Request sent to Y. By comparing the received sequence number with the one contained in its RTD table, X detects this route request is generated in response to its original request.

Similarly, for the path $B{\rightarrow}A$, B floods a DSR Route Request that is translated into an AODV RREQ by scar-zone nodes. These scar-zone nodes will receive RREPs that will be translated back to DSR Route Replies.

5.2   AODV ↔ OLSR

Here, we consider that C_1 runs AODV while C_2 runs OLSR. This case is more complex to treat since these protocols belong to different families. More elaborated node interactions are then required.

For the path from A to B, $A{\rightarrow}B$, node A floods a RREQ in its cell. Nodes of the scar-zone cannot perform any direct mapping of an AODV RREQ into an OLSR control packet. In Figure 4, node X forwards the RREQ to its neighbors. Node Y belonging to the OLSR cell will receive this message. Y's routing translator daemon catches this packet, from the context transmitted by its NRPDP, determines that it is in the scar-zone, and looks up its OLSR routing table for the destination. With the retrieved information, the translator daemon replies with a RREP including the number of hops obtained from the OLSR routing table. Node A will receive several RREPs and choose the shortest path to start the communication. From now on, A has a path to join B but B does not have a reverse path to A. There are several solutions to this problem. As soon as the RREQ is received, X sends a TC − message (Topology Declaration, *i.e.* advertisement of LS) announcing that A can be reached through X.

Now, if we consider the case where B wants to launch a communication with node A, the process is more complex since A's address will not be in the routing table of the OLSR nodes. If we consider that the AODV cell uses the same address prefix, nodes of the scar-node can send a Host/Network Association (HNA) message indicating they are gateways for the announced prefix. Since this assumption may not hold in ad hoc networks where addresses can be randomly chosen, another solution is required. We propose that the translator routing daemon of OLSR nodes in S_{C_1,C_2} uses LS update packets to inform I_{C_2} nodes that

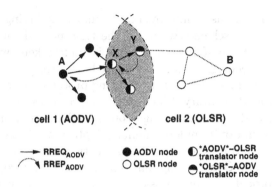

Figure 4. Translation AODV ↔ OLSR.

another network is reachable through them. Upon the reception of this information, the routing daemon of OLSR nodes adds a default route in their OLSR routing table. This entry will be used when forwarding a packet to a node in the other cell. When an OLSR scar-zone node receives a packet for node A, it passes the packet to an AODV node that initiates a RREQ and waits to receive the RREP before forwarding the packet.

Note that there is no need in these examples for OLSR nodes to implement the AODV protocol and inversely. They only require to *understand* routing control packets of other protocols and take appropriate action defined by a set of rules depending on the context.

6. Conclusion

The case of merging networks is an area with many open questions. Little attention has been given by the ad hoc routing community on issues related to heterogeneity involving layers above PHY and MAC. Our proposal is one of the first attempts to consider routing protocol heterogeneity. This paper proposes a novel framework for merging heterogeneous cells. Contrary to previous proposals, we consider the case where cells use distinct routing schemes. One of our main goals is to enable a smooth and flexible cell interoperability whatever routing protocols are involved. We proposed a new routing approach, *i.e.* a context-aware routing based on neighborhood capabilities. We designed NRPDP, a new protocol for discovering neighbor routing capabilities, and an associated routing daemon (the Routing Translator Daemon).

The advantage of our solution is to enable flexible interoperation between merging cells. We showed that three of the main candidates for standardization can easily interoperate without modifying their specifications. With our scheme, we avoid the time needed for a complete reconfiguration. Even if reconfiguration is undertaken, we avoid the stoppage before reconfiguration is achieved. As well, envisioning the benefits of our approach to adaptive routing protocols, our solution can be viewed as complementary. Since adaptive protocols borrow concepts from proactive and reactive routing protocols, the framework presented in this paper would only undergo minor adaptations. In fact, only new translation rules and protocol-specific interactions must be added to our translator daemon. Generalizing our approach to an interoperability of all routing protocols, we are considering the development of an inter-network routing protocol that will be used as a common routing protocol between wired and wireless heterogeneous networks. This work is also motivated by the fact that, similarly to wired networks, bandwidth (*i.e.*, radio) is a scarce resource as well as energy that cannot be managed without introducing some policy and negotiation between merging networks.

As a conclusion, if the current trend followed by the IETF's MANET Working Group is verified, we believe several ad hoc routing protocols will be standardized. These protocols would benefit from our approach since they would become *compliant* instead of concurrent.

References

[Clausen and Jacquet, 2003] Clausen, T. and Jacquet, P. (2003). Optimized Link State Routing Protocol OLSR. RFC. RFC-3626.

[D.B. Johnson and Hu, 2003] D.B. Johnson, D.A. Maltz and Hu, Y-C. (2003). The Dynamic Source Routing protocol for Mobile ad hoc networks (DSR). Internet-Draft. draft-ietf-manet-dsr-09.txt - Work-in-progress.

[Haas, 1997] Haas, Z. (1997). A new routing protocol for the reconfigurable wireless networks. In *Proc. of the IEEE International Conference on Universal Personal Communications*.

[JSAC, 1999] JSAC, IEEE (1999). Special issue on software radios. In *Mobile Computing*, volume 17-4. IEEE.

[Lu and Bhargava, 2001] Lu, Y. and Bhargava, B. (2001). Achieving flexibility and scalability: A new architecture for wireless network. In *Proc. of International Conference on Internet Computing (IC'2001)*, pages 1105–1111, Las Vegas.

[Nesargi and Prakash, 2002] Nesargi, S. and Prakash, R. (2002). Manetconf: Configuration of hosts in a mobile ad hoc network. In *Proc. of Infocom'02*, New York.

[Perkins et al., 2003] Perkins, C. E., Belding-Royer, E. M., and Das, S. (2003). *Ad hoc On-Demand Distance Vector AODV Routing*. IETF RFC 3661.

[R. Ogier and Lewis, 2004] R. Ogier, F. Templin and Lewis, M. (2004). Topology Dissemination Based on Reverse-Path Forwarding (TBRPF). RFC. RFC-3684.

PERFORMANCE IMPACT OF MOBILITY IN AN EMULATED IP-BASED MULTIHOP RADIO ACCESS NETWORK

Philipp Hofmann, Christian Bettstetter, Jeremie Wehren*,
and Christian Prehofer
DoCoMo Euro-Labs, Future Networking Lab, Munich, Germany
lastname@docomolab-euro.com

Abstract   This paper investigates the performance of a multihop radio access network. In our testbed, nodes communicate to one access point using IEEE 802.11b and AODV routing. We measure the average packet delay and delivery ratio, if the node movement is emulated employing the random waypoint and random direction model, respectively. We find that random waypoint mobility yields up to 100 % better results. This shows that the testbed performance is highly sensitive to the mobility model, even if comparable mobility behavior is assumed.

Keywords: Multihop radio access, routing performance, mobility modeling, testbed, ad hoc routing, AODV, random waypoint model, random direction model, network emulation.

1.   Introduction

The paradigm of a *multihop radio access network* (MRAN) is that mobile nodes serve as "wireless routers" to extend the coverage of fixed access points (see Fig. 1). This new network architecture is especially interesting for mobile communication systems in which the cell size is very small, e.g., due to operation at high frequencies. Although the design of such networks is still in the research phase, basic functionalities can already be shown today through the interworking of IP-based ad hoc networking protocols and fixed IP networks [1–6].

This paper studies the packet-level performance of a single-cell MRAN, based on measurements in a WLAN testbed with one access

---

*Jeremie Wehren studies at the Institut Eurécom, Sophia-Antipolis, France, and Ecole Polytechnique Fédérale de Lausanne (EPFL), Switzerland.

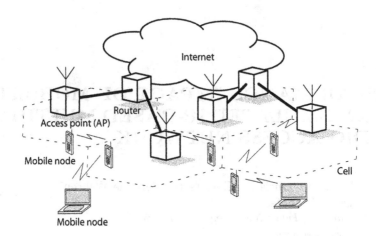

Figure 1.   A multihop radio access network (MRAN) connected to a fixed network

point and seven mobile nodes. Each of the nodes runs the *ad hoc on-demand distance vector (AODV)* routing protocol [7] over IEEE 802.11b in ad hoc mode. The mobile network is emulated by creating a virtual dynamic topology: the mobile nodes move virtually according to a given mobility model, and packets between nodes that are currently not within transmission range are filtered out.

We are especially interested in the impact of the mobility model on the network performance. To investigate this issue, we compare two well-known models: the *random waypoint* (RWP) model [8] and the *random direction* (RD) model [9, 10]. Although the two models are intuitively very similar, we show that the resulting performance is very different. Whereas the qualitative difference can be explained with recent theoretical results on both models [10–16], the quantitative difference is significant: using the RD model, the average packet delay is about twice as high as that of the RWP model; the packet delivery fraction of the RD model is almost 30 % lower.

The remainder of this paper is organized as follows: Section 2 describes our testbed and emulation environment for MRANs. Section 3 explains the used mobility models and their inherent properties. Next, Section 4 presents and interprets the performance results. Section 5 addresses related work, before Section 6 concludes.

2.   Description of the Testbed

As illustrated in Figure 2, our MRAN testbed consists of seven laptops and one desktop computer. The laptops represent the mobile nodes; the

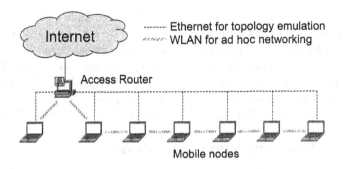

Figure 2.   Testbed topology

desktop computer combines the functionality of an access point (AP) and access router (AR). All computers are equipped with IEEE 802.11b wireless interfaces. These interfaces run in ad hoc mode and are used for the transmission of payload and routing protocol packets. As routing protocol, we employ the AODV IPv4 implementation of [17]. The laptops are positioned in a way that each of them can establish a direct link to any other laptop, i.e., we have a fully-meshed network topology.

The movement of the laptops and the resulting connectivity graph is emulated: Each node moves virtually according to a mobility trace that is stored in a simple text file (ns-2 mobility file). Packets transmitted between laptops that are currently not within transmission range of each other (according to their virtual position) are discarded. This virtual dynamic topology is controlled by the desktop. Via an Ethernet connection, it informs the laptops about their current (virtual) neighbors and sends commands to set the packet filters accordingly. The actual filtering is achieved on the MAC layer using the tool *iptables*. The setup is implemented using the *NRL mobile network emulator* (MNE) [18] running on the desktop.

As the network topology is emulated while all nodes are physically linked together, the nodes share a single channel. On the one hand, this reduces the available bandwidth per node; on the other hand, hidden or exposed node problems do not occur. As we focus on the performance of the network layer, we ignore these link layer issues.

3.   Mobility Models

There is an ongoing discussion in the research community as how to model the mobility of nodes in performance studies. Purely stochastic mobility models can be implemented quickly, and they yield "averaged" performance values, which can be easily reproduced by other researchers. Scenario-based studies — which include obstacles [19, 20] and/or use real-world mobility traces [21] — are more realistic but produce results that are only valid in this particular scenario. In this paper, we focus on purely stochastic mobility models, namely on the RWP and RD model. As mentioned above, we use pre-recorded mobility files to determine the movement of the nodes in the network emulation. These files have been generated from stochastic simulations of the two used models.

3.1   Random Waypoint Model

The most popular mobility model in current research on ad hoc networks is the RWP model. It describes the movement of a node in a bounded area as follows (see Fig. 3(a)): A node randomly chooses a destination point ("waypoint") $\mathbf{P}$ and moves with speed V on a straight line to this point. After waiting a pause time T_p, it chooses a new destination and speed, moves to this destination, and so on. This movement can be described as a sequence of random variables

$$\{(\mathbf{P}_i, V_i, T_{p,i})\}_{i \in \mathbb{N}} = \{(\mathbf{P}_1, V_1, T_{p,1}), (\mathbf{P}_2, V_2, T_{p,2}), \ldots\}, \qquad (1)$$

where an additional waypoint $\mathbf{P}_0$ is needed for initialization. Each destination point $\mathbf{P}_i$ is randomly chosen from a uniform distribution over the area. The movement from one waypoint to the next waypoint is called a movement transition. The speed as well as the pause times are chosen from an arbitrary probability distribution with well defined expected values $E\{V\}$ and $E\{T_p\}$, respectively.

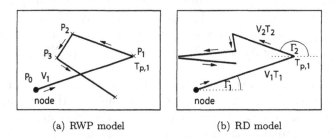

(a) RWP model   (b) RD model

Figure 3.   Illustration of mobility models and random variables

Recently, it has been shown that this model has the following unintended properties:

1 *Non-uniform node distribution.* The probability density of a moving mobile node's location is non-uniform with a maximum oc-

currence probability in the middle of the system area and zero occurrence probability at the very border [10, 11, 13–16].

2 *Speed decay.* The ensemble average of the nodes' speed decreases over time until a steady-state is achieved. If the speed V is chosen between $[0, v_{max}]$ in each waypoint, the network may converge toward a non-mobile network [22].

These properties will also influence the MRAN performance results given later in this paper.

3.2 Random Direction Model

In the RD model, a node chooses a destination direction Γ (an angle) and a transition period T, rather than a destination point (see Fig. 3(b)).[1] After moving in a straight line in the given direction for the given time, it pauses a period T_p, and then chooses a new direction, period, and speed. The complete movement process of a node can be described as

$$\{(\Gamma_i, V_i, T_i, T_{p,i})\}_{i \in \mathbb{N}} = \{(\Gamma_1, V_1, T_1, T_{p,1}), (\Gamma_2, V_2, T_2, T_{p,2}), \ldots\}. \quad (2)$$

The directions are chosen from a uniform random distribution on the interval $[0, 2\pi[$. In this paper, the transition times are chosen from an exponential probability distribution with an expected value $E\{T\}$.

Whenever a node reaches the border of the area, it "bounces back" following the rule that the incoming angle is equal to the outgoing angle (see Fig. 3(b)). Using this border rule, a "uniform" spatial distribution of the nodes can be achieved [10].

The RD model is very popular in the research on cellular networks (see, e.g., [9, 23]), but it has gained surprisingly little attention in the research community on ad hoc networking. The RWP model, on the other hand, is almost solely used in the research community on ad hoc networks. Many papers in this domain use the RWP model.

4. Performance Evaluation

4.1 Setup

The mobile nodes move independently on a square system area of length a. Each of them sends UDP packets with a constant bit rate κ to the access router using multihop forwarding. All nodes have the same transmission range r_0 and equal stochastic movement parameters. The access router is located in the middle of this area.

Our performance parameters are as follows:

- *Average packet delay.* The packet delay is the time that a packet needs to be transmitted from the mobile node to the access router. The average packet delay is the mean of the delays of all packets arriving at the access router. Only successfully transmitted packets are considered.

- *Packet delivery fraction (PDF).* The packet delivery fraction is the percentage of packets that are successfully transmitted from the mobile nodes to the AR.

To avoid transient effects, the emulation runs for a certain warm-up period. After this time, we start measuring the performance for a duration of 0.5 h. The same experiment is repeated Ω times, each time with a different mobility trace. The parameters are shown in Table 1.

Table 1. Parameters for Performance Evaluation

Basics	
Number of mobile nodes	$n = 7$
Constant bit rate of each node	$\kappa = 30\,\text{kbit/s}$
Square area with side length	$a = 600\,\text{m}$
Virtual radio transmission range	$r_0 = 200\,\text{m}$
Measurements	
Warm-up period (no measurement)	0.5 h
Measurement period	0.5 h
Number of independent experiments	$\Omega = 10$
Movement	RWP or RD
Speed of nodes	constant v
Pause time	$T_p = 0$
RWP: uniformly distributed waypoints	
RD: Γ uniformly distributed	
$\quad$ T exponentially distr. with mean according to (7)	

4.2   Movement Parameters

Let us now describe the parameters of the mobility models used for our performance analysis. The "level of mobility" of the RWP and RD model can be determined in terms of the nodes' speed, pause time, and direction change frequency $\lambda = \frac{\#\ \text{direction changes}}{\text{measurement time}}$, i.e., its expected number of direction changes per unit time. For reasonable comparison, we would like to set all parameters to be the same in both models. We set the speed of all nodes to a constant value over the entire emulation, i.e., $V_i = v \forall i$ for all nodes in both models. The pause time is set to zero in both models. The only parameter that needs to be determined is the direction change frequency, so that $\lambda_{\text{RWP}} = \lambda_{\text{RD}}$ holds.

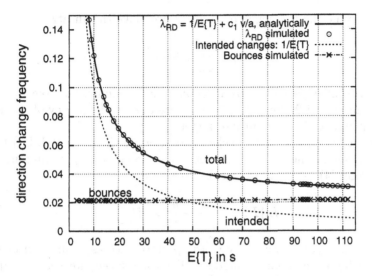

Figure 4.   Direction change frequency of a mobile RD node with speed $v = 10\,\text{m/s}$ on a square area of length $a = 600\,\text{m}$

Using the RWP model, the direction change frequency is

$$\lambda_{\text{RWP}} = \frac{1}{E\{T\} + E\{T_p\}} = \frac{1}{E\{T\}}, \tag{3}$$

where T is the duration of one transition, i.e., the time of a movement between two waypoints. Its expected value can be written as $E\{T\} = \frac{1}{v} E\{L\}$, with L being the Euclidean distance between two waypoints. Since the waypoints are uniformly distributed in a square area, we can borrow well-known results from geometric probability theory [24] and write $E\{L\} = 0.5214\,a$. In summary, this yields

$$\lambda_{\text{RWP}} = c_0 \frac{v}{a} \quad \text{with } c_0 = 1.918. \tag{4}$$

Using the RD model, two types of direction changes occur: (i) intended direction changes after each transition period T and (ii) "forced" direction changes upon each bounce back from the border of the system area. The frequency of intended direction changes is implicitly given by $1/E\{T\}$. The frequency of bounces is denoted by $1/E\{T_{bounce}\}$. The total direction change frequency of the RD model is then

$$\lambda_{\text{RD}} = \frac{1}{E\{T\}} + \frac{1}{E\{T_{bounce}\}}. \tag{5}$$

To determine the frequency of bounces, we make the ansatz

$$\frac{1}{E\{T_{bounce}\}} \approx c_1 \frac{v}{a}. \tag{6}$$

The constant c_1 can be derived empirically by simulation (see Fig. 4). We obtain $c_1 \approx 1.29$ for $E\{T\} \geq 3\,\text{s}$. Note that the frequency of bounces is independent of the input parameter $E\{T\}$.

We now have sufficient knowledge to compute the value of $E\{T\}$ that must be set in the RD model, such that $\lambda_{\text{RWP}} = \lambda_{\text{RD}}$ is guaranteed. Combining (4), (5), and (6) yields

$$E\{T\} \approx \frac{1}{c_0 - c_1} \frac{a}{v} \approx 1.59 \frac{a}{v}. \tag{7}$$

In our experiments, we measure the performance as a function of the speed v for both mobility models. For each speed, we set the input parameter $E\{T\}$ of the RD model according to (7).

4.3   Results and Interpretation

Figure 5(a) shows the average packet delay over v. The vertical bars show the value range of the Ω independent experiments, i.e., the interval between the minimum and maximum average delay. We observe that the delay for the RD model is always higher than that for the RWP model. This is a consequence of the statistical spatial node distribution of the mobility models, as explained in Section 3: Since the AR is located in the middle, the number of hops between a mobile RWP node and the AR is, on average, lower than that of an RD node. This shorter hop distance results in shorter delays in the RWP model. While we did expect this qualitative behavior, the quantitative difference is indeed surprising: Independent of v, the RD model yields a 100 % higher average delay. This is a specific example showing that an understanding of mobility modeling is very important to interpret simulation results.

Let us analyze in more detail the causes for lower packet delays in the RWP model. The main part of the delay is caused by the route discovery process, since AODV establishes routes only on demand. We explain this in the following: A node wants to send a payload packet to the AR but has no appropriate entry in its routing table. The routing protocol will recognize that a route to the AR must be found. Hence, it buffers the packet and initiates a route discovery process by broadcasting a *route request* (RREQ) packet with the address of the AR. If the AR receives the RREQ, it sends back a *route reply* (RREP) packet to the mobile node. When the mobile node receives the RREP, the route is established and the

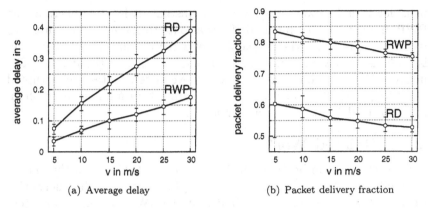

(a) Average delay   (b) Packet delivery fraction

Figure 5.   Network performance over node speed v

payload packet can be transmitted. The duration of how long packets are buffered also depends on the number of hops between the mobile node and the AR. The reason for this is that AODV uses the *expanding ring search technique* for the route discovery. If an RREQ is sent for the first time, the time-to-live of this packet is set to one hop. If no RREP is received within a certain period, a second RREQ is sent with an increased time-to-life. This process is repeated until either a route is found or the maximum limit for the number of RREQs has been reached. In the latter case, no route has been found and the buffered packets are discarded.

In summary, the higher delay of the RD model is mainly caused by the on-demand nature of AODV, in particular its expanding ring search technique. Initial measurements with a proactive routing protocol, such as OLSR [25], indicate that the packet delay is much smaller here, and the difference between the two mobility models is marginal. The reason is that proactive protocols do not buffer packets during the route discovery process.

The relation between route discovery and delay also explains the impact of speed. A higher speed of the nodes results in more frequent route disruptions, which in turn necessitates more route discoveries. Thus, the average delay of the packets is increased for a higher speed.

Finally, we analyze the packet delivery fraction (PDF), which is shown in Figure 5(b). Packets get lost either if a node can not find a route to the AR or if an established route is disrupted during the transmission of a packet. The PDF for the RD model is almost 30 % lower than that for the RWP model. For example, with $v = 5$ m/s, the RWP model yields a reasonable PDF of 83 %. With the RD model, however, only 60 % of the packets arrive at the AR. We conjecture that the reason for this

behavior is as follows: at a given time instant, it is more likely in the
RWP model that a mobile node finds a route to the AR. Note that this
is not obvious, since the probability of a RWP node to be completely
isolated is higher than in the RD model (see [26] without AR). The
impact of v is as follows: If the nodes move faster, routes are disrupted
more frequently, and packets get lost during transmission more often.
Hence, the PDF decreases for increasing v.

5.   Related Work

Related work can be found in simulation-based studies on the impact
of mobility modeling on performance of (stand-alone) ad hoc networks.
Belding-Royer et al. [11] recognized that the RWP model causes "fluctuations" in the number of neighbors of a node, which in turn has effect
on the routing performance. Bai et al. [27] did a performance study of
AODV, comparing the RWP, group, freeway, and Manhattan grid mobility. Yoon et al. [22] demonstrated by simulations that the speed decay
of the RWP model (Property 2) has significant impact on performance
of routing protocols. The inhomogeneous node distribution of the RWP
model (Property 1), has been regarded from a theoretical view in [14, 26,
28]; its impact on routing performance has not been elaborated so far.

Related work also comprises simulation-based studies of ad hoc networks with AODV over 802.11b using the RWP model [29]. Moreover,
descriptions of other ad hoc networking testbeds can be found in [6, 30–
32] and references in [7]. Finally, performance studies of MRANs are
given in [6, 33].

6.   Conclusions and Further Work

Our performance analysis of an MRAN testbed has shown that we
must be very careful in making statements about the (absolute) performance of a routing protocol, since it is *highly sensitive* to the used mobility model. In fact, understanding the impact of mobility on network-
level performance is a non-trivial task. The inherent characteristics of
a mobility model have direct influence on the network topology graph,
namely its static and dynamic connectivity. This connectivity, in turn,
has impact on the performance of medium access control and routing.
This paper has demonstrated that the routing performance difference
between two common and intuitively very similar stochastic mobility
models can be very high. In conclusion, we believe that the research
community would greatly benefit from defining a set of "benchmark scenarios," i.e., a well-defined collection of mobility traces that can be used
to evaluate new protocols. Such a collection may consist of real-world

traces (e.g., WLAN users on a campus, cars in downtown) and purely stochastic models (such as the RWP and RD model).

Notes

1. The RD model described here is a generalization of the *random direction model* in [11]. In that paper nodes change direction only at the area border.

References

[1] J. Broch, D. A. Maltz, and D. B. Johnson, "Supporting hierarchy and heterogenous interfaces in multi–hop wireless ad hoc networks," in *Proc. Workshop on Mobile Computing*, (Perth, Australia), June 1999.

[2] U. Jönsson, F. Alriksson, T. Larsson, P. Johansson, and G. Q. Maguire, "MIPMANET: Mobile IP for mobile ad hoc networks," in *Proc. ACM MobiHoc*, (Boston, USA), Aug. 2000.

[3] Y. Sun, E. M. Belding-Royer, and C. E. Perkins, "Internet connectivity for ad hoc mobile networks," *Intern. J. of Wireless Inform. Netw.*, vol. 9, Apr. 2002.

[4] J. Xi and C. Bettstetter, "Wireless multi-hop Internet access: Gateway discovery, routing, and addressing," in *Proc. 3Gwireless*, (San Francisco, USA), May 2002.

[5] R. Wakikawa, J. T. Malinen, C. E. Perkins, A. Nilsson, and A. J. Tuominen, "Global connectivity for IPv6 mobile ad hoc networks." IETF Draft, Oct. 2003.

[6] N. Bayer, B. Xu, and S. Hischke, "An architecture for connecting ad hoc networks with the IPv6 backbone (6Bone) using a wireless gateway," in *Proc. European Wireless (EW)*, (Barcelona, Spain), Feb. 2004.

[7] E. M. Belding-Royer and C. E. Perkins, "Evolution & future directions of the ad hoc on-demand distance vector routing protocol," *Ad Hoc Netw.*, July 2003.

[8] D. B. Johnson and D. A. Maltz, *Mobile Computing*, ch. Dynamic Source Routing in Ad Hoc Wireless Networks. Kluwer, Feb. 1996.

[9] R. A. Guérin, "Channel occupancy time distribution in a cellular radio system," *IEEE Trans. Veh. Technol.*, vol. 36, Aug. 1987.

[10] C. Bettstetter, "Mobility modeling in wireless networks: Categorization, smooth movement, and border effects," *ACM Mobile Comp. Commun. Rev.*, July 2001.

[11] E. M. Royer, P. M. Melliar-Smith, and L. E. Moser, "An analysis of the optimum node density for ad hoc mobile networks," in *Proc. IEEE ICC*, (Helsinki, Finland), June 2001.

[12] C. Bettstetter and C. Wagner, "The spatial node distribution of the random waypoint mobility model," in *Proc. German Workshop on Mobile Ad Hoc Networks (WMAN)*, (Ulm, Germany), Mar. 2002.

[13] D. M. Blough, G. Resta, and P. Santi, "A statistical analysis of the long-run node spatial distribution in mobile ad hoc networks," in *Proc. ACM MSWiM*, (Atlanta, USA), Sept. 2002.

[14] C. Bettstetter, G. Resta, and P. Santi, "The node distribution of the random waypoint mobility model for wireless ad hoc networks," *IEEE Trans. Mobile Comput.*, vol. 2, July 2003.

[15] C. Bettstetter, H. Hartenstein, and X. Pérez-Costa, "Stochastic properties of the random waypoint mobility model," *ACM/Kluwer Wireless Netw.*, Sept. 2004.

[16] W. Navidi and T. Camp, "Stationary distributions for the random waypoint mobility model," *IEEE Trans. Mobile Comput.*, vol. 3, Feb. 2004.

[17] "AODV-UU." http://www.docs.uu.se/docs/research/projects/scanet/aodv.

[18] W. Chao, J. P. Macker, and J. W. Winston, "NRL Mobile Network Emulator," Tech. Rep. NRL/FR/5523–03-10,054, Naval Research Laboratory, Jan. 2003.

[19] A. Jardosh, E. M. Belding-Royer, K. C. Almeroth, and S. Suri, "Towards realistic mobility models for mobile ad hoc networks," in *Proc. ACM MobiCom*, (San Diego, USA), Sept. 2003.

[20] J. Kammann, M. Angermann, and B. Lami, "A new mobilty model based on maps," in *Proc. IEEE VTC*, (Orlando, USA), Oct. 2003.

[21] J. G. Jetcheva, Y.-C. Hu, S. PalChaudhuri, A. K. Saha, and D. B. Johnson, "Design and evaluation of a metropolitan area multitier wireless ad hoc network architecture," in *Proc. IEEE WMCSA*, (Monterey, USA), Oct. 2003.

[22] J. Yoon, M. Liu, and B. Noble, "Random waypoint considered harmful," in *Proc. IEEE Infocom*, (San Francisco, USA), Apr. 2003.

[23] C. Hartmann and J. Eberspächer, "Adaptive radio resource management in F/TDMA cellular networks using smart antennas," *Eur. Trans. Telecom.*, Sept. 2001.

[24] B. Ghosh, "Random distances within a rectangle and between two rectangles," *Bull. Calcutta Math. Soc.*, vol. 43, 1951.

[25] T. Clausen and P. Jacquet, "Optimized Link State Routing Protocol (OLSR)." IETF RFC 3626, Oct. 2003.

[26] C. Bettstetter, "Topology properties of ad hoc networks with random waypoint mobility," in *Proc. ACM MobiHoc*, (Annapolis, USA), June 2003. Short paper.

[27] F. Bai, N. Sadagopan, and A. Helmy, "IMPORTANT: a framework to systematically analyze the impact of mobility on performance of routing protocols for ad hoc networks," in *Proc. IEEE Infocom*, (San Francisco, USA), Mar. 2003.

[28] P. Santi, "The critical transmitting range for connectivity in mobile ad hoc networks," *IEEE Trans. Mobile Comput.*, 2004. To appear.

[29] S. R. Das, C. E. Perkins, E. M. Royer, and M. K. Marina, "Performance comparison of two on-demand routing protocols for ad hoc networks," *IEEE Personal Commun. Mag.*, Feb. 2001.

[30] D. A. Maltz, J. Broch, and D. B. Johnson, "Lessons from a full-scale multihop wireless ad hoc network testbed," *IEEE Personal Commun. Mag.*, Feb. 2001.

[31] D. S. J. DeCouto, D. Aguayo, B. A. Chambers, and R. Morris, "Performance of multihop wireless networks: Shortest path is not enough," in *Proc. Workshop on Hot Topics in Networking (HotNets)*, (Princeton, USA), Oct. 2002.

[32] M. Möske, H. Füßler, H. Hartenstein, and W. Franz, "Performance measurements of a vehicular ad hoc network," in *Proc. IEEE VTC*, (Milan, Italy), May 2004.

[33] M. Ghassemian, P. Hofmann, C. Prehofer, V. Friderikos, and H. Aghvami, "Performance analysis of Internet gateway discovery protocols in ad hoc networks," in *In Proc. IEEE WCNC*, (Atlanta, USA), Mar. 2004.

Broadcast Services and Topology Control in Ad-Hoc Networks

Flaminio Borgonovo, Matteo Cesana, Luigi Fratta

DEI, Politecnico di Milano

borgonov, cesana, fratta@elet.polimi.it

Abstract

The design of effective Medium Access Control protocol for ad hoc network is a highly challenging task due to the characteristics of such networks. The medium is shared and unreliable by definition, thus the access mechanism should be robust to channel errors and collisions. Furthermore, all the coordination functions must be totally distributed in a pure ad hoc environment.

In this paper we address the issue of designing an effective MAC protocol for Inter Vehicular Communications (IVC) networks with a particular focus on broadcast services and topology control capabilities.

1 Introduction

The design of ad hoc networks has recently attracted a lot of attention, mainly because many characteristics of such networks, especially in a highly mobile environment, make the design of a prompt, efficient, flexible, and reliable MAC very difficult.

Applications of mobile ad hoc networks can range from military ones, where networks need to be deployed immediately without the support of either base stations or fixed network infrastructures, to inter-vehicle communications, designed for both traffic safety enhancement and entertainment purposes. The inter-vehicle communications application poses the most stringent requirements, due to a highly variable topology and to the need to provide a continuous exchange of broadcast information to support traffic control applications [1].

Within this application field, to achieve a reduction of 50% in road accidents within 2010 [2], the European Commission has promoted an integrated safety program, including: traffic management, driver assistance, passive safety, and emergency management. To this end, a special interest exists in developing Intelligent Transportation System (ITS) services based on wireless vehicular communications both among vehicles (Inter Vehicles Communication, IVC) and between vehicles and fixed infrastructure at the side of the road (Vehicles to Road Communication, VRC) [3] [4].

Typical applications of IVC and VRC include:

- *Information and Warning Functions (IWF)*: this service requires to transmit alarm information regarding stopped vehicles, traffic accidents, traffic intensity and congestion, road surface conditions. Such an information should be promptly available at the designated receivers and therefore requires the transmission of short aperiodic broadcast packets with high priority and stringent delay requirements.

- *CoOperative Driver Assistance (CODA)*: this application is helpful for handling and managing merging situation, overtaking and lane changing manoeuvres. A direct addressing of neighboring vehicles and a reliable unicast communication link with low latency are needed. Broadcast functionality might be useful for the CODA functions as well, e.g. to send a gap request to all nearby vehicles.

Most of the applications in IVC networks, as those mentioned above, require the vehicles to continuously "broadcast" some background information, such as cruise parameters [5]. This information is intrinsically broadcast because directed mostly to neighboring vehicles. In such a scenario, the MAC level should be able to provide the upper layers with prompt and reliable broadcast channels.

Almost none of the standardized wireless technologies, like the IEEE802.11, seems to be well suited for IVC applications, since the broadcast service they implement is highly unreliable.

Furthermore, due to the variable environment, in mobile ad hoc networks all protocols and coordinating functions must be totally distributed. Since the topology of the network has a dramatic impact on its performance, an effective topology control algorithm should be integrated at MAC level in order to guarantee good performance in terms of throughput and network connectivity with dynamic and variable network topologies. These kinds of algorithms often operate in a fully distributed way by modulating the transmitting power of each terminal according to the network situation it is experiencing.

In this paper we address the issue of designing an effective MAC protocol for IVC networks with a particular focus on broadcast services and topology control capabilities. In Section 2 we briefly discuss the effectiveness of IEEE802.11 as far as the broadcast service is concerned. Further on, we highlight the issue of designing an effective broadcast support at MAC level by reviewing some of the existing research results. Section 3 comments the need of implementing a topology control in an ad hoc network environment. In Section 4 we briefly describe the features of the ADHOC MAC proposed in [6, 7], to support a reliable broadcast service and we present a novel and effective topology control algorithm which can be easily integrated within ADHOC MAC. Section 5 presents numerical results on th performance of broadcast and topology control algorithm. In Section 6 some concluding remarks are given.

2   MAC Design and Broadcast services for Ad Hoc Networks

Since the wireless medium is inherently a shared resource, the effectiveness of the mechanism that controls the access to the channel determines the capacity of the network and has a dramatic impact on system complexity and cost.

The ad hoc networks architecture is completely distributed, that is, no logical distinction among the entities composing the network exists. No central stations, like in cellular networks, can control and rule the channel access, nor coordinate the operations of the end users. Everybody in a pure ad hoc network is contemporary base station and end user.

MAC protocols for ad hoc networks must cope with some problems typical of a wireless environment:

- *Unreliable Links*: wireless media is unreliable since any transmission can suffer a high bit error rate, due to multi-path, fading and shadowing phenomena. Furthermore, wireless nodes can move around at non negligible speed.

- *Collision*: wireless media is shared, so some kind of access control should be implemented at MAC level in a distributed way.

- *Power Consumption*: power saving issues should be further taken into account in MAC design since wireless nodes are often battery powered.

The existing MAC protocols for ad hoc networks can be classified in two categories: *contention free* and *contention based*. The contention free protocols avoid collisions during access phase through a radio resource scheduling algorithm which assigns each users its own transmission slot on a time division basis [8] [9], thus they are based on some kind of Time Division Multiple Access (TDMA) protocol. These solutions work with a synchronous communication scenario where the radio channel is divided into several time slots, and each wireless station has knowledge of this assignment. This solution reduces bandwidth wastage due to collision but it needs a complex slot scheduling algorithm, which has to be totally distributed.

On the other hand, the basic principle of contention based protocols is competition, i.e., wireless nodes compete to gain the access to the channel, and possible collisions are resolved by randomized retransmission procedures. These protocols are based on ALOHA and on CSMA [10, 11]. Strength points of these protocols are simplicity and good performance in low traffic situations. Unfortunately, these solutions doesn't scale well at high loads since collisions tend to become systematic and affect the network performance. For this reason, mechanism of collision avoidance are often adopted in contention based access protocols.

The Distributed Coordination Function (DCF) mechanism adopted in the IEEE802.11 standard is the most used contention based MAC protocol [12]. It implements some concepts proposed in MACA (Multiple Access with Collision Avoidance) protocol [13] and in MACAW (Multiple Access with Collision Avoidance for Wireless networks) protocol [14]. In details, after a physical carrier sensing phase, a handshaking phase can be used in order to alleviate hidden nodes interference and minimize the number of exposed nodes. A station willing to transmit sends out a *Request To Send* (RTS) to probe the availability of the destination. The destination, if available, answers with a *Clear To Send* (CTS) packet. All the stations overhearing one of the two control packets are prevented from accessing the channel. The RTS/CTS exchange is preceded with a physical carrier sensing phase, in order to limit collisions among control packets.

As far as the broadcast service is concerned, two concepts of broadcasting can be defined: *single-hop broadcast* and *multi-hop broadcast*. The former lets a packet to be correctly received by any of the recipients within the transmitting range of the transmitter. All the wireless transmissions can be seen as a single hop broadcast, provided that no collision can happen at one or more of the designated recipients. Due to radio range limitations, *single-hop broadcasting* does not cover all terminals of the network. The *multi-hop broadcast*, requires a mechanism that enables a packet to be reached by all the nodes in the network. Relaying and forwarding functions must be performed by "intermediate" terminals.

In IEEE 802.11 MAC protocol, no RTS/CTS exchange is used for the transmission of broadcast frames, regardless of their length. Further, no acknowledgement (ACK) is transmitted by any of the recipients of the broadcast frame. Therefore, the *single hop broadcast* service implemented in IEEE802.11 is highly unreliable since the sender cannot detect possible collisions [12]. Many efforts have been done in order to enhance the reliability of the IEEE802.11 *single-hop broadcast* service [15] [17] [18], but no definite clear solution has been proposed.

Most of the advanced multi hop broadcast protocols, such as the tree-based protocol in [19], do not work well for ad hoc networks due to the dynamic nature of the network topology. Hence, the flooding approach and its variants, have been proposed as the preferred means for multi-hop broadcast service [20]. In flooding, each station retransmits a broadcast packet just once until all terminals are reached. Such a procedure is highly inefficient in networks that present an high degree of connectivity since it generates many redundant retransmissions. In addition this procedure suffers from the broadcast storm problem if integrated with random access procedures [21], since neighbor nodes are likely to re-transmit a broadcast packet almost at the same time, causing massive collisions.

As already pointed out in the previous section, the above drawback is very critical in vehicular control applications, where vehicles continuously "broadcast" some background information, such as cruise parameters [5]. This information, furthermore, is intrinsically single-hop broadcast because directed mostly to neighbor vehicles, and a flooding procedure would saturate the whole network with useless information [20].

3 Topology Control in Ad Hoc Networks

Power control procedures that adapt the terminals transmitted power according to network needs are widely and effectively used in wireless systems.

For instance, cellular systems often resorts to centralized power control procedure to provide a satisfactory Signal to Noise Ratio (SIR) to all the connections and more generally to enhance the system performance.

On the other hand, in ad hoc networks the motivations for adopting power control procedures are even stronger and manifold.

Firstly, power control schemes are adopted for power and energy saving reasons, which are a must because of the limited amount of energy available at the battery-powered wireless terminals [22] [23] [24].

Secondly, power control procedures can be used to control the topology of the network with the purpose of improving the shared resource reuse [25], limiting the collisions among concurring transmissions [26] and providing guaranteed connectivity [27].

Topology control capabilities are of utmost importance in ad hoc applications such as IVC, where the mobile terminals rapidly move. Think of the situation of a car driving from a low populated countryside area into an urban densely populated one. The TX/RX device transmission range, that must be large in the countryside to provide connectivity, needs to be drastically reduced in the urban area to limit the number of neighboring terminals. This can be achieved by controlling the transmitted power which should be set in order to optimize the trade off between network capacity and connectivity. The higher is the transmitted power the higher is the network connectivity and the smaller is the network capacity due to the reduced radio resource reuse.

The basics of many of the proposed algorithm is to control the number of each terminal's neighbors in a distributed way [28, 29, 30, 31]. Namely, each terminal gathers information on the number of neighboring stations and adapts the level of transmitted power to maintain almost constant this number. The algorithms available in the literature mainly differ on the methods to get the information of the neighboring terminals and on the steps of the power update procedure.

In the next section we present a topology control algorithm which can be easily implemented with the information at layer 2 provided by ADHOC MAC briefly reviewed in the next section.

4 The ADHOC-MAC protocol

ADHOC MAC operates with a time slotted structure, where slots are grouped into virtual frames (VF) of length N, and no frame alignment is needed. The slotting information can be explicitly provided by external sources, such as GPS, or implemented in a distributed way. In the following, we assume an environment where the terminals can be grouped into clusters such that all the terminals of a cluster are interconnected by broadcast radio communication. Such a cluster is defined as One-Hop (OH). A terminal can belong to more than one OH-cluster, leading to the case of non disjoint clusters. The union of OH-clusters with a common subset is called a Two-Hop (TH) cluster. An example of such an environment is shown in Figure 1 where seven terminals are grouped into three OH-clusters forming one TH-cluster.

4.1 RR-ALOHA

For a correct operation, the ADHOC MAC needs that each active terminal has assigned a Basic CHannel (BCH), corresponding to a slot in the VF, which is a reliable single hop broadcast channel not suffering from the hidden-terminal problem. This is obtained in a distributed way by the RR-ALOHA protocol, where, as in R-ALOHA, contention is used to get access to an available slot in the frame and, upon success, the same slot is reserved in the following frames and no longer accessed by other terminals until it is released. Since the ad hoc environment is

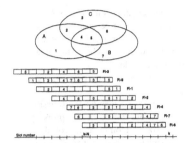

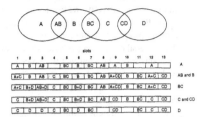

Figure 1: *Example of the FI information propagated by the terminals 1-7 in the one-hop clusters A, B, and C represented by ellipses.*

Figure 2: *An Example of how terminals belonging to the labelled disjoint areas perceive reservation of slots.*

not fully broadcast, the information needed for the RR-ALOHA correct operation is provided to all terminals by means of the BCHs.

Each transmission on the BCH contains, besides the data, a control field named Frame Information (FI). The FI is a vector with N entries specifying the status of each of the N slots preceding its transmission, as observed by the transmitting terminal itself. The slot status can be either BUSY or FREE: it is BUSY if a packet has been correctly received or transmitted by the terminal, otherwise it is FREE. In the case of a BUSY slot the FI also contains the identity of the transmitting terminal.

Based on received FIs, each terminal marks a slot, say slot k, either as RESERVED, if slot k-N is coded as BUSY in at least one of the FIs received in the slots from k-N to k-1, or as AVAILABLE, otherwise.

As in R-ALOHA an available slot can be used for new access attempts. Upon accessing an AVAILBLE slot, terminal j will recognize after N slots (a time frame) its transmission either successful, if the slot is coded as "BUSY by terminal j" in all the received FIs, or failed, otherwise.

The correctness of the above procedure in setting up a channel has been proved in [6] for a general network configuration, showing that no attempt at access can be made in a slot already in use when such access would cause a collision.

For the sake of simplicity, we just consider here the single TH-Cluster network of Figure 1. According to the RR-ALOHA procedure we can state the following. All terminals belonging to the same TH-cluster mark the slots in the same way. In fact, all the terminals receive the FI generated by the terminals belonging to the TH-cluster common set (terminals 4 and 5 in Figure 1) and such FI concerns all the transmissions in the TH cluster itself. A single BUSY code is enough to force a RESERVED slot. Similarly, any slot signaled as BUSY is recognized as RESERVED by all the terminals in the TH cluster and therefore, since a RESERVED slot cannot be accessed, no other terminal within the TH-cluster can transmit, and no collision will occur.

It is worth noticing that the transmission of terminals belonging to disjoint OH-clusters cannot collide, and the slots can be reused. However, since OH-Clusters can overlap, the slot reuse in an OH-cluster can still be constrained if the same slot is in use in the non disjoint OH cluster.

To illustrate this effect Figure 2 shows an example of the slot marking, observed at different terminals, where the area labels also denote terminals in the area. For the sake of clarity the RESERVED slots are represented by the corresponding terminal identity. Note that some terminals can receive FIs denoting the same BUSY slot for different terminals. This is the case of terminal B: it receives FI from AB, denoting slot 1 BUSY to terminal A, but also receives

FI from terminal BC, denoting slot 1 BUSY to terminal C (the reservation is marked as A+C). In fact, quite correctly, the two terminals in A and C can transmit in slot 1 because A does not detect the FI from BC and C does not detect the FI from AB. Similarly, terminals lying at least three hops apart can reuse the same slot as they can never collide with each other. This shows that the goal of avoiding collisions is achieved with an optimal use of slots.

Thus it can be seen that the procedure achieves the proposed goal of setting up channels with no interference if the cluster configuration does not change.

However, when the clusters merge because of terminal migration or activation, transmission collision in established channels can still occur. In fact, transmissions in a given slot, properly reserved for different terminals belonging to disjoint clusters would, upon merging, collide at the new common terminals. The RR-ALOHA procedure enables the colliding terminals to become aware of any collision. When a collision is discovered, the slot is released and a new set up procedure is started. The frequency of this situation depends on the variations of the network topology due to the activation of new terminals and the mobility of the active ones.

The BCH gained by R-ALOHA can also be used to assure additional MAC services such as a fast and reliable single hop broadcast channel, and a signaling channel to dynamically reserve additional channels with the bandwidth and priority needed to fulfill any QoS requirement. Furthermore, the FI can be used to safely set up point-to-point channels ([7, 6]), and to implement an efficient muli-hop broadcast mechanism, as described next.

4.2 Multi-Hop Broadcast

The ADHOC-MAC operation can be extended to implement multi-hop broadcast services over the whole network. The required relaying function, usually implemented at the network layer, can be effectively implemented at MAC layer using the connectivity information provided by the FI. Note that an effective multi-hop broadcast support at layer 2 can ease the design of routing protocols at the network layer.

As with flooding, the broadcast packets in the ADHOC-MAC network need to be numbered and the relaying procedure is applied only the first time the broadcast packet is received by a terminal. Let C_i be the set of neighbors of terminal i, i.e. all the terminals in the same OH-cluster, and C_j^i, for any $j \in C_i$, the sets of neighbors' neighbors. Given that terminal i receives a broadcast packet from terminal z in slot k, we define the set of neighbors that have not received the packet in slot k by $S_i \subset C_j^i$. All these sets are identified by terminal i through the information carried by the FIs received in the N slots following slot k.

At slot $k + N$, terminal i recognizes whether or not it needs to relay the broadcast packets. Terminal i does not relay the packet if $S_i = \phi$ or if, for at least one $j \in (C_j^i - S_i)$ the following condition is satisfied:

$$S_i \subseteq C_j^i \text{ AND } \left(|C_j^i| > |C_i| \text{ OR } (|C_j^i| = |C_i| \text{ AND } ID_j > ID_i) \right).$$

where ID_i denotes the address of terminal i. Basically, terminal i does not relay the packet if set S_i is a subset of C_j^i and if either C_j^i has higher cardinality than C_i or, having the same cardinality, the address of j is higher than the address of i.

4.3 Topology Control in ADHOC MAC

The need for an effective network topology control has been widely discussed in Section 3 and will be further supported by the numerical results presented in the next section.

Hereafter we present a topology control algorithm that can be easily implemented using the distributed information provided by the FI in ADHOC MAC.

The algorithm works in a fully distributed way by updating at each *Power Control Period (PCP)* the transmitting power of each terminal in order to keep active a given number of bidirectional links between each terminal and its neighbors. Two terminals are connected by a bidirectional link if and only if they are within the transmitting range one another. The parameters of the algorithm are: K, the target number of bidirectional links, the value of the PCP and the power ramping step P_{step} which represents the minimum amount of power increase.

Using the information contained in the FI, each terminal updates a list α of neighboring nodes ordered according to the received power, from the higher to the lower. In order to reach this goal, each terminal should insert in its FI the information on its transmission power level.

At the end of each PCP a terminal collects all the FIs received in the previous frame. If N_r is the number of these received FIs, the terminal adjusts its transmitting power according to the following rules:

Rule 1 : *If $N_r \leq K$, the terminal increments its transmission power of P_{step}, and signals that $N_r \leq K$ setting the* connection flag *to 0 in its own FI.*

The transmitted power cannot exceed a maximum value P_{max}. The *connection flag* in the FIs signals the connectivity status of the transmitter to its neighbors.

Rule 2 : *If $N_r = K$, the terminal sets its transmitting power so that to reach the K-th neighbor in its list. It signals $N_r = K$ by setting the* connection flag *to 1 in its FI.*

Rule 3 : *If $N_r > K$, for each of the last $N_r - K$ neighbors in list α it adds the neighbor to a subset F if in the neighbor's FI the* connection flag *is set to 0 or is set to 1 and the ID of the terminal itself is signalled. If subset F is empty the terminal sets its transmission power to reach the K-th neighbor of the ordered list and sets the* connection flag *in its FI to 1, otherwise it sets its transmission power so to reach the most distant neighbor belonging to subset F and sets the* connection flag *in its FI to 2.*

The rationale of the algorithm is to maintain the number of bidirectional links equal to K for each terminal. A terminal is allowed to have more than K bidirectional links (Rule 3) only to keep the connectivity status of its neighbors above the quality constraint ($N_r \geq K$).

In order to implement the proposed algorithm in ADHOC MAC the information carried by the FI should be enhanced. As already pointed out, each terminal should include in its FI its transmitting power level, which can be coded using 1 byte. Furthermore, a two bits *connection flag* must be added.

5 Performance Evaluation

In this section we present preliminary performance evaluations on the effectiveness of ADHOC MAC with respect to the single-hop broadcast service (Section 5.1) and multi-hop broadcast service (Section 5.2). Furthermore, we study the impact of the topology control algorithm presented in the previous section on the dynamics of ADHOC MAC (Section 5.3).

All the numerical results have been obtained by simulating the ADHOC MAC protocol in a network scenario where the terminals are dynamically generated according to a Poisson process and are placed in a square area with side length $A = 1Km$ according to a uniformly distributed probability density. To avoid border effects, the square area is shaped like a wraparound surface.

Upon generation, each terminal tries to acquire its BCHs and, after $L = 300$ frames have elapsed, it dies whether it has acquired the BCH or not. Each terminal has a transmission radius R and each frame is composed by N transmission slots.

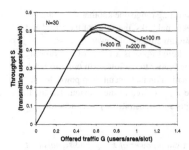

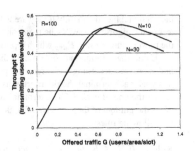

Figure 3: *The throughput of BCH channels versus the channel load for a torus surface, different coverage radii and frai*

Figure 4: *The throughput of BCH channels versus the channel load for a torus surface, for different :izes N.*

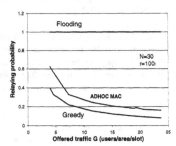

Figure 5: *Packet relaying probability versus the average number of neighbors in a square region with edge = $1Km$ and with coverage radius = $100m$.*

5.1 Single Hop Broadcast Efficiency

According to ADHOC MAC, a terminal can support the single-hop broadcast service once it has acquired a BCH channel. To evaluate the efficiency of such a procedure we measure the throughput of BCH defined as the average number of terminals that use the channel per slot per unit area. In general, this figure is function of the offered traffic and depends on system parameters such as N and L.

Figure 3 shows the throughput S versus the offered traffic G for $N = 30$, and $R = 100, 200, 300m$. The throughput reaches a maximum value S_{max} and shows an unstable behavior due to congestion caused by repeated access attempts.

We have observed that the S_{max} is much smaller than 1 because the protocol prevents the use of some slots to avoid hidden terminal collisions, thus limiting the radio resources reuse. This is the overhead one must pay to guarantee a reliable single hop broadcast. This overhead, as observed in Figure 3, changes with R. In fact, as R decreases, the probability of slot reuse in disjoint OH-Clusters increases, but reaches a saturation value for $A/R \geq 10$.

On the other hand, if R is comparable with A, S_{max} increases again because the number of terminals outside the OH-Cluster reduces. In the unrealistic case of $R = A$, S_{max} would be equal to 1 since all terminals are within a single OH-Cluster.

Figure 4 shows S versus G for $N = 10, 30$. In general, a larger frame size should favor the throughput since it easier to find a free slot when a new channel is added at a given throughput S. However, a larger frame means proportionally larger users densit, which reduces reuse.

In fact, adding users in the plane can transform non-overlapping OH-clusters into overlapping ones and users that originally used the same slot must use different slots when merged into the TH-cluster. When $G < G_{max}$ the former effect is predominant, whereas the second prevails in the congested zone. A further effect, though with less impact on throughput, is the reduction of the overhead due to collisions when a large frame length is adopted.

5.2 Multi-Hop Broadcast efficiency

As described in section 4.2, ADHOC-MAC provides a simple method to attain an effective multi-hop broadcast service, exploiting the distributed information transmitted in the FI fields. To quantitatively evaluate such an efficiency, we measured the number of retransmissions to ensure that a broadcast packet reaches all destinations. The numerical results obtained by simulation are presented in Figure 5 that shows the probability that a terminal retransmits a broadcast packet, i.e., the fraction of terminals involved in the broadcast relaying function, versus G, for $N = 30$ and $R = 100m$.

For comparison purposes, we show the performance of the flooding technique and a centralized greedy approach. The former, adopted in IEEE 802.11, requires each terminal to retransmit a broadcast packet once and thus represent an upper bound for the fraction of terminals involved in the relaying function. The lower bound, more interesting for actual efficiency evaluation, is provided by the solution of the broadcast retransmissions minimization problem, which has been proven to be NP-hard [19]. Instead, we have used a suboptimal solution obtained by a centralized greedy procedure.

In the heuristic we used, the terminal with the higher number of neighbors still missing the broadcast packet is selected to perform the retransmission at each iteration step. The procedure terminates when all the terminals have received the packet. Such a procedure would require at each terminal the full knowledge of the network topology and a central controller should further coordinate the relaying phase.

The numerical results of Figure 5 have been obtained assuming a network topology composed by a regular square grid of nodes at a distance equal to terminal radio range, so that each node has four grid neighbors. The grid is introduced to guarantee the complete network connectivity. Additional nodes are generated randomly in order to change the offered traffic.

With both the greedy algorithm and the ADHOC MAC multi-hop algorithm the relaying probability decreases as the load increases. In fact, as G increases, the average number of neighbors increases and thus the fraction of terminals that need to retransmit decreases.

We note that ADHOC MAC performs closely to the greedy algorithm and, for a high degree of connectivity, ADHOC MAC provides a gain of 85% in terms of overhead reduction with respect to flooding.

5.3 Topology Control Algorithm Efficiency

The performance analysis of the topology control algorithm has been split into two traffic configurations: static and dynamic. In the static configuration the number of the terminals within the network is fixed throughout the simulation. In this scenario we validated the correctness of the topology control algorithm and we tested its convergence time. To do this, we randomly generated M terminals within the network at the same time and we run ADHOC MAC with the topology control algorithm.

In the dynamic scenario the terminals enter and leave the network randomly. In this configuration we evaluated the impact of the topology control algorithm on the performance of ADHOC MAC when varying the traffic intensity and the parameter K.

The *standard setting* of the parameters used in the following simulations is: $L = 300 frames$, $N = 30 slots/frame$, $Step = 1m$, $PCP = 5 frames$ and P_{max} is set to reach a distance of $250m$ in square simulation area with edge equal to $1Km$.

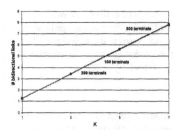

Figure 6: *Number of bidirectional links versus K for three cases of users' density in the* standard setting.

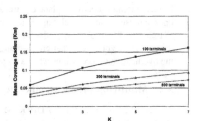

Figure 7: *Average coverage radius versus K for three cases of users' density in the* standard setting.

5.3.1 Static Analysis

Figure 6 gives the average number of the actual installed bidirectional links versus K for different values of terminals density ($\sigma = 100, 300, 500$ terminals/Km^2). As expected, the topology control algorithm adapts the transmitted power according to the terminals' density and provides in all the cases the same number of bidirectional links.

Figure 7 shows the average value of the terminals' coverage radius versus K. For a given K the average value of the coverage radius grows as the terminals' density decreases. The required power to reach K neighbors is inversely proportional to the network density. On the other hand, when fixing the network density, the value of the coverage radius increases with the parameter K, i.e., the average transmitting power grows if the number of required neighbors grows.

5.3.2 Dynamic Analysis

The throughput of the single-hop broadcast service provided by ADHOC MAC that has been measured with a fixed transmission radius shows an unstable behavior when the offered traffic increases, as observed in Figure 3. If the topology control algorithm behaves correctly, the throughput degradation as G increases should disappear. This behavior has been verified by the numerical results shown in Figure 8 which gives the network throughput S Net (number of transmissions per slot all over the network) versus the network offered traffic G Net for the two cases with and without the topology control algorithm. The throughput measured for $K = 5, 7, 9$ is the same and linearly increases with the offered traffic.

The higher throughput with respect to the case of constant $R = 100m$ is mainly due to the the topology control algorithm that, by reducing the coverage radius as the offered traffic increases, favors the radio resource reuse and limits the access collisions.

These two phenomena have been monitored in our simulations.

Figure 9 shows the average coverage radius versus the G Net. The topology control shrinks the terminals' coverage radius when G Net increases, with a consequent higher possibility of reusing the same slot all over the network.

The access collision probability versus G Net is shown in Figure 10. With the fixed coverage radius a sharp increase in the collisions is observed for high G Net values (corresponding to the unstable conditions of Figure 8). On the other hand, the topology control algorithm practically maintain the collision probability constant.

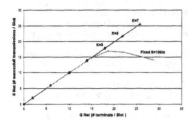

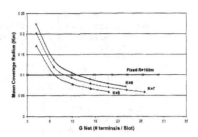

Figure 8: *S Net versus G Net for the cases with (K=3,5,7) and without (coverage radius R=100m) topology control in the* standard setting.

Figure 9: *Average value of the coverage radius versus G Net for the cases with (K=3,5,7) and without (coverage radius R=100m) topology control in the* standard setting.

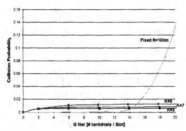

Figure 10: *Percentage of collisions versus G Net for the cases with (K=3,5,7) and without (coverage radius R=100m) topology control in the* standard setting.

6 Conclusions

The provision of effective broadcast services within vehicles is a central point for designing applications like traffic control and accident signaling in IVC. The ADHOC MAC is one of the few MAC protocols able to provide a reliable single hop broadcast channel within mobile terminals, according to which any broadcast transmission is guaranteed to be correctly received by all the neighboring nodes of the transmitter. Furthermore, thanks to the connectivity information exchange within ADHOC MAC, effective algorithms for multi-hop broadcast transmissions and topology control purposes can be designed.

In this paper we have reviewed the capabilities of ADHOC MAC in providing broadcast services whose performances have been thoroughly analyzed through simulation. Further on, we have proposed a novel topology control algorithm able to dynamically adapt to network changes due to terminal mobility which can be easily integrated within ADHOC MAC framework. We have validated the algorithm by testing its behavior in static traffic environment and finally we have shown its good performance with dynamic traffic.

References

[1] *"Special issue on intelligent vehicle highway systems"*, IEEE Transactions on Vehicular Technology, vol. 40, no. 1, 1991.

[2] European Transport Safety Council (ETSC), *"Intelligent Transportation Systems and Road Safety Report"*, Brussels, Belgium, 1999.

[3] M. Aoki, *"Inter-vehicle communication: technical issues on vehicle control applications"*, IEEE Communication Magazine, vol. 34, pp. 90-93, Oct. 1996.

[4] I. Chisalita, N. Shahmehri, *"A novel architecture for supporting vehicular communica-tion"*, in Proc. of IEEE VTC 2002-Fall, vol. 2, pp. 1002-1006, Sept. 2002.

[5] R. Hager, R. Mathar, and J. Mattfeldt, *"Intelligent cruise control and reliable communica-tion of mobile stations"*, IEEE Transactions on Vehicular Technology, vol. 44, no. 3, pp. 443-448, 1995.

[6] F. Borgonovo, A. Capone, M. Cesana, L. Fratta, *"ADHOC MAC: new MAC architec-ture for ad hoc networks providing efficient and reliable point-to-point and broadcast ser-vices"*, ACM Wireless Networks, to appear.

[7] F. Borgonovo, A. Capone, M. Cesana, L. Fratta, *"ADHOC MAC: a new, flexible and reliable MAC Architecture for ad hoc Networks"*, In Proc. of IEEE WCNC, vol. 2, pp. 965-970, March 2003.

[8] I. Chlamtac and A. Farago, *"Making Transmission Schedules Immune to Topology Changes in Multihop Packet Radio Networks"*, IEEE/ACM Transactions on Networking, vol. 2, no. 1, pp. 23-29, Feb. 1994.

[9] Chenxi Zhu , M. S. Corson, *"A Five-Phase Reservation Protocol (FPRP) for Mobile Ad Hoc Networks"*, Wireless Networks, vol. 7, no. 4, September 2001.

[10] L. Roberts, *"ALOHA Packet System With and Without Slots and Capture"*, Computer Com-munications Reviews, vol. 5, no. 2, pp. 28-42, April 1975.

[11] L. Kleinrock and F. Tobagi, *"Packet Switching in Radio Channels. I. Carrier Sense Multi-ple Access Models and Their Throughput Delay Characteristics"*, IEEE Transanctions on Communications, vol. COM23, no. 12, pp. 1400-1416, Dec. 1975.

[12] IEEE 802.11 Working Group, *"Part 11: Wireless LAN Medium Access Control (MAC) and Physical Layer (PHY) Specifications"*, ANSI/IEEE Std. 802.11, Sept. 1999.

[13] P. Karn, *"MACA, A New Channel Access Protocol for Packet Radio"*, in Proc. of ARRL/CRRL Amateur Radio 9th Computer Networking Conference, pp. 134-140, 1990.

[14] V. Bharghavan et al., *"MACAW: A Media Access Protocol forWireless LAN's"*, Computer Communications Reviews, vol. 24, no. 4, pp. 212-225, Oct. 1994.

[15] M. Impett, M.S. Corson, *"A Receiver Oriented Approach to Reliable Broadcast in Ad Hoc Networks"*, in Proc. of IEEE WCNC 2000, Chicago, IL , USA, Sept. 2000

[16] S.T. Sheu, Y. Tsai, J. Chen, *"A Highly Reliable Broadcast Scheme for IEEE 802.11 Multi-hop Ad Hoc Networks"*, in Proc. of IEEE ICC 2002, New York, NY, USA, April, 2002.

[17] K. Tang, M. Gerla, *"Mac Layer Broadcast Support in 802.11 Wireless Networks"*, In Proc. of IEEE MILCOM 2000, 2000.

[18] I. Chlamtac, S. Kutten, *"Tree-based Broadcasting in Multihop Radio Networks"*, IEEE Transactions on Communications, Oct 1987.

[19] Brad Williams, Tracy Camp, *"Comparison of broadcasting techniques for mobile ad hoc networks"*, in Proc. of ACM MobiHOC, June 2002.

[20] Y. C. Tseng , S. Y. Ni , Y. S. Chen and J. P. Sheu, *"The broadcast storm problem in a mobile ad hoc network"*, Wireless Networks, vol. 8, no. 2/3, March 2002.

[21] S. Agarwal, S. Krishnamurthy, R. H. Katz, S. K. Dao, *"Distributed Power Control in Ad-hoc Wireless Networks*, in Proc. of IEEE PIMRC 2001.

[22] E.S. Jung, N. H. Vaidya, *A Power Control MAC Protocol for Ad Hoc Networks*, in Proc. of ACM MOBICOM 2002, Atlanta, USA, September 2002.

[23] J.-P. Ebert, B. Stremmel, E. Wiederhold, A. Wolisz, *An Energy-efficient Power Control Approach for WLANs*, Journal of Communications and Networks (JCN), vol. 2, no. 3, pp. 197-206, Sept. 2000.

[24] J. P. Monks, V. Bharghavan, W. W. Hwu, *A power controlled multiple access protocol for wireless packet networks*, In Proc. of IEEE INFOCOM, April 2001

[25] S. Wu, Y. Tseng, J. Sheu, *Intelligent Medium Access for mobile ad hoc networks with busy tones and power control*, IEEE Journal on Selected Areas in Communications, vol. 18, no. 9, pp. 1647-1657, Sept. 2000.

[26] R. Ramanathan, R. Rosales-Hain, *Topology control of multihop wireless networks using transmit power adjustment*, in Proc. of IEEE INFOCOM 2000, vol. 2, pp. 404-413, March 2000.

[27] M. Gerharz, C. deWaal, P. Martini, P. James, *A cooperative nearest neighbours topology control algorithm for wireless Ad Hoc networks*, in Proc. of ICCCN 2003, pp. 412-417, October 2003.

[28] D. M. Blough, M. Leoncini, G. Resta, P. Santi, *The k-neigh protocol for symmetric topol-ogy control in Ad Hoc networks*, in Proc. of ACM MobiHoc 2003, pp. 141-152, June 2003.

[29] J. Liu, B. Li, *Mobile Grid: capacity-aware topology control in mobile Ad Hoc networks*, in Proc. of ICCCN 2002, pp. 570-574, Oct. 2002.

[30] R. Wattenhofer, L. Li, P. Bahl, Y. M. Wang, *Distributed topology control for power effi-cient operation in multihop wireless Ad Hoc networks*, in Proc. of IEEE INFOCOM 2001, vol.3, pp. 1388-1397, April 2001.

SPACE AND TIME CURVATURE IN INFORMATION PROPAGATION IN MASSIVELY DENSE AD HOC NETWORKS

Philippe Jacquet
INRIA
Domaine de Voluceau
78150 Rocquencourt
France
philippe.jacquet@inria.fr

Abstract   Gupta and Kumar have shown that effective wireless range decreases in inverse function of local traffic density. We show that a variable traffic density impacts the curvature of paths in a dense wireless ad hoc network the same way a variable optical density bends light paths. We set up the general laws that paths must satisfy in presence of traffic flow density. Introducing Time constraint in packet delivery, we generalize this curvature problem to a space-time problem with mobile networks.

1.   Introduction

Mobile ad hoc networks involve nodes that are moving on a network domain and communicate via radio means. The domain of a network can be indifferently a battlefield, a urban quarter, a building floor, etc. Most papers in the litterature take as an assumption models where the distribution of traffic and nodes are uniform over their network domain. This basic model leads to fundamental results, as the illuminating result of Gupta and Kumar **?** which states that the maximum capacity per node in a flat domain is in $O(\sqrt{\frac{1}{N \log N}})$. In this paper we will depart from the uniform model and assume that the traffic density varies with node location. We will provide results on how shortest path are affected by traffic density gradients. In particular we will show that in asymptotic conditions the routing paths obey to similar laws as in non linear optic.

We also generalize this result to the case where nodes are mobile and can hold the packet for some time before retransmitting it. Relaxing time constraint in the packet delivery and introducing mobility pattern depending on node position, we generalize the equation we obtained for a stationary network and prove that mobility can actually increase the network capacity.

In their reference paper on the capacity of mobile ad hoc networks, Gupta and Kumar ? showed that in presence of traffic density of λ bit per time unit per square unit area, the typical radius of correct reception decays in $O(\frac{1}{\sqrt{\lambda}})$. This result assumes an uniform density model and quantity λ is the density of traffic *including* the load generated by packet that are retransmitted on their way to their destination on multihop paths. To our view, this estimate is the most fundamental result. As a direct consequence the average number of hops needed to connect two arbitrary points in a bounded domain is therefore $O(\sqrt{\lambda})$ since the distance must be divided by the radio ranges. As pointed out by Gupta and Kumar, this property has a strong implication in the evaluation of the maximum capacity attainable by a random node when the node density increases. If C is the capacity generated by each node and N is the number of nodes in the network, Gupta and Kumar found that the maximum bandwidth attainable is $C_N = \frac{\alpha}{\sqrt{N \log N}}$, quantity α depending on propagation models. However the order of magnitude is easy to get: the density of traffic generated per unit square are is $O(CN)$. Let $r(C, N)$ be the typical radio range, thus the number of retransmissions needed to route a packet from its source to its destination is $O(\frac{1}{r(C,N)})$. The latter estimate, in turn, yields a traffic density (including retransmissions) of $O(\frac{CN}{r(C,N)})$. Therefore $r(C, N) = O(\sqrt{\frac{r(C,N)}{CN}})$, namely $r(C, N) = O((CN)^{-1})$. The average number of neighbor per node is $O(\pi r(C, N)^2 N)$; it should be larger than $\log N$ in order to guarantee connectivity, which leads to the estimate $C = O((N \log N)^{-1/2})$.

If instead we consider that the network spans on a domain in dimension D, then the radius of correct reception decays in the inverse of the Dth root of emitter density, which impacts the maximum capacity (replacing the exponent $-1/2$ by $-1/D$).

This paper addresses the case where the traffic pattern is not uniform but varies as a continuous function of node location. We investigate the case where the traffic and node densities are large enough to make the efficient radio ranges infinitely small (compared to traffic density gradient and domain size). In this perspective, shortest paths (in number of hops) from sources to destinations look like continuous lines. In the sequel we call lines generated by shortest path routes, *propagation lines*. In the uniform model propagation lines are expected to be straight lines. In this paper we limit our investigation to the propagation lines and we analyze how the latter are affected by the variation of traffic density. We summarize our findings into macroscopic differential equations involving propagation lines curvatures.

The paper is organized as follows. The second section investigates more thoroughly a model of mobile ad hoc network in order to provide more accurate estimates about Gupta and Kumar results. The model assumes a 2D domain under slotted time with a density of λ emitters per slot and square area unit. Under this simple model we will provide quantitative confirmation of Gupta and Kumar results. The third section introduces the concept of massively dense networks and how propagation lines are affected by variable traffic density. At a macroscopic level the variable traffic density acts like a variable optical index and curves the propagation lines from sources to destinations. The fourth section introduces the time component and node mobility in the problem and shows that larger packet delivery delay can reduce the number of hops.

2.   Quantitative results on time slotted networks

Quantification of the problem

Kumar and Gupta estimates were originally derived from information theory considerations and are not related to any particular network implementation. If we assume a specific implementation, then there will be a quantity β such that the typical radius of correct reception of a packet is equal to $\frac{\beta}{\sqrt{\lambda}}$. By typical radius we mean the radius below which probability of correct reception of a packet is above a given threshold. The quantity β will depend on many parameters such as the probability threshold, the attenuation coefficient of wave propagation and the minimum signal-over-noise ratio required for correct reception. Notice that the typical disk of correct reception contains in average a finite number of transmitters per slot, since the area is proportional to $\frac{1}{\lambda}$.

If we consider a network dispatched in a domain of dimension D then the estimate of the radius will be $\frac{\beta}{\lambda^{1/D}}$. In the sequel we will look at 2D domains generalizing occasionnaly the results on other dimensions.

When the density λ increases in a fixed domain, then the minimum number of hops connecting two points A, B tends to be equivalent to $d(A,B)\frac{\sqrt{\lambda}}{\beta}$ where $d(A,B)$ denotes the euclidian distance between mobile node A and mobile node B. Meanwhile, the increase of the number of relays naturally increases the traffic density. If ν is the actual traffic density generation per unit area, *i.e.* the traffic locally generated on mobile nodes, not the traffic relayed by the mobile nodes, then the average density traffic will satisfy the identity: $\lambda = \nu\bar{d}\frac{\sqrt{\lambda}}{\beta}$ where $\bar{d}$ is the average euclidian distance between two end points in a connection.

This previous identity assumes that the pattern of path between points covers the domain in an uniform manner so that the traffic density, generated and relayed, is constant on the whole domain. In this case the path that connect two points with the minimum number of hops is very close to the straight line. But the question arises about the shape of the shortest path when the traffic density is not uniform. We will show that when the density increases while keeping proportional to a given continuous function, then the propagation paths tend to conform to continuous line, that we call propagation lines. Under these assumptions we will provide the general equations that the propagation lines must satisfy. We will show that variable traffic densities affect shortest path the same way as variable optical indices affect light path in a physical medium.

Propagation model

We consider the following model: time is slotted, all mobile nodes are synchronized, transmissions on begining of slots. We consider an area of arbitrary size $\mathcal{A}$ (we will ignore border effect). N transmitters are distributed according to a Poisson process. We call λ the density of transmitter per slot and per square area unit. We have $\lambda = fN/\mathcal{A}$ where f is the rate of packets transmission per slot and per node.

Let a node X at a random position (we ignore border effects). We assume that all nodes transmit at the same nominal power. The reception signal at distance r is $P(r) = r^{-\alpha}$ with $\alpha > 2$. Typically $\alpha = 2.5$. Let W the signal intensity received by node X at a random slot. Quantity W is a random variable: let $w(x)$ its density function. In **?** it is shown that the Laplace transform of $w(x)$, $\tilde{w}(\theta) = \int w(x)e^{-x\theta}dx$ satisfies the identity:

$$\tilde{w}(\theta) = \exp(2\pi\lambda \int_0^\infty (e^{-\theta r^{-\alpha}} - 1)rdr) . \tag{1}$$

Using standard algebra we get

$$\tilde{w}(\theta) = \exp(-\lambda\pi\Gamma(1 - \frac{2}{\alpha})\theta^{2/\alpha}) \tag{2}$$

If the node location domain was a line instead of an area (consider a sequence of mobiles nodes on a road) then we would have

$$\tilde{w}(\theta) = \exp(-\lambda\Gamma(1 - \frac{1}{\alpha})\theta^{1/\alpha}) \tag{3}$$

If, instead the location domain was a volume (consider aircrafts network), then

$$\tilde{w}(\theta) = \exp(-\frac{4}{3}\lambda\pi\Gamma(1 - \frac{3}{\alpha})\theta^{3/\alpha}) \tag{4}$$

In the following we restrict ourselves on a 2D domain.

Neighbor model

A node is considered neighbor of another node if the probability of receiving packets from each other is greater than a certain threshold p_0. For example $p_0 = 1/3$. Under this model, we can affect to p_0 the value which optimizes the distance traveled by a packet per transmission as in **?**. We assume that the slotted system contains an acknowledgment mechanism so that each succesful transmission does not triger any new retransmission for the same hop. In this case the distance travelled by the packet is equal to the distance from the transmitter to the receiver. When the transmission fails then the distance is zero and the node reschedule a new retransmission at a random time (we assume that λ involve the load due to retransmissions).

We assume that a packet can be decoded if its signal-over-noise is greater than a given threshold K. Typically $K = 10$. Therefore another node is neighbor if its distance r is such that $P(W < r^{-\alpha}/K) > p_0$, i.e. when $r < r(\lambda)$ where $r(\lambda)$ is the critical radius such that $\int_0^{r(\lambda)} w(x)dx = p_0$. By simple algebra it comes that $r(\lambda) = \lambda^{-1/2}r(1)$. This result confirms the result of Gupta and Kumar in this very specific network model. We find $\beta = r(1)$. The surface covered by the radius $r(\lambda)$ is the neighborhood area $\sigma(\lambda) = \frac{\sigma(1)}{\lambda}$ and $\sigma(1) = \pi r(1)^2$.

Computation of β, $r(1)$ and $\sigma(1)$

In order to simplify the presentation we set $C = \pi\Gamma(1 - \frac{2}{\alpha})$ and $\gamma = \frac{2}{\alpha}$. By application of reverse Laplace we have:

$$P(W < x) = \frac{1}{2i\pi} \int_{-i\infty}^{+i\infty} \frac{\tilde{w}(\theta)}{\theta} e^{\theta x} d\theta \tag{5}$$

Expanding $\tilde{w}(\theta) = \sum_n \frac{(-C)^n}{n!}\theta^{n\gamma}$, we get

$$P(W < x) = \frac{1}{2i\pi} \sum_n \frac{(-C)^n}{n!} \int_{-i\infty}^{+i\infty} \theta^{n\gamma-1} e^{\theta x} d\theta \tag{6}$$

By bending the integration path toward the negative axis we get

$$\frac{1}{2i\pi} \int_{-i\infty}^{+i\infty} \theta^{n\gamma-1} e^{\theta x} d\theta = \frac{\sin(\pi n\gamma)}{\pi} \int_0^{\infty} \theta^{n\gamma-1} e^{-\theta x} d\theta$$

$$= \frac{\sin(\pi n\gamma)}{\pi} \Gamma(n\gamma)x^{-n\gamma}$$

Figure 1 shows the plot of $P(W < x)$ versus x for $\alpha = 2.5$ and $\lambda = 1$. Notice that $P(W < x)$ reaches $p_0 = 1/3$ close to $x = x_0 = 20$. Therefore $\beta = r(1) = (x_0 K)^{-1/\alpha} \approx 0.12$. Therefore $\sigma(1) \approx 0.045$.

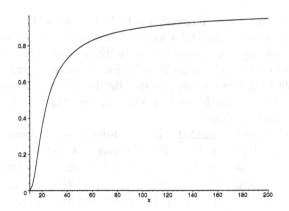

Figure 1.   Quantity $P(W < x)$ versus x for $\alpha = 2.5$, no fading.

Modeling of fading

Signals propagating through random obstacles experience random fadings. An usual modeling of fading consists into introducing a random factor F to signal attenuation at distance r: $r^{-\alpha}$. For example $\log F$ is uniform on $[-v, v]$. In this case we have a new expression of $\tilde{w}(\theta)$:

$$\tilde{w}(\theta) = \exp(-\pi \lambda \Gamma(1 - \frac{2}{\alpha}) \phi(-\frac{2}{\alpha}) \theta^{2/\alpha}) \tag{7}$$

with $\phi(s) = E(F^{-s})$, the Dirichlet transform of the fading. When fading is uniform on $[-v, v]$ we have $\phi(s) = \frac{\sinh(sv)}{sv}$. For any given real number x we also have $P(W < xF)$ equaling

$$\sum_n \frac{(-CF(-\gamma))^n \sin(\pi n \gamma)}{\pi n!} \Gamma(n\gamma) \phi(n\gamma) x^{-n\gamma} \tag{8}$$

which helps the computation of $\sigma(1)$ with fading.

3.   Massively dense networks

We now consider massively dense networks on a 2D domain. We denote by $\lambda(x, y)$ the traffic density at the point of coordinate (x, y) on the domain. We suppose that function $\lambda(x, y)$ is continuous in (x, y), or at least Lebesgue integrable. When $\lambda(x, y)$ are uniformly large, the results of Gupta Kumar together with the result of the previous section state that the radio ranges tend to be "microscopic" and routes can be considered as continuous lines between nodes. Packets travelling on a route $\mathcal{C}$ passing on the point of coordinate (x, y) will experience hops of length $\frac{\beta}{\sqrt{\lambda(x,y)}}$ passing in the vicinity of point (x, y).

Let $n(x,y) = \frac{\sqrt{\lambda(x,y)}}{\beta}$. The number of hops that a packet will experience on route C is something close to $\int_C n(x(s), y(s))ds$ where s is a curvilign absciss on route C.

In the sequel we are looking for route with the shortest hop number. Searching the path that minimize the hop number between two points A and B is therefore equivalent for looking for the path light between A and B in a medium with non-uniform optical index $\lambda(x, y)$. There is a known result about the optimal path that minimize a path integral $\int_C nds$.

THEOREM 1 *The optimal path satisfies on each of its point* $\mathbf{z}(s) = (x(s), y(s))$ *such that s is acurvilign absciss* $(ds = \sqrt{(dx)^2 + (dy)^2} = |d\mathbf{z}|)$:

$$\frac{d}{ds}(n(\mathbf{z}(s))\frac{d\mathbf{z}(s)}{ds}) = \nabla n(\mathbf{z}(s)) \qquad (9)$$

where ∇ *is symbol of gradient vector.*

The proof is classical. If we consider a small perturbation C^* of optimal path C where $\mathbf{z}^*(s) = \mathbf{z}(s) + \delta\mathbf{z}(s)$ we should have $\int_{C^*} nds - \int_C nds = \delta[\int_C nds] = 0$. We have $\delta[\int_C nds] = \int_C \delta[nds]$. Since $\delta[n] = \nabla n.\delta\mathbf{z}(s)$ and $\delta[ds] = \frac{d\mathbf{z}}{ds}.d\delta\mathbf{z}$ we get

$$\delta[\int_C nds] = \int_C \nabla n.\delta\mathbf{z}(s)ds + \int_C n\frac{d\mathbf{z}}{ds}.\frac{d\delta\mathbf{z}}{ds}ds$$

Integrating by part the second right hand side integral of the above and assuming that both C and C^* share the same end points (*i.e.* $\delta\mathbf{z}(s) = 0$ at both ends), we get:

$$\delta[\int_C nds] = \int_C (\nabla n - \frac{d}{ds}(n\frac{d\mathbf{z}}{ds})).\delta\mathbf{z}(s)ds.$$

Since $\delta\mathbf{z}(s)$ can be arbitrary, and that in all case $\delta[\int_C nds] = 0$, then $\nabla n - \frac{d}{ds}(n\frac{d\mathbf{z}}{ds}) = 0$ on the optimal path.

Therefore finding the optimal path is just an application of geometric optics. Notice that when $\nabla n = 0$ (uniform traffic density) propagation lines are straight lines (no curvature).

However we face a major problem in the fact that the distribution of path is actually impacting traffic density. This lead to an egg-and-chicken problem which may not that easy to solve. We call $\Phi(x, y)$ the flow density of information transiting in the vicinity of point (x, y). Quantity $\Phi(x, y)$ is expressed in bit per meter, since it expresses the flow of packet crossing a virtual unit of segment of length of 1 meter centered on point (x, y). This flow impact the traffic density by the fact that each packet must be relayed every $\beta/\sqrt{\lambda(x, y)}$ meter in the vicinity of point (x, y). Therefore locally:

$$\lambda(\mathbf{z}) = \Phi(\mathbf{z})\frac{\sqrt{\lambda(\mathbf{z})}}{\beta} \qquad (10)$$

In other words $\lambda(\mathbf{z}) = (\frac{\Phi(\mathbf{z})}{\beta})^2$ and

$$n(\mathbf{z}) = \frac{\Phi(\mathbf{z})}{\beta^2} \tag{11}$$

When considering domain of dimension D we have $\lambda = \Phi^{\lambda^{1/D}}\frac{}{\beta}$ and $n = (\frac{\Phi}{\beta D})^{\frac{1}{D-1}}$. Notice that the equations are singular when $D = 1$.

As an example we can assume a planar domain massively and uniformly filled with mobile nodes and gateway nodes. We denote by μ_G the spatial density of gateways. We assume that the mobile nodes are much more dense than the gateways. We denote by ν the traffic density generated in any point. ν is expressed in bits per square meters per slot. The flow density Φ is constant in the domain and is equal to $\nu\bar{d}$. We suppose that mobile nodes sends and receives flows from their closest gateway. Therefore $\bar{d} = \int_0^\infty \exp(-\pi\mu_G r^2)dr = \frac{1}{2\sqrt{\mu_G}}$:

$$\Phi = \frac{\nu}{2\sqrt{\mu_G}} \tag{12}$$

But in this case $\nabla n = 0$ and propagation lines are straight lines. The document ? provides non trivial examples where the propagation lines are curved and can be exactly calculated. This is the cases when traffic is generated toward a central gateway or when the traffic is generated toward gateways regularly spaced on a circle circonference.

Practical implementation of shortest path protocol

Implementing a routing protocol that follows the geodesic lines is not a difficult task. Indeed there is no need that nodes monitor their local traffic density n nor to advertize the gradient vector. In fact a shortest path algorithm, such as OLSR ?, will suffice. Of course one will need to limit the neighborhood of the node to those whose hello success rate exceeds p_0. To this end we make use of Hysterisis Strategy in *advanced link sensing* option and set up HYST-THRESHOLD-HIGH parameter to value p_0 that provides the best success rate, hop distance compromize. Tuned that way OLSR will automatically provide the shortest path that fit the traffic density gradient curvature.

4.   Introduction of time component

In the previous section we were assuming very strict timing constraints so that packets are forwarded like *hot potatoes* without any pause between retransmissions. Recently Grossglauser and Tse ? showed that mobility increases the capacity of wireless ad hoc networks. This due to the fact when nodes are moving one just wait that nodes come closer instead of immediately starting relaying when nodes are far apart. In particular the increase in capacity can be dramatic (in $O(1)$ instead of $O(\lambda^{-1/2})$) if one consider ergodic mobility patterns. Unfortunately the delay for packet becomes unbounded when the density increases. The aim of this section is to quantify the gain in retransmissions number while we let the time constraint on delay delivery T vary.

Although basic sensors networks or other smart dust are not expected to be mobile, However one can imagine more sophisticated sensors produced by via nanotechnology that can be mobile by themselves. The sensors can also travel because they are embedded in a mobile device or because the background medium is mobile (think about sensor in a river stream) Interestingly intermediate nodes can choose to store packet while moving instead of immediately retransmitting it. As long they move to the right direction this may considerably reduce the average number of retransmissions between source and destination. Of course the consequence is that packet delay delivery will considerably increase. This may be a solution for non urgent background traffic to take advantage of mobility and therefore have much less impact on network global load. Therefore we will model this very property by introducing space-time considerations in the framework presented in this paper.

Throughout this section we will assume that a node has a packet (or a sequence of packets) to transmit to a destination node with a time delivery constraint of T. In words, each packet should arrive to its destination no later than after a delay T. It basically means that we add the time dimension to our 2D problem. A path now contains the time dimension and will connect a source space-time point (z_0, t_0) to a destination space-time point (z_1, t_1), given that $t_1 = t_0 + T$. When $T = 0$, and neglecting propagation delays and processing in relay nodes, we get to our previous analysis restricted to space components. Our aim is to show that with some mobility models, when T tends to infinitely, the number of retransmissions needed to connect point (z_0, t_0) to point (z_1, t_1) tends to be negligible compared to the number of retransmissions needed to connect point (z_0, t_0) to point (z_1, t_0), (*i.e.* same spatial point but zero delay).

In order to set up notations and convention for this very general problem, we will first start with an unrealistic mobility model.

A simplistic mobility model

In this first example, we make the assumption that a node which has a packet to transmit or to relay can also trave with its packet at speed v. We also make the unrealistic hypothesis that the node that carries the packet can travel in any direction it wishes and that it makes the decision according to the destination of the packet. Therefore at any time the node that carries the packet has to make the decision of either transmitting it to the next hop or to travel with it on the propagation path. When the node chooses to hold the packet, we say that the packet is in hold state. We consider the optimal path C when $T = 0$ which connects z_0 to z_1, we assume that the space time path will be the path C plus a time component. In order to avoid too many notations we will still denote the space-time path by C.

With these very hypotheses, the number of actual hops the packet will experience during its propagation on path C is equal to $\int_C n|dz - vdt|$ where $\mathbf{v}$ is the vector speed at point $\mathbf{z}$. Since we assume that the speed can be made colinear with $d\mathbf{z}$ then the number of hops is equal to $\int_C n|1 - v\frac{dt}{|dz|}| \times |dz|$. In the following we call $h = \frac{dt}{|dz|}$ the average (local) packet holding time per distance unit, or we will denote by $\gamma = \frac{h}{v}$ the average fraction of distance traveled in hold state by a packet per distance unit. We therefore has $T = \int_C h|dz| = \int_C \frac{\gamma}{v}|dz|$.

Under this hypothesis it is clear that when $\gamma \to 1$, then the number of hops tends to be negligible compared to $\int_C n|dz|$ (zero delay case). In this case we don't need to have T unbounded, since in the most extreme case the node that generated the packet can simply drive his way to the destination and deliver the packet when in the neighborhood of the destination without transmitting it to intermediate nodes.

However this model is far from realistic. There is no reason that the mobility pattern of a node could depend that crucially on the destinations location of the packet it is holding. In fact a relay node can hold several packet to different destination at the same time and the node will have no way to split itself in several parts in order to move toward these different destinations at the same time. In the following subsection we consider a more realistic mobility model where the nodes are subject to random walks that are independent to data traffic conditions.

The random walk model

In this model we assume that at any time node travels at a random speed toward a random direction and keep its speed and heading during a time duration τ. After time τ it randomly change speed and heading. This like a particle in motion in a gas. Quantity τ refers to the free space motion delay during which the particle moves in straight line. At the end of period τ the particle experiences a collision that changes its motion vector in a random way. Notice that τ can be made random as well (we may assume that it is exponentially distributed). We assume that the expectation of speed vector $E[\mathbf{v}] = 0$. We also assume that the speed vectors have isotropic direction and $c_2\mathbf{I}$ is the covariance matrix, $\mathbf{I}$ being the identity matrix. We could accept some un-isotropic aspects so that covariance could depart from colinearity with identity matrix, but we will not do it for the sake of presentation.

Quantity $\sqrt{c_2}\tau$ is the standard deviation of node location after one free space travel. We assume that this quantity is of the same order as of hop distance r

(remaining that $r = \frac{1}{n}$. In other words the free space travel distance is of the same order as the hop distance. It is also instrumenetal in our proof that the speed is distributed on values that are not bounded by a finite number.

When a packet arrive in a mobile node, the router has to select whether it will transmit the packet to the next hop or keep it in hold state. We define a decision process which is based on the localisation of the next hop and the speed vector of the host node. If the node decides set the packet in hold state it will keep it as long its speed does not change. Therefore a hold state will last at least one τ period. The decision making automaton use a parameter x that is a positive real number and which depends on the delivery delay constraint T of the packet. Let θ be the angle made by the direction to the next hop. The node decide to immediately transmit its packet to the next hop iff the two following conditions hold:

1 its speed is larger than parameter x;

2 the speed direction angle is contained in interval $[-\theta - \frac{\pi}{1+x}, -\theta + \frac{\pi}{1+x}]$;

otherwise the packet stays in hold state. If the node keeps the packet in hold state it will keep it to its next motion vector change. At this moment it will proceed to a new packet state decision according to its new motion vector and to the localisation of the curent potential next hop. If the node has been decided to be transmitted immediately then the reciever will also proceed to a state selection. A packet may be transmitted over several hops before returning back in hold state. Basically when $x = 0$, then the packet is always immediately retransmitted to its next hops as with $T = 0$; and when $x \to \infty$, then the packet is less likely retransmitted and stay longer in hold state. The probability that a node chooses to immediately transmit the packet is $p(x) = \frac{1}{1+x}P(|\mathbf{v}| > x)$ and let $v(x) = E[|\mathbf{v}|, |\mathbf{v}| > x]\frac{1}{\pi p(x)}\sin(\frac{\pi}{1+x})$, it is clear that $v(x) > \frac{x}{\pi}\sin(\frac{\pi}{1+x})$. since the computation is done on speed greater than x with direction uniformly distributed in $[-\theta - \frac{\pi}{1+x}, -\theta + \frac{\pi}{1+x}]$. The average motion vector when the packet is in hold state is colinear with the direction to the next hop and has modulus equal to $p(x)v(x)$. According to hypothesis we have $\lim_{x \to \infty} p(x) = 0$ and $\lim_{x \to \infty} v(x) = \infty$.

The average distance the packet will travel before a new decision has to be taken (either in hold state or in immediate retransmission) is $p(x)r + (1 - p(x))p(x)v(x)\tau$ with variance $v_2(x)$. Notice that $\lim_{x \to \infty} v_2(x) = c_2\tau^2$. It comes that the average fraction of hold state travel per unit distance is

$$\gamma = \frac{(1 - p(x))v(x)\tau}{r + (1 - p(x))v(x)\tau} \tag{13}$$

We have clearly $\lim_{x \to \infty} \gamma = 1$ since $v(x)$ tends to infinity, in this case we also have

$$T = \int_C \frac{(1 - p(x))v(x)\tau^2}{(r + (1 - p(x))v(x)\tau)^2 p(x)} |d\mathbf{z}| \qquad (14)$$

which also tends to infinity since the product $p(x)v(x)$ tends to zero. However we cannot be sure that the packet actually reaches its destination. We know that the packet is *on average* on the path C. In order to check how far from path it is actually we have to look at the variance of packet localization. After each decision step, the packet travels in average a distance $p(x)r + (1 - p(x))p(x)v(x)\tau$ with a variance $v_2(x)$. Therefore in order to travel a distance of one unit the packet will have go through an average number of decision steps equal to $\frac{1}{p(x)r + (1-p(x))p(x)v(x)\tau}$. Therefore the variance of its position is close to $\frac{v_2(x)}{p(x)r + (1-p(x))p(x)v(x)\tau}$. In order to be safe we have to prove that this variance is small, so that the packet does not evade too far from the path and that when time limit will be critical the number of hops it will have to travel in emergency to reach its destination won't be too long. Since $v_2(x) \sim c_2 \tau^2$ which is of order of r^2 the variance is at most equal to $\frac{c_2 \tau^2}{p(x)r}$ which is order $r/p(x)$ which is small. In other words the number of hops is (using identity $r = \frac{1}{n}$):

$$\int_C \frac{n}{1 + n(1 - p(x))v(x)\tau n} |d\mathbf{z}| + O(|d\mathbf{z}|\sqrt{\frac{n}{p(x)}}) \qquad (15)$$

Similarily the delivery delay $T = \int_C \frac{(1-p(x))v(x)\tau^2 n^2}{(1+(1-p(x))v(x)\tau n^2)^2 p(x)} |d\mathbf{z}|$. When the parameter $v(x)\tau$ is large compared to $1/n$ then the number of hops is equivalent to $\int_C \frac{|d\mathbf{z}|}{v(x)\tau}$ and $T \sim \int_C \frac{|d\mathbf{z}|}{p(x)v(x)}$. Notice that the optimal path may vary when T changes, for example if mobility model is uniform on the network domain then when T is large optimal path will be straight lines. In other word the curvature of optimal paths may also depend on the time component. Of course all the quantities x, $v(x)$ and $p(x)$ may also vary on the spatial domain leading to further optimization.

5. Conclusion and perspectives

It seems that propagation lines don't change when the route optimization criterium changes. For example if hop number is changed in packet total delay time, the route should basically remains the same. The reason for this conjecture is that the condition of traffic at any given point in the network location is basically the same *modulo* an homothetic factor $\lambda(\mathbf{z})$. The only aspect which changes is the distance travelled by the packet per hop, but the delay per hop will be the same in distribution. We have a similar point about bandwidth allocation criterium. The transmission at any point will take the same amount of bandwidth, only the per hop distance travelled by the packet will differ. Under this perspective simple shortest path algorithms such as OLSR ? should be asymptotically close to optimal.

References

P. Jacquet, "Element de théorie analytique de l'information, modelisation et évaluation de performances," INRIA Research Report RR-3505, http://www.inria.fr/rrrt/rr-3505.html, 1998.

P. Gupta and P.R. Kumar. Capacity of wireless networks. Technical report, University of Illinois, Urbana-Champaign, 1999. http://citeseer.nj.nec.com/gupta99capacity.html

F. Baccelli, B. Blaszczyszyn, P. Mühlethaler, "A spatial reuse Aloha MAC protocol for Multihop wireless mobile networks," INRIA Research Report RR-4955, 2003. http://www.inria.fr/rrrt/rr-4955.html

M. Grossglauser, D. Tse, "Mobility increases the capacity of ad hoc wireless networks," INFO-COM 2001, citeseer.nj.nec.com/grossglauser01mobility.html.

T. Clausen, P. Jacquet, "The Optimized Link State Routing protocol (OLSR)," IETF RFC 3626, 2003.

P. Jacquet, "Geometry of information propagation in massively dense ad hoc networks," INRIA Research Report RR-4992, 2003.

CLUSTER-BASED LOCATION-SERVICES FOR SCALABLE AD HOC NETWORK ROUTING

S. Sivavakeesar and G. Pavlou
Center for communication systems research, University of Surrey, Guildford, Surrey GU2 7XH, United Kingdom

Abstract: We propose a location-service to assist location-based routing protocols, realized through our Associativity-Based Clustering protocol. The main goal of our scheme, which employs hierarchical principles, is to minimize the control traffic associated with location-management. In location-based routing protocols, the control traffic is mainly due to location-updates, queries and responses. Our scheme employs a novel geographically-oriented clustering scheme in order to minimize control traffic without impairing performance. In our location management scheme, nodes are assigned home-zones, and are required to send their location-updates to their respective home-zones through a dominating-set. This strategy, unlike similar location-management approaches, minimizes inevitable superfluous flooding by every node, and prevents location updates and queries from traversing the entire network unnecessarily, hence conserving bandwidth and transmission power. We evaluate our proposed scheme through simulations, and the results indicate that our protocol scales well with increasing node-count.

Key words: Ad hoc Networks, Scalable Location Management, Hierarchical Clustering Protocol, Scalable Routing.

1. INTRODUCTION

A mobile ad hoc network (MANET) is a network without any pre-existing infrastructure, and this paper considers the problem of routing in such networks of large-scale. In multihop MANETs, the routing protocol is the key to efficient operation. However, the design of an effective and efficient routing protocol in MANETs is extremely challenging because of mobility, limited battery energy, unpredictable behavior of radio channels, and time-varying bandwidth[2]. The absence of fixed infrastructure means that the mobile nodes (MNs) communicate directly with one another in a peer-to-peer fashion, and requires routing over multihop wireless paths. The main difficulty arises from the fact that multihop paths consist mainly of wireless links whose endpoints are likely to be moving independently of one another. Consequently, node mobility causes the frequent failure and re-activation of links, effecting a reaction from the network's routing algorithm to the changes in topology, thus increasing network control traffic and contributing to congestion. This routing task becomes extremely challenging when the network grows in size, and when two problems such as increasing node-density and large number of nodes have to be tackled. High node-density, where a node is within a radio-range of a large number of neighbors, often leads to superfluous forwarding of routing related control traffic, and large network size necessitates the maintenance of large routing tables. These two features are inter-related, and often affect the routing protocol scalability.

A considerable body of work has addressed on routing in MANETs, including a new generation of on-demand and efficient pro-active routing approaches[7,8]. These routing algorithms, however, tend to use flooding or broadcasts for route computation. While they can operate well in small networks, they incur heavy control traffic for discovery and maintenance of end-to-end routes, which forms a major bottleneck for large networks having node membership in the order of thousands over a large geographical area. In addition, flooding in MANETs does not work well due to the presence of hidden and exposed-terminals, and does not scale[10]. In recent years, a new family of protocols has been introduced for large-scale ad hoc networks that make use of the approximate location of nodes in the network for geography-based routing[3,4,7]. The amount of state information that needs to be stored by nodes in this case is minimal, because location-based routing does not use pre-computed routes for packet forwarding. As a result, link-breakage in a route does not affect the end-to-end session. These protocols, however, often need proper location-services, and hence location-management plays a vital role. Previous work in this area has shown that the asymptotic overhead of location-management is heavily dependent on the service primitives (location updates or registration, maintenance and discovery) supported by the location-management protocol of the location-service[3,4]. However, the location-registration or update cost normally dominates other costs for all practical purposes, and thus novel schemes are required to limit this control traffic. In our location-management scheme, we try to achieve this with an introduction of stable geographically-oriented clustering protocol, which we name Associativity-based Clustering protocol. This protocol does not involve any extra control traffic, and periodic *HELLO* messages as in AODV or other location-service approaches[1,8]. We use the concept of *virtual-clusters* we introduced in[1], and each *virtual-cluster* functions as a home-zone for a set of nodes. Nodes that reside within a *virtual-cluster* thus maintain approximate location information of a set of nodes, which select that *virtual-cluster* as their home-zone, in a distributed fashion. Since mobility is the main cause of uncertainty in ad hoc networks, our clustering protocol and algorithm takes this as the main criterion in order to select a relatively long-lived cluster head (CH) in each *virtual-cluster*[1]. Our strategy is to address the scalability issue both in dense and in large-scale networks. Scalability in dense networks is addressed efficiently by allowing only a few dominating set of nodes to make "summarized" composite periodic location-updates on behalf of a set of dominated nodes (i.e., CHs handle the location-updates on behalf of their members). This is to minimize superfluous flooding by every node to the entire network, as in other location-services[5]. Scalability in large-scale networks is addressed by strictly using geo-forwarding-based (location-based) unicasting as opposed to flooding even for location-registration process, and prevent location-updates, queries and replies from arbitrarily traversing unnecessary parts of the ad hoc network. As a result, the performance of our geo-forwarding-based routing strategy hardly degrades due to excessive control traffic, and from poor route convergence and routing-loops resulting from mobility. This paper thus deals with the problem of designing a location-update scheme in a scalable way to provide accurate destination information in order to enable efficient routing in mobile ad hoc networks.

The rest of this paper is organized as follows. Section II examines related previous work, and presents our motivation. The novel associativity-based clustering protocol and the proposed location-service technique are described in section III. Section IV evaluates the proposed scheme through simulations, and demonstrates that our location-service results in less signaling traffic in comparison to other similar methods. Section V presents our conclusions and future work.

2.   RELATED WORK AND OUR MOTIVATION

This section briefly explains the basic principles and problems associated with location-based routing protocols. It is then followed by a brief review of key location-service techniques and popular clustering protocols in the literature.

2.1   Basic Principles of Location-Based Routing

In location-based routing, the forwarding decision by a node is primarily
based on the position of a packet's destination and the position of the node's
immediate one-hop neighbours[7]. The position of the one-hop neighbors is
typically learned through one-hop broadcasts realized through periodic
beaconing or *HELLO* transmissions. In order to learn the current position of
a specific node, the help of a location-service is needed. MNs first identify
their location-servers, and register their current position with these servers
through location-update packets. When a node needs to know the location of
its desired destination partner, it contacts the appropriate location-server and
obtains this information. A location-query packet is used by the querying
node, and results in a location-response packet. There are three main packet
forwarding strategies for location-based routing: greedy forwarding,
restricted-directional flooding, and hierarchical approaches. In the first two
strategies, a node forwards a given packet to only one (greedy-forwarding)
or to more (restricted directional flooding) one-hop neighbors that are
located closer to the destination than the forwarding node itself. Recovery
strategies are needed in these two approaches, when there is no one-hop
neighbor that is closer to the destination than the forwarding node itself. The
third approach uses hierarchical principles for scalability reasons. There are
four key location-service strategies found in the literature: Distance routing
effect algorithm for mobility (DREAM) approach, quorum-based location-
service, grid location-service (GLS), and home-zone based location-service.
The details of these location-service approaches are briefly described next.

2.2   Related Work on Location-service

According to the DREAM approach, each node tries to maintain location
information about each other node in the network. This approach can be
regarded as an *all-for-all* approach, whereby all nodes are involved, and
every node maintains the location of all nodes. Due to the communication
complexity and periodic flooding of location-updates, DREAM is considered
the least scalable location-service technique, and thus inappropriate for
large-scale MANETs[7]. In the quorum-based location system, a subset of all
mobile nodes is chosen to host location databases. A virtual backbone is then
constructed between the nodes of the subset, using a non-location-based ad
hoc routing mechanism. This scheme, however, does not specify as to how
the virtual backbone nodes are selected and managed. Further, the quorum
system depends on a non-location-based ad hoc routing protocol for the
virtual backbone, which tremendously increases the implementation
complexity. In the case of grid location-service (GLS), the area that contains

the ad hoc network is divided into a hierarchy of squares[5]. In this hierarchy, n-order squares contain exactly four (n-1)-order squares, forming quadtrees. Since GLS requires all nodes to store information about some other nodes, it can be classified as an *all-for-some* approach. In GLS, the location updates of a particular node have to traverse the entire network, as the location-servers of a given node are spread throughout the network. In addition, whenever a querying node contacts the nearest location-server of another far-away node, whose location information is requested for, that query needs to be forwarded to the node being queried. This forwarding is based on the location information maintained by a far-away server, which is nearest to the querying node. In this case, the freshness of the information obtained is questionable, as nodes are required to update far-away location servers less frequently. This greatly depends on how quickly a particular entry in the location-server times-out or becomes stale. If, on the other hand, the frequency of updates increases, this may have a serious impact on scalability. Further, due to mobility, the role of location-servers for a particular node will keep on changing, and as a result the location-update and query packets may find it difficult to detect the appropriate location-servers. In the case of home-zone-based location-service, the position C of the home-zone for a node can be derived by applying a well-known hash-function to the node identifier. All MNs within a circle with radius R centered at C have to maintain location information for the nodes. The home-zone approach is also an *all-for-some* approach. If the home-zone is sparsely populated, R may have to be increased, resulting in several tries with increasing R for updates as well as queries. In this way, increasing or decreasing the R depending on node-density is very complex when it comes to practical implementation. Although our scheme makes use of similar home-zone strategy, the way in which the location-service is realized is both simple and scalable, as it will be explained in section III.

2.3   Related Work on Clustering

The purpose of clustering is two-fold: the first is to create a network of hierarchy, and the second is to select a dominating-set of nodes i.e. the cluster heads. This results in scalable network, where the location-server functionality is equitably distributed among network nodes. Choosing CHs optimally is an NP-hard problem[1]. Thus existing solutions to this problem are based on heuristic (mostly greedy) approaches and none of them attempts to retain the topology of the network[1]. Also, almost none of them consider node mobility as the main criterion in the clustering process. As a result, they fail to guarantee a stable cluster formation. In a MANET that uses cluster-based services, network performance metrics such as throughput and delay are tightly coupled with the frequency of cluster reorganization. Therefore, stable cluster formation is essential for better Quality of Service (QoS). The most popular clustering approaches in the literature are the lowest identifier (Lowest-ID) and maximum-connectivity. But these two, along with others, do not provide a quantitative measure of cluster stability.

2.4   Our Motivation

As mentioned before, the location-based routing strategy is chosen in order to improve scalability. On the other hand, the location management cost should not be high. The adoption of a hierarchical strategy together with the use of a dominating-set demonstrates as to how the control traffic is minimized without compromising route computation accuracy. For analysis purposes, the ad hoc network is represented as an undirected graph G = (V, E), where V is the set of nodes in the graph, and E is the set of edges in the graph.

3.   HOME-ZONE BASED HIERARCHICAL LOCATION MANAGEMENT

3.1   Associativity-based Stable Clustering

In order to make our clustering mechanism scalable, we make use of the notion of *virtual-clusters* we introduced in[1]. The idea is that a geographical area is divided into equal regions of circular shape in a systematic way so that each MN can determine the circle it resides in if location information is available. In our scheme, each *virtual-cluster* has a unique identifier based on the geographic location, which can be calculated using a publicly known function. Each MN should have a complete picture of the locations of these *virtual-clusters* and their centres[1].

Our clustering protocol does not involve any extra control traffic; instead, periodic *HELLO* message – as in other similar location-management approaches – is enough. This clustering facilitates electing a dominating set i.e. the cluster heads (CHs). In order to maintain stable clusters, a new associativity-based criterion is used to elect CHs[9]. A node is elected as a cluster head, if it has the highest associativity-state with respect to its present *virtual-cluster*, and stays nearer to its *virtual-cluster*-centre (VCC), in comparison to other nodes in the same cluster. This implies spatial, temporal, and connection stability. Each MN periodically monitors its current speed, and whenever its speed is zero, it starts measuring its "associativity". This is because typically after an unstable migration period, there exists a period of stability, where a mobile node spends some "dormant" or "residence time" within the *virtual-cluster* before it starts moving again[9].

When a MN is stationary, it measures its associativity with respect to a particular *virtual-cluster* by periodic "ticks" that takes place every ASSOCIATIVITY_TICK_PERIOD. This process however does not involve any transmissions. In this way, any node X within the kth *virtual-cluster* that has its total number of ticks (n_{xk}) greater than $A_{threshold}$, will exhibit higher degree of "associativeness", and hence have greater "dormant time". If, however, the speed of the MN is monitored to be greater than μ_{TH} (a system parameter), its number of ticks immediately becomes zero. The heuristic used by our clustering scheme is given by equation (4). Any node X determines the criterion value (Ω_{xk}) in k^{th} cluster by calculating the following:

1. Its distance from the centre (VCC) of a particular *virtual-cluster*. Assuming node X, whose location co-ordinates at time 't' are ($x_{xk}(t)$, $y_{xk}(t)$), in the kth *virtual-cluster*, whose center's Cartesian co-ordinates are (x_{ck}, y_{ck}), its distance at time 't' can be calculated by:

$$d_{xk}(t) = \sqrt{(x_{xk}(t) - x_{ck})^2 + (y_{xk}(t) - y_{ck})^2} \qquad (1)$$

2. Each node stores the "residence" or "dormant time" in the last m number of clusters it has visited. This is basically the time period from the instance at which the MN's velocity is zero within a particular *virtual-cluster* and the instance at which it is more than μ_{TH}. Then the mean "dormant-time" in terms of number of "ticks" (N_{mean_x}) is calculated as follows. Assuming that the "dormant-time" of node X in the jth *virtual-cluster* is R_{X_j}. Then node X's mean "dormant-time" (R_{mean_x}) is determined by considering its "dormant-times" in the last *m* number of *virtual-clusters* as given by equation (2). The mean "dormant-time" in terms of number of "ticks" is derived from equation (3). With this, the clustering criterion value (Ω_{xj}) for node X in *virtual-cluster* j is determined from equation (4).

$$R_{mean_x} = \frac{\sum_{j=1}^{j=m} R_{X_j}}{m} \qquad (2)$$

$$N_{mean_X} = \frac{R_{mean_X}}{ASSOCIATIVITY\_TICK\_PERIOD} \tag{3}$$

$$\Omega_{xj} = \begin{cases} \dfrac{N_{mean_X} - n_{xj}}{d_{xj}(t)} & \forall d_{xj}(t) \neq 0, n_{xj} \neq 0 \\[2mm] \dfrac{N_{mean_X} - n_{xj}}{d_{min}} & \forall d_{xj}(t) = 0, n_{xj} \neq 0 \\[2mm] \dfrac{1}{d_{xj}(t)} & \forall d_{xj}(t) \neq 0, n_{xj} = 0 \\[2mm] \dfrac{1}{d_{min}} & \forall d_{xj}(t) = 0, n_{xj} = 0 \end{cases} \tag{4}$$

Accordingly, any node X that has the highest value for the clustering criterion Ω_x is elected in either a centralized way or in a distributed way, depending on whether the present CH is available or not[1]. Equation (4) tries to ensure that the resulting clusters are more stable, and have uniform coverage by the respective CHs. If the CH lies very near to a VCC, it can have a uniform coverage, and hence ensures that all member nodes of a *virtual-cluster* are connected to this CH directly or via k-hops, where k is bound by $D/(2R_{TX})$, and D is the diameter of the *virtual-cluster*, with R_{TX} being the transmission radius of a node. Ω_x is proportional to the expected "residence time", and inversely proportional to the distance from the respective VCC. The system parameter d_{min} ($\neq 0$) is the minimum value that $d_{xk}(t)$ can take. The structure of *HELLO* packet has been modified to include this criterion value, so that any node can know the neighbors' Ω-values. Depending on the current state and circumstances, any node can disseminate one of the following four different packet types: *JOIN*, *HELLO_CH*, *HELLO_NCH*, and *SUCCESSOR*[1]. These control packets, except periodic *HELLO* packets by CHs, are relayed by intermediate MNs only within the *virtual-cluster*, where they have originated from. The distributed operation and the bootstrapping process of this clustering protocol are similar to those we proposed in[1]. Whenever a present CH knows that it is going to leave the *virtual-cluster* it is currently serving, it will select a member node that has the highest value for the criterion value as its successor, and inform about it the cluster members through the SUCCESSOR packet. In this case, the CH is elected in a centralized manner. The present CH can decide it is going to leave the cluster, when its monitored speed at a moment exceeds μ_{TH}. Unlike in any other clustering algorithm, our algorithm has another unique feature in that whenever a CH leaves the *virtual-cluster*, it will loose its CH status. In this way this algorithm ensures that no other visiting MN can challenge an existing CH within a particular *virtual-cluster*, and thus causing transient instability. All aspects of our strategy ensure that stable CHs are elected, and thus results in stable clustering. This clustering scheme is thus fully distributed, where all the nodes share the same responsibility and act as CHs depending on the circumstances.

In our location management scheme, we maintain a two-level hierarchical topology, where elected cluster heads at the lowest level (level-0) become members of the next higher level (level-1)[8,11]. In our scheme, the CHs are needed, and thus are elected using the virtual clustering concept only at level-0. CH election is not triggered in level-1, where only the cluster-membership detail is maintained. Level-0 hierarchy is used for efficient location-updating, while level-1 hierarchy is used for resilience as explained in section III.B.

3.2   Homezone-based Hierarchical Location-Service

We make the following assumptions in our model: 1) the area to be covered is heavily populated with mobile nodes, 2) Heavy-traffic is expected within the network (i.e., multiple simultaneous communications among nodes are possible), 3) every node is equipped with GPS (Global Positioning System) capability that provides it with its current location, and 4) there exists a universal hash-function that maps every node to a specific home-zone based on the node's identifier[3,4].

A home-zone is basically a *virtual-cluster* that has a unique identifier. Any node that is present in that *virtual-cluster* is responsible for storing the current locations of all the nodes that select this cluster area as their home-zone, and hence functions as a location-server. Since our location-management scheme requires that all nodes store the location information on some other nodes, it can be classified as an *all-for-some* approach, which can scale well[7]. The static mapping of our hash-function, as given by equation (5), is to facilitate simplicity and distributed operation. This hash-function has to be selected such that, i) all MNs should be able to use the same function to determine the home-zone of a specific node, ii) every *virtual-cluster* has the same number of nodes for which the MNs residing within that cluster should maintain location information, iii) and the mapping functionality has to be time-invariant.

$$hf(NodeIdentifier) \rightarrow Home - Zone \tag{5}$$

In any location-service, nodes are required to update their location information depending on their mobility. In our scheme, the update generation is mobility-driven as well as time-driven. The time-driven approach is to make sure that even if a node is stationary, periodical update is made to its home-zone. The unique aspect of our location-management scheme is that it tries to minimize the location-update cost with the help of dominant-set elected at level-0 using our clustering protocol. Dominant-set is basically a set of CHs elected at level-0 using our virtual-clustering principles. Accordingly, each node maintains four different table types: neighbor-table, location-cache, location-register, and forwarding-pointer-table. Each node has two different neighbor-tables maintained separately at each hierarchy level, and one of each of the other three types of tables maintained at level-0 only. Only the CH at level-0 has entries in its neighbor-table maintained at level-1. Neighbor-table at level-0 is used by every node to maintain the members of a particular *virtual-cluster* together with its one-hop neighbors, irrespective of their cluster identity. The periodic *HELLO* messages within a specific *virtual-cluster* are used to maintain this neighbor-table at level-0. As mentioned before, nodes of a specific cluster can relay *HELLO* packets of another node only when the latter resides within the same *virtual-cluster*[1]. However, when a bordering node receives a *HELLO* from a node of a different *virtual-cluster*, the former maintains the details of the latter in the neighbor-table for geo-forwarding purposes. Periodic *HELLO* messages by CHs have to be unicast by gateways between CHs of adjacent *virtual-clusters* to an extent that can be limited for scalability. This *HELLO* dissemination among neighboring CHs at level-1 facilitates maintenance of neighbor-table by CHs. Location-register of a node within a specific *virtual-cluster* has the location information of MNs whose home-zone is identical to the *virtual-cluster* of the former. The location-cache of a node is updated, whenever that node happens to know the location information of another non-member and non-home-zone node, for example, during location-discovery or data handling.

From the neighbor-table at level-0, any CH knows about its members, which may have different home-zones. In our scheme, the cluster head gathers the location information of its member nodes that have a common home-zone. Unless these nodes are highly mobile, the CH generates a "summarized" single location-update towards that common home-zone, on behalf of its member nodes. As a result, the need for every node, especially in a high-density network, to generate individual location-update packet is minimized. On the other hand, any node with high-mobility has to make its own location-updates, if it has not become a member of any cluster. In our scheme, it is assumed that whenever any node's speed increases beyond μ_{TH}, it will not be considered as a member of any cluster, and thus required to initiate its own location-updates. As specified in[1], in addition to its mere presence in a *virtual-cluster*, any node has to be included within the neighbor-table of the respective CH. In location-updates and data handling, the absolute locations of nodes are not needed; instead, the *virtual-cluster* id (VID) is enough. This is possible because of two reasons: 1) since these IDs are unique, from the VID any forwarding node we can obtain the co-ordinates of the corresponding VCC, and use it for geo-forwarding, 2) only for inter-cluster packet forwarding location-based routing is used, while at the local cluster-level proactive distance vector routing is used. Whenever a node in a home-zone receives the location-update which is meant for that home-zone, it can stop geo-forwarding that packet any more. Instead, it updates its location-register and informs other nodes within the same home-zone (i.e. *virtual-cluster*) through a periodic *HELLO* packet, which includes the location-register maintained by that node, so that other member nodes can update their location-registers. Within a *virtual-cluster*, efficient broadcast is utilized as opposed to flooding. This broadcast is based on reliable unicast realized through constant interaction between MAC-level and routing-level as used in "core-broadcast"[10]. A node whose speed exceeds μ_{TH} makes its own updates. As long as such a node's total number of *virtual-cluster* boundary-crossing so far is less than a threshold, it makes use of "forwarding-pointer" concept for correct data forwarding[4,5]. Accordingly, whenever such node moves out of its present *virtual-cluster*, it leaves a "forwarding-pointer" in the previous cluster without initiating a location-update up to its home-zone. Hence, any packet that has been geo-forwarded based on the old cluster ID can still traverse the chain of forwarding pointers to locate the user at its present location in a different cluster.

As in any location management approach, whenever a node needs location information of another node, the former has to first find the location-server (home-zone) of the latter and initiate the location-discovery process. In our strategy, the querying node uses the same mapping hash-function to determine the home-zone of its desired communicating partner. It then geo-forwards the location-query packet to that home-zone. Location-response can be initiated by the node being queried or any intermediate node as long as it contains the "fresh" information about the node being queried, or any node in the home-zone of the node being queried. In order to enable "freshness" of location information, each entry in any of the four tables maintained by each node is subject to a time-out mechanism. The use of sequence numbers achieves the same effect, in addition to avoiding routing-loops that may be introduced by mobility[8]. In addition, due to the way location-servers (home-zones) are maintained in our location strategy, the location-update or location-query packets can be unicast using geo-forwarding principles as opposed to flooding as used in other similar approaches (for e.g. in GLS[5]). This, in turn, prevents location-update and query packets from traversing unnecessary parts of the MANET. This minimizes the control traffic, and conserves scarce bandwidth and transmission energy. In the worst case, when a querying MN has not received any location-response within the LOCATION_RESPONSE_TIME_OUT period, after having tried for MAX_LOC_QUERY_RETRIES, it will start gradually flooding its location-query in the network.

The maintenance of level-1 hierarchy is for resilience purposes. Accordingly, the level-1 cluster hierarchy members periodically update each other with the fresh location information of the nodes maintained in the four different tables. This update is subject to different scopes for scalability, as in the case of fisheye state routing (FSR)[8]. This is beneficial, in case there are no nodes in a specific *virtual-cluster*. The nodes that select such a *virtual-cluster* as their home-zone do not necessarily know about its emptiness, and they may continue to send location-updates either through their level-0 CHs or by themselves. By exchanging such information by CHs, adjacent clusters may maintain location-information (in location-cache) for nodes that select the empty *virtual-cluster* as their home-zone. With this approach, any location-query that is directed to the empty *virtual-cluster* for location information about nodes whose home-zone happens to be the empty *virtual-cluster*, can still receive location-response from adjacent clusters.

4.   EVALUATION THROUGH SIMULATION

We chose GLS – another similar location-management approach – in our attempt to compare the performance of our strategy. For this purpose, we implemented our location-management strategy based on associativity-based clustering protocol and GLS in GloMoSim[1,12]. The distance between any two VCCs in our scheme is 200m. Each node moves using a random waypoint model, with a constant speed chosen uniformly between zero and maximum speed, which is here taken as 10 ms^{-1}. Each scenario was run for a 300 simulated seconds. Due to space limitation, we analyze the scalability of our location-service only in terms of increasing node-count. In order to properly model increasing network sizes, the terrain-area is also increased with an increase in the number of nodes |V| so that the average node-density (γ) is kept constant. The number of nodes is varied from 20, 80, 180, 320, 500 and 720. The terrain-area size is varied so that the average node-degree remains the same and accordingly 200X200 m^2, 400X400 m^2, 600X600 m^2, 800X800 m², 1000X1000 m² and 1200X1200 m² were selected for each run. The transmission range of each node is 100 m, and the wireless link capacity 2 Mbps. Traffic is generated using random CBR connections having a payload size of 512 bytes. These CBR connections are randomly generated so that at any moment the total number of source-destination pairs is kept constant – and each session lasts for a time-period which is uniformly distributed between 40 and 50 seconds.

Fig. 1 shows the average control cost (routing related cost) incurred per node for different CBR traffic sessions. Here the control cost includes periodic location-updates, *HELLO* packets, location queries and responses. As it can be seen, the control cost is lower in the case of our location-service, and hence our scheme outperforms GLS. Fig. 2 demonstrates the fact that in our location-management strategy the location-update packets don't traverse the unnecessary parts of the network. This is determined by considering the average number of location-updates that are relayed by each node in the network. As it can be seen from Fig. 2, although location-updates are much lower than that of GLS, the average number of location-update packets relayed by any node in our scheme slightly increases with the number of nodes. This can be attributed to the hash-function selected for simulation purposes. As the number of nodes within the network increases, the terrain-area size is also increased. With this increase, the total number of *virtual-clusters* or home-zones within the considered area also increases. As a result, some nodes may select far-away *virtual-clusters* as their home-zones.

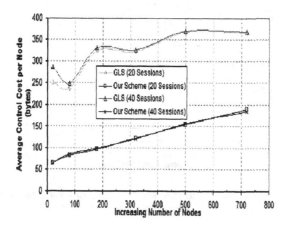

Figure 1. Average control cost incurred per node as a function of increasing number of node-count

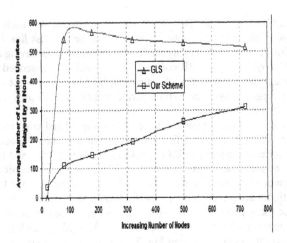

Figure 2. Average number of location-updates packets relayed by a node as a function of increasing number of nodes

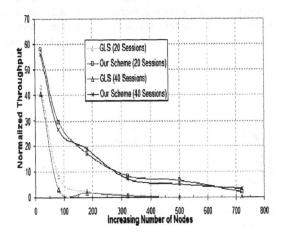

Figure 3. Normalized throughput as a function of increasing number of nodes and increasing number of sessions

Fig. 3 depicts the normalized throughput of both schemes as a function of increasing number of nodes under different traffic scenarios. The normalized throughput is defined as the total number of packets actually delivered to their respective destinations divided by the total number of packets generated within the whole network. Although the throughput in our scheme is higher than that of GLS, the throughputs in both schemes tend to decrease as the number of nodes increases. This is due to the fact the link capacity was 2 Mbps, and it poses the main bottleneck in the scenarios considered.

5. CONCLUSIONS AND FUTURE WORK

In this paper we presented the design and performance of an efficient location-service based on our novel associativity-based clustering strategy. By employing the dominating-set (CHs) to perform periodic location-updates on behalf of other nodes, we have demonstrated that our scheme leads to less control traffic when compared to the GLS. In addition, our location-management strategy conserves scarce resources such as battery energy and wireless bandwidth by preventing the location-updates, queries and responses from traversing the unnecessary parts of the ad hoc network. Mathematical analysis and simulation results confirmed the performance advantages of our scheme. In our future work, we have decided to construct longevity routes based on the proposed location-service as our next step in realizing quality of service routing. We plan to report such findings in future papers.

ACKNOWLEDGEMENT

This work was undertaken in the context of the UK Engineering and Physical Sciences Research Council (EPSRC) Programmable Ad-hoc Networks (PAN - GR/S02129/01) research project.

REFERENCES

1. S.Sivavakeesar, and G.Pavlou, "Stable Clustering Through Mobility Prediction for Large-Scale Multihop Intelligent Ad Hoc Networks", IEEE WCNC, Mar. 2004.
2. X.Hong, M.Gerlo, Y.Yi, K.Xu, and T.J.Kwon, "Scalable Ad Hoc Routing in Large, Dense Wireless Networks Using Clustering and Landmarks", *Proc. Int'l Conf. on Communications.* (ICC 2002), vol. 25, no. 1, Apr. 2002, pp. 3179 – 3185.

3. S-C.M.Woo, and S.Singh, "Scalable Routing Protocol for Ad Hoc Networks", Wireless Networks, Kluwer Academic Publishers, vol. 7, no. 5, Sep. 2001, pp. 513 – 529.

4. S.J.Philip, and C.Qiao, "ELF: Efficient Location Forwarding in Ad Hoc Networks", *Proc. Global Telecommunications Conference*. (Globecom 2003), vol. 22, no. 1, Dec. 2003, pp. 913 – 918.

5. J.Li, J.Jannoti, D.S.J.De Couto, D.R.Karger, and R.Morris, "A Scalable Location Service for Geographic Ad Hoc Routing", *Proc. 6th Int'l Conf. on Mobile Computing and Networking* (Mobicom 2000), Aug. 2000, pp. 120 – 130.

6. J.Sucec and I.Marsic, "Location Management for Hierarchically Organised Mobile Ad hoc Networks", Proc. IEEE WCNC 2002, March 2002, pp. 603-607.

7. M.Mauve, J.Widmer, and H.Hartenstein, "A Survey on Position-Based Routing in Mobile Ad Hoc Networks", IEEE Network, vol. 15, no. 6, Nov./Dec. 2001, pp. 30 – 39.

8. X.Hong, K,Xu, and M.Gerla, "Scalable Routing Protocols for Mobile Ad Hoc Networks", IEEE Network, vol. 16, no. 4, Jul./Aug. 2002, pp. 11 – 21.

9. C-K.Toh, G.Lin, and M.Delwar, "Implementation and Evaluation of an Adaptive Routing Protocol for Infrastructureless Mobile Networks", *Proc. Int'l Conf. on* Computer Communications & Networks (IEEE IC3N), Las Vegas, 2000.

10. R.Sivakumar, P.Sinha, V.Bharghavan, "CEDAR: A Core-Extraction Distributed Ad hoc Routing Algorithm", IEEE Journal on Selected Areas in Communications, vol. 17, no. 8, August 1999, pp 1 – 12.

11. J.Sucec, and I.Marsic, "Clustering Overhead for Hierarchical Routing in Mobile Ad Hoc Networks", Proc. IEEE INFOCOM'2002, June 2002, pp. 1698 – 1706.

12. X.Zengu, R.Bagrodia, and M.Gerla, "GloMoSim: A Library for Parallel Simulations of Large-scale Wireless Networks", Proceedings of the 12th Workshop on Parallel and Distributed Simulations, May 1998.

ON SELECTING NODES TO IMPROVE ESTIMATED POSITIONS

Erwan Ermel,[1,2] Anne Fladenmuller,[1] Guy Pujolle,[1] and André Cotton,[2]

[1]*Laboratoire d'Informatique de Paris 6*
Université Pierre et Marie Curie
Paris, France
firstname.lastname@lip6.fr

[2]*Thales Communication*
BGCOM/TCF/SEA/TAI
Colombes, France
firstname.lastname@fr.thalesgroup.com

Abstract

We consider node localization problems in ad hoc wireless networks in which two types of nodes are considered: nodes with self-locating capability like GPS and nodes with no self-locating capability. Our simple algorithm for improving the position accuracy consists of selecting and processing only the nodes that are likely to enhance the position estimation. We focus our approach on defining a hull of neighboring nodes as key of position accuracy enhancement.

1. Introduction

Recent developments of wireless technologies have allowed new types of networks to be envisaged. Mobile *ad hoc* networks (MANETs), which consist of wireless hosts establishing multi-hops communication with each others in the absence of fixed infrastructure, represent one of the current challenge of the networking research community. Many applications are considered for these new mobile ad hoc networks, for instance military ones to establish communications in war theatres, for disasters relief or more basically to set a network in large zones when it is difficult to install a cable-based infrastructure.

Routing in such conditions is quite challenging. For instance, due to the unpredictable mobility of nodes, the network topology is inherently unstable. Important aspects to consider include, the robustness of the solution, its compatibility with existing solutions in wired environments, and its availability of routes. Most existing routing proposals are topology-based routing protocols such as DSR [Johnson et al., 2001], AODV [Perkins, 2001], OLSR [Clausen and Jacquet, 2003]. However, with these schemes, when the number of nodes increases, the size of the routing tables grows considerably, they become difficult to maintain and therefore their scalability quickly becomes a complex issue. Another approach has been proposed which overcomes this problem: position-based routing protocols, such as GPSR [Karp and Kung, 2000] based on the Greedy-Face-Greedy algorithm [Bose et al., 2001], and Geocasting [Navas and Imelinski, 1997]. Such an approach does not require any routing table as routing is based on the geographical position of nodes. Nevertheless, such an approach requires that each node has its own position coordinates to be part of the routing protocol. This can be achieved, if all nodes are equipped of positioning equipments such as a GPS type systems but this hypothesis can be seen as restrictive. As a matter of fact, for cost reasons, it is more than likely that ad hoc networks will be composed of heterogeneous nodes: some will have self-locating capability such as GPS [Hofmann-Wellenhof et al., 1997] or Galileo [Galileo,], whereas some others, that we'll refer to as *simple* nodes, will have to estimate their position.

Position estimation methods are mostly based on geometrical computations like triangulation and trilateration. To evaluate the distance between two nodes, four classes of methods can be defined. The first one consists of estimating the distance by estimating the *time-of-flight* of a signal between two anchors. Time Of Arrival (TOA) is used in [Capkun et al., 2002, Werb and Lanzl, 1998] and in all radar systems while the Time Difference Of Arrival (TDOA) technique is used in [Savvides et al., 2001, Ward et al., 1997] (differential time from two anchors or from two different signal like radio and ultra-sound). The second class is based on the strength of the signal. For a given emitting power, the distance can be estimated as the signal strength decreases with the distance. Radars equipments are also based on this technique [Savarese et al., 2001, Beutel, 1999]. The third class is based on triangulation. The Angle of Arrival (AoA) estimates the direction of an incoming signal from several anchors, and then estimates the position. This method is used in VOR systems [vor,]. The fourth and last class merges all the remaining techniques like connectivity based approach [Doherty et al., 2001][Niculescu and Nath, 2003][Bulusu et al., 2000], Indoor Localization Systems [Want et al., 2000].

In [Ermel et al., 2004], we propose a simple convex hull selection to enhance the accuracy of the position estimation process. We extends in this paper this approach by selecting nodes by their position and also their position accuracy.

In this paper, we propose to enhance the accuracy of the estimated position without adding any new constraint on the environment so this proposal could be used for each of the position estimation method stated above. The structure of this paper is done as follow. We first present the assumptions and the definition made in this paper. Section 3 details two hull methods to select anchors within neighbors nodes, followed in Section 4 by our simulation results. Section 5 concludes the paper.

2.   Assumptions and definitions

We limit our approach to a one-hop anchor selection but the technique is also feasible for n-hops node selection. No distance measurement is to be used between nodes to estimate the position of a simple node. Therefore a simple node only exploits its own connectivity to other nodes in its direct neighborhood.

Let S be a *simple* node. Let S_{est} be the estimated position and S_{real} be the coordinates of the real location of S. As S_{real} is to be estimated by S_{est}, S_{real} information is only used by simulations.

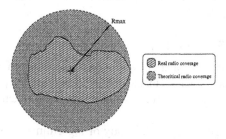

Figure 1.   Theoretical and real radio coverage.

Let R_{max} be the maximum theoretical transmission range of node S (Fig. 1). We also define the accuracy of the node position C_{acc}, which is function of the localization error represented by the distance between S_{real} and S_{est}:

$$C_{acc} = 1 - \frac{\left|\overline{S_{real}S_{est}}\right|}{R_{max}} \tag{1}$$

where $|\overline{AB}|$ is the distance between A and B. By construction, $0 \leq C_{acc} \leq 1$. The position estimation process may badly estimate S_{est} and somehow an impossible geometrical case occurs: S_{est} doesn't lie into the radio coverage area of S i.e. $\left|\overline{S_{real}S_{est}}\right| \leq R_{max}$. Thus a minimum function has to be used to solve this problem:

$$C_{acc} = 1 - \frac{min\left(\left|\overline{S_{real}S_{est}}\right|, R_{max}\right)}{R_{max}} \tag{2}$$

Self-locating nodes with accurate position, like position given by a GPS have a position accuracy of 1. Simple nodes which have to estimate their position have a position accuracy $0 \leq C_{acc} < 1$. As real positions are unknown for simple nodes, simple nodes have to estimate their position but also the precision of their position. Several methods to estimate the position accuracy, like statistical approach and area computing approach, are detailed in [Ermel et al., 2003].

In following sections, we assume that the nodes can estimate their position accuracy C_{acc} as a function of their estimated position and their neighborhood. We also define R_{error} as the maximum radius of the radio coverage of a node that takes into account the accuracy of its estimated position:

$$R_{error} = R_{max} + R_{max}(1 - C_{acc}). \qquad (3)$$

For example, for a self-locating node, $R_{error} = R_{max}$ while for simple nodes $R_{max} < R_{error} \leq 2R_{max}$.

3.   Anchors selection

Our main goal in this paper is to enhance the accuracy of an estimated position by selecting only anchors that are likely to improve the position estimation process. We detail in this section two hull selections schemes to achieve our goal.

Computational geometry is the study of algorithms for solving geometric problems on a computer. These problems are for example selecting a convex hull among a list of nodes, Voronoï diagrams, geometric searching. These algorithms are well detailed in [Preparata and Shamos, 1991, de Berg et al., 1997, O'Rourke, 1998, Lemaire, 1997].

The main idea of using a convex hull as a selection method among nodes is to choose only nodes that have the greatest distance between anchors. As the position estimation process is based on trilateration, the further the anchors are from each other, the better will be the accuracy of the estimated position.

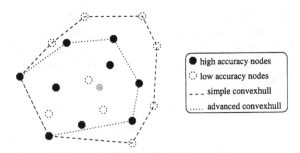

Figure 2.   Convex hull: simple convex hull considers only the distance metric to elect hull nodes while in advanced hull distance and position accuracy of the nodes are taken into account.

Only the hull nodes are taking into account in the position estimation process. The remaining nodes are simply discarded. A convex hull example is shown in Fig. 2.

We detail here two hull methods: the simple convex hull and the advanced convex hull.

3.1 Simple convex hull

In the plane, the convex hull of S, a set of points, can be visualized as the shape assumed by a rubber band that has been stretched around the set S and released to conform as closely as possible to S.

The simple convex hull selects only the nodes for their physical position whatever their position accuracy C_{acc} is.

An example of this simple convex hull definition is shown in Fig. 2.

3.2 Advanced hull

The simple hull selection phase chooses the border nodes of the neighborhood regardless of their position accuracy C_{acc}. Thus it seemed interesting to improve this process by selecting the nodes with the highest accuracy the closest to the ones chosen by the simple hull method. While the simple convex hull is defined with a single distance metric, we then defined an advanced hull with two metrics: the position and the position accuracy of the nodes.

As shown in fig. 2, if a simple greedy advanced convex hull is used, the two resulting hulls are quite different. As we worked from the assumption that the further the nodes are, the better the resulting position accuracy is, the use of accurate nodes appears not that far from being the best solution. Nevertheless, the selection of the inner nodes of the hull, but the closer from the hull border, is not a simple task.

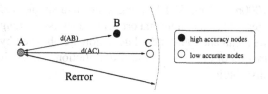

Figure 3. Virtual accuracy definition: accurate nodes may be chosen as an anchor despite their distance from the source A

Therefore, we define a Virtual Accuracy parameter V_{acc} of a node as:

$$V_{acc}^B = \frac{|\overline{AB}|}{R_{error}^A} C_{acc}^B \qquad (4)$$

where $|\overline{AB}|$ is the distance between a node A and an anchor B, R_{error}^A the maximum transmission range of A defined in Section 2, and C_{acc}^B the position accuracy of B.

The Virtual Accuracy V_{acc} merges the distance and the position accuracy metrics defined in the advanced hull into a simple metric. The advanced hull selection is divided into two steps:

- A simple convex hull selection to determine the nodes belonging to the convex hull.

- The selecting of the appropriate nodes as a function of their virtual accuracy V_{acc} parameter.

The second step consists of selecting nodes close to the hull with the highest virtual accuracy. Nodes that do not belong to the simple hull compare their virtual position accuracy to their nearest hull nodes; then the highest virtual accuracy node is selected as a member of the advanced hull. If the virtual accuracy of the nodes is the same, the node with the highest accuracy C_{acc} is selected. Example of this case is shown in Fig. 3.

Note that by selecting nodes with a higher virtual accuracy than hull nodes, the hull is no longer convex.

The simple convex hull selection is based only on a simple distance metric whereas the advanced hull is based not only on a distance metric but also on the position accuracy of the nodes. We will study in Section 4 the performances of these two hull selection methods.

4.   Simulation Results

The simulations were performed under Java. 50 nodes were randomly placed in a 1000m x 1000m square. Self-locating nodes and simple nodes were also randomly selected. The maximum theoretical transmission range R_{max} was set to 170m. The mean and the standard deviation are obtained by a maximum likelihood estimation of the mean of the β-distribution. Matlab has been used to solve the β-distribution.

4.1   Evaluation of the hull selection

We first evaluate the *fairness* of the selection between the simple nodes (sn) and self-locating nodes (sln). The selection may favor high accurate nodes to the detriment of others. De facto, our selection can be called unfair as the different classes of nodes are not equal to the selection process.

Let i (respectively j) be the number of sln nodes before the selection (respectively sn nodes). We store into a selection matrix SM_{sln} (respectively

SM_{sn}) the number of remaining sln nodes (respectively sn nodes) after the selection step as a function of i and j. $SM_{sln}(i,j)$ is the number of remaining sln nodes after the selection. We define also two other matrices FM_{sln} and FM_{sn} (Fairness Matrices):

$$FM_{sln}(i,j) = \frac{SM_{sln}(i,j)}{i} \text{ and } FM_{sn}(i,j) = \frac{SM_{sn}(i,j)}{j}$$

$FM(i,j)$ is the fraction of the number of nodes of a class after the selection by the initial number of neighbor nodes of the same class. Let X_{sln}^k and X_{sn}^k be a set of value defined as:

$$X_{sln}^k = \{FM_{sln}(i,j) \text{ with } i+j = k; \forall i \neq 0; \forall k \leq n\}$$
$$X_{sn}^k = \{FM_{sn}(i,j) \text{ with } i+j = k; \forall j \neq 0; \forall k \leq n\}$$

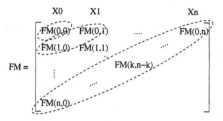

with for example

$$X_{sn}^1 = \{FM_{sn}(0,1)\}$$
$$X_{sln}^n = \{FM_{sln}(1,n) \ldots FM_{sn}(k,n-k) \ldots FM_{sn}(n,0)\}$$

At last, X_{sln}^k and X_{sn}^k are merged into a single term X^k as:

$$X^k = X_{sln}^k \cup X_{sn}^k$$

The *fairness index* **F** defined in [Jain, 1992] is now used on the X^k sets to estimate the fairness of the selection among the sln nodes and the sn nodes:

$$F_{index}(X^k) = \frac{\left(\sum_{i=1}^n x_i\right)^2}{n \sum_{i=1}^n x_i^2} \text{ with } x_i \in X^k. \qquad (5)$$

For all values of x_i, the *fairness index* **F** always remains between 0 and 1. If the selection is fair among the different classes of nodes, **F** is equal to 1.

The results of the fairness index are shown in Fig. 4 and Fig. 5

Fig 4 shows the fairness of the hull selection among the self-locating nodes and the simple nodes. From 1 to 3 neighbor nodes, no distinction is made in the selection process between the self-locating nodes and the simple nodes. These results are indeed expected as no selection is done under 3 neighbor nodes.

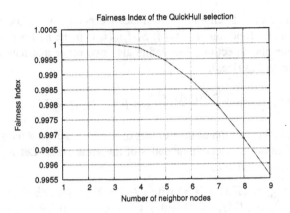

Figure 4.   Fairness index of the hull selection among the neighbor nodes.

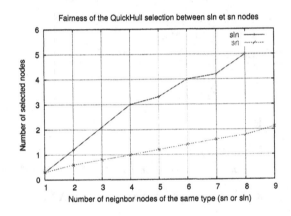

Figure 5.   The selection advantages the sln nodes to the detriment of sn nodes.

Above 3 neighbors, the fairness index decreases and proves that our selection is no fairer. However **F** only gives an idea of the fairness of the selection between the different nodes classes but gives no clues on which class is privileged i.e. we don't know if the sln or the sn nodes are privileged. As the values used in **F** are already averages of data, **F** decreases slightly from 1 to 0.9955. These values seem very low, but by construction of the averages, high variations of the fairness index can't be seen in a global view. But if comparisons are made between simple cases, the variations of **F** are noticeable.

As for Fig 5 the selection ratio between the sln nodes and the sn nodes is shown. For example, with 4 sln neighbor nodes, 3 are kept by the selection while for 4 sn nodes, only one of them is kept by the selection. Thus our selection really favors sln nodes.

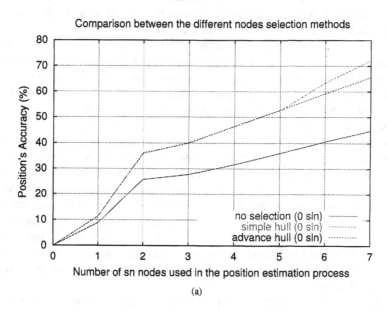

(a)

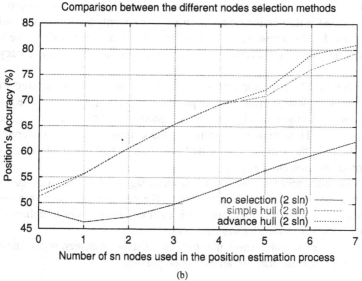

(b)

Figure 6. Evaluation of the effect of the nodes selection on the accuracy of the estimated position of the nodes.

Fig.6 shows the impact of our selection on the accuracy of an estimated position. To compare our method, we choose as the reference model the greedy scheme: all the neighbor nodes are selected in the position estimation process. None are discarded. The estimated position is obtained by a simple centroid formula, where all the nodes are given the same weight.

Fig. 6(a) shows the resulting accuracy of an estimated method when no self-locating nodes are present in the neighborhood (sln=0) while in Fig. 6(b) two sln nodes are present. In both figures, the simple selection gives in every cases better position accuracy than the greedy approach does. The selection enhances the position accuracy up to 20% for the simple hull selection.

While the simple hull selection enhances the position accuracy of an estimated position for every case, the advanced hull selection gives different results depending on the number of sln nodes selected in the position estimation process. When no sln nodes are used, like in Fig. 6(a), the resulting position accuracy is even worst than the one given buy the greedy method. These results are due to the fact that the advanced hull selection uses the position accuracy of the nodes. But in that case, there is not trustworthy node. Therefore the results show a *snow ball effect* on the position accuracy: position errors are widely propagated along the position estimation process. Thus the advanced hull selection is not suitable when no trustworthy nodes are present in the neighborhood.

Fig. 6(b) shows that the advanced hull selection gives better results than in Fig. 6(a). In this case, 2 sln nodes are used in the position estimation process. The advanced hull selection do not improve significantly the simple hull selection in low density networks (only 1 to 2 %). In high density cases, like with at least 7 neighbors, the advanced selection improves the position accuracy up to 5%.

The complexity of a simple convex hull is $o(n.ln(n))$ of its one step while advance hull selection need two steps, the first is a simple hull selection , $o(n.ln(n))$, and the second for searching good virtual nodes is performed in $o(n(n-1))$. Thus from the results and the complexity of the different algorithms, only the simple convex hull selection is really suitable as a good position estimation enhancement.

The position accuracy mainly takes advantages of the distance between nodes and of the network density. Using the position accuracy as a selection factor in a node selection process is then not a good way to enhance the position accuracy in low density networks.

The simulation results show that only the distance between nodes have an importance in the accuracy of an estimated position.

5.   Conclusion

We present and compare in this paper two simple methods to select anchors in a wireless network to enhance the position estimation of simple nodes, with no self-locating capabilities.

The first proposal consists of defining a convex hull among neighbor nodes while our second proposal extends the simple convex hull selection by using in addition to the distance metric the position accuracy of neighbor nodes . Whatever the hull method used to select the nodes, the resulting position accuracy is enhanced from a greedy scheme up to 20%. We also show that the accuracy of an estimated position only takes advantage of the distance between the hull nodes, whatever their position accuracy. Thus the simple hull selection gives better results than the advanced hull selection in low density networks, while in high density networks, the advanced hull selection improves slightly the simple selection.

Our next step consists of implementing such selection algorithms in a global geographical routing protocol in a heterogeneous network.

References

[vor,] *Flight Training Handbook*. Aviation Book Co.

[Beutel, 1999] Beutel, J. (1999). Geolocalisation in a picoradio environment. Master's thesis, ETH Zurich, Electronics Lab.

[Bose et al., 2001] Bose, P., Morin, P., Stojmenovic, I., and Urrutia, J. (2001). Routing with guaranteed delivery in ad hoc wireless networks. volume 7, pages 609–616.

[Bulusu et al., 2000] Bulusu, N., Heidemann, J., and Estrin, D. (2000). Gps-less low cost outdoor localization for very small devices. *IEEE Personal Communication, Special Issue on Smart Spaces and Environment*, 7(5):28–34.

[Capkun et al., 2002] Capkun, S., Hamdi, M., and Hubaux, J. P. (2002). Gps-free positioning in mobile ad-hoc networks. *Cluster Computing*, 5(2).

[Clausen and Jacquet, 2003] Clausen, T. and Jacquet, P. (2003). Optimized link state routing (olsr) protocol. *Internet Draft*.

[de Berg et al., 1997] de Berg, M., Kreveld, M., Overmars, M., and Scharzkopf, O. (1997). *Computational Geometry, Algoritms and Application*. Springer.

[Doherty et al., 2001] Doherty, L., Pister, K. S. J., and Ghaoui, L. E. (2001). Convex optimization methods for sensor node position estimation. In *Proceedings of IEEE INFOCOM'2001*, Anchorage.

[Ermel et al., 2003] Ermel, E., Fladenmuller, A., Pujolle, G., and Cotton, A. (2003). Estimation de positions dans des reseaux sans-fil hybrides. In *CFIP 2003*.

[Ermel et al., 2004] Ermel, E., Fladenmuller, A., Pujolle, G., and Cotton, A. (2004). Improved position estimation in wireless heterogeneous networks. In *Networking 2004*.

[Galileo,] Galileo. http://europa.eu.int/comm/dgs/energy_transport/galileo/.

[Hofmann-Wellenhof et al., 1997] Hofmann-Wellenhof, B., Lichtenegger, H., and Collins, J. (1997). *Global Positioning System : Theory and Practice*. Springer-Verlag.

[Jain, 1992] Jain, R. (1992). *The Art of Computer Systems Performance Analysis*. Wiley.

[Johnson et al., 2001] Johnson, D. B., Maltz, D. A., Hu, Y.-C., and Jetcheva, J. G. (2001). The dynamic source routing protocol for mobile ad hoc networks. *Internet Draft*.

[Karp and Kung, 2000] Karp, B. and Kung, H. T. (2000). Gpsr : Greedy perimeter stateless routing for wireless networks. In *Proceedings of ACM/IEEE MOBICOM'00*.

[Lemaire, 1997] Lemaire, C. (1997). *Triangulatation de Delaunay et arbres multidimensionnels*. PhD thesis, Ecole des mines de Saint-Etienne.

[Navas and Imelinski, 1997] Navas, J. and Imelinski, T. (1997). Geocast - geographic addressing and routing. In *Proceedings of ACM/IEEE MOBICOM'97*, volume 3, pages 66–76.

[Niculescu and Nath, 2003] Niculescu, D. and Nath, B. (2003). Dv based positioning in ad hoc networks. *Telecommunication Systems*, 22(1):267–280.

[O'Rourke, 1998] O'Rourke, J. (1998). *Computational Geometry in C*. Cambridge University Press.

[Perkins, 2001] Perkins, C. E. (2001). Ad hoc on demand distance vector (aodv) routing. *Internet Draft*.

[Preparata and Shamos, 1991] Preparata, F. P. and Shamos, M. I. (1991). *Computational Geometry: An Introduction*. Springer Verlag.

[Savarese et al., 2001] Savarese, C., Rabaey, J. M., and Beutel, J. (2001). Localization in distributed ad-hoc wireless sensor networks. *Proceedings of the ICASSP*.

[Savvides et al., 2001] Savvides, A., Han, C.-C., and Strivastava, M. B. (2001). Dynamic fine-grained localization in ad-hoc networks of sensors.

[Want et al., 2000] Want, R., Hopper, A., Falcao, V., and Gibbons, J. (2000). The active badge location system. *ACM Transaction on Information Systems*, 10(1):91–102.

[Ward et al., 1997] Ward, A., Jones, A., and Hopper, A. (1997). A new location technique for the active office. *IEEE Personal Communications*, 4(5):42–47.

[Werb and Lanzl, 1998] Werb, J. and Lanzl, C. (1998). A positioning system for finding things indoors. *IEEE Spectrum*, 35(9):71–78.

ENERGY-EFFICIENT MULTIMEDIA COMMUNICATIONS IN LOSSY MULTI-HOP WIRELESS NETWORKS

Al Harris, Cigdem Sengul, Robin Kravets, and Prashant Ratanchandani
Department of Computer Science
University of Illinois at Urbana-Champaign
{aharris,sengul,rhk}@cs.uiuc.edu

Abstract   A key concern in multi-hop wireless networks is energy-efficiency due to battery-power constrained mobile nodes. The network interface is a significant consumer of energy [7, 8, 15] causing a substantial amount of energy to be wasted by sending packets that cannot be used by the receiver. Given the small MAC layer packet sizes of wireless channels as compared to multimedia application data frames, inter-packet dependencies are formed (*i.e.*, the loss of a single packet renders a group of packets useless). In this paper, we present an application-aware link layer protocol to reduce the energy wasted by sending such useless data in lossy networks.

Keywords: Energy-efficient design, application-aware MAC, multi-hop wireless networks

1.   Introduction

The increase in the availability of mobile computing devices has led to the design of communication protocols for multi-hop wireless networks. However, wireless networking poses implementation challenges due to characteristics such as resource constraints (*e.g.*, battery power) and lossy links. Our research addresses the issue of energy-efficient multimedia communication in such wireless networks by revisiting per-hop mechanisms.

The key observation for wireless communication is that energy consumption is affected by decisions at all layers. Essentially, although high quality communication is an end-to-end issue, lossy communication channels are a MAC layer issue. Therefore, a cross-layer approach is necessary in multimedia communication where there are dependencies between packets. Blindly processing each packet at the MAC layer may waste energy from the transmission of unusable data. The majority of current research in multi-hop wireless networks assumes that each packet is an independent application layer frame, and so the loss of one packet does not impact the correctness of others. However, the small size of MAC layer packets (1.5KB for IEEE 802.11 [11]) does not support the framing of larger application layer data frames (*e.g.*, on the order of 10KB for MPEG I-frames). This mismatch forces applications to fragment their data across multiple MAC layer packets, creating dependencies between such packets (*i.e.*, the loss of one packet implies the loss of the entire frame). Therefore, while information about link quality is essential to achieve effective communication, the type and characteristics of application data also impact the success of energy conservation techniques.

The contribution of our research is an investigation of the impact of per-hop mechanisms on end-to-end communication in the presence of lossy communication channels. Energy savings are achieved by exposing application layer framing [5] information to the link layer. Our approach is based on the observation that data transmitted to the receiving application, but not usable by the application, represents wasted energy. To prevent the transmission of partial frames to the receiver, we propose the use of a link layer mechanism that tracks the transmission of individual packets of an application layer frame at each hop, dropping all the packets of a frame if a single packet is lost. This mechanism is based on the idea that a partial frame should be dropped as soon as possible and only packets belonging to complete frames should be forwarded. Additionally, combined with a hop-by-hop reliability mechanism (*e.g.*, as in IEEE 802.11 [11]), the proposed approach also compensates for transmission-based losses and increases energy efficiency.

This paper is organized as follows. Section 2 presents the motivation for developing energy-efficient protocols. Section 3 defines two performance metrics: *effectiveness* and *energy-efficiency*. Section 4 presents the application-aware link layer protocol in detail and Section 5 evaluates the effectiveness of our protocol via a simulation study. Finally, Section 6 presents conclusions and future directions.

2.   Energy Management in Multi-Hop Wireless Networks

Battery capacity is not increasing in step with the energy requirements of new mobile computing technology. Therefore, methods of saving energy must be designed to enable the use of mobile computing devices. Our approach uses application framing-aware link layer mechanisms to reduce the number of incomplete frames transmitted through the network. In this section, we discuss our approach in the context of current research in energy management and dropping policies in single-hop and multi-hop wireless networks.

2.1   Energy-Aware Communication

The concern for saving energy has spawned a large body of work on energy-saving routing protocols. In general, energy-aware routing protocols for multi-hop wireless networks focus on load balancing based on energy consumption and routing efficiency [3, 4, 23, 25]. However, such work does not consider efficient transmission of data once the energy efficient route has been found. Essentially, energy-efficient routing and our data-centric approach are orthogonal, so their benefits can potentially be combined.

FEC-based error recovery solutions [1, 10, 12, 14, 19, 24, 26] add overhead in the form of extra data needed to recover from errors and in the form of computation necessary to generate the codes for such recovery. For an FEC-based solution to be optimal, the code must match the error rate of the link. If the error rate is over-estimated, too much extra data will be included, wasting bandwidth and adding to the delay at each hop. However, if the error rate is under-estimated, all errors will not be corrected. For the estimation to be accurate across a multi-hop wireless environment, the FEC-code must be re-calculated at each hop, due to the fact that each hop has varying error rates.

Another energy-saving is reducing the transmitter energy [9, 18, 20], which reduces the amount of energy used during transmissions at the cost of reducing the effective transmission range, rate, and/or reliability. This type of solution has no concern for data relations between packets or for loss recovery.

2.2   Supporting End-to-End Communication with Hop-by-Hop Mechanisms

The judicious use of hop-by-hop mechanisms has been discussed in the context of both congestion control and support for end-to-end communication in last-hop wireless environments. While research in both these areas does not address energy management, the techniques are similar.

The problem of fragmentation wasting resources has been known for a long time: *The loss of any one fragment means that the resources expended in sending the other fragments of that datagram are entirely wasted [13].* While such waste could be reduced by ensuring the transmission of one application layer frame per link layer packet, such intelligent fragmentation may not always be possible. For example, the problem of having larger application data frames than ATM cells has been explored by Floyd [22]. In general, *fragmentation-based congestion collapse* is caused by bandwidth being wasted through the transmission of fragments of packets that cannot be reassembled at the receiver into valid packets due to the loss of some of the fragments during a congested period of the network. ATM with partial packet discard was shown to be helpful in combating such fragmentation-based congestion collapse [2, 16, 17]. One major difference between the partial packet discard approach and our approach is the desire to eliminate the transmission of partial frames. With partial packet discard, the initial packets (or cells) are transmitted until a loss is encountered. These packets traverse the entire path and are discarded at the receiver. While the latter packets do not add to congestion, these initial packets are not dropped. In our approach, we add a minimal amount of buffering which allows us to delay sending packets from any frame in which some packets are missing to prevent energy consumption from the initial packets.

In a similar vein, [21] proposed dropping subsequent packets of a video frame after the loss of one of the packets containing a fragment of the video frame. The dropping mechanism presented in Section 4 is an application of this technique to multi-hop wireless networks. Essentially, the focus of previous applications of packet discard techniques has been to avoid congestion collapse, however, even in the absence of congestion, when energy conservation is also a concern, packet discard techniques can be used to achieve energy savings, as will be shown in Section 5.

3.   Protocol Effectiveness and Energy Efficiency

The difficulty in designing MAC protocols is finding a balance between effectiveness and energy-consumption. We define *effectiveness* to be the goodput, or percentage of the data received at the end host that is usable by the end host application. It is clearly possible for an energy-conserving protocol to save energy but at the cost of very poor protocol effectiveness. For example, if an energy-aware protocol indiscriminately drops packets based on battery levels of the mobile nodes and a certain application has a frame size that spans two packets, it is possible that many of the packets that are received at the end host are unusable due to the fact that half of the frame was dropped by the protocol. In this situation, many of the received packets actually constitute wasted energy since the data in the packets is useless to the receiving application.

To factor the effect of useless packets into energy consumption analysis, we use an *energy efficiency* metric. Energy-efficiency (E_P) is the ratio of the number of usable packets (P_G) received at the end-host to the total energy (E_T) used in the transmission of a data stream ($E_P \equiv \frac{P_G}{E_T}$). Energy-efficiency ($E_P$) can be used to evaluate the effect of an energy-aware protocol on the application as well as its effect on the total energy consumption of the network and so, deciding whether or not the protocol is actually useful.

4.   Application-Aware Link Layer Protocol

Our approach to providing energy-efficient transport of data in a multi-hop wireless network uses information about application layer framing at the link layer. Through the use of knowledge about application layer framing, the link layer makes intelligent decisions to improve protocol effectiveness and energy-efficiency. To this end, we use two mechanisms, an intelligent dropping mechanism described in Section 4.2 and a link layer retransmission mechanism described in Section 4.3. The intelligent dropping mechanism achieves better energy-efficiency by reducing the number of unusable packets transmitted by dropping partial frames at each hop in the path. Furthermore, enabling retransmissions at the link layer achieves better energy-efficiency by reducing the number of unusable packets received at the end host (*e.g.*, as in IEEE 802.11b [11]). The transport protocol used on top of the proposed link layer protocol is described in Section 4.1.

The parameters that affect the performance of the link layer mechanisms are: the application layer frame size, the loss rate of the links, the hop count, and the mobility rate. The effects of variations in these parameters are explored in the rest of this section and in Section 5. Essentially, the proposed mechanisms achieve protocol effectiveness and energy-efficiency at the expense of an increase in delay and a small amount of extra buffer space at each node. These details are explored further in the remainder of this section.

4.1 Transport Protocol Support

To develop an energy-efficient link layer protocol for wireless multi-hop networks, some information about application layer frames needs to be exposed to the link layer. To this end, a two field header is added to each packet, which is filled at the transport layer. The first field, *frame_no*, contains the application frame number of the data. The second field, *frame_size*, contains the number of packets for this application layer frame. These fields contain the main parameters that are used by the link layer mechanisms. It is important to note that if the fields are not present, the link layer performs like a traditional link layer. This provides backwards compatibility. Also, packets sent with *frame_size* equal to one are treated as if being sent by a regular link layer and are transmitted immediately.

4.2 Intelligent Dropping Mechanism

The dropping mechanism is based on the simple idea that if any part of a frame is lost, the entire frame is useless. Therefore, each node only sends packets containing fragments of complete frames by using a simple mechanism previously proposed for wired networks for fragmented packets. Essentially, an extra buffer capable of holding two frames of data is kept at each node, where packets are buffered until a full frame is received. Once all the packets of a frame have been received, the packets in the frame are placed at the the tail of the send queue. Using this mechanism, no node sends any packets of a frame for which it is not in possession of the entire frame. This method dramatically reduces the number of incomplete frames received at any node in the path at the cost of increased delay from buffering the packets of a frame until all are received.

The link layer using this mechanism does not fail to send any packets that will be usable by the end host. Therefore, the only decrease in transmissions is due to the reduction in useless packets sent. There are two costs that are incurred by this protocol. The first is a delay cost due to the fact that packets are only sent after an entire frame has been buffered. The application layer frame size and the hop count directly affect this cost. As the application layer frame size increases the delay increases due to the fact that larger application layer frames are fragmented into a greater number of link layer packets that need to be buffered. Furthermore, as the hop count increases, the delay increases due to gathering all packets before forwarding. If the number of hops is k, the number of flows is f, and the number of packets in one application layer frame is N, for a shared channel we have the following: Normal Delay $\equiv f \times (N + k - 1)$ and Dropping Delay $\equiv f \times (N \times k)$. Therefore, the dropping mechanism creates a multiplicative delay, while a non-buffering link layer protocol incurs a linear delay with respect to the hop count and frame size. However, Normal Delay assumes perfect pipelining of packets at each hop, which may not be achievable in practice due to contention between nodes at both ends of the link. The increase due to the dropping mechanism is due to the collection of all packets in a frame at each hop, which may also potentially reduce such contention between nodes. The second cost incurred is the need for more buffer space at the nodes. The dropping mechanism requires enough buffer space to hold two frames. Therefore, the needed buffer size consequently increases as the frame size increases.

The higher the loss rate of each link, the more likely that partial frames will be created (through the loss of some of the packets in each frame). Therefore, as the loss rate increases, better energy-efficiency gains are expected. The overall results of this mechanism in Section 5 show not only better energy-efficiency than a standard link layer, but also lower total energy expenditures. As the rate of mobility increases, the chance of one half of a frame traversing one path, and another traversing another path increases. However, since this protocol is not currently mobility-aware, this causes frames that could have been transmitted successfully to be dropped. However, the benefits return once the flow settles to a new route.

4.3   The Retransmission Mechanism

To increase the number of usable frames received at the end host, the proposed link layer protocol utilizes link layer retransmissions. Essentially, MAC layer information about transmission errors is exposed to the link layer. When a transmission error is detected by the MAC layer, the packet involved in the error is retransmitted. An additional parameter for the retransmission mechanism is the number of times to attempt to retransmit a packet lost due to transmission error. For this paper each packet retransmission at the link layer is only attempted once. The packet retransmission limit is set to 7 for short packets and to 4 for long packets in IEEE 802.11b standard [11]. However, only one retransmission attempt is performed by our approach since multiple retransmissions cause additional delay and increase the power expenditure. If the retransmission of a packet fails, the rest of the remaining packets in the frame are dropped.

The cost of link layer retransmissions is extra delay added onto the delay incurred by the intelligent dropping mechanism. This delay is bound by the total number of retransmission attempts r for all flows: Retransmission Delay $\equiv$ $(f+r)(N*k)$. The effects of the parameters considered on this mechanism are essentially the same as the effects noticed for the intelligent dropping mechanism. As the hop count and the application layer frame size increases, the delay cost increases. The retransmission mechanism is also not mobility-aware and, therefore, performance degrades as mobility speeds increase. As the loss rate increases, the total number of retransmissions also increase. This leads to an interesting difference. Namely, while the total energy used by a link layer protocol using retransmissions may be higher than a standard link layer protocol, the energy-efficiency improves due to higher goodput performance (See Section 5). Therefore, a link layer protocol using retransmissions is more effective than a standard link layer protocol.

5. Evaluation

In this section, we present results from our simulation of three different link layer protocols. The first protocol is the *standard* link layer which does no buffering, dropping or retransmissions. The second is a application-aware link layer protocol, which implements the intelligent *dropping* mechanism. The third is a link layer protocol with a *retransmission* mechanism. All are simulated with the ns-2 network simulator [6]. For the application data, we use a CBR stream modified to include the two field header described in Section 4.1. All simulations use the IEEE 802.11 MAC layer. The simulations run in a $1000 \times 1000m^2$ area with 50 nodes. AODV is used for routing. Additionally, we use *random waypoint* mobility model. The effect of three parameters are tested in the simulations: the application layer frame size, the mobility rate, and the link error rate. The link error rate is modeled using a random probability of each packet being dropped. Our simulation results represent an an average of five runs with identical traffic models but different randomly generated network topologies. The effectiveness of the protocols are evaluated using the following metrics: the number of complete frames received at the end host, the number of partial frames received at the end host, the average end to end delay, the total number of MAC layer transmissions, and the average hop count. Additionally, we provide comparisons based on the *energy-efficiency* metric described in Section 3.

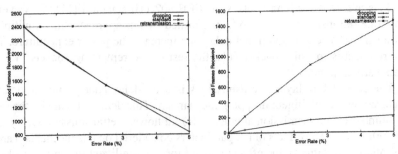

| Figure 1. | Complete frames at receiver. | Figure 2. | Partial frames at receiver. |

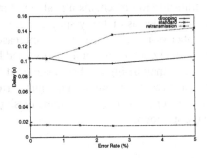

Figure 3.   End-to-end Delay

5.1   Effects of Error Rate on Performance

To evaluate the impact of error rate, simulations were run that maintain an average hop count of five, a frame size of four, and a mobility rate of 0.1 m/s. The error rates used are 0%, 0.5%, 1.5%, 2.5%, and 5%. The simulation results show that the dropping mechanism delivers slightly fewer complete frames to the receiver than the standard link layer (see Figure 1). Since nodes do not forward partial frames, any loss on each link is from complete frames. As expected, the link layer with retransmissions delivers significantly more complete frames as the loss rate increases. However, the dropping mechanism successfully limits the number of partial frames delivered to the receiver (see Figure 2). The partial frames that do get to the receiver are due to errors on the last link. These can obviously not be avoided by a dropping mechanism.

The main cost of the proposed mechanism is the delay incurred by buffering. The results, as expected, show that the delay is proportional to the frame size (see Figure 3). This is due to the fact that at each hop, before the first packet in a frame can be sent, the rest of the packets in the frame must be received. For the 5 hop network with a frame size of 4, the delay cost is 80ms. As expected, our experiments show that the delay increases linearly with the size of the

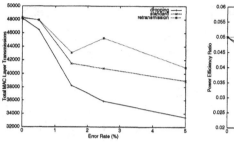

Figure 4.   MAC layer transmissions.

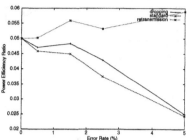

Figure 5.   Power-efficiency.

application layer frames. As the error rate increases, the link layer using the retransmission mechanism incurs more delay as the need for retransmissions increases.

To evaluate the energy consumption of each protocol, the total number of MAC layer transmissions is used (since this metric translates directly to energy consumption). Figure 4 illustrates that the standard link layer falls in between the augmented link layers. The dropping mechanism has significantly fewer transmissions than the others due to the fact that it drops any incomplete frame. It is important to combine these results with the results depicted in Figure 1 to see that while many packets were dropped, this is not noticed by the end application since the number of usable frames remains approximately the same. Obviously, the retransmission mechanism uses more transmissions. Although this may be interpreted as failure of the retransmission mechanism to achieve the goal of being energy-efficient, the comparisons based on the energy-efficiency metric prove otherwise. Based on the definition of the energy-efficiency metric, higher energy-efficiency means a more efficient protocol. It is observed that the dropping mechanism achieves a high energy-efficiency by eliminating useless transmissions. The retransmission mechanism shows the best energy-efficiency ratio. This is accomplished by retransmitting lost packets and dropping frames that cannot be rebuilt.

5.2   Effects of Mobility on Performance

To evaluate the effects of mobility, the error rate is held constant at 1.5%, the average hop count is 3.8 and the frame size is four. Because neither of the link layer mechanisms are mobility-aware, we expect performance degradation as mobility increases. As the mobility rate increase to around 6 m/s, the dropping mechanism begins to drop frames that have parts that travel different paths (see Figure 6). The retransmission mechanism sustains its performance longer due to the fact that link breaks are fixed by the time the retransmission

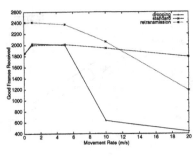
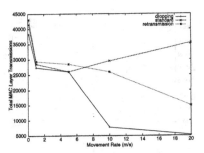

Figure 6. Mobility vs. goodput.   *Figure 7.* Mobility vs. MAC layer transmissions.

is attempted. However, the retransmission mechanism begins to deliver fewer complete frames to the end host than the standard link layer as the mobility rate increases beyond 11 m/s. Because the retransmission mechanism only attempts one retransmission, when that transmission fails, the simple dropping mechanism takes over.

Figure 7 compares the number of MAC layer transmissions in the presence of mobility. The link layer using the dropping mechanism has a significant drop in transmissions as it begins to drop most of the frames. However, these savings in energy are offset by the ineffectiveness of the protocol. The retransmission mechanism achieves fewer transmissions as the retransmissions begin to fail, and the simple dropping mechanism takes over. Again, the savings in energy is overshadowed by the ineffectiveness of the protocol.

6.   Conclusions

Energy conserving protocols are essential for the operation of multi-hop wireless networks due to resource constraints (*e.g.*, battery power). However, such protocols should not only focus on saving energy without any concept of the effect on the application. We define a energy-saving protocol as effective if it maximizes the percentage of usable data received at the end host. Furthermore, we present a new metric for the energy-efficiency defined as the ratio of the number of usable packets received by the end host to the total energy used in transmission. Essentially, the energy-efficiency of a protocol takes into account not only the energy consumption of the protocol but also the effectiveness of the protocol. To this end, an application-aware link layer mechanism has been presented. This mechanism combined with a link layer retransmission mechanism makes use of application layer framing information to achieve significant gains in energy-efficiency. Future work involves adding mobility-awareness into the MAC layer mechanisms to handle route breaking and recovery.

References

[1] S. Aramvith and M.-T. Sun. A coding scheme for wireless video transport with reduced frame skipping. In *SPIE Visual Communication and Image Processing Conference*, 2000.

[2] G. Armitage and K. Adams. Packet reassembly during cell loss. *IEEE Network Magazine*, 7(5):26–34, September 1993.

[3] S. Banerjee and A. Misra. Minimum energy paths for reliable communication in multi-hop wireless networks. In *3rd ACM International Symposium on Mobile Ad Hoc Networking and Computing (MobiHoc)*, 2002.

[4] J.-H. Chang and L. Tassiulas. Energy conserving routing in wireless ad-hoc networks. In *IEEE INFOCOM*, 2000.

[5] D. D. Clark and D. L. Tennenhouse. Architectural considerations for a new generation of protocols. In *ACM Symposium on Communications Architectures and Protocols*, 1990.

[6] K. Fall and K. Varadhan. *ns Notes and Documentation*. The VINT Project.

[7] L. M. Feeney and M. Nilsson. Investigating the energy consumption of a wireless network interface in an ad hoc networking environment. In *IEEE INFOCOM*, 2001.

[8] P. Gauthier, D. Harada, and M. Stemm. Reducing power consumption for the next generation of PDAs: It's in the network interface! In *Workshop on Mobile Multimedia Communications MoMuc*, 1996.

[9] J. Gomez, A. T. Campbell, M. NaghShineh, and C. Bisdikian. PARO: supporting dynamic power controlled routing in wireless ad hoc networks. *winet*, 9:443–460, 2003.

[10] R. Han and D. Messerschmitt. A progressively reliable transport protocol. *Multimedia Systems*, 1999.

[11] IEEE 802 LAN/MAN Standards Committee. Wireless LAN medium access control (MAC) and physical layer (PHY) specifications. IEEE Standard 802.11, 1999.

[12] P. Karn. Toward new link layer protocols. In *Tucson Amateur Packet Radio (TAPR) symposium*, 1994.

[13] C. Kent and J. Mogul. Fragmentation considered harmful. In *ACM SIGCOMM '87*, August 1987.

[14] B. Kim, Z. Xiong, and W. Pearlman. Progressive video coding for noisy channels. In *Proceedings ICIP 98*, 1998.

[15] R. Kravets and P. Krishnan. Application-driven power management for mobile communication. *Wireless Networks*, 6(4):263–277, 2000.

[16] M. Labrador and S. Banerjee. Enhancing application throughput by selective packet dropping. In *IEEE ICC*, 1999.

[17] L. Leslie and D. McAuley. Fairisle: An ATM network for the local area. In *ACM SIGCOMM '91*, 1991.

[18] J. Monks, V. Bharghavan, and W. Hwu. A power controlled multiple access protocol for wireless packet networks. In *IEEE INFOCOM*, 2001.

[19] C. Parsa and J.J. Garcia-Luna-Aceves. Improving TCP performance over wireless networks at the link layer. *Mobile Networks and Applications*, 1:57–71, 2000.

[20] R. Ramanathan and R. Rosales-Hain. Topology control of multihop wireless networks using transmit power adjustment. In *IEEE INFOCOM*, 2000.

[21] S. Ramanathan, P. Rangan, and H. Vin. Frame-induced packet discarding: An efficient strategy for video networking. In *Fourth International Workshop on Networking and Operating System Support for Digital Audio and Video*, 1993.

[22] A. Romanow and S. Floyd. Dynamics of TCP traffic over ATM networks. *IEEE Journal on Selected Areas in Communications*, 13(4):633–641, May 1995.

[23] S. Singh, M. Woo, and C. S. Raghavendra. Power-aware routing in mobile ad hoc networks. In *4th Annual International Conference on Mobile Computing and Networking (MobiCom)*, 1998.

[24] S. Tabbane. *Handbook of Mobile Radio Networks*. Artech House Mobile Communications Library, 2000.

[25] C.-K. Toh. Maximum battery life routing to support ubiquitous mobile computing in wireless ad hoc networks. *IEEE Communications Magazine*, 39(6):138–147, 2001.

[26] M. Zorzi and R. R. Rao. Error control and energy consumption in communications for nomadic computing. *IEEE Transactions on Computers*, 46:279–289, 1997.

ANALYZING THE ENERGY CONSUMPTION OF IEEE 802.11 AD HOC NETWORKS

Daniel de O. Cunha, Luís Henrique M. K. Costa, and Otto Carlos M. B. Duarte
Grupo de Teleinformática e Automação - PEE/COPPE - DEL/POLI
*Universidade Federal do Rio de Janeiro - Rio de Janeiro, Brazil**
doc,luish,otto@gta.ufrj.br

Abstract   This paper analyzes the energy consumption of ad hoc nodes using IEEE 802.11 interfaces. Our objective is to provide theoretical limits on the lifetime gains that can be achieved by different power saving techniques proposed in the literature. The evaluation takes into account the properties of the medium access protocol and the process of forwarding packets in ad hoc mode. The key point is to determine the node lifetime based on its average power consumption. The average power consumption is estimated considering how long the node remains sleeping, idle, receiving, or transmitting.

Keywords: Energy Conservation, Wireless Communication, Ad Hoc Networks

1.   Introduction

A critical factor of the wireless ad hoc network operation is the energy consumption of the portable devices. Typically, wireless nodes are battery-powered and the capacity of these batteries is limited by the weight and volume restrictions of the equipments. Consequently, it is important to reduce the energy consumption of the nodes in the ad hoc network. Moreover, in multihop ad hoc networks each node may act as a router. Thus, the failure of a node due to energy exhaustion may impact the performance of the whole network.

Most works on ad hoc networks assume the use of IEEE 802.11 wireless LAN interfaces. Nevertheless, IEEE 802.11 interfaces operating in ad hoc mode have some peculiarities that are frequently disregarded. Chen *et al.* [1] analyzed the energy consumption of the IEEE 802.11 MAC protocol in infrastructured mode. Feeney and Nilsson [2] measured the energy consumption of IEEE 802.11 interfaces in ad hoc mode and showed that the idle cost is relatively high, since the nodes must constantly sense the medium in order to

*This work has been supported by CNPq, CAPES, FAPERJ, and RNP/FINEP/FUNTTEL.

identify the transmissions addressed to them. Monks *et al.* [3] analyzed the effect of transmission power control on the energy consumption of the nodes. Singh and Raghavendra [4] analyzed the potential gain of their PAMAS protocol, but ignored the power consumption of idle interfaces. Bhardwaj *et al.* [5] derived upper bounds on the lifetime of sensor networks considering the collaborative profile of such networks. The derived bounds relate to the network as a whole and not to specific nodes.

In this paper, we analyze the energy consumption of ad hoc nodes taking into account the interactions of the IEEE 802.11 MAC protocol and the packet forwarding performed on the ad hoc multi-hop networks. This is done based on the fraction of time that the interfaces spend in each operational state and on the capacity of the ad hoc networks. Finally, we analyze the potential gain of different power saving techniques. The theoretical limits of each technique can be used as guidelines in the development of novel power-saving schemes.

This paper is organized as follows. Section 2 analyzes the effects of ad hoc packet forwarding on the node energy consumption. The potential gains of different power saving techniques are obtained in Section 3. Finally, Section 4 concludes this work.

2.   Energy Consumption of the Nodes

The analyses presented in this section assumes the use of IEEE 802.11b interfaces operating in ad hoc mode at 11Mbps using the Distributed Coordination Function (DCF), with RTS/CTS handshake [6]. We can model the average power (P_m) consumed by the interface as

$$P_m = t_{Sl} \times P_{Sl} + t_{Id} \times P_{Id} + t_{Rx} \times P_{Rx} + t_{Tx} \times P_{Tx} \; , \qquad (1)$$

where t_{Sl}, t_{Id}, t_{Rx}, and t_{Tx} are the fractions of time spent by the interface in each of the possible states: Sleep, Idle, Receive, and Transmit, respectively. These fractions of time satisfy the condition $t_{Sl} + t_{Id} + t_{Rx} + t_{Tx} = 1$. Analogously, P_{Sl}, P_{Id}, P_{Rx}, and P_{Tx} are the powers consumed in the four states. Considering P_m and the initial energy of the node (E), we can calculate the node lifetime (T_v), which represents the time before the energy of the node reaches zero, as

$$T_v = \frac{E}{P_m} \; . \qquad (2)$$

The lifetime analysis presented here takes into account only the energy consumption of the wireless interfaces, ignoring the energy consumed by the other circuits of the equipment. Initially, we assume the absence of any power-saving strategy, which implies $t_{Sl} = 0$. With this restriction, the maximum lifetime of a node is achieved with the node permanently in Idle state, as Eq. 3 shows.

$$T_{idle} = \frac{E}{P_{Id}} \; . \qquad (3)$$

In order to evaluate the effect of DCF over the energy consumption, we first analyze two nodes in direct communication. This scenario enables maximum transmission capacity because there is no contention. Then, we analyze the effect of ad hoc forwarding in the energy consumption.

Direct Communication

This scenario consists of two nodes separated by a distance that allows direct communication. The maximum utilization is achieved if the source always has a packet to transmit when the medium is free. In this case, t_{Id}, t_{Rx}, and t_{Tx} are the fractions of time the node spends in each state to transmit one data frame, according to DCF operation. Ignoring the propagation delay, the transmission time of a data frame is divided as shown in Figure 1.

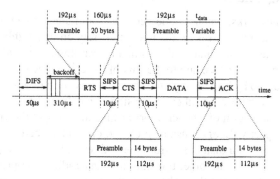

Figure 1.   Total transmission time of a data frame.

The backoff time is uniformly distributed between 0 and 31 (CW_{min}) slots of 20μs each. The average backoff is 15.5 slots, or 310μs per frame. The interframe spaces are SIFS = 10μs and DIFS = 50μs. Moreover, a preamble is sent before each frame. This preamble can be long, lasting for 192μs, or short, lasting for 96μs [7]. Our analysis considers the long IEEE 802.11 preamble, since the short preamble is not compatible with old interfaces. The RTS, CTS, and ACK control frames are transmitted in one of the IEEE 802.11 basic rates. We assume a basic rate of 1Mbps. Thus, the 20 bytes of the RTS are transmitted in 160μs, while the 14 bytes of CTS and ACK take 112μs. The data frame includes a 34-byte MAC header in addition to the data payload, which includes any overhead added by upper layers. Therefore, the transmission time is

$$T_{frame} = backoff + 4 \times t_{pr} + 3 \times SIFS + DIFS +$$
$$t_{RTS} + t_{CTS} + t_{data} + t_{ACK} . \tag{4}$$

We can obtain t_{Id}, t_{Rx}, and t_{Tx} for the emitter and destination nodes as a function of the packet length used. The Backoff, DIFS, and SIFS are periods where both nodes stay idle. During the periods corresponding to the RTS and data packets the emitter is in Tx and the destination in Rx state. The opposite situation occurs during the CTS and ACK periods.

Based on the results for the emitter and destination nodes, we can also calculate t_{Id} and t_{Rx} for "overhearing" nodes, which is necessary to the forwarding chain analysis. Overhearing nodes do not take part in the point-to-point communication but they are in the range of the emitter and/or the receiver. Thus, these nodes spend energy receiving frames addressed to other nodes. There are three kinds of overhearing node: a node that only overhears traffic originated from the emitter, $overhearing_e$, a node that only overhears traffic originated from the destination, $overhearing_d$, and a node that overhears traffic originated from both the emitter and the destination, $overhearing_{ed}$.

Forwarding Chain

In ad hoc networks, when a node needs to communicate with someone out of its direct transmission range, the node must rely on its neighbors to deliver the packets. The intermediate nodes form a forwarding chain with its extremities connected to the source and to the sink of the communication. The packets are forwarded hop by hop through the chain. In this configuration, consecutive packets compete with each other, increasing contention. Li et $al.$ [8] showed that the ideal utilization of a generic forwarding chain is $\frac{1}{4}$ of the one-hop communication capacity. Li et $al.$ used a propagation model where a packet can be correctly received at a distance r from the emitter and where the packet transmission can interfere with other transmissions in a radius of approximately $2r$. We assume in this paper that when a node is overhearing a communication from a distance d such that $r < d < 2r$, the signal strength is still able to change the state of the interface to Rx. Even if the correct reception is impossible, the interface tries to receive the frames.

Therefore, a node in an ideal forwarding chain spends $\frac{1}{4}$ of the time as an emitter, $\frac{1}{4}$ of time as a destination and $\frac{1}{2}$ as an $overhearing_{ed}$ node. Then, the average power consumption of a node in the forwarding chain is

$$P_m = \frac{1}{4} \times P_e + \frac{1}{4} \times P_d + \frac{1}{2} \times P_{o_{ed}} , \qquad (5)$$

where P_e, P_d, and $P_{o_{ed}}$ are, respectively, the average power consumed by a node spending all the time as an emitter, a destination, and an $overhearing_{ed}$ node.

Quantitative Analysis

In order to provide a quantitative analysis, we adopt the measurements by Feeney and Nilsson [2] for IEEE 802.11b interfaces operating at 11Mbps. Table 1 presents an approximation of their results. To ease the comparison with

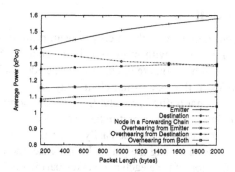

Figure 2. Average power for nodes in different situations.

the maximum lifetime with no energy saving of Eq. 3, Table 1 also shows the consumption of the four states relatively to the Idle consumption (P_{Id}). Based

Table 1. IEEE 802.11b interface energy consumption.

State	Consumption (W)	Ratio
Sleep	0.050	$0.07P_{Id}$
Idle	0.740	P_{Id}
Rx	0.900	$1.2P_{Id}$
Tx	1.350	$1.8P_{Id}$

on Table 1, and on Eqs. 1, 2, and 5, we obtain the average power and lifetime of nodes in different situations as a function of P_{Id} (Figure 2). As Figure 2 shows, for emitters and nodes taking part of forwarding chains the average power consumed, P_m, increases as the data length increases. This is due to the increasing of t_{Tx} with the augmentation of the data frame, and implies in a reduction of the node lifetime. Nevertheless, the node lifetime (Eq. 2) decreases slowly with the packet size comparing to the maximum throughput achievable. Thus, it is possible to transmit more data using large packets. Moreover, nodes in the $overhearing_{ed}$ condition have a lifetime from 13% (for 160-byte packets) to 15% (for 2000-byte packets), shorter than idle nodes.

3. Power Saving Techniques

This section analyzes three major power saving techniques for ad hoc networks. The first one uses the remaining energy as routing metric [9]. The idea is to avoid the continuous use of the same nodes to forward packets. The second approach is transmission power control [3]. Due to the attenuation of RF signals, it may be interesting to reduce the distance of a communication, even if it increases the total number of hops. The third technique is the transition to low power mode [4]. The objective is to maximize the time a node spends at the low power state. The following sections detail each technique.

Energy-Aware Routing

Energy-aware routing balances the energy consumption of the nodes by selecting routes through nodes with more remaining energy. Since the source and the sink of a communication are fixed, these nodes do not benefit from this technique. The nodes in the forwarding chain may save energy. The following analysis considers that traffic is evenly distributed among n disjoint paths. Thus, each intermediate node takes part in the active forwarding chain $\frac{1}{n}$ of the time. Nevertheless, the analysis can be easily extended to the case where the traffic is unevenly divided among the paths. In this case, the fraction $\frac{1}{n}$ should be replaced by the fraction of time each node takes part in the forwarding chain.

In order to evaluate the gain achievable by this technique, we use the consumption of the node when continuously forwarding packets, and the consumption of the node when not taking part of the forwarding chain. The average power consumption can be expressed as

$$P_{m_{bal}} = \frac{P_{fc}}{n} + \frac{(n-1)P_{\overline{fc}}}{n}, \tag{6}$$

where P_{fc} is the average power consumption in the active forwarding chain, whereas $P_{\overline{fc}}$ is the average power consumed when not forwarding. While P_{fc} is plotted in Figure 2, we have two limit cases for $P_{\overline{fc}}$. In the best case, the node that leaves the active forwarding chain does not overhear the traffic of the new active chain, and thus consumes P_{Id}. In the worst case, however, the node continuously overhears the traffic of the active forwarding chain, thus consuming $P_{o_{ed}}$ (Figure 2). In this case, the node is close enough to the active forwarding chain, being in the interference range of the forwarding nodes.

Using the average power consumptions, we obtain the limit lifetime gain of this technique. Figure 3(a) shows the limit gain as the number of used paths increases ($\frac{1}{n} \rightarrow 0$) and as a function of the packet length for the two cases discussed above. Note that the packet length has a small effect on these limits, since they depend on the relation between P_{Id}, $P_{o_{ed}}$, and P_{fc}. While the values showed in Figure 3(a) are limits when $\frac{1}{n} \rightarrow 0$, Figure 3(b) plots the variation of this gain with n for the case of 2000-byte packets. With $n = 4$, at least 66% of the maximum lifetime gain is achieved, for both situations. The important result is that energy-aware routing achieves significant gains using few paths.

Transmission Power Control

Min and Chandrakasan [10] analyzed the conditions under which it is advantageous to use two hops instead of one, by reducing the transmission power. They model the energy consumption as $\alpha + \beta d^n$, where α is the distance-independent, and βd^n is the distance-dependent term. The coefficient n represents the path loss and is typically between 2 and 6 [3]. Min and Chandrakasan claim that the use of two hops is profitable when the reduced distance-

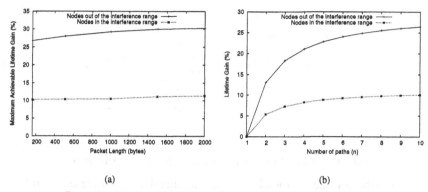

(a)   (b)

Figure 3.   Limits of the lifetime gains with the energy-aware routing.

dependent consumption is higher than the fixed cost associated to the inclusion of an additional hop. The variable portion of the energy consumption of the wireless interface is due to the RF amplifier. Assuming that all the difference between P_{Tx} and P_{Rx} is due to the power amplifier, the lower limit of P_{Tx} is P_{Rx} and all the additional consumption scales with the distance, as modeled by βd^n. Hence, given the values adopted in our analysis (Table 1), the distance-dependent consumption ($P_{Tx} - P_{Rx}$) is equal to $0.6P_{Id}$ for $d = r$. Moreover, assuming no power saving, the interface consumes at least P_{Id}. Thus, the fixed cost of the communication can be estimated by the difference between P_{Id} and P_{Rx}, which is $0.2P_{Id}$. Let T_{Tx} and T_{Rx} be, respectively, the amount of time that the emitter stays in the Tx and Rx states during the transmission of one packet. The terms α and βd^n for $d = r$ for the emitter, destination, and *overhearing$_{ed}$* nodes are shown in Table 2.

Table 2.   Packet transmission costs for different node types.

Node	α	βd^n
Emitter	$(T_{Tx} + T_{Rx})\,0.2P_{Id}$	$T_{Tx} \times 0.6P_{Id}$
Destination	$(T_{Tx} + T_{Rx})\,0.2P_{Id}$	$T_{Rx} \times 0.6P_{Id}$
Overhearing$_{ed}$	$(T_{Tx} + T_{Rx})\,0.2P_{Id}$	0

Under these conditions and ignoring overhearing nodes, the per-packet cost of direct communication is $2(T_{Tx} + T_{Rx})0.2P_{Id} + (T_{Tx} + T_{Rx})0.6P_{Id}$, while the two-hop communication cost with $d = \frac{r}{2}$ is $4(T_{Tx} + T_{Rx})0.2P_{Id} + 2\beta\frac{r}{2}^n$, where $\beta\frac{r}{2}^n$ is the distance-dependent cost of one hop communication at a distance $d = \frac{r}{2}$. Thus, the use of two hops is advantageous if the resulting $\beta\frac{r}{2}^n$ is lower than $(T_{Tx} + T_{Rx})0.1P_{Id}$, i.e., if the resulting power consumption of the Tx state, P_{Tx}, for the communication is lower than $1.3P_{Id}$. This indicates that for channels with a path loss coefficient (n) higher than 2.58 the use of two hops instead of one is advantageous.

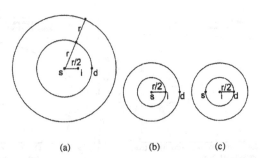

Figure 4.   Transmission and interference ranges of the communications.

Nevertheless, the overhearing nodes can significantly increase the overall energy consumed. Supose the situation of Figure 4, where the source, s, wants to communicate with the destination, d, at a distance r from s and there is a third node, i, between s and d, at a distance $\frac{r}{2}$ from the source, which can be used as an intermediate hop. Considering only these three nodes and the propagation model where the interference range is twice the transmission range, the use of two hops instead of one is not profitable because nodes in the interference range overhear the transmissions. The use of a second hop causes two transmissions of the same packet, with half the original range. In direct communication, s would use a range of r, resulting in a interference range of $2r$ (Figure 4(a)). Node d can correctly receive the packet and i is an overhearing node. Using two-hop communication, node s uses a transmission range of $\frac{r}{2}$ in order to node i be able to receive the packet. The $\frac{r}{2}$ range implies a interference range of r, making d an overhearing node for this transmission (Figure 4(b)). After the first transmission, i sends the packet to d. In this second transmission s is an overhearing node (Figure 4(c)). Considering the overhearing nodes, the per-packet cost of direct communication is $3(T_{Tx} + T_{Rx})0.2P_{Id} + (T_{Tx} + T_{Rx})0.6P_{Id}$, while the two-hop communication cost with $d = \frac{r}{2}$ is $6(T_{Tx} + T_{Rx})0.2P_{Id} + 2\beta\frac{r}{2}^{n}$. Note that $\beta\frac{r}{2}^{n}$ is always positive, which means that the two-hop communication of Figure 4 always consumes more than direct communication, independently of the path loss coefficient.

Nevertheless, the two-hop communication with a range of $\frac{r}{2}$ covers an area four times smaller than the area covered by the direct communication with range r. In general, there are other nodes near the three nodes of Figure 4 that will be overhearing. If we assume an uniform distribution of overhearing nodes, each transmission of the two-hop scenario implies $\frac{1}{4}$ of the overhearing nodes of direct communication. Accounting for the two transmissions of the two-hop scenario, the total number of overhearing nodes is half the number of overhearing nodes in the single transmission of direct communication. Therefore, as the number of overhearing nodes (N) per communication range (given by a πr^2 area) increases, the ratio between the total energy consumed in

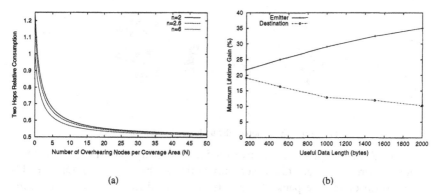

Figure 5. Energy conservation with transmission power control.

the two-hop scenario and the total energy consumed in direct communication approaches 0.5.

Figure 5(a) plots the ratio between the total energy consumed in the two-hop scenario and in direct communication, i.e., the two hops relative consumption, with varying density of overhearing nodes, for different path loss coefficients (n). When there is no overhearing node near the communication, the two-hop consumption tends to the consumption of one-hop communication as n increases, and even a low density of overhearing nodes can result in significant energy savings using two hops (Figure 5(a)). Even for $n = 2$, the two hops relative consumption is around 0.7, assuming four overhearing nodes per communication range.

Considering only direct communication, the reduction of P_{Tx} to the lowest possible value is attractive, because all the reduction is converted into lifetime gain. Figure 5(b) shows the limit of the lifetime gain for the emitter and the destination nodes for different packet lengths, as $P_{Tx} \rightarrow P_{Rx}$ (and the distance between emitter and destination tends to zero). In this case, there is a significant difference in the gain for different packet lengths. As the length increases, the emitter gain increases while the destination gain decreases. As the packet length increases, the fraction of time spent by the emitter in Tx increases, and the time spent by the destination in Tx decreases.

Transition to Sleep State

The significant difference of consumption between the Idle and Sleep states makes the transition to sleep state profitable. Nevertheless, due to the distributed nature of ad hoc networks, the use of this technique is limited. A sleeping node must rely on its neighbors to store eventual packets addressed to it. Moreover, as the node may be asleep at packet arrival, the network latency increases. Thus, most works on this technique admit larger delays.

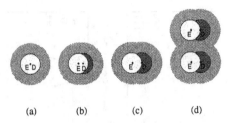

Figure 6.   Situation of the nodes at different distances.

The PAMAS protocol [4] aims to reduce the energy consumption without latency increase. Nodes fall asleep only at times when they would not be able to transmit or receive packets. This is the case when a node is overhearing the communication of two other nodes. This approach reduces the time nodes spend in the Idle state, as well as reduces the periods in which the nodes are consuming energy by overhearing the communication of other nodes.

PAMAS uses a separate signaling channel to decide when nodes must fall asleep. Nevertheless, we can adopt a PAMAS-like technique over IEEE 802.11. In the IEEE 802.11 standard, when a node receives a RTS or CTS frame, the node sets its NAV (Network Allocation Vector) according to the virtual carrier sense mechanism. In practice, a node that overhears the RTS/CTS exchange will not be able to transmit or receive packets for the period specified in the NAV, therefore this node can fall asleep during that period, without affecting the network performance.

As Figure 6 shows, the nodes in the range of the emitter (white area) can sleep just after the end of the RTS transmission, while the nodes in the range of the destination (dark-gray area) can sleep only after the transmission of the CTS frame. We refer to the union of these two areas as the *Power Saving area* (PS-area). The nodes in the interference range (light-gray area) of both the emitter and the destination are unable to fall asleep since they can not correctly receive the RTS or CTS frames. They are overhearing nodes. Depending on the distance, d, between the emitter and the destination, the fraction of nodes that are in each situation changes. Figs. 6(a) and 6(c) show the limit situations where the emitter and the destination are at distances $d = 0$ and $d = r$, respectively. Figure 6(b) shows an intermediate situation: the distance $d = 0.7r$ is the radius of a circle with half the area of the original circle of radius r. The power saved increases as the fraction of overhearing nodes decreases. Therefore, we consider two neighbor PS-areas that are as close as possible, i.e., two PS-areas with overlapping interference areas (the light-gray portion in Figure 6(d)). Then, we assume that each PS-area is responsible for only half the adjacent interference area. The average power consumption of the nodes that fall asleep after the transmission of the RTS and of the CTS frames are P_{rts} and P_{cts}, respectively, and N is the average number of nodes in the commu-

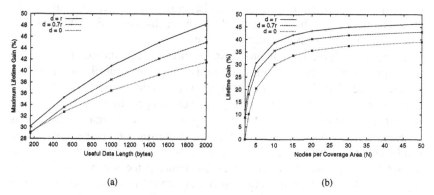

(a)   (b)

Figure 7.   Limits of the lifetime gains of the PAMAS-like power saving scheme.

nication range, given by a πr^2 area. Assuming the fair sharing of the channel among all nodes and that the cost of the transition to sleep state is negligible, the average power consumptions can be computed by weighting the average power of nodes in the different possible situations based on the involved areas, and consequently the number of nodes in each situation. The average powers consumed for the different distances discussed above are

$$P_{m_{d=0}} = \frac{P_e + P_d + (N-2)P_{rts} + 1.5P_{o_{ed}}}{2.5N} \,, \tag{7}$$

$$P_{m_{d=0.7r}} = \frac{P_e + P_d + (N-2)P_{rts} + 0.44NP_{cts} + 1.82NP_{o_{ed}}}{3.26N}, \; and \tag{8}$$

$$P_{m_{d=r}} = \frac{P_e + P_d + (N-2)P_{rts} + 0.61NP_{cts} + 1.83NP_{o_{ed}}}{3.44N} \,. \tag{9}$$

The limit gain achievable by this technique (when $N \to \infty$), as a function of the packet length, is shown in Figure 7(a). The maximum gain is achieved with the limit distance $d = r$. Moreover, the gain using large frames is 50% higher than using small frames. As Figure 7(b) shows for the gain using 2000-byte frames, with a node density of 10 nodes per communication range, more than 70% of the maximum gain is achieved.

4.   Conclusions

This paper analyzed the energy consumption of ad hoc nodes considering the interactions of the IEEE 802.11 MAC protocol and the ad hoc packet forwarding. Our goal was to provide theoretical gain limits which help the development of power-saving schemes. The use of larger packets increases the fraction of time spent by the interface in the Tx state, reducing the node lifetime. Nevertheless, our results show that the lifetime reduction is compensated by the higher throughput achieved using larger packets. Therefore, large packets are more energy efficient. Our analysis shows that an overhearing node has a lifetime up to 15% smaller than idle nodes.

Then, we analyzed the potential gains of three widely studied power saving techniques: energy-aware routing, transmission power control, and transition to sleep state. The limit gain of energy-aware routing varies from 11%, for nodes in disjoint paths with overlapping radio ranges, to 30%, for nodes in isolated forwarding chains. Moreover, up to 66% of the maximum gain is achieved using only four disjoint paths. Using transmission power control, the results show that the use of two hops instead of one can save up to 50% of the total energy consumed per packet, by reducing the number of overhearing nodes. Additionally, transmission power control increases the lifetime of nodes in direct communication from 21%, for small packets, to 35%, for large packets. Destination nodes can also benefit from this technique. Finally, a PAMAS-like power saving scheme, which uses the transition to sleep state, achieves up to 48% lifetime gain. More importantly, more than 70% of the possible gain is achieved with a density of 10 nodes per communication range.

References

[1] J.-C. Chen, K. M. Sivalingam, P. Agrawal, and S. Kishore, "A comparison of MAC protocols for wireless local networks based on battery power consumption," in *IEEE INFOCOM*, Mar. 1998, pp. 150–157.

[2] L. M. Feeney and M. Nilsson, "Investigating the energy consumption of a wireless network interface in an ad hoc networking environment," in *IEEE INFOCOM*, Mar. 2001.

[3] J. Monks, J.-P. Ebert, W.-M. W. Hwu, and A. Wolisz, "Energy saving and capacity improvement potential of power control in multi-hop wireless networks," *Computer Networks*, vol. 3, no. 41, pp. 313–330, Feb. 2003.

[4] S. Singh and C. Raghavendra, "PAMAS: Power aware multi-access protocol with signalling for ad hoc networks," *ACM Computer Communication Review*, vol. 28, no. 3, pp. 5–26, July 1998.

[5] M. Bhardwaj, A. Chandrakasan, and T. Garnett, "Upper bounds on the lifetime of sensor networks," in *IEEE ICC*, June 2001, pp. 785–790.

[6] IEEE, "Wireless LAN medium access control (MAC) and physical layer (PHY) specifications," IEEE Standard 802.11, 1999.

[7] ——, "Wireless LAN medium access control (MAC) and physical layer (PHY) specifications: Higher-speed physical layer extension in the 2.4 GHz band," IEEE Standard 802.11b, 1999.

[8] J. Li, C. Blake, D. S. J. D. Couto, H. I. Lee, and R. Morris, "Capacity of ad hoc wireless networks," in *ACM MOBICOM*, July 2001, pp. 61–69.

[9] S. Singh, M. Woo, and C. S. Raghavendra, "Power-aware routing in mobile ad hoc networks," in *ACM MOBICOM*, Oct. 1998, pp. 181–190.

[10] R. Min and A. Chandrakasan, "Top Five Myths about the Energy Consumption of Wireless Communication," ACM MOBICOM, Sept. 2002, poster.

ENERGY-EFFICIENT RELIABLE PATHS FOR ON-DEMAND ROUTING PROTOCOLS

Tamer Nadeem[1], Suman Banerjee[2], Archan Misra [3], Ashok Agrawala[1]

[1]*Dept. of Computer Science, University of Maryland, College Park, USA.*
Emails: {nadeem,agrawala}@cs.umd.edu

[2]*Dept. of Computer Science, Wisconsin University, USA. Email: suman@cs.wisc.edu*

[3]*IBM T.J. Watson Research Center, Hawthorne, USA. Email: archan@us.ibm.com*

Abstract   We define techniques to compute energy-efficient reliable paths within the framework of on-demand routing protocols. The choice of energy-efficient reliable paths depend on link error rates on different wireless links, which in turn depend on channel noise. We show how our scheme accounts for such channel characteristics in computing such paths. Additionally, we perform a detailed study of the AODV protocol and our energy-efficient variants, under various noise and node mobility conditions. Our results show that our proposed variants of on-demand routing protocols can achieve orders of magnitude improvement in energy-efficiency of reliable data paths.

1.   Introduction

Battery-power is typically a scarce and expensive resource in wireless devices. Therefore, a large body of work has addressed energy-efficient link-layer forwarding techniques [1–5] and routing mechanisms [6–11] for multi-hop wireless networks. However, an end-to-end reliability requirement can significantly affect the choice of data paths in both these objectives. In particular, the choice of energy-efficient routes should take into account the channel noise in the vicinity of these nodes. Such noise would lead to transmission errors and consequent re-transmissions, thus increasing the energy costs for reliable data delivery.

In this paper we describe how such minimum energy end-to-end reliable paths can be calculated *and implemented* for reactive (on-demand) routing protocols. We have experimented with the Ad-hoc On-demand Distance Vector Routing protocol (AODV) [12]. It should, however, become obvious from our description that our technique can be generalized to alternative on-demand routing protocols (e.g., DSR [13] and TORA [14]). Through our experimentation, we perform a detailed study of the AODV protocol and our energy-efficient variants, under various noise and node mobility conditions.

2.   Related Work

A large number of researchers have addressed the energy-efficient data transfer problem in the context of multi-hop wireless networks. They can be classified into two distinct categories. One group focuses on protocols for minimizing the energy requirements over end-to-end paths. Typical solutions in this approach have ignored the retransmission costs of packets and have therefore chosen paths with a large number of small hops [5, 7]. For example, the proposed protocol in [5] is one such variable energy protocol using a modified form of the Bellman-Ford algorithm, where the nodes modify their transmission power based on the distance to the receiver, and where this variable transmission energy is used as the link cost to effectively compute minimum energy routes.

An alternative approach focuses on algorithms for increasing the lifetime of wireless nodes, by attempting to distribute the forwarding load over multiple paths. This distribution is performed by either intelligently reducing the set of nodes needed to perform forwarding duties, thereby allowing a subset of nodes to sleep over idle periods or different durations (e.g, PAMAS [6], SPAN [10], and GAF [11]), or by using heuristics that consider the residual battery power at different nodes [9, 8, 15] and route around nodes nearing battery exhaustion.

In [16] we formulated the link cost considering the error rates. We showed how can we approximate those costs to fit the proactive routing protocols that use Bellman-Ford algorithm in calculating the best paths. However, in this paper: 1) We study two different mechanisms for bit error rates estimation in practice, 2) We define and implement the modifications needed to compute energy efficient routes in the AODV as an example of the reactive protocols, and 3) We extensively study the performance benefits of our proposed schemes under various wireless environment conditions.

3.   Minimum Energy Reliable Paths

To compute minimum energy paths, we need to evaluate candidate paths not merely based on the energy spent in a single transmission attempt across the wireless hops, but rather on the total energy required for packet delivery, *including potential retransmissions due to errors and losses* on the link.

For any particular link (l) between a transmitting node and a receiving node, let Pt denote the transmission power and p_l represent the packet error probability. Assuming that all packets are of a constant size, the energy involved in each transmission attempt across of a packet, E_l, is simply a function of Pt.

Any signal transmitted over a wireless medium experiences two different effects: attenuation due to the medium, and interference with ambient noise at the receiver. Due to the characteristics of the wireless medium, the transmitted signal suffers an attenuation proportional to a function of the distance between the receiver and the transmitter. The ambient noise at the receiver is independent of the distance between the source and distance, and depends purely on the operating conditions at the receiver. We consider two scenarios: 1) Fixed Transmission Power: in which each node chooses the transmission power to be a fixed constant and independent of the link distance, and 2) Variable Transmission Power: where a transmitter node adjusts Pt to ensure that the strength of the (attenuated) signal received by the receiver is a constant (independent of D) and is minimally above a certain threshold level, P_{Th}.

We consider two different operating models for end-to-end reliable paths:

3.1   Hop-by-Hop Retransmissions (HHR):

Each individual link provides reliable forwarding to the next hop using link-layer retransmissions. The *mean* number of individual packet transmissions for the successful transfer of a single packet is $1/(1-p_l)$. Therefore, the mean energy cost, C_l, required for the successful transmission of this packet is:

$$C_l = \frac{E_l}{1 - p_l} \tag{1}$$

3.2   End-to-End Retransmissions (EER):

The individual links do not provide link-layer retransmissions and error recovery. Reliable packet transfer is achieved only via retransmissions initiated by the source node. The actual end-to-end cost energy, C, requirements for a given path with nodes $0, \ldots, n$ in sequence for the EER case is given by:

$$C = \frac{\sum_{i=0}^{n-1} E_{i,i+1}}{\prod_{i=0}^{n-1} (1 - p_{i,i+1})} \qquad (2)$$

where, $E_{i,i+1}$ is the energy required for a single transmission across the link $\langle i, i+1 \rangle$ and $p_{i,i+1}$ is the packet error probability of the link.

The above formulas assume unlimited number of retransmissions for each packet. In practice, wireless cards limit the retransmissions number for HHR case, where as protocols such as UDP don't provide packet reliability. However, those formulas are still valid because: 1) many wireless cards use large number of retransmission trials in case of HHR, and 2) we consider connections in which the sources have infinite or very large number of packets. Also, we do not consider the cost of the control packets, e.g.,IEEE 802.11 RTS/CTS/ACK frames, since the cost of the data packets dominates other costs.

4.   Estimating Link Error Rate

In order to implement our proposed mechanism, it is sufficient for each node to estimate only the bit error rate, BER, on its incoming wireless links. In this section we discuss the following two possible mechanisms:

4.1   BER using Radio Signal-to-Noise Ratio

The bit error rate, BER, of a wireless channel depends on the received power level, Pr, of the signal. The exact relationship between BER and Pr depends on the choice of the signal modulation scheme. However, in general, several modulation schemes exhibit the generic relationship: BER $\propto erfc(\sqrt{\frac{constant \times Pr}{N}})$, where N is the noise density (noise power per Hz) and $erfc(x)$ is defined as the complementary function of $erf(x)$ and is given by: $erfc(x) = 1 - (2/\sqrt{\pi}) \int_0^x \exp^{-t^2} dt$. For the case of BPSK (Binary Phase-Shift Keying) and QPSK (Quadrature Phase-Shift Keying) the bit error is obtained by [17]

$$\text{BER} = 0.5 \times erfc(\sqrt{\tfrac{Pr \times W}{N \times f}}) \qquad (3)$$

where f is the transmission bit rate and W is the channel bandwidth (in Hz). Note that the CCK (Complementary Code Keying) used by IEEE 802.11b to achieve the 11 Mbps, which we assume in this paper where the bit rate f is 11 Mbps and the channel bandwidth W is 2 MHz, is modulated with the QPSK technology.

Most wireless interface cards typically measure the signal-to-noise ratio (SNR) for each received packet. SNR is a measure of the received signal strength relative to the background noise and is often expressed in decibels as: $SNR = 10 \log \frac{Pr}{N}$. From the SNR value measured by the interface card we can calculate the ratio Pr/N in Equation 3 for each received packet.

This SNR-based error rate estimation technique is useful specially primarily in free space environments where such error models are applicable. Consequently this is not applicable for indoor environments. For such environments we use an alternative technique that is based on empirical observations of link error characteristics, which we describe next.

4.2   BER **using Link Layer Probes**

In this empirical mechanism, we estimate the BER of the incoming links by using link layer probe packets. Each node periodically broadcasts a probe packet within its local neighborhood. Each such packet has a local sequence number which is incremented with each broadcast. Each neighbor of this node receives only a subset of those probes due to channel errors. Each node stores the sequence number of the last correctly received probe from each of its neighbors. On the reception of the next (i^{th}) probe from a node, the receiving node can calculate s_i, the number of probes lost since the last received probe. The packet error rate (p) for a probe packet of length $prob\_size$ bits (assuming independent bit errors) is: $p = 1 - (1 - \text{BER})^{prob\_size}$. Given the packet error rate for probes since the last probe reception as $p = \frac{s_i}{s_i+1}$, the receiving node can compute the incoming link BER as:

$$\text{BER} = 1 - \exp(\frac{\log(1 - \frac{s_i}{s_i+1})}{prob\_size}) \tag{4}$$

In wireless environments, broadcast packets are more prone to losses due to collision than unicast packets that use channel contention mechanisms, e.g. the RTS/CTS technique employed in IEEE 802.11. Therefore our probe-based BER estimation technique can potentially over-estimate the actual BER experienced by the data packets. However, the probe-based mechanism is still applicable for these reasons:

1 BER is over-estimated in parts of the wireless network with high traffic load. Since our route computation technique is biased against high BER, routes will naturally avoid these areas of high traffic load. This will lead to an even distribution of traffic load in the network, increasing network longevity and decreasing contention.

2 The criteria of selecting optimum route in our algorithm is based on the relative relations between costs of the candidate routes not the actual costs. As the traffic load on the network gets evenly balanced over links, the BER will be equally over-estimated for all the links. This implies that the costs relative ordering of the candidate routes is largely unaffected.

Generally, Pr and N in Equation 3 vary with time: N varies due to the environment conditions, and Pr, which changes with distance, varies due to the nodes mobility. Consequently, we can not base the SNR-based BER estimation on a single measurement. Therefore, SNR-based mechanism uses probes packets to calculate those parameters as function of several measurements over a window of time, in order to capture the dynamics of the network. For both the SNR-based and probe-based schemes, each node broadcasts probe packet, at an average period t (one second in the implementation). To avoid accidental synchronization and consequently collisions, t is jittered by up to $\pm0.25t$. Each node continuously updates its estimate of the BER using an exponentially weighted moving average of the sampled BER values. As in all such averaging techniques, the estimate can be biased towards newer samples depending on the rate at which the noise conditions changes, i.e. increasing node mobility.

4.3   BER **Estimation for Variable Power Case**

For the fixed transmission power case, both probe and data packets are transmitted with the same constant power. Therefore, the BER experienced by data packets is the same as probe packets However, the same is not true for the variable power transmission case. In the variable power case, the transmission power used for a given data packet depends on the link's length. However, probe packets are broadcasted to all its possible neighbors and is transmitted with the fixed maximum transmission power (Pt_{max}). Equation 3 implies that packets received at a higher power (e.g.

probe packets) will experience lower BER than the packets received at lower power (e.g. data packets). Therefore a suitable adjustment is required for BER estimation for data packets.

Since a node, in the variable power case, chooses the transmission power level such that the power level of the received data packet at the receiver is P_{Th}. Since the noise N can be calculated using the SNR and Pr values measured by the wireless interface card, we can estimate the BER for data packets using the SNR-based technique by substituting P_{Th} for Pr in Equation 3.

A related but different correction scheme is applied for the BER estimation of data packets when using the probe-based technique. We omit the details due to space constraints [18].

5. AODV and its Proposed Modifications

AODV builds routes using a route request / route reply query cycle. When a source node desires a route to a destination for which it does not already have a route, it broadcasts a route request (RREQ) packet across the network. Nodes receiving this packet update their information for the source node and set up backwards pointers to the source node in the route tables. A node receiving the RREQ may send a route reply (RREP) if it is either the destination or if it has a fresh route to the destination. Otherwise, it rebroadcasts the RREQ. If a node receives a RREQ which they have already processed, it discards the RREQ and do not forward it. As the RREP propagates back to the source, nodes set up forwarding pointers to the destination. Once the source node receives the RREP, it may begin to forward data packets to the destination. Details of the AODV protocol can be found in [12]. Our proposed modifications adhere to the on-demand philosophy, i.e. paths are still computed on-demand and as long as an existing path is valid, we do not actively change the path.

5.1 AODV Messages and Structures

To perform energy efficient route computation for reliable data transfer, we need to exchange information about energy costs and loss probabilities between nodes that comprise the candidate paths. This information exchange is achieved by adding additional fields to existing AODV messages and structures (RREQ, RREP, Broadcast ID table, and Routing table) and does not require the specification of any new message. Due to space constrain, the description of the relevant changes are left to [18].

5.2 Route Discovery

5.2.1 Route Request Phase. The source node triggers the route discovery by initializing the new added fields in a RREQ message as $C_{req} = 0$, $E_{req} = 1$, and $Q_{req} = 1$ (the latter two are valid for the EER case). RREQ messages are transmitted at the node's maximum power level in order to reach all legitimate one hop neighbors. When an intermediate node n_i receives RREQ message from a previous node n_{i-1}, it calculates the energy (E_l) consumed by node n_{i-1} in a single transmission attempt of a data packet over the link $l = \langle i - 1, i \rangle$ as a function of the the transmission power, Pt of node n_{i-1}.

For the fixed power case, this transmission power, Pt is a globally known constant. In the variable power case, the control messages, e.g. probe packets and RREQ messages, are sent with a fixed maximum transmission power Pt_{max}, which is globally known. Therefore, node n_i can calculate the transmit power to be used by node n_{i-1} for data packets as: $Pt = P_{Th} \times \frac{Pt_{max}}{Pr_{max}}$ where that Pr_{max} is the power level at which control messages from n_{i-1} are received at n_i.

Subsequently, node n_i updates fields in the RREQ message as: $C_{req}=C_{req} + \frac{E_l}{1-p_l}$ for the HHR case, and as: $E_{req}=E_{req} + E_l$, $Q_{req}=Q_{req} \times (1 - p_l)$, and $C_{req}=\frac{E_{req}}{1-Q_{req}}$ for EER case. The packet error

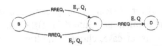

Figure 1.   Calculating the energy cost for the EER case.

rate (p_l) is calculated by node n_i using the data size known in advance and the estimated BER_l as $p_l = 1 - (1 - \text{BER}_l)^{data\_size}$.

Node n_i examines the broadcast identification number stored in the RREQ message to check if it has seen any previous RREQ message belongs to the same route request phase or not. If this is the first instance for this RREQ, node n_i adds a new entry in its *Broadcast ID* table and initializes its values as $H_{bid} = H_{req}$, $C_{bid} = C_{req}$, $E_{bid} = E_{req}$, $Q_{bid} = Q_{req}$, and $Prev_{bid} = n_{i-1}$ where H_{req} is the number of hops traversed by this RREQ messages. Otherwise a previous RREQ message has been seen by the node n_i. In this case it compares the updated cost value in the RREQ message with that stored in the table entry. In the HHR case, if the boolean expression

$$(C_{req} < C_{bid}) \text{ OR } (C_{req} = C_{bid} \text{ AND } H_{req} < H_{bid}) \tag{5}$$

is true, then this RREQ message is further forwarded. Otherwise the currently best known route has lower cost than the new route discovered by this RREQ message, and so is discarded.

For a correct formulation in the EER case, the comparison rule (Expression 5) does not apply. This is because the cost function is not linear in the EER case. Consider Figure 1 in which node n receives two RREQ messages through two different paths from the source. The end-to-end energy costs for the two paths are $\frac{E_1+E}{Q_1 \times Q}$ and $\frac{E_2+E}{Q_2 \times Q}$ respectively. The node n should choose the path defined by $RREQ_1$ if and only if, $\frac{E_1+E}{Q_1 \times Q} < \frac{E_2+E}{Q_2 \times Q}$. However, at the node, n, information on E and Q are not available and so this inequality cannot be evaluated. Therefore, to optimally compute energy-efficient routes in the EER case, each separate RREQ message needs to be forwarded towards the destination. To do this, we need to maintain a separate entry in the Broadcast ID table for each RREQ message. Those entries can potentially lead to an exponential growth in the size of the table, and hence is not practical. Therefore, in practice we propose the same forwarding mechanism as used in the HHR case. To improve the quality of the chosen paths, we increase the state maintained in the Broadcast ID table to store information about the five RREQ messages with the lowest costs at each node.

As described in our modification, the intermediate nodes may broadcast multiple RREQ messages for the same route request phase, as an opposite to a single RREQ message in original AODV.

5.2.2   Route Reply Phase.   When the destination node, or an intermediate node that has information about the partial route to the destination and its corresponding partial cost, receives the first RREQ, it stores locally the calculated total cost as the minimum cost of the route and send back the route reply (RREP) message. For any further RREQ reception of the same route request phase, the node updates the minimum stored cost and send back RREP message if and only if the calculated total cost of RREQ message is lower than the minimum stored one. Therefore, nodes may forward multiple RREP messages in response to better routes found by successive RREQ messages.

RREQ messages propagate back on the same path traversed by RREQ using the $Prev_{bid}$ fields stored at the intermediate nodes. At each intermediate node, the RREP fields are updated in a manner similar to the RREQ messages updates to reflect the cost of the route from the current

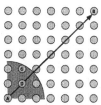

Figure 2. The 49-node grid topology. The shaded region marks the maximum transmission range for the node, *A*. $A \rightarrow B$ is one of the example flows used on this topology.

intermediate node to the destination. Each intermediate node store this calculated cost for further route requests for the same destination from any node [18].

6. Simulation Experiments and Performance Evaluation

In this section, we report on extensive simulation-based studies on the performance of the AODV protocol, both with and without our energy-aware modifications. The performance comparisons were done using the *ns-2* simulator, enhanced with the CMU-wireless extensions (the underlying link layer is IEEE 802.11b with 11 Mbps data rate).

We have studied the performance of the different schemes for both HHR and EER cases, under both fixed and variable transmission power scenarios. In this paper, we will, however, focus only on the HHR case. This is because all practical link-layer protocols for multi-hop wireless attempt to provide some degree of reliable forwarding through the use of retransmissions or error control coding strategies.

To study the performance of our suggested scheme, we implemented and observed three separate routing schemes:

a) The Shortest-Delay (**SD**): The original AODV routing protocol that selects the route with the minimum latency.

b) The Energy-Aware (**EA**): Enhances the AODV protocol by associating a cost with each wireless link which is the energy required for a single packet transmission without the retransmission considerations. In this formulation of wireless link cost, the link error rates are ignored.

c) Our Retransmission-Energy Aware (**RA**): Which enhances the AODV protocol as described in this paper. As discussed previously, the link cost now considers the impact of retransmissions necessary for reliable packet transfer.

6.1 Network Topology and Link Error Modeling

For our experiments we use different topologies having 49 nodes randomly distributed over a 700×700 square region. The maximum transmission radius of a node is 250 units. We present results for three different topology scenarios:

- *Static Grid:* Nodes are immobile and equi-spaced along each axis as in Figure 2.

- *Static Random:* Nodes are immobile and uniformly distributed over the region.

- *Mobile Random:* Initially, nodes are distributed uniformly at random over the region. During the simulation, nodes move around the region using the random waypoint model [13] with zero pause time.

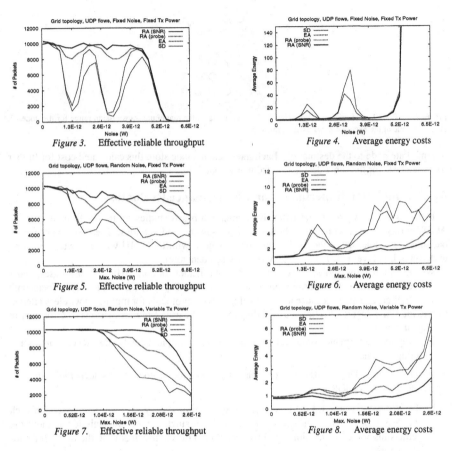

Figure 3. Effective reliable throughput

Figure 4. Average energy costs

Figure 5. Effective reliable throughput

Figure 6. Average energy costs

Figure 7. Effective reliable throughput

Figure 8. Average energy costs

In all our simulations we had a set of 12 flows that were active over the duration of the experiment. We use both TCP and UDP flows for different experiments. For the TCP flows, we use its NewReno variant. For the UDP flows, we choose the traffic sources to be constant bit rate (CBR) sources at rate of 5 packets per second. The UDP packets and TCP segments were 1000 bytes each. Each of the simulations was run for a fixed duration of 250 seconds. Each point in the results is the average of 10 runs. When hop-by-hop reliability (HHR) is used, the UDP and TCP flows have similar performance patterns as was confirmed in our experiments. Therefore, we only present the results for the UDP flows.

All the control packets, e.g., probe packets, RREQ, RREP messages, IEEE 802.11 RTC/CTS/ACK frames, as well as the data packet experience the same bit error rate (BER) of a wireless link which depends on the ambient noise level as shown in Equation 3. We partitioned the entire square region into small square grids (50 × 50 units each). We model the ambient noise of each of these small square regions as independent identically distributed white Gaussian noise of μ mean and standard deviation σ. The noise mean μ for the different small square grids was chosen to vary between two configurable parameters, N_{min} and N_{max} corresponding to minimum and maximum noise respectively, while the noise standard deviations σ was chosen to be equal to $(0.1 \times \mu)W$. We used different distributions for the μ over the entire region for different experiments. In this paper, we focus only on the following extreme cases:

1 *Fixed noise environment:* N_{min} is equal to N_{max} and their values vary between $0.0W$ and $13.0 \times 10^{-12}W$.

2 *Random noise Environment:* We fix N_{min} to $0.0W$ and vary N_{max} between $0.0W$ and $13.0 \times 10^{-12}W$.

6.2   Metrics

We observed two different metrics:

1 **Average energy:** Computed per data packet by dividing the total energy expenditure (over all the nodes) by the total data units (sequence number for TCP and packets for UDP) received correctly at destinations. Note that this cost includes energy consumption due to control packets as well as the data packets. This metric is measured in Jouls.

2 **Effective Reliable Throughput:** This metric counts the number of the reliably transmitted packets from the source to the destination, over the simulated duration.

We choose 0.282W to be the transmission power for the fixed transmission power experiments, and to be the maximum transmission power in case of the variable transmission power experiments. Similarly, $100\mu J$ is chosen to be the energy cost for single attempt transmission of a bit a over a link for the fixed transmission power experiments, and to be the maximum energy cost corresponding to the maximum transmission power in case of the variable transmission power experiments. As implemented in ns-2, we use Friis and Two-ray ground models and their corresponding parameters values as our propagation models. Due to space constraints, we will show performance comparisons for the variable power case for only a few sample experiments.

6.3   Static Grid Topologies

Figures 3 and 4 show the effective reliable throughput and the average energy cost for experiments with fixed noise environments for UDP flows. Note that each data point on the plot corresponds to an experiment with a specified fixed noise value for the entire square region. Clearly for very low noise environments, all schemes are equivalent. However, as the noise in the environment starts to increase, the RA schemes show significant benefits. The SNR-based link error estimation has a superior performance than the probe-based technique. It is interesting to note that for both EA and SD schemes, the effective reliable throughput does not decrease monotonically. Instead at certain intermediate noise values (e.g. $1.3 \times 10^{-12}W$ and $3.12 \times 10^{-12}W$) the throughput goes to zero. This is an interesting phenomena that is related to the relative size of the RREQ and the data packets.

Consider the flow $A - B$ in Figure 2. Both SD and EA schemes for the fixed transmission power case chooses a path with minimum number of hops. Therefore, the first hop for this flow will be the link $\langle A, C \rangle$. For a static link , the BER is constant and depends on the noise value and the received power. but the packet error rates (PER) is not. PER depends on packet size and is smaller for RREQ packets than the data packets. When the noise is $1.3 \times 10^{-12}W$, the BER for the $\langle A, C \rangle$ link is 0.0008. The corresponding PER for RREQ packets is about 0.5. Therefore RREQ packets sent by node A is correctly received at C in about 50% of the cases and the link $\langle A, C \rangle$ is chosen by both SD and EA scheme using their usual cost metrics. However, the PER experienced by the data packets on the same link is nearly 1. This causes significant losses for

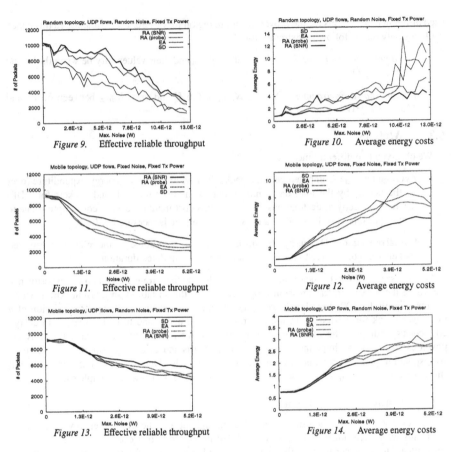

Figure 9.   Effective reliable throughput

Figure 10.   Average energy costs

Figure 11.   Effective reliable throughput

Figure 12.   Average energy costs

Figure 13.   Effective reliable throughput

Figure 14.   Average energy costs

data packets and therefore the throughput achieved is close to zero. However, when the noise level increases (i.e. say $1.56 \times 10^{-12}W$), the BER on the link goes up (i.e. to 0.00186). This causes the PER for RREQ packets to increase to 0.8. Therefore most of these RREQ packets get lost across link $\langle A, C \rangle$. Consequently both SD and EA schemes start to shift to paths with shorter hops with lower BER and their performance starts to approach the RA scheme.

The RA scheme do not suffer from this anomalous behavior. This is because the RA schemes choose routes based on the BER. Therefore, it automatically avoid links with high PER for data packets. This behavior is clearly visible in the grid topology, since the number of alternatives routes are discrete and few. Figures 5 and 6 show the corresponding plots for the random noise environment. The EA and SD schemes consume about 140% more energy per successfully transferred data unit than the RA schemes, when the maximum noise in the environment is $> 3.9 \times 10^{-12}W$ and still achieves only half the throughput of the RA schemes.

For the sake of completeness, we present sample results for the variable transmission power case. In this experiment, we chose a power threshold at the receiver, P_{Th}, to be 1.0×10^{-9} W. The transmission power needs to be chosen such that the receiving node receives the packet with this power. Figures 7 and 8 show the effective reliable throughput and the average energy costs

Max. speed (m/s)	Fixed Noise Env. Noise: 2.0×10^{-12} W		Random Noise Env. Max. noise: 2.0×10^{-12} W	
	$\dfrac{T_{probe}}{T_{SNR}}$	$\dfrac{E_{probe}}{E_{SNR}}$	$\dfrac{T_{probe}}{T_{SNR}}$	$\dfrac{E_{probe}}{E_{SNR}}$
1	0.90	1.08	0.90	1.07
5	0.84	1.19	0.95	1.09
10	0.72	1.29	0.85	1.26
15	0.74	1.32	0.89	1.29

Table 1. Comparison of SNR-based and the probe-based versions of the RA scheme. T_{probe} and T_{SNR} are the reliable throughput of the probe-based and the SNR-based schemes respectively. Similarly E_{probe} and E_{SNR} are the average energy for the two schemes.

for UDP flows on a grid topology in a random noise environment for the variable transmission power case. As before, we can see performance benefits of the RA schemes over the EA and SD schemes.

6.4 Static Random Topologies

We next present results of the randomly generated static topologies (Figures 9 and 10). As before, the RA schemes provide significant performance benefits over the SD and EA schemes. For low noise environments, the SD and EA schemes consume 50-100% more energy for each reliably delivered data unit, while for high noise environments the energy requirements for the SD and EA schemes are about 2-3 times higher.

6.5 Mobile Topologies

In Figures 11 and 12 we show the effective reliable throughput and the average energy per reliable delivered data unit respectively in the fixed noise environment. Figures 13 and 14 are the corresponding plots for the random noise environment. In both these cases, the maximum speed nodes is 10 m/s.

We observe that the impact of mobility increases with increase in the channel noise. For example, in absence of channel noise, the throughput achieved for the mobile topologies is about 10% lower than the corresponding static topologies. As the channel noise increases the data throughput achieved for the mobile topologies is significantly lower.

From the energy consumption plots (Figures 12 and 14) it is clear that the RA scheme that uses the SNR-based link error estimation performs significantly better than all the other schemes. This is because mobility impacts the link error rates, and the SNR-based scheme is able to quickly adapt its estimates of the continuously changing link error rates.

This effect is clearly visible in Table I where we compare the performance of the two RA schemes over different mobile topologies with varying speeds. As the mobility speed increases, we can see that the RA (SNR) scheme is more energy-efficient in comparison to the RA (probe) scheme (the benefits increase from 7% to 29% as the maximum speed of nodes increase from 1 m/s to 15 m/s in the random noise environment). It also achieves higher reliable throughput.

7. Conclusions

In this paper we have extensively studied the performance of the AODV protocol under varying wireless noise conditions. Our simulation studies show that the energy-aware modification of AODV can result in a significant (sometimes orders of magnitude) reduction in total energy consumption per packet, with the added benefit of higher throughput as well. In essence, the higher overheads of our energy-aware route establishment process (e.g., forwarding of multiple RREQs) are more than compensated for by the lower energy consumed in data forwarding. We presented two schemes to estimate the link error rates based on the characteristics of the environment. Our simulations show that the performance gains are impressive for both RA (SNR) and RA (probe) schemes compared with other schemes.

Acknowledgments

This work was supported in part by the Maryland Information and Network Dynamics (MIND) Laboratory, its founding partner Fujitsu Laboratories of America, and by the Department of Defense through a University of Maryland Institute for Advanced Computer Studies (UMIACS) contract.

References

[1] W. Ye, J. Heidemann, and D. Estrin, "An energy efficient mac protocol for wireless sensor networks," in *Proceedings of Infocom*, June 2002.

[2] B. Prabhakar, E. Uysal-Biyikoglu, and A. El Gamal, "Energy-efficient Transmission over a Wireless Link via Lazy Packet Scheduling," in *Proceedings of IEEE Infocom*, Apr. 2001.

[3] J.H. Gass Jr., M.B. Pursley, H.B. Russell, and J.S. Wysocarski, "An adaptive-transmission protocol for frequency-hop wireless communication networks," *Wireless Networks*, vol. 7, no. 5, Sept. 2001.

[4] A. El Gamal, C. Nair, B. Prabhakar, E. Uysal-Biyikoglu, and S. Zahedi, "Energy-efficient Scheduling of Packet Transmissions over Wireless Networks," in *Proc. of IEEE Infocom*, 2002.

[5] K. Scott and N. Bambos, "Routing and channel assignment for low power transmission in PCS," in *Proceedings of ICUPC*, Oct. 1996.

[6] S. Singh and C.S. Raghavendra, "Pamas-power aware multi-access protocol with signaling for ad hoc networks," in *ACM Comm. Review*, 1998.

[7] J. Gomez-Castellanos, A. Campbell, M. Naghshineh, and C. Bisdikian, "PARO: A power-aware routing optimization scheme for mobile ad hoc networks," *ACM/Baltzer Journal on Mobile Networks*, 2002.

[8] J.-H. Chang and L. Tassiulas, "Energy conserving routing in wireless ad-hoc networks," in *Proceedings of Infocom*, Mar. 2000.

[9] C.K. Toh, H. Cobb, and D. Scott, "Performance evaluation of battery-life-aware routing schemes for wireless ad hoc networks," in *Proceedings of ICC*, June 2001.

[10] B. Chen, K. Jamieson, H. Balakrishnan, and R. Morris, "Span: An Energy-Efficient coordination Alogrithm for Topology Mainte nance in Ad Hoc Wireless Networks," *ACM Wireless Networks Journal*, Sept. 2002.

[11] Y. Xu, J. Heidemann, and D. Estrin, "Geographically-informed Energy Conservation for Ad Hoc Routing," in *Proc. of ACM Mobicom*, July 2001.

[12] C.E. Perkins and E.M. Royer, "Ad-hoc on-demand distance vector routing," in *Proceedings of the 2nd IEEE Workshop on Mobile Computing Systems and Applications*, Feb. 1999.

[13] D. Johnson and D. Maltz, "Dynamic source routing in ad hoc wireless networks," in *Mobile Computing*, 1996, pp. 153–181.

[14] V. Park and S. Corson, "Temporally-ordered routing algorithm (tora) version 1: Functional specification, draft-ietf-manet-tora-spec-04.txt," .

[15] A. Misra and S. Banerjee, "MRPC: Maximizing network lifetime for reliable routing in wireless environments," in *Proceedings of WCNC*, Mar. 2002.

[16] S. Banerjee and A. Misra, "Minimum energy paths for reliable communication in multi-hop wireless networks," in *Proc. of Mobihoc*, June 2002.

[17] J.G. Proakis, "Digital communications," *Third Edition, McGraw-Hill, Inc., New York*, 1995.

[18] T. Nadeem, S. Banerjee, A. Misra, and A. Agrawala, "Energy-Efficient Reliable Paths for On-Demand Routing Protocols," Tech. Rep. UMIACS-TR-2004-25 and CS-TR-4582, University of Maryland, April 2004.

MINIMUM POWER SYMMETRIC CONNECTIVITY PROBLEM IN WIRELESS NETWORKS: A NEW APPROACH

Roberto Montemanni, Luca Maria Gambardella

Istituto Dalle Molle di Studi sull'Intelligenza Artificiale (IDSIA)
Galleria 2, CH-6928 Manno-Lugano, Switzerland
{roberto, luca}@idsia.ch

Abstract   We consider the problem of assigning transmission powers to the nodes of a wireless network in such a way that all the nodes of the network are connected by bidirectional links and the total power consumption is minimized.

A new exact algorithm, based on a new integer programming model, is described in this paper together with a new preprocessing technique.

Keywords: Wireless networks, minimum power topology, exact algorithms.

1.   Introduction

Ad-hoc wireless networks have been significantly studied in the past few years due to their potential applications in battlefield, emergency disasters relief, and other application scenarios (see, for example, Singh et al., 1999, Ramanathan and Rosales-Hain, 2000, Wieselthier et al., 2000, Wan et al., 2001 and Lloyd et al., 2002). Unlike wired networks of cellular networks, no wired backbone infrastructure is installed in ad-hoc wireless networks. A communication session is achieved either through single-hop transmission if the recipient is within the transmission range of the source node, or by relaying through intermediate nodes otherwise.

We consider the problem of minimum transmit power bidirectional topology in multi-hop wireless networks where individual nodes are typically equipped with limited capacity batteries and therefore have a restricted lifetime. Topology control is one of the most fundamental and critical issues in multi-hop wireless networks which directly affect the

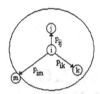

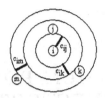

Figure 1.   Wireless communication model.

Figure 2.   Costs for the mathematical formulation IP.

network performance. In wireless networks, topology control essentially involves choosing the right set of transmitter power to maintain adequate network connectivity. In energy-constrained networks, where replacement or periodic maintenance of node batteries is not feasible, the issue is very critical since it directly impacts the network lifetime.

In Ramanathan and Rosales-Hain, 2000 the problem of controlling topology using transmission power control in wireless networks is firstly approached in terms of optimization. It is showed that a network topology which minimizes the maximum transmitter power allocated to any node can be constructed in polynomial time. This is a critical criterion in battlefield applications since using higher transmitter power increases the probability of detection by enemy radar. In this paper, we focus on the minimum power topology problem in wireless networks with omnidirectional antennae. It has been shown in Clementi et al., 1999 that this problem is NP-complete. Related work in the area of minimum power topology construction include Wattenhofer et al., 2001 and Huang et al., 2002, which propose distributed algorithms.

Unlike in wired networks, where a transmission from i to m generally reaches only node m, in wireless networks with omnidirectional antennae it is possible to reach several nodes with a single transmission (this is the so-called *wireless multi-cast advantage*, see Wieselthier et al., 2000). In the example of Figure 1 nodes j and k receive the signal originated from node i and directed to node m because j and k are closer to i than m, i.e. they are within the transmission range of a communication from i to m. This property is used to minimize the total transmission power required to connect all the nodes of the network. For a given set of nodes, the *Minimum Power symmetric Connectivity (MPC) problem* is to assign transmission powers to the nodes of the network in such a way that all the nodes are connected by bidirectional links and the total power consumption over the network is minimized. Having bidirectional links simplifies one-hop transmission protocols by allowing acknowledgement messages to be sent back for every packet (see Althaus et al., 2003). It is assumed that no power expenditure is involved in reception/processing activities, that a complete knowledge of pairwise distances between nodes is available, and that there is no mobility.

2. Problem description

In order to formalize the problem, a model for signal propagation has to be selected. We adopt the model presented in Rappaport, 1996. Signal power falls as $\frac{1}{d^\kappa}$, where d is the distance from the transmitter to the receiver and κ is an environment-dependent coefficient, typically between 2 and 4 (we will set $\kappa = 4$). Under this model, and adopting the usual convention (see, for example, Althaus et al., 2003 and Montemanni et al., 2004) that every node has the same transmission efficiency and the same detection sensitivity threshold, the power requirement for supporting a link from node i to node j, separated by a distance d_{ij}, is then given by

$$p_{ij} = (d_{ij})^\kappa \tag{1}$$

It is important to notice that the results presented in this paper are still valid in case more complex signal propagation models more complex are taken into account.

We assume that there is no constraint on maximum transmission powers of nodes. However, the algorithm we discuss in this paper can be extended straightforwardly to the case when this assumption does not hold. If, for example, node i cannot reach node j even when it is transmitting to its maximum power (i.e. $d_{ij}^\kappa >$ maximum power of node i), then p_{ij} can be redefined as $+\infty$.

MPC can be formally described as follows:

Given the set V of the nodes of the network, a *range assignment* is a function $r : V \to \mathcal{R}^+$. A *bidirectional link* between nodes i and j is said to be established under the range assignment r if $r(i) \geq p_{ij}$ and $r(j) \geq p_{ij}$. Let now $B(r)$ denote the set of all bidirectional links established under the range assignment r. *MPC* is the problem of finding a range assignment r minimizing $\sum_{i \in V} r(i)$, subject to the constraint that the graph $(V, B(r))$ is connected.

As suggested in Althaus et al., 2003, a graph theoretical description of *MPC* can be given as follows:

Let $G = (V, E, p)$ be an edge-weighted graph, where V is the set of vertices corresponding to the set of nodes of the network and E is the set of edges containing all the possible (unsorted) pairs $\{i, j\}$, with $i, j \in V$, $i \neq j$. A cost p_{ij} is associated with each edge $\{i, j\}$. It corresponds to the power requirement defined by equation (1).

For a node i and a spanning tree T of G (see, for example, Kruskal, 1956), let $\{i, i_T\}$ be the maximum cost edge incident to i in T, i.e. $\{i, i_T\} \in T$ and $p_{i i_T} \geq p_{ij} \ \forall \{i, j\} \in T$. The *power cost* of a spanning tree T is then $c(T) = \sum_{i \in V} p_{i i_T}$. Since any connected graph contains a spanning tree, and a broadcast tree must be connected, MPC can be described as the problem of finding the spanning tree T with minimum power cost $c(T)$. This observation is at the basis of the integer programming formulation which will be presented in Section 3.

3.   An integer programming formulation

A weighted, directed graph $G' = (V, A, p)$ is derived from G by defining $A = \{(i, j), (j, i) | \{i, j\} \in E\} \cup \{(i, i) | i \in V\}$, i.e. for each edge in E there are the respective two (oriented) arcs in A, and a dummy arc (i, i) with $p_{ii} = 0$ is inserted for each $i \in V$. p_{ij} is defined by equation (1) when $i \neq j$. In order to describe the new integer programming formulation for MPC, we also need the following definition.

Given $(i, j) \in A$, we define the *ancestor* of (i, j) as

$$a_j^i = \begin{cases} i & \text{if } p_{ij} = \min_{\{i,k\} \in E}\{p_{ik}\} \\ arg\max_{k \in V}\{p_{ik} | p_{ik} < p_{ij}\} & \text{otherwise} \end{cases} \quad (2)$$

According to this definition, (i, a_j^i) is the arc originated in node i with the highest cost such that $p_{i a_j^i} < p_{ij}$. In case an *ancestor* does not exist for arc (i, j), vertex i is returned, i.e. the dummy arc (i, i) is addressed.

In formulation IP a spanning tree (eventually augmented) is defined by z variables: $z_{ij} = 1$ if edge $\{i, j\}$ is on the spanning tree, $z_{ij} = 0$ otherwise. Variable y_{ij} is 1 when node i has a transmission power which allows it to reach node j, $y_{ij} = 0$ otherwise.

$$(IP) \quad \text{Min} \sum_{(i,j) \in A} c_{ij} y_{ij} \quad (3)$$

$$\text{s.t. } y_{ij} \leq y_{i a_j^i} \qquad \forall (i, j) \in A, a_j^i \neq i \quad (4)$$

$$z_{ij} \leq y_{ij} \qquad \forall \{i, j\} \in E \quad (5)$$

$$z_{ij} \leq y_{ji} \qquad \forall \{i, j\} \in E \quad (6)$$

$$\sum_{i \in S, j \in V \setminus S, \{i,j\} \in E} z_{ij} \geq 1 \qquad \forall S \subset V \quad (7)$$

$$z_{ij} \in \{0, 1\} \qquad \forall \{i, j\} \in E \quad (8)$$

$$y_{ij} \in \{0, 1\} \qquad \forall (i, j) \in A \quad (9)$$

In formulation IP an incremental mechanism is established over y variables (i.e. transmission powers). The costs associated with y variables in the objective function (3) are given by the following formula:

$$c_{ij} = p_{ij} - p_{ia_j^i} \quad \forall (i,j) \in A \tag{10}$$

c_{ij} is equal to the power required to establish a transmission from node i to node j (p_{ij}) minus the power required by node i to reach node a_j^i ($p_{ia_j^i}$). In Figure 2 a pictorial representation of the costs arising from the example of Figure 1 is given.

Constraints (4) realize the incremental mechanism by forcing the variable associated with arc (i, a_j^i) to assume value 1 when the variable associated with arc (i,j) has value 1, i.e. the arcs originated in the same node are activated in increasing order of p. Inequalities (5) and (6) connect the spanning tree variables z to transmission power variables y. Basically, given edge $\{i,j\} \in E$, z_{ij} can assume value 1 if and only if both y_{ij} and y_{ji} have value 1. Equations (7) state that all the vertices have to be mutually connected in the subgraph induced by z variables, i.e. the (eventually augmented) spanning tree. Constraints (8) and (9) define variable domains.

3.1 Valid inequalities

A set of valid inequalities is proposed in Montemanni and Gambardella, 2003 for a formulation described in the same paper. Most of these inequalities can be easily adapted to formulation IP, and the remainder of this section is devoted to their description in terms of formulation IP.

In order to describe these valid inequalities, we will refer to the subgraph of G' defined by the y variables with value 1 as G_y. Formally, $G_y = (V, A_y)$, where $A_y = \{(i,j) \in A | y_{ij} = 1$ in the current solution of $IP\}$.

Connectivity inequalities: since graph G_y must be connected by definition, each node i must be able to communicate with at least another node. Its transmission power must then be sufficient to reach at least the node j which is closest to it. This can be expressed through the following set of inequalities:

$$y_{ij} = 1 \quad \forall (i,j) \in A \text{ s.t. } a_j^i = i \tag{11}$$

Bidirectional inequalities 1: for each arc $(i,j) \in A$, if $y_{ij} = 0$ and $y_{ia_j^i} = 1$ then the transmission power of node i is set to reach node a_j^i and nothing more. The only reason for node i to reach node a_j^i and nothing more is the existence of a bidirectional link on edge $\{i, a_j^i\}$ in

G_y. Consequently $y_{a_j^i i}$ must be equal to 1. This is what the following set of constraints states.

$$y_{a_j^i i} \geq y_{ia_j^i} - y_{ij} \quad \forall (i,j) \in A \text{ s.t. } a_j^i \neq i \tag{12}$$

Notice that if $y_{ij} = 1$ then $y_{ia_j^i} = 1$ because of inequalities (4) and consequently in this case the constraint does not give any new contribution. If $y_{ij} = 0$ and $y_{ia_j^i} = 0$ then again the constraint does not give any new contribution.

Bidirectional inequalities 2: consider arc $(i,j) \in A$, where j is the farthest node from i (i.e. $\nexists (i,k) \in A, a_k^i = j$) and suppose $y_{ij} = 1$. The only reason for node i to reach node j is the existence of a bidirectional link on edge $\{i,j\}$ in G_y. Consequently y_{ji} must be equal to 1, as stated by the following set of constraints.

$$y_{ji} \geq y_{ij} \quad \forall (i,j) \in A \text{ s.t. } \nexists (i,k) \in A, a_k^i = j \tag{13}$$

Notice that if $y_{ij} = 0$ the constraint does not give any contribution to formulation IP.

Tree inequality: in order to be strongly connected, the directed graph G_y must have at least $2(|V| - 1)$ arcs, as stated by the following constraint.

$$\sum_{(i,j) \in A} y_{ij} \geq 2(|V| - 1) \tag{14}$$

Reachability inequalities 1: in order to define this set of valid inequalities, we need the following definitions.

$G_a = (V, A_a)$ is the subgraph of the complete graph G' such that $A_a = \{(i,j) | a_j^i = i\}$. Notice that $|A_a| = |V|$ by definition.

$\mathcal{R}_i = \{j \in V | j \text{ can be reached from } i \text{ in } G_a\}$.

The inequalities are based on the consideration that, since graph G_y must be strongly connected, it must be possible to reach every node j starting from each node i. This implies that at least one arc must exist between the nodes which is possible to reach from i in G_a (i.e. $\mathcal{R}_i$) and the other nodes of the graph (i.e. $V \backslash \mathcal{R}_i$). The following set of inequalities arises:

$$\sum_{(k,l) \in A, \, k \in \mathcal{R}_i, l \in V \backslash \mathcal{R}_i} y_{kl} \geq 1 \quad \forall i \in V \tag{15}$$

Reachability inequalities 2: in order to define this set of valid inequalities, we need the following definition.

$\mathcal{Q}_i = \{j \in V | i \text{ can be reached from } j \text{ in } G_a\}$.

These inequalities are based on the idea that, since graph G_y must be strongly connected, it must be possible to reach every node i from every other node j of the graph. This means that at least one arc must exist between the nodes which cannot reach i in G_a (i.e. $V \backslash \mathcal{Q}_i$) and the other nodes of the graph (i.e. $\mathcal{Q}_i$). The following set of constraints arises:

$$\sum_{(l,k)\in A, l\in \mathcal{Q}_i, \, k\in V\backslash \mathcal{Q}_i} y_{lk} \geq 1 \quad \forall i \in V \tag{16}$$

In the remainder of this paper we will refer to formulation IP reinforced with inequalities (11)-(16) as IP_R. We will use this last, reinforced formulation because the extra inequalities strictly constrain y variables to assume quasi-feasible values (in terms of IP^{PMB}) only and this leads to much shorter solving times (see Montemanni and Gambardella, 2003).

4.   Preprocessing procedure

The theoretical result described in this section is used to reduce the number of edges of a problem (and consequently the number of variables of formulation IP).

Given a problem, we suppose we have an heuristic solution, heu, with cost $cost(heu)$ for it. Given a node i, all its transmission power levels that, if implemented, would induce a cost higher than $cost(heu)$ can be ignored. More formally:

Theorem 1. *If the following inequality holds*

$$2p_{ij} + \sum_{k\in V\backslash\{\{i\}\cup\{j\}\}, \, a_l^k=k} p_{kl} > cost(heu) \tag{17}$$

then edge $\{i,j\}$ can be deleted from E.

Proof. If p_{ij} is the power of node i in a solution, this means that the power of node j must be greater than or equal to $p_{ji}(= p_{ij})$, i.e. arc (j,i) must be in the solution, because otherwise there would be no reason for node i to reach node j. The sum in the left hand side of the inequality represents a lower bound for the power required by nodes different from i and j to maintain the network connected. The left hand side of inequality (17) represents then a lower bound for the total power required in case node i transmits to a power which allows it to reach node j and nothing farther. For this reason, if inequality (17) holds, edge $\{i,j\}$ can be deleted from E. $\qquad\qquad\square$

It is important to notice that once edge $\{i,j\}$ is deleted from E, the value of the ancestor of arc (i,k) $((j,l))$ with $a_k^i = j$ $(a_l^j = i)$ has to be updated to a_j^i (a_i^j).

5.   The iterative exact algorithm IEX

In this section we describe an algorithm which solves to optimality formulation IP (i.e. the minimum power symmetric connectivity problem).

It is very difficult to deal with constraints (7) of formulation IP_R in case of large problems. For this reason some techniques which leave some of them out have to be considered. We present an iterative algorithm (IEX) which in the beginning does not consider constraints (7) at all, and then adds them step by step only in case they are violated.

In order to speed up the approach, the following inequality should also be added to the initial integer problem IP_R:

$$\sum_{\{i,j\}\in E} z_{ij} \geq |V| - 1 \qquad (18)$$

Inequality (18) forces the number of active z variables to be at least $|V| - 1$ - this condition is necessary in order to have a spanning tree - already at the very first iterations of the algorithm.

The integer program defined as IP_R without constraints (7) but with inequality (18), is solved and the values of the z variables in the solution are examined. If the edges corresponding to z variables with value 1 form a spanning tree then the problem has been solved to optimality, otherwise constraints (19), described below, are added to the integer program and the process is repeated.

At the end of each iteration, the last available solution is examined and, if edges corresponding to z variables with value 1 generate a set CC of connected components with $|CC| > 1$, then the following inequalities are added to the formulation:

$$\sum_{i\in C, j\in V\backslash C, \{i,j\}\in E} z_{ij} \geq 1 \quad \forall C \in CC \qquad (19)$$

Inequalities (19) force z variables with value 1 to connect the (elsewhere disjoint) connected components in CC to each other.

6.   Computational results

In this section we present some experiments aiming to evaluate the performance of the preprocessing technique described in Section 4 and of the exact algorithm presented in Section 5.

Tests have been carried out on problems randomly generated as described in Althaus et al., 2003. For each problem of size $|V|$ generated, $|V|$ points - they are the nodes of the network - have been chosen uniformly at random from a grid of size 10000×10000.

The preprocessing technique and the IEX algorithm have been implemented in ANSI C, and the callable library of ILOG CPLEX 6.0 (see *http://www.cplex.com*) has been used to solve the integer programs encountered during the execution algorithm IEX. Tests have been carried out on a SUNW Ultra-30 machine.

In this paper consider networks with up to 50 nodes, but it is important to notice that in case the algorithm is used within a distributed/dynamic environment, the typical local vision of a node can be estimated in a few tens of nodes, that may be reflected into a much larger global network.

6.1   Preprocessing procedure

In order to apply the preprocessing procedure described in Section 4, a heuristic solution to the problem must be available. For this purpose we use one of the simplest algorithms available, MST, which works by calculating the *Minimum Spanning Tree T* (see Prim, 1957) on the weighted graph with costs defined by equation (1), and by assigning the power of each transmitter i to p_{ii_T}, as described near the end of Section 2. More complex algorithms, which guarantee better performance, have been proposed (see, for example, Althaus et al., 2003). It is worth to observe that if these algorithms have had been adopted, also the preprocessing technique would have produced better results than those reported in the remainder of this section.

In Table 1 we present, for different values of $|V|$, the average percentage of arcs deleted by the preprocessing procedure over fifty runs.

Table 1 suggests that the preprocessing technique we propose dramatically simplifies problems. It is also interesting to observe that the percentage of arcs deleted considerably increases when the number of nodes ($|V|$) increases. This means that, when dimensions increase, the extra complexity induced by extra nodes is partially mitigated by the increased efficiency of the preprocessing technique. This should help to contain the complexity explosion faced when the number of nodes goes up.

6.2 IEX algorithm

In Table 2 we present the average computation times required by different exact algorithms to solve to optimality problems for different values of $|V|$. Fifty instances have been considered for each value of $|V|$.

The results in the second column of Table 2 are those presented in Althaus et al., 2003 (obtained on an AMD Duron 600MHz PC) multiplied by a factor of 3.2 (as suggested in Dongarra, 2003). This makes

Table 1. Preprocessing technique. Average performance.

$|V|$	Arcs deleted (%)
10	57.556
15	63.781
20	66.526
25	70.393
30	72.464
35	74.647
40	76.106
45	77.568
50	78.688

Table 2. Exact algorithms. Average computation times (sec).

$|V|$	Althaus et al., 2003	Montemanni and Gambardella, 2003	IEX
10	2.144	0.192	0.052
15	18.176	0.736	0.196
20	71.04	8.576	0.601
25	188.48	33.152	2.181
30	643.2	221.408	13.481
35	2278.4	1246.304	28.172
40	15120	9886.08	79.544

them comparable with the other results of the table. In the third column the results presented in Montemanni and Gambardella, 2003 are summarized. Table 2 shows that algorithm IEX outperforms, in terms of time required to retrieve optimal solutions, the other exact methods. In particular it is important to observe that the gap between the computational times of this algorithm and those of the other methods tends to increase when the number of nodes considered increases.

7.   Conclusion

In this paper we have considered the problem of assigning transmission powers to the nodes of a wireless network in such a way that all the nodes are connected by bidirectional links and the total power consumption is minimized.

We have presented a new integer programming formulation for the problem and a new algorithm, based on this formulation, which retrieves optimal solutions according to the model considered. A preprocessing technique has been also proposed.

We are currently researching on a framework where the algorithm we propose is used locally at each node of a distributed/dynamic network. The objective will be to evaluate the behavior of our novel approach in this context.

Acknowledgments

The work was partially supported by the FET unit of the European Commission through project BISON (IST-2001-38923).

References

Althaus, E., Călinescu, G., Măndoiu, I.I., Prasad, S., Tchervenski, N., and Zelikovsky, A. (2003). Power efficient range assignment in ad-hoc wireless networks. In *Proceedings of the IEEE WCNC 2003 Conference*, pages 1889–1894.

Clementi, A., Penna, P., and Silvestri, R. (1999). Hardness results for the power range assignment problem in packet radio networks. *LNCS*, 1671:195–208.

Dongarra, J.J. (2003). Performance of various computers using standard linear algebra software in a fortran environment. Technical Report CS-89-85, University of Tennessee.

Huang, Z., Shen, C.-C., Srisathapornphat, C., and Jaikaeo, C. (2002). Topology control for ad hoc networks with directional antennas. In *Proceedings of the ICCCN 2002 Conference*.

Kruskal, J.B. (1956). On the shortest spanning subtree of a graph and the travelling salesman problem. *Proceedings of AMS*, 7:48–50.

Lloyd, E., Liu, R., Marathe, M., Ramanathan, R., and Ravi, S. (2002). Algorithmic aspects of topology control problems for ad hoc networks. In *Proceedings of the ACS MobiHoc 2002 Conference*, pages 123–134.

Montemanni, R. and Gambardella, L.M. (2003). An exact algorithm for the min-power symmetric connectivity problem in wireless networks. Technical report, Istituto Dalle Molle di Studi sull'Intelligenza Artificiale (IDSIA).

Montemanni, R., Gambardella, L.M., and Das, A.K. (2004). The minimum power broadcast tree problem in wireless networks: a simulated annealing approach. Submitted for publication.

Prim, R.C. (1957). Shortest connection networks and some generalizations. *Bell System Technical Journal*, 36:1389–1401.

Ramanathan, R. and Rosales-Hain, R. (2000). Topology control of multihop wireless networks using transmit power adjustment. In *Proceedings of the IEEE INFOCOM 2000 Conference*, pages 404–413.

Rappaport, T. (1996). *Wireless Communications: Principles and Practices*. Prentice Hall.

Singh, S., Raghavendra, C., and Stepanek, J. (1999). Power-aware broadcasting in mobile ad hoc networks. In *Proceedings of the IEEE PIMRC 1999 Conference*.

Wan, P.-J., Călinescu, G., Li, X.-Y., and Frieder, O. (2001). Minimum energy broadcast routing in static ad hoc wireless networks. In *Proceedings of the IEEE INFOCOM 2001 Conference*, pages 1162–1171.

Wattenhofer, R., Li, L., Bahl, P., and Wang, Y.M. (2001). Distributed topology control for power efficient operation in multihop wireless ad hoc networks. In *Proceedings of the INFOCOM 2001 Conference.*

Wieselthier, J., Nguyen, G., and Ephremides, A. (2000). On the construction of energy-efficient broadcast and multicast trees in wireless networks. In *Proceedings of the IEEE INFOCOM 2000 Conference*, pages 585–594.